Biological Field Emission Scanning Electron Microscopy

Current and Future Titles in the Royal Microscopical Society—John Wiley Series

Published

Principles and Practice of Variable Pressure/Environmental Scanning Electron Microscopy (VP-ESEM) – Debbie Stokes

Aberration-Corrected Analytical Electron Microscopy – Edited by Rik Brydson

Diagnostic Electron Microscopy – A Practical Guide to Interpretation and Technique – Edited by John W. Stirling, Alan Curry and Brian Eyden

Low Voltage Electron Microscopy – Principles and Applications – Edited by David C. Bell and Natasha Erdman

Standard and Super-Resolution Bioimaging Data Analysis: A Primer – Edited by Ann Wheeler and Ricardo Henriques

Electron Beam-Specimen Interactions and Applications in Microscopy – Budhika Mendis

Biological Field Emission Scanning Electron Microscopy, 2 Volume Set – Edited by Roland A. Fleck and Bruno M. Humbel

Forthcoming

Understanding Practical Light Microscopy – Jeremy Sanderson

Correlative Microscopy in the Biomedical Sciences – Edited by Paul Verkade and Lucy Collinson

The Preparation of Geomaterials for Microscopical Study: A Laboratory Manual – Owen Green and Jonathan Wells

Electron Energy Loss Spectroscopy – Edited by Rik Brydson and Ian MacLaren

Biological Field Emission Scanning Electron Microscopy

Volume II

Edited by

Roland A. Fleck

Centre for Ultrastructural Imaging
King's College London
United Kingdom

Bruno M. Humbel

Imaging Section
Okinawa Institute of Science and Technology
Japan

WILEY

This edition first published 2019

Registered Offices
John Wiley & Sons, Inc., 111 River Street, Hoboken, NJ 07030, USA
John Wiley & Sons Ltd, The Atrium, Southern Gate, Chichester, West Sussex, PO19 8SQ, UK

Editorial Office
John Wiley & Sons Ltd, The Atrium, Southern Gate, Chichester, West Sussex, PO19 8SQ, UK

For details of our global editorial offices, customer services, and more information about Wiley products visit us at www.wiley.com.

Wiley also publishes its books in a variety of electronic formats and by print-on-demand. Some content that appears in standard print versions of this book may not be available in other formats.

Library of Congress Cataloging-in-Publication Data

Names: Fleck, Roland A., 1970- editor. | Humbel, Bruno M., editor.
Title: Biological field emission scanning electron microscopy / edited by
Roland A. Fleck, Bruno M. Humbel.
Description: First edition. | Hoboken, NJ : Wiley, 2019. | Series: Royal
Microscopical Society - John Wiley series | Includes bibliographical
references and index. |
Identifiers: LCCN 2018021231 (print) | LCCN 2018021980 (ebook) | ISBN
9781118663240 (Adobe PDF) | ISBN 9781118663264 (ePub) | ISBN 9781118654064
(hardback)
Subjects: | MESH: Microscopy, Electron, Scanning–methods | Microscopy,
Electron, Scanning–instrumentation
Classification: LCC QH212.S3 (ebook) | LCC QH212.S3 (print) | NLM QH 212.S3 |
DDC 502.8/25–dc23
LC record available at https://lccn.loc.gov/2018021231

Cover Design: Wiley
Cover Image: Courtesy of Roland Fleck with coloring by Willy Blanchard, Electron Microscopy Facility, University of Lausanne, Lausanne Switzerland

Set in 10/12pt SabonLTStd by SPi Global, Chennai, India
Printed and bound in Singapore by Markono Print Media Pte Ltd

10 9 8 7 6 5 4 3 2 1

Contents to Volume I

About the Editors		xix
List of Contributors		xxi
Foreword		xxv

1 Scanning Electron Microscopy: Theory, History and Development of the Field Emission Scanning Electron Microscope 1
David C. Joy

1.1 The Scanning Electron Microscope 1
1.2 The Thermionic Gun 3
1.3 The Lanthanum Hexaboride ("LaB_6") Source 4
1.4 Other Enhanced "Higher Brightness" Sources 4
1.5 The Twenty-First Century SEM 6
1.6 The Future for Ion Beam Imaging – Above and Beyond 6
References 6

2 Akashi Seisakusho Ltd – SEM Development 1972–1986 7
Michael F. Hayles

2.1 Introduction 7
2.2 TEM Development 7
2.3 SEM Development, with TEM Repercussions 8
2.4 TEM Again, BUT SEM Lives On 8
2.5 MSM (Mini Scanning Microscope) Table-Top Series 9
2.5.1 Alpha Bench-Top Series 9
2.5.2 SMS Bench-Mounted Series 9
2.5.3 ISI Series 10
2.5.4 DS130 (W) Tungsten 11
2.6 Sigma (SS) Series 15
2.6.1 Vari-Zone Lens Modes of Operation 15
2.6.2 Environmental SEM 17
2.6.3 Integrated Circuit (IC) Orientated SEM 18
2.6.4 Dedicated Integrated Circuit (IC) SEM 18
2.6.5 Measurement Notation and Bar 20
Acknowledgements 23
References 23

3 **Development of FE-SEM Technologies for Life Science Fields** 25
Mitsugu Sato, Mami Konomi, Ryuichiro Tamochi and Takeshi Ishikawa

3.1 Introduction 25
3.2 Principle of SEM and Mechanism of Resolution 26
3.2.1 Principle of SEM 26
3.2.2 Mechanism of SEM Resolution 27
3.3 Commercialization of FE-SEM and the Impact of its Application 30
3.3.1 Commercialization of FE-SEM 30
3.3.2 Impact of FE-SEM on Applications 31
3.4 Development of In-Lens FE-SEM and its Impact 32
3.4.1 Development of the In-Lens FESEM 32
3.4.2 Adoption of the High Sensitivity Backscatter Electron Detector (YAG Detector) 34
3.4.3 Expansion of a Low Accelerating Voltage Application 34
3.5 Introduction of Semi In-Lens FE-SEM 36
3.5.1 Objective Lens of the Semi In-Lens System 36
3.5.2 Advance of the Signal Detection System 37
3.5.3 Applications 39
3.6 Resolution Improvement by the Deceleration Method 42
3.6.1 Retarding Method and Boosting Method 42
3.6.2 Resolution Improvement Effect by the Deceleration Method 42
3.6.3 Expanded Applications with Ultralow Accelerating Voltage 43
3.7 Popularization of a Schottky Emission Electron Source and Progress of a Cold FE Electric Gun 43
3.7.1 Popularization of a Schottky Emission Electron Source 43
3.7.2 Progress of the CFE Electron Gun 44
3.8 Advent of Truly “Easy to Use” FE-SEM 45
3.8.1 Outline 45
3.8.2 No Expertise Necessary 45
3.8.3 Simplified Beam Alignment 46
3.8.4 Coaching for Skill Improvement 46
3.9 Closing Remarks 46
Acknowledgements 47
References 48

4 **A History of JEOL Field Emission Scanning Electron Microscopes with Reference to Biological Applications** 53
Kazumichi Ogura and Andrew Yarwood

4.1 The First JEOL Scanning Electron Microscopes 53
4.2 The First Cryo-SEM 55
4.3 Development of JEOL Field Emission SEMs 57
4.4 In-Lens Field Emission SEM Development 59
4.5 Introduction of the JEOL Semi In-Lens 61
4.6 Evolution of the JEOL Semi In-Lens FE-SEMs 64
4.7 Development of Beam Deceleration and the JEOL Energy Filter 66
4.8 A Unique Aberration Corrected FE-SEM 70
4.9 Ongoing Semi In-Lens FE-SEM Development 71
4.10 JEOL Introduces the Super Hybrid Lens 72
4.11 Development of the JEOL Gentle Beam System 74
4.12 Conclusion 77
References 77

5 TESCAN Approaches to Biological Field Emission Scanning Electron Microscopy 79
Jaroslav Jiruše, Vratislav Košťál and Bohumila Lencová

5.1 Historical Introduction 79
5.2 Biological Samples In SEM 80
5.2.1 Why FE-SEM? 81
5.2.2 SEM Optics and Displaying Modes 81
5.2.3 Ultra-high Resolution Microscopy 84
5.2.4 Beam Deceleration Mode 85
5.2.5 In-Flight Beam Tracing 87
5.2.6 Detection System 87
5.3 Methods for Imaging Biological Samples 89
5.4 Imaging of Sensitive Samples 90
5.4.1 Imaging at Low Energies 90
5.4.2 Low Vacuum Operation 91
5.4.3 Observation of Biological Samples Without any Preparation 92
5.5 Advanced FE-SEM Techniques in Biology 94
5.5.1 Correlative Microscopy 94
5.5.2 Integration of the Confocal Raman Microscope in SEM 95
5.5.3 FIB-SEM Instrumentation 96
5.5.4 Preparation of Cross-sections and 3D Tomography in FIB-SEM 99
Conclusions 100
Acknowledgements 100
References 100

6 FEG-SEM for Large Volume 3D Structural Analysis in Life Sciences 103
Ben Lich, Faysal Boughorbel, Pavel Potocek and Emine Korkmaz

6.1 Introduction 103
6.2 High Resolution SEM Imaging at Low Accelerating Voltage 104
6.3 Spot Size 104
6.4 Beam Penetration 106
6.5 Contrast and Signal-to-noise Ratio 108
6.6 Serial Block Face Imaging 109
6.7 Challenges of SBFI 110
6.8 Recent Advances 110
6.9 Multienergy Deconvolution 111
6.10 Stability 111
6.11 Speed 113
6.12 Automated Acquisition, Reconstruction, and Analysis 114
6.12.1 Maps 114
6.13 Conclusion 114
References 115

7 ZEISS Scanning Electron Microscopes for Biological Applications 117
Isabel Angert, Christian Böker, Martin Edelman, Stephan Hiller, Arno Merkle and Dirk Zeitler

7.1 Biological Imaging using Zeiss Technology 119
7.1.1 ZEISS GEMINI® Technology 119
7.1.2 Helium Ion Microscopy in Biology 127
7.1.3 Multibeam SEM 128

7.2 3D Imaging – Live Happens in 3D 129
7.2.1 GEMINI®: One Technology for Several 3D Approaches 129
7.3 Correlative Solutions – for a Deeper Insight 133
7.3.1 Shuttle & Find – The Interface for Easy Correlation 134
7.3.2 Correlative Microscopy Going 3D 135
7.3.3 New Possibilities in Correlative Microscopy 137
Acknowledgement 139
References 139

8 SEM Cryo-Stages and Preparation Chambers 143
Robert Morrison

8.1 Overview 143
8.2 History 145
8.3 Types of Cooling 145
8.3.1 Braid Cooling 145
8.3.2 Gas Cooling 145
8.4 Location of the Preparation Chamber 147
8.4.1 On-Column Preparation Chamber (Figure 8.5) 147
8.4.2 Off-Column Preparation Chamber (Figure 8.6) 147
8.5 Location of the Cooling Dewar 149
8.5.1 On-Column Cooling (Figure 8.7) 149
8.5.2 Off-Column Cooling (Figure 8.8) 150
8.6 Sample Preparation 150
8.7 Freezing Mehods 151
8.7.1 Slushed Nitrogen Freezing 151
8.7.2 Propane Jet Freezing (Moor, Kistler and Müller, 1976) 151
8.7.3 Ethane Plunging (Dubochet *et al.*, 1988) 152
8.7.4 Slam Freezing (Dempsey and Bullivane, 1976) 152
8.7.5 High-Pressure Freezing (Dahl and Stachelin, 1989) 152
8.8 Mounting Methods 154
8.8.1 Surface Mounting 154
8.8.2 Edge Mounting 154
8.8.3 Filter Mounting 154
8.8.4 Hole Mounting 155
8.8.5 Liquid Film Mounting 155
8.8.6 Rivet Mounting 157
8.9 Fracturing 157
8.10 Sublimation 158
8.11 Coating 159
8.11.1 Metal Sputtering 160
8.11.2 Carbon Coating 160
8.12 More Advanced Techniques and Equipment 161
8.12.1 CryoFIB Lift-Out and On-Grid Thinning 161
8.12.2 Cryo Rotate Stages 163
8.12.3 SEM Stage Bias 163
8.12.4 Cryo-STEM in SEM 164
8.12.5 Cryo-EDS 165
8.13 Conclusion/Summary 165
References 165

9 **Cryo–SEM Specimen Preparation Workflows from the Leica Microsystems Design Perspective** **167**
Guenter P. Resch

9.1 Introduction 167
9.2 Specimen Fixation 168
9.2.1 Ambient Pressure Freezing Methods 168
9.2.2 High Pressure Freezing 169
9.3 The Vacuum Cryo-Transfer Shuttle 172
9.4 Freeze Fracture and Freeze Etching 176
9.4.1 Freeze Etching 181
9.5 Cryo-Planing 181
9.6 Coating 183
9.7 The Leica EM VCT500 Cryo-Sem Set 185
9.8 Summary 186
Acknowledgements 188
References 188

10 **Chemical Fixation** **191**
Bruno M. Humbel, Heinz Schwarz, Erin M. Tranfield and Roland A. Fleck

10.1 Introduction 191
10.2 Aldehydes 192
10.2.1 Formaldehyde (FA) 192
10.2.2 Glutaraldehyde (GA) 194
10.3 Acrolein 195
10.4 Osmium Tetroxide 195
10.5 Uranyl Acetate 198
10.6 Less Common Fixatives 198
10.6.1 Malachite Green 198
10.6.2 Ruthenium Red 199
10.7 Mixtures of Fixtures 200
10.7.1 Formaldehyde and Glutaraldehyde 200
10.7.2 Buffered Formaldehyde and Picric Acid 201
10.7.3 Glutaraldehyde and Osmium Tetroxide 201
10.7.4 Osmium Tetroxide–Potassium Ferrocyanide Staining 202
10.7.5 Osmium Tetroxide–Thiocarbohydrazide–Osmium Tetroxide (OTO) 202
10.7.6 Osmium Tetroxide and Tannic Acid 202
10.8 Summary of Action of Fixatives 203
10.9 Buffers 203
10.10 Water Source 204
10.11 SEM Preparation 205
10.11.1 Critical-Point Drying and Room Temperature Preparation 205
10.11.2 Ionic Liquids 206
10.11.3 Resin 208
10.12 Some Thoughts on the Preparation Protocols for Volume Microscopy 209
10.13 Conclusion 211
References 211

11 A Brief Review of Cryobiology with Reference to Cryo Field Emission Scanning Electron Microscopy **223**
Roland A. Fleck, Eyal Shimoni and Bruno M. Humbel

11.1 Cryopreservation and Brief History of Low Temperature Biology 223
11.2 History of Freezing for Electron Microscopy Observation 224
11.3 Temperature and the Condensed Phases of Water 227
11.3.1 Supercooling 227
11.3.2 Homogeneous Ice Nucleation 228
11.3.3 Heterogeneous Ice Nucleation 228
11.3.4 Post-Nucleation 229
11.3.5 Vitrification 230
11.3.6 Thawing 231
11.3.7 Irruptive Recrystallisation 231
11.3.8 Migratory Recrystallisation 231
11.3.9 Spontaneous recrystallisation 232
11.4 Freeze Drying 232
11.5 Mechanisms of Low Temperature Damage and Injury 234
11.6 Cryoprotectants 237
11.7 Cryopreservation of Biological Systems 239
11.8 Vitrification the 'Key' to Cryo-FEGSEM 244
11.9 Conclusion 253
Acknowledgements 253
References 253

12 High-Resolution Cryo-Scanning Electron Microscopy of Macromolecular Complexes **265**
Sebastian Tacke, Falk Lucas, Jeremy D. Woodward, Heinz Gross and Roger Wepf

12.1 Summary 265
12.2 Introduction 265
12.3 Prerequisites for High-Resolution Sem (HRSEM) 266
12.3.1 Macromolecular Structure Preservation 266
12.3.2 Controlling the Freeze-Drying Process: Partial Freeze-Drying 269
12.4 A Versatile High-Vacuum Cryo-Transfer System 271
12.5 Blurring, Noise and other Artefacts 273
12.5.1 Mass Loss During Imaging – Beam Damage 274
12.5.2 Contamination During Imaging 274
12.5.3 The Sample Support: Strategies for HRSEM for 'Beam Transparent' Samples 277
12.5.4 Low Dose Imaging versus Signal-to-Noise Ratio 278
12.6 All About Coating 279
12.6.1 Signal Enhancement – Contrasting Techniques for HRSEM 280
12.6.2 Coating Techniques 282
12.6.3 Tungsten Planar Magnetron Sputtering versus e-Beam Evaporation 284
12.6.4 Coating Film Thickness 284
12.6.5 Elevation Angle 286
12.6.6 Which Metal? 288
12.6.7 Example: Comparison of W, Cr and Pt/Ir/C Coating of Protein 2S Crystal Layer 289
12.7 Final Resolution Obtainable During HRSEM Work from Metal Coating 290
12.8 Molecular HRSEM Imaging on Frozen Hydrated Bulk Tissue 292
12.9 Conclusion 294
Acknowledgements 295
References 295

Contents to Volume II

About the Editors xix
List of Contributors xxi
Foreword xxv

13 FESEM in the Examination of Mammalian Cells and Tissues 299
Andrew Forge, Anwen Bullen and Ruth Taylor

Acknowledgements 308
References 308

14 Public Health/Pharmaceutical Research – Pathology and Infectious Disease 311
Paul A. Gunning and Bärbel Hauröder

14.1 Introduction 311
14.2 Biological Tissue Specimen Preparation 312
14.2.1 Fixation Considerations 312
14.2.2 Dehydration Considerations 314
14.2.3 Alternative Preparation Methods 315
14.3 Biomaterials Sample Preparation 317
14.3.1 Cross-sections; Soft Tissues and Soft/Semi-soft Biomaterials 318
14.3.2 Cross-sections; Hard Tissues and Hard Biomaterials 319
14.4 Determination of an Optimal Instrumental Approach 321
14.4.1 Example 1. Nanocrystalline Silver Coated Wound Dressing 323
14.4.2 Example 2. Biofilms on Fibrous Gauze Wound Dressing 323
14.4.3 Example 3. Resorbable Polymer Microspheres 324
14.4.4 Example 4. Gel Networks 325
14.5 Energy Dispersive X-Ray Microanalysis 326
14.5.1 EDS Example 1. Change Control 327
14.5.2 EDS Example 2. Contaminant Identification 329
14.5.3 EDS Example 3. Organic Materials 329
14.5.4 EDS Example 4. Detection of Asbestos and Mineral Fibres in Lung Tissue 331
14.5.5 EDS Example 5. Detection and Analysis of Foreign Bodies and Inclusions in Tissue 333
14.6 Quality Control of Biomedical Products 336
14.6.1 EDS Example 6. 'White Powder' as a Potential Biohazards Material 336
14.7 Summary 340
Acknowledgements 341
References 341

15 Field Emission Scanning Electron Microscopy in Cell Biology Featuring the Plant Cell Wall and Nuclear Envelope 343
Martin W. Goldberg

15.1 Introduction 343
15.2 Plant Cell Wall 346
15.3 Plasmodesmata: Connecting Plant Cells 350
15.4 Nuclear Envelope In Plants, Animals and Fungi 352
15.5 Conclusion 358
Acknowledgements 358
References 358

16 Low-Voltage Scanning Electron Microscopy in Yeast Cells **363**
Masako Osumi

16.1 Introduction 363
16.2 Development and Outline of Ultrahigh Resolution Low-Voltage Scanning Electron Microscopy (UHR LVSEM) 364
16.2.1 Resolution of UHR LVSEM 364
16.2.2 Dynamics of the Ultrastructure During Cell Wall Formation in *Schizosaccharomyces pombe* 365
16.2.3 Identification of Cell Wall Components 367
16.3 Development and Outline of Ultralow-Temperature Low-Voltage Scanning Electron Microscopy (ULT LVSEM) 371
16.3.1 High-pressure Freezing Method for ULT LVSEM 371
16.3.2 Freeze-fracture Method of Sample Preparation for TEM 371
16.3.3 Resolution of ULT LVSEM 372
16.3.4 *In situ* Observation of High-Pressure Frozen *S. pombe* Cells by ULT LVSEM 372
16.3.5 3D ULT LVSEM Images of Septum Formation 372
16.4 *In situ* Localization of the Cell Wall Component *a*-1,3-Glucan and Its Synthase During Septum Formation 375
16.4.1 Immunoelectron Microscopy (IEM) 375
16.4.2 Freeze-fracture Replica Labeling Method for TEM 375
16.4.3 Analysis of Localization of the Cell Wall Components and its Syntase by IEM 376
16.5 Discussion and Summary 379
Acknowledgements 380
References 380

17 Field Emission Scanning Electron Microscopy in Food Research **385**
Johan Hazekamp and Marjolein van Ruijven

17.1 Introduction 385
17.2 A Closer Look at Food Microstructure 385
17.3 Cryo-Preparation and Observation 386
17.4 Applications of Food Microscopy 389
17.4.1 Ice Cream 389
17.4.2 Foams 389
17.4.3 Monoglyceride Networks 390
17.4.4 Gelling Agents 392
17.5 Facts and Artefacts 392
17.6 Concluding Remarks/Summary 395
Acknowledgements 395
References 395

18 Cryo-FEGSEM in Biology **397**
Paul Walther

18.1 Introduction 397
18.2 Cryo-Preparation and Cryo-FEGSEM 398
18.2.1 Freezing 398
18.2.2 Cryo-Stage and Cryo-Transfer to the SEM 400
18.2.3 Cryo-Fracturing 400
18.2.4 Beam Sensitivity and Coating 400

18.2.5 Cryo-planing 405
18.2.6 Partial Freeze Drying 408
18.3 Discussion and Outlook 408
18.4 Materials and Methods 409
Acknowledgements 411
References 411

19 Preparation of Vitrified Cells for TEM by Cryo-FIB Microscopy 415
Yoshiyuki Fukuda, Andrew Leis and Alexander Rigort

19.1 Introduction 415
19.2 Operating Principle of the FIB Instrument 417
19.2.1 Interactions of the Ion Beam and Sample – Sputtering and Milling 417
19.2.2 Beam Generation and Shaping in the Ion Column 417
19.2.3 Ion Beam-Induced Damage and Artefacts 419
19.3 Cryo-FIB Applications in Biology 420
19.4 Instrumentation for Cryo-FIB Milling 421
19.5 The Manufacture of *In Situ* Cryo-TEM Lamellae from Vitrified Cells on EM Grids 423
19.5.1 FIB Protocol for Cryo-TEM Lamella Preparation 423
19.6 Electron Cyro-Microscopy and Tomography of Large FIB-Milled Windows 426
19.6.1 Providing Windows into the Cell's Cytoplasm 426
19.6.2 Study of FIB-Prepared Primary Neuronal Cells 428
19.6.3 *In Situ* Mapping of Macromolecular Complexes 429
19.7 Outlook: Enabling Structural Biology *In Situ* by Cryo-FIB Preparation 432
Acknowledgements 435
References 435

20 Environmental Scanning Electron Microscopy 439
Rudolph Reimer, Dennis Eggert and Heinrich Hohenberg

20.1 Introduction 439
20.2 Signal Generation in the ESEM 441
20.2.1 Imaging of Non-conductive Samples 442
20.2.2 Imaging of Hydrated Samples 443
20.3 Radiation Damage 444
20.4 Application of ESEM For Investigation of Hydrated Biomedical Samples 445
20.4.1 Water in Biomedical Material: Hydrated Samples in the ESEM 445
20.4.2 Biofilms 448
20.4.3 Lipids 449
20.4.4 Tissue Surfaces 450
20.4.5 Tissue Microanatomy 451
20.4.6 Specific Staining 452
20.4.7 Correlative Fluorescence Light and Environmental Scanning Electron Microscopy 453
20.4.8 Living Specimens 454
20.5 Preparation Steps for ESEM 454
20.5.1 Investigation of Living Specimens 454
20.5.2 Investigation of Hydrated Tissues 455
20.6 Summary and Outlook 455
Acknowledgements 458
References 458

21 Correlative Array Tomography **461**
Thomas Templier and Richard H.R. Hahnloser

21.1 Introduction 461
21.1.1 Array Tomography and Its Tradeoffs 461
21.1.2 Volumetric Electron Microscopic Imaging: To Handle, Stain, and Store Hundreds of Ultrathin Sections 462
21.1.3 Correlative Light and Electron Microscopy 462
21.1.4 Workflow 463
21.2 Cat Sample Preparation Protocols 463
21.2.1 Fixation and Embedding 463
21.2.2 Section Cutting and Collection 467
21.2.3 Postembedding On-Section Immunohistochemistry 469
21.2.4 Data Acquisition 472
21.3 Application: Identification of Projection Neuron Type in Ultrastructural Context 476
21.4 Conclusion 479
Acknowledgements 479
References 479

22 The Automatic Tape Collection UltraMicrotome (ATUM) **485**
Anwen Bullen

22.1 Introduction 485
22.2 Sample Preparation for Atum 487
22.2.1 Staining 487
22.2.2 Sectioning 487
22.2.3 Mounting 488
22.2.4 Imaging 489
22.2.5 Alignment and Segmentation 490
Acknowledgements 493
References 493

23 SBEM Techniques **495**
Christel Genoud

23.1 Introduction 495
23.2 Serial Sections with TEM 496
23.3 Development of Environmental SEM 498
23.3.1 Detection Modes in SEM Compatible with Resin Blocks 499
23.3.2 Insertion of a Microtome in the SEM Chamber 500
23.3.3 Sample Preparation 502
23.3.4 Pre-embedding Immunolabelling 504
23.3.5 Sample Mounting 504
23.3.6 Image Acquisition 505
23.3.7 Post-processing and Analysis of Data 507
23.4 Results Obtained With This Technique 507
23.4.1 Neuroscience 507
23.4.2 Cell Biology 508
23.4.3 Organs and Tissues 508
23.4.4 Organism 508
23.4.5 Material Science 509
23.5 Conclusion 509
References 509

24 FIB-SEM for Biomaterials **517**
Lucille A. Giannuzzi

24.1 Introduction and FIB Basics 517
24.2 Geometry for 2D Sectioning And Imaging 520
24.3 Geometry for 3D Sectioning and Imaging 522
24.4 Applications of 2D Sectioning and Imaging with FIB-SEM 523
24.5 3D FIB-SEM Tomography 524
24.6 3D FIB-SEM Tomography with Multi-signal SEM Acquisition 526
24.7 TEM Specimen Preparation with FIB-SEM 528
24.8 Summary 530
Acknowledgements 530
References 531

25 New Opportunities for FIB/SEM EDX in Nanomedicine: Cancerogenesis Research **533**
Damjana Drobne, Sara Novak, Andreja Erman and Goran Dražić

25.1 Introduction 533
25.2 Materials and Methods 534
25.3 Results 537
25.4 Discussion 540
Acknowledgement 542
References 542

26 FIB-SEM Tomography of Biological Samples: Explore the Life in 3D **545**
Caroline Kizilyaprak, Damien De Bellis, Willy Blanchard, Jean Daraspe and Bruno M. Humbel

26.1 Introduction 545
26.2 Focus on Sample Preparation 548
26.2.1 Chemical Fixation 549
26.2.2 High-Pressure Freezing and Freeze-Substitution 551
26.2.3 Resin Embedding 554
26.3 Focus on the Geometry of the Sample in Relation to the Ion and Electron Beam 557
26.3.1 Geometry of the Instrument 557
26.3.2 Geometry of the Resin Block 558
26.4 Beauty of the FIB-SEM Investigation 561
26.4.1 Example of the Mitochondrial Network in Mouse Liver 561
26.5 Conclusion 561
References 562

27 Three-Dimensional Field-Emission Scanning Electron Microscopy as a Tool for Structural Biology **567**
J.D. Woodward and R.A. Wepf

27.1 Introduction 567
27.2 Theory 570
27.2.1 Backprojection 570
27.2.2 Image Formation in the SEM 571
27.2.3 SEM Backprojection 571
27.3 Tomographic 3DSEM 573
27.3.1 Tilt Strategies 574
27.3.2 Signal-to-Noise Ratio 576
27.3.3 Resolution 576

27.4 Single-Particle Reconstruction 578
27.5 In Practice 580
27.5.1 Sample Preparation 580
27.5.2 Tomographic Reconstruction 580
27.5.3 Single-Particle Reconstruction 581
27.6 Applications 581
27.6.1 Handedness 581
27.6.2 Bridging the Resolution Gap 582
References 584

28 Element Analysis in the FEGSEM: Application and Limitations for Biological Systems 589
Alice Warley and Jeremy N. Skepper

28.1 Introduction 589
28.2 Specimen Preparation 590
28.2.1 Chemical Fixation, Critical Point Drying, Freeze-Drying or Resin-Embedding 590
28.2.2 Cryo-Immobilisation and Freeze-Drying 591
28.2.3 Preparation for Analysis in the Frozen-Hydrated State 592
28.2.4 Choice of Coating Material 593
28.3 Production of X-Rays and X-Ray Detection 593
28.3.1 Production of X-Rays 593
28.3.2 X-Ray Detection 594
28.4 Interaction Volume 595
28.4.1 Size of Interaction Volume in Biological Specimens 597
28.5 EDS Using Low Voltages 598
28.5.1 Low Excitation of X-Rays 598
28.5.2 Voltage Instability 598
28.5.3 Low Beam Current 598
28.5.4 Surface Coating 601
28.6 Peak Identification 601
28.7 Quantification 601
28.7.1 ZAF 604
28.7.2 XPP 604
28.7.3 Phi-Rho-Zed 604
28.7.4 Continuum Normalisation 604
28.7.5 Peak-to-Local Background 605
28.8 Summary 605
Acknowledgements 605
References 606

29 Image and Resource Management in Microscopy in the Digital Age 611
Patrick Schwarb, Anwen Bullen, Dean Flanders, Maria Marosvölgyi, Martyn Winn, Urs Gomez and Roland A. Fleck

29.1 Introduction 611
29.2 Resource Management 612
29.3 Resource Discoverability 612
29.4 Usage Optimization 613
29.5 Utilization Monitoring 614
29.6 Data Management 615
29.7 Project Management 615
29.8 Image Acquisition 615

29.9 Multimedia Data in Science 616
29.10 Benefits of Digital Management Systems 616
29.11 Image Database – A System-Relevant Component in Science 617
29.12 Image Database – An Integrated Tool in Today's Laboratory Work 618
29.13 From Technical Limitations to Solutions 619
29.14 Data Volume on Acquisition 619
29.15 Review of Data 620
29.16 Store Once, View Multiple Times 620
29.17 3D Image Processing 621
29.18 3D Volume Processing 622
References 623

30 Part 1: Optimizing the Image Output: Tuning the SEM Parameters for the Best Photographic Results 625
Oliver Meckes and Nicole Ottawa

30.1 Image Adjustments 625
30.1.1 Sharpness/Astigmatism 625
30.1.2 Brightness/Contrast 625
30.1.3 Integration Time/Noise Reduction 629
30.2 Empty Magnification and Useful Scan Size 631
30.3 SAVE: 8 Bit, 16 Bit, and The Whole Image Formats … 633
30.3.1 The Various Image Formats 633

Part 2: Post-Processing of the Photomicrograph 637
30.4 Optimization and Colourization of Sem Images 640
30.4.1 Common Digital Formats 641
30.4.2 Filters 641
30.5 Colouring of SEM Images 647
30.5.1 Creating Masks 647
30.5.2 From the Black-and-White Image with Alpha Channel to Colour Image 651
30.5.3 Blending Various Detector Signals 653
30.6 Conclusion 657
References 657

31 A Synoptic View on Microstructure: Multi-Detector Colour Imaging, nanoflight® 659
Stefan Diller

31.1 Introduction 659
31.2 Black and White versus Colour Imaging 661
31.2.1 Electron Microscopists and Coloured Imaging 661
31.2.2 Colour 663
31.2.3 Cinematographers Camera Tricks 666
31.2.4 Motion Picture and the SEM 667
31.3 Materials and Methods 670
31.3.1 Nanoflight Hardware System Setup 670
31.3.2 Nanoflight Software Setup 672
31.4 Conclusion 677
Acknowledgements 677
References 678

Index 679

About the Editors

ROLAND A. FLECK, PhD, FRCPath, FRMS, is a Professor in Ultrastructural Imaging and Director of the Centre for Ultrastructural Imaging at King's College London. Having specialised in basic research into cellular injury at low temperatures and during cryo-preservation regimes he has developed specialist knowledge of freeze fracture/freeze etch preparation of tissues and wider cryo-microscopic techniques. As director of the Centre for Ultrastructural Imaging he supports advanced three dimensional studies of cells and tissues by both conventional room temperature and cryo electron microscopy. He is a visiting Professor of the Faculty of Health and Medical Sciences, University of Copenhagen and Professor of the UNESCO Chair in Cryobiology, National Academy of Sciences of Ukraine, Institute for Problems of Cryobiology, Kharkiv, Ukraine.

BRUNO M. HUMBEL, Dr. sc. nat. ETH, is head of the Imaging Section at the Okinawa Institute of Science and Technology, Onna son, Okinawa, Japan. He is awarded a research professorship at Juntendo University, Tokyo, Japan. He got his PhD at the Federal Institute of Technology, ETH, Zurich, Switzerland, with Prof. Hans Moor and Dr. Martin Müller, both pioneers in cryo-electron microscopy (high-pressure freezing, freeze-fracturing, freeze-substitution and low-temperature embedding, cryo-SEM, cryo-sectioning). His research focuses on sample preparation for optimal, life-like imaging of biological objects in the electron microscope. The main interests are preparation methods based on cryo-fixation applied in Cell Biology. From here, hybrid follow-up methods like freeze-substitution or freeze-fracturing are used. He is also involved in immunolabelling technology, e.g., ultra-small gold particles and has been working on techniques for correlative microscopy and volume microscopy for a couple of years. He teaches cryo-techniques and immunolabelling and correlative microscopy in international workshops and has professional affiliations with Zhejiang University, Hangzhou, People's Republic of China as a distinguished professor and co-director of the Center of Cryo-Electron Microscopy and with the Federal University of Minas Gerais, Belo Horizonte, Brazil, as a FAPEMIG visiting professor at the Centro de Microscopia da UFMG.

List of Contributors

Isabel Angert, Carl Zeiss Microscopy GmbH, Oberkochen, Germany

Christian Böker, Carl Zeiss Microscopy GmbH, Oberkochen, Germany

Damien De Bellis, Electron Microscopy Facility, University of Lausanne, Lausanne, Switzerland

Willy Blanchard, Electron Microscopy Facility, University of Lausanne, Lausanne, Switzerland

Faysal Boughorbel, Thermo Fisher Scientific, Eindhoven, The Netherlands

Anwen Bullen, UCL Ear Institute, University College London, United Kingdom

Jean Daraspe, Electron Microscopy Facility, University of Lausanne, Lausanne, Switzerland

Stefan Diller, Scientific Photography, Wuerzburg, Germany

Goran Dražić, National Institute of Chemistry, Laboratory for Materials Chemistry, Ljubljana, Slovenia

Damjana Drobne, Department of Biology, Biotechnical Faculty, University of Ljubljana, Slovenia

Martin Edelman, Carl Zeiss Microscopy GmbH, Oberkochen, Germany

Dennis Eggert, Heinrich Pette Institute, Leibniz Institute for Experimental Virology, Hamburg, Germany

Andreja Erman, Institute of Cell Biology, Faculty of Medicine, University of Ljubljana, Slovenia

Dean Flanders, The Friedrich Miescher Institute for Biomedical Research, Basel, Switzerland

Roland A. Fleck, Centre for Ultrastructural Imaging, King's College London, UK

Andrew Forge, UCL Ear Institute,University College London, United Kingdom

Yoshiyuki Fukuda, Department of Structural Biology, Max Planck Institute of Biochemistry, Martinsried, Germany and Department of Cell Biology and Anatomy, Graduate School of Medicine, University of Tokyo, Tokyo, Japan

Christel Genoud, Friedrich Miescher Institute for Biomedical Research, Basel, Switzerland

Lucille A. Giannuzzi, L.A. Giannuzzi & Associates LLC, Fort Myers, FL, USA and EXpressLO LLC, Lehigh Acres, FL, USA

Martin W. Goldberg, Science Laboratories, School of Biological and Biomedical Sciences, Durham University, Durham, United Kingdom

Urs Gomez, Imagic Bildverarbeitung AG, Glattbrugg, Switzerland

Heinz Gross, ETH ScopeM, Swiss Federal Institute of Technology, ETH-Hönggerberg, Zurich, Switzerland

Paul Gunning, Smith & Nephew Advanced Wound Management, Hull, UK

Richard H.R. Hahnloser, Institute of Neuroinformatics, University of Zurich and ETH Zurich, Neuroscience Center, Zurich, Switzerland

Bärbel Hauröeder, Zentrales Institut des Sanitätsdienstes der Bundeswehr, Koblenz, Germany

Mike F. Hayles, Cryo-FIB-SEM Technology, Eindhoven, The Netherlands

Johan Hazekamp, Unilever R&D Vlaardingen, Vlaardingen, The Netherlands

Stephan Hiller, Carl Zeiss Microscopy GmbH, Oberkochen, Germany

Heinrich Hohenberg, Heinrich Pette Institute, Leibniz Institute for Experimental Virology, Hamburg, Germany

Bruno M. Humbel, Electron Microscopy Facility, University of Lausanne, Switzerland and Imaging Section, Okinawa Institute of Science and Technology, Onna-son, Okinawa, Japan

Takeshi Ishikawa, Science and Medical Systems Business Group, Hitachi High-Technologies Corporation, Minato-ku, Tokyo, Japan

Jaroslav Jiruše, TESCAN, Brno, Czech Republic

David C. Joy, 232 Science and Energy Research, Facility and Department of Materials Science and Engineering, The University of Tennessee, Knoxville, TN, USA

Caroline Kizilyaprak, Electron Microscopy Facility, University of Lausanne, Lausanne, Switzerland

Vratislav Košťál, TESCAN, Brno, Czech Republic

Mami Konomi, Science and Medical Systems Business Group, Hitachi High-Technologies Corporation, Minato-ku, Tokyo, Japan

Emine Korkmaz, Thermo Fisher Scientific, Eindhoven, The Netherlands

Andrew Leis, Bio21 Molecular Science and Biotechnology Institute, The University of Melbourne, 30 Flemington Road, Parkville, Victoria 3010, Australia

Bohumila Lencová, TESCAN, Brno, Czech Republic

Ben Lich, Thermo Fisher Scientific, Eindhoven, The Netherlands

Falk Lucas, ETH ScopeM, Swiss Federal Institute of Technology, ETH-Hönggerberg, Zurich, Switzerland

Maria Marosvölgyi, arivis AG, Business Unit arivis Vision, Rostock, Germany

Oliver Meckes, Eye of Science, Reutlingen, Germany

Arno Merkle, Carl Zeiss Microscopy GmbH, Oberkochen, Germany

Robert Morrison, Quorum Technologies, Ltd, Laughton, UK

Sara Novak, Department of Biology, Biotechnical Faculty, University of Ljubljana, Slovenia

Kazumichi Ogura, JEOL Ltd, Akishima, Tokyo, Japan

Masako Osumi, Japan Woman's University, Tokyo, Japan and NPO Integrated Imaging Research Support, Tokyo, Japan

Nicole Ottawa, Eye of Science, Reutlingen, Germany

Pavel Potocek, Thermo Fisher Scientific, Eindhoven, The Netherlands

Rudolf Reimer, Heinrich Pette Institute, Leibniz Institute for Experimental Virology, Hamburg, Germany

Guenter P. Resch, Nexperion e.U. – Solutions for Electron Microscopy, Wien, Austria

Alexander Rigort, Thermo Fisher Scientific, FEI Deutschland GmbH, Germany

Marjolein van Ruijven, Unilever R&D Vlaardingen, Vlaardingen, The Netherlands

Mitsugu Sato, Research and Development Division, Hitachi High-Technologies Corporation, Hitachinaka, Ibaraki-ben, Japan

Patrick Schwarb, Imagic Bildverarbeitung AG, Glattbrugg, Switzerland

Heinz Schwarz, Max Planck Institute for Developmental Biology, Tübingen, Germany

Eyal Shimoni, Department of Chemical Research Support, Weizmann Institute of Science, Rehovot, Israel

Jeremy Skepper, Cambridge Advanced Imaging Centre, University of Cambridge, Cambridge, United Kingdom

Sebastian Tacke, ETH ScopeM, Swiss Federal Institute of Technology, ETH-Hönggerberg, Zurich, Switzerland

Ryuichiro Tamochi, Science and Medical Systems Business Group, Hitachi High-Technologies Corporation, Minato-ku, Tokyo, Japan

Ruth Taylor, UCL Ear Institute, University College London, United Kingdom

Thomas Templier, Institute of Neuroinformatics, University of Zurich and ETH Zurich, Neuroscience Center, Zurich, Switzerland

Erin M. Tranfield, Institute Gulbenkian de Ciência,Oeiras, Portugal

Paul Walther, Central Facility for ElectronMicroscopy, Ulm University, Ulm, Germany

Alice Warley, Centre for Ultrastructural Imaging, King's College London, United Kindom and Visiting Professor, Department of Histology and Cell Biology Faculty of Medicine, University of Granada, Granada, Spain

Roger Wepf, ETH ScopeM, Swiss Federal Institute of Technology, ETH-Hönggerberg, Zurich, Switzerland and UQ, CMM University of Queensland, Brisbane, Australia

Martyn Winn, STFC Daresbury Laboratory, Warrington, United Kingdom

Jeremy Woodward, SBRU, University of Cape Town, South Africa

Jeremy D. Woodward, Department of Integrative Biomedical Sciences in the division of Medical Biochemistry & Structural Biology and Structural Biology Research Unit, University of Cape Town, Cape Town, South Africa

Andrew Yarwood, JEOL (UK) Ltd, Welwyn Garden City, Hertfordshire, UK

Dirk Zeitler, Carl Zeiss Microscopy GmbH, Oberkochen, Germany

Foreword

The task of microscopy in biology is to provide the structural information for correlation of structure and function in complex biological systems. Specimen preparation and imaging techniques should be directed towards the preservation and imaging of the smallest possible significant details in order to fully exploit this unique, integrating feature of biological microscopy, complementing the more linear biochemical procedures. High spatial and temporal resolution is required to describe an aqueous dynamic biological system closely related to the living state. A quantitative description of biological structures down to molecular dimensions may remain a dream, but one must nevertheless attempt to realize it.

With the advent of the field emission electron source paired by important progress in related electron optics, scanning electron microscopy has reached a level of structural and analytical resolution comparable to transmission electron microscopy. Cryoimmobilization techniques, in addition, can rapidly arrest the living processes.

The main problems during the preparation of biological material for electron microscopy arise from the necessity to transform the living sample into a solid state in which it can resist the physical impact of the electron microscope. Basically, preparative procedures are identical for TEM and SEM. In SEM the much higher dose and the necessity to localize the signal at the specimen surface demand additional solutions.

Biological electron microscopy laboratories are often established and financed to service biological research projects, with the biological question being academically the only relevant part. The efforts, however, to investigate new preparative ideas and to set up, modify and/or improve existing approaches are high. Sound methodological research directed to the solution of relevant biological questions helps to educate greater understanding of the physical and chemical changes a sample may encounter during processing and imaging. Thus, advancing electron microscopy into a source of primary information not only for systems of reduced complexity, such as vitrified layers of macromolecules, will enable microscopists to discuss with bioscientists directly linkages between structure and function and for micrographs to be more often selected for the data they contain rather than for mere illustrative purposes.

Biologists normally choose the electron micrographs according to their expectation and thanks to the introduction of more stable and user friendly instruments many biologists routinely operate advanced instrumentation, however, high resolution (highly significant) information is not obtained only because a field emission SEM (FEGSEM) is used. All the steps involved in specimen preparation and imaging must be carefully understood in order to appreciate the possible level of primary information.

Sound basic research has been possible only in a few laboratories and the present books (Volume I and II) aim to collate knowledge of sample preparation procedures and electron optics to promote best practice in the use of the FEGSEM to answer basic biological questions.

Martin Müller

13

FESEM in the Examination of Mammalian Cells and Tissues

Andrew Forge, Anwen Bullen and Ruth Taylor
UCL Ear Institute, London, UK

Vertebrate tissues and the cells of which they are comprised are structurally highly organised. The topographical distribution of molecules, the arrangement of macromolecular assemblies and the disposition of subcellular organelles underlie the way in which a cell works, and different cell types are structurally organised in particular ways to enable them to perform their specific functions. The proper, often precise, arrangement of different cell types in a tissue is also essential to the maintenance of physiological activity. Architectural details are therefore crucial to understanding how cells and tissues perform and how they malfunction when disrupted during disease. FESEM provides a number of advantages for detailed assessment of cell and tissue architecture. In this chapter we focus on some of the procedures for preparing samples for FESEM imaging under non-cryogenic conditions and the kind of information that can be obtained, illustrated by examples from our own work on the inner ear of a variety of vertebrates, as well as from studies of mammalian cells in culture.

One advantage of scanning electron microscopy (SEM) is the ability to examine easily large fields of cells and large samples. This enhances assessment of the relative incidence of features of interest, as well as the context of those features in relation to other cells, the tissue, organ (Figure 13.1A to G) or even the entire organism. The details of cellular organisation within a tissue and changes that may occur with development or pathology can also be identified (Figure 13.1H and I). With FESEM, the high resolution capability potentially allows assessment of organisation from the macromolecular level upwards and the low electron beam energies employed for imaging reduce the chances of beam-induced damage to the specimen, thereby enhancing preservation of delicate structures (Figure 13.1D to G). Most commonly with SEM it is the surfaces of samples that are examined, but this can provide a wealth of information. Different cells may have differing and distinguishing

Biological Field Emission Scanning Electron Microscopy, First Edition.
Edited by Roland A. Fleck and Bruno M. Humbel.

surface features, for example the presence or absence of microvilli or of endocytotic vesicles that are characteristic of particular cell types and therefore provide identifying markers (Figure 13.1D). The surfaces of individual cells may have different characteristics depending on their locations; for example, in epithelial cells, the luminal (apical) surface across which secretion, absorption or environmental interaction may occur differs from that of the lateral surface with, or across, which there may be interactions with adjacent cells in the tissue (Figure 13.1H and I). The basal surface, where there may be interactions with extracellular matrices or with other cell types such as neurones, may also be distinguished. Cell surfaces may be covered by some kind of, usually organised, macromolecular complex, and the surface will contain particular molecules that may be evenly distributed or clustered into specific areas (Figure 13.1E–G). Cell surface features may be lost or altered in pathology (Figure 13.2A an B). There are also a number of methods that have been developed to examine the internal structure of cells and tissues by SEM. There is thus the possibility of exploring architectural organisation from the level of the intracellular macromolecular organisation within a much broader context of the organised tissue. With the ability to identify molecular complexes and the application of backscattered electron detection to localise proteins labelled by immunogold techniques, FESEM offers opportunities to make comprehensive analyses of normal cell and tissue architecture and of the causes and consequences of pathological processes.

The ability to image samples without a conductive coat, made possible by the use of a low voltage electron beam in FESEM, potentially also allows assessment of tissues at high resolution in a 'close-to-life' state in which there has been minimal preparation of the sample prior to examination in the microscope. This can be accomplished using techniques such as rapid freezing to preserve structure in samples that are then examined on a temperature

→

Figure 13.1 Structural details of sensory epithelia of the inner ear at different magnifications. (A) Intact spiral of the organ of Corti, the auditory sensory epithelium, of the mouse. Arrows denote the location of the thin strip of sensory epithelium. Anaglyph prepared from stereoimages reveals the 3D form of the organ of Corti spiral. (B) The organisation of sensory cells – the hair cells – along the organ of Corti spiral is displayed (guinea pig). The hair cells derive their name from the organised bundle of projections from their apical surface. There is a single row of inner hair cells (ihc) and three rows of outer hair cells (ohc) along the sensory strip. Each hair cell is separated from its neighbour by intervening supporting cells. (C) At higher magnification (mouse organ of Corti) details of the surface characteristics of hair and supporting cells are revealed. The hair bundle is composed of 'stereocilia' in rows of ascending height, the bundle 'polarisation' thereby defined is the same for all hair cells. The sterocilia of ihc are wider than those of ohc. There are variations in the surface characteristics of the supporting cells. (D) Endocytotic vesicle openings at the surface of some cells (arrows), while other cells are characterised by long microvilli (guinea pig organ of Corti). (E, F, G) Details of specialised cell surface coat around stereocilia: (E) lateral links between stereocilia (arrows) (vestibular sensory epithelium of mouse); (F) tip links (arrows) at the tip of shorter stereocilium linking to longer ones behind in the direction of bundle polarisation (newt saccular macula); (G) high power of tips links suggests bifurcation at the upper end (ohc of mouse). (H) Cellular organisation within the body of the sensory epithelium (mouse organ of Corti). The supporting cells have an enlarged cell body that encloses the base on an ohc (arrow) and a thin phalangeal process that rises at an angle to the surface. The bodies of the ohc are angled in the opposite direction. (I) Reorganisation of the epithelium with loss of hair cells. The supporting cells have enlarged and are upright rather than angled. The characteristics of the apical surface of the cells – numerous short microvilli – are distinct from the lateral membrane of the cell – distributed short projections. Scale bars: A, 100 μm; B, 5 μm; C, 2.5 μm; D, 1 μm; E, 200 nm; F, 100 nm; G and H, 5 μm.

controlled cold stage in the microscope (Echlin, 1978; Read and Jeffree, 1991). However, such procedures require specialised equipment and particular expertise and are not easily applicable to most vertebrate tissues, although isolated cells and cell cultures may be prepared in this way (Osumi *et al.*, 2006; Walther, 2008) (and see Chapter 16 by Osumi and Chapter 18 by Walther). Tissue processing procedures appropriate for the study to be undertaken are therefore often applied prior to examination by FESEM. This inevitably leads to some compromises; for example, techniques that provide optimal structural preservation may not always be compatible with the protocols required for preservation of antigens or the exposure of the epitopes necessary for antibody recognition and labelling with gold particle probes.

Nevertheless, however the sample is to be examined, it must be prepared in some way to be compatible with the environment inside the instrument. This inevitably involves drying, which can induce distortion and shrinkage of vertebrate tissue so some form of fixation is essential (although of itself that does not prevent tissue shrinkage (Boyde and Maconnachie, 1980; Kaab, Notzil and ap Gwynn, 1998). In most cases, fixation is best achieved with glutaraldehyde. Although the use of low voltages in FESEM can allow examination

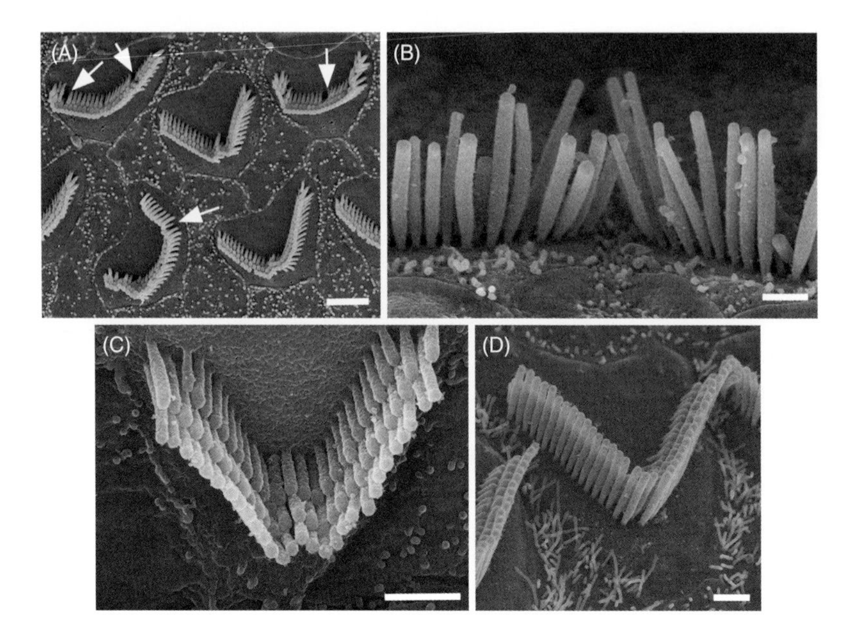

Figure 13.2 Details of pathologies (A, B) and effects of different preparation techniques (C, D). (A) Hair bundles on outer hair cells are mis-shapen and individual sterocilia are missing from bundles (arrows). The outline of the surface of each hair cell is resolved and the shape of the apical surface is seen to conform to the shape of the bundle (Bardet-Biedl syndrome (BBS)-6 KO mouse (May-Simera *et al.*, 2009)). (B) Variations in lengths and widths of stereocilia within a single hair bundle of an inner hair cell (Plastin-1 KO mouse; Taylor *et al.*, 2015a). (C) Wrinkling of surface membranes and some bending of stereocilia in a sample prepared conventionally. (D) Smooth surface membranes and upright stereocilia in a sample prepared using the OTOTO procedure. Scale bars: A, 2 μm; B, 1 μm; C, 1 μm; D, 1 μm.

of samples with no conductive coat, the thickness, complexity and irregularity of many vertebrate tissues means that it is often difficult to establish a conductive path through a non-coated sample that enables current to flow to ground. Several procedures for incorporating heavy metals directly into samples to provide an endogenous conductive pathway have now been described. These usually involve the binding of osmium to provide conduction, either through the use of tannic acid (Jongebloed *et al.*, 1999) or thiocarbohydrazide (TCH) (Davies and Forge, 1987; Malik and Wilson, 1975). In our hands, the use of tannic acid often results in deposits upon the sample surface so TCH is preferred. In this protocol (Malik and Wilson, 1975), following an initial post-fixation with OsO_4 the tissue is exposed to TCH, then OsO_4, and these two steps are then repeated (i.e. Os-TCH-Os-TCH-Os (OTOTO)). Extensive washing (with water) between each step to remove unbound TCH or OsO_4 is necessary as both chemicals are relatively insoluble and any that is unbound in the tissue can precipitate on the surface if not washed away. After dehydration and drying, the sample is mounted on a holder using conductive silver paint (e.g. fast drying silver suspension, Agar Scientific) as adhesive. This provides the continuity between the specimen and ground.

The protocol results in the binding of osmium to enhance conductivity throughout the tissue, not just at the surface, and since the sample can be examined in the instrument without a sputtered or evaporated heavy metal coating, the resolution at high magnifications of the grains resulting from the condensation of deposited metal is avoided. We have

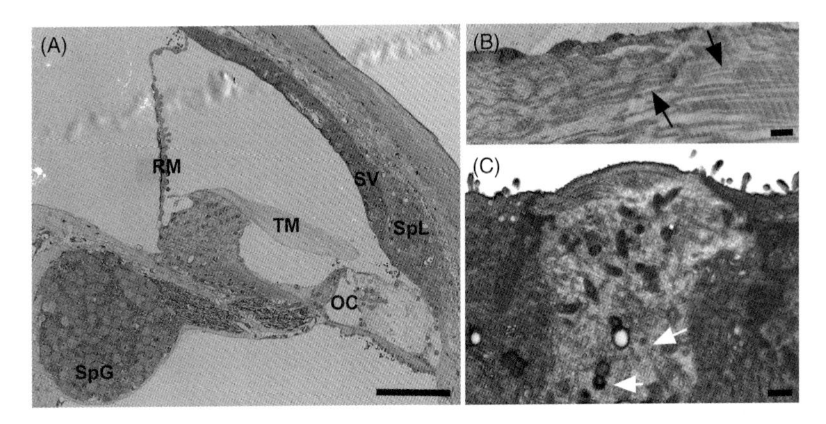

Figure 13.3 Thin sections imaged by backscatter detection in the SEM. (A) Low power image of a large area in which the high resolution of the imaging reveals subcellular details. Single cochlear turn after conventional preparation for thin sections and lead citrate and uranyl acetate staining of the section. The cellular composition of all the tissues that comprise the cochlea are visible. OC = organ of Corti; TM = tectorial membrane; SpG = spiral ganglion containing the cell bodies of the auditory nerves; SpL = spiral liagament; SV = stria vascularis (ion transporting epithelium of the cochlea); RM = Reissner's membrane. (B) Fine details in conventionally prepared thin-section: tectorial membrane in which collagen banding (arrows) is resolved. (C) Thin section of sample prepared by the OTOTO procedure that was first examined by SEM, then removed from SEM stub and embedded in plastic. High resolution capability enables identification of fine subcellular details such as microtubules (arrows). Scale bars: A, 100 μm; B, 2 μm; C, 0.5 μm.

also found that the use of the TCH-Os- repeated protocol (OTOTO) confers a degree of rigidity on tissues that lessens the distortion or collapse of structure during drying (Davies and Forge, 1987) (Figure 13.2C and D), although fine detail may be preserved even with glutaraldehyde-only fixation followed by dehydration and drying (Figure 13.2C). In addition, samples prepared in this way can be processed after examination by SEM for thin-sectioning (Hunter-Duvar, 1978; Hunter-Duvar and Mount, 1978; Taylor *et al.*, 2015b) (Figure 13.3C).

Routine thin sections of plastic embedded tissue as normally prepared for TEM can be examined directly with the FESEM. FESEM instruments are capable of producing high emission currents at low voltages, compared to microscopes employing other emission sources. The probe current, and therefore the backscatter electron signal from a sample, can be increased by this method without increasing the spot size of the beam. These conditions improve the signal to noise ratio of backscatter imaging and are particularly useful in imaging biological specimens (Richards, Owen and ap Gwynn, 1999). The heavy metals used in the tissue processing and/or bound to the sections by post-section staining provide the moieties from which backscattering derives and their differential binding to structural elements in the tissue provides the contrast. The relatively low energy at the sample surface in normal FESEM operating conditions lessens the heating and, thus, distortion of the plastic section under electron bombardment, and a coating of carbon over the section can enhance conductivity. Sections may also be placed on solid conductive supports, such as silicon wafers, as opposed to the grids used in TEM. Examination of thin sections under these conditions by FESEM (see Chapter 22 by Bullen on ATUM) means that very large areas of tissue can be examined without the interruptions of grid bars, as would happen with TEM, over a wide range of magnifications, from those that cover low power light microscopy to ultrastructural resolutions normally only available with TEM (Figure 13.3A,B andC). Impregnating tissues with heavy metals during processing, for example using OTOTO procedures, prior to embedding in plastic and sectioning (Figure 13.3C) is now becoming increasingly used for the examination of serial sections to enable three-dimensional reconstruction of tissue. One way this is achieved is with a microtome inside the SEM. As each successive section is cut off, images of the exposed block face are obtained by backscattered electron imaging (scanning block face SEM -SBFSEM) (Denk and Horstmann, 2004) (see Chapter 23 by Genoud). The relatively low energy of the electron beam in an FESEM reduces effects on the embedding plastic that could lead to distortion and enhances resolution as there is relatively low beam penetration into the sample and thus less beam broadening than would occur at higher beam energies. Potential charging of the non-conductive plastic can be accommodated with a thin heavy metal coating over the external surface of the plastic block. This procedure essentially requires a dedicated SEM since the microtome takes the place of the normal specimen holder inside the instrument so needs to be removed for other types of SEM imaging. An alternative to this is so-called 'serial section scanning electron microscopy (S3EM)', also known as 'array tomography' (Horstmann *et al.*, 2012) (see Chapter 21 by Templier and Hahnloser). Here long ribbons of serial thin sections of tissue, which can be prepared conventionally or with additional heavy metal incorporation, are collected on to a silicon wafer. The sections are conventionally labelled with uranium and/or lead salts and then, after drying, are coated with evaporated carbon to assist conduction. The section ribbons are then imaged by backscatter or secondary electron detection in the SEM (see Chapter 22). Software programs (e.g. Atlas 5 from Zeiss or SEM Supporter from JEOL) drive the microscope stage and perform image recognition and automatic image recording so that the same area of interest on each successive section is imaged and recorded. The software for array tomography is still in development but this technique affords opportunities to achieve low and high resolution 3D reconstruction of the same tissue with relative ease because an individual sample can be re-imaged several times so examination of the 3D structure over a range of magnifications is possible. All these procedures can potentially be used with tissue sections in which proteins have been immunolabelled using gold particles, which are detectable in the backscatter mode of the instrument (Schwarz and Humbel, 2014).

Immunogold labelling applied directly to samples for examination by backscatter electron detection in the SEM allows exploration of the location and distribution of surface antigens on individual cells (Müller *et al.*, 1989; Walther and Müller, 1986; Walther *et al.*, 1984). For such studies cells or tissues are best fixed with formaldehyde at 2–4% to preserve antigenicity (which is adversely affected by the concentrations of glutaraldehyde normally used for tissue fixation), with addition of a low concentration –0.025%-glutaraldehyde to partially ameliorate the suboptimal structural preservation during processing for SEM that results from the use of formaldehyde alone. Following fixation and immunogold labelling by routine methods, samples are dehydrated, dried and then provided with a conductive coating of carbon, but it is also possible after the immunogold labelling to process samples with osmium and thiocarbohyrazide, as outlined above, to enhance conductivity without losing the ability to detect the gold label.

As examples, good preservation of the morphology of single cells such as lymphocytes can be obtained after fixation in formaldehyde (Figure 13.4A) and the immunolabelling reveals the protein of interest to be located exclusively at the tips of microvilli (Figure 13.4B). Cell cultures grown on a glass coverslip, or similar substrate, are easy to handle for processing and differences between cells in culture can be identified; in the sample shown (Figure 13.4C and D), the culture has been transfected with the gene encoding a protein (espin), which enhances the growth of elongated microvilli, together with that of a gene for a protein of

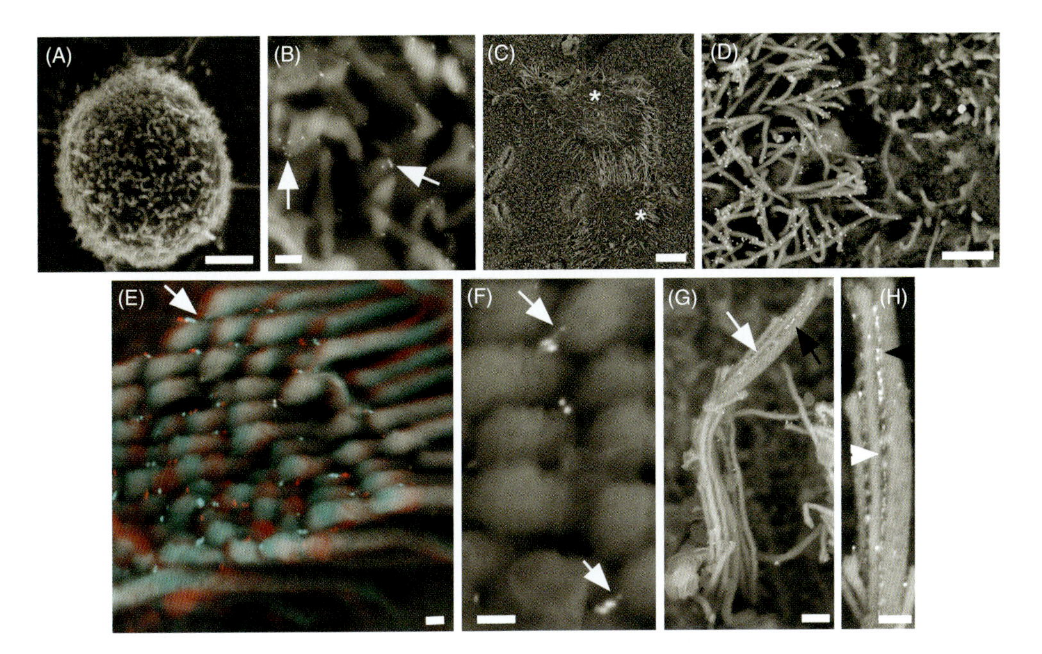

Figure 13.4 Immunogold labelling of cells and tissues. (A and B) Surface protein on lymphocyte. (A) Lymphocyte retains spherical shape with microvilli distributed across the surface. (B) Backscatter image of sample in A reveals 10 nm gold particles (white spots, arrowed) at tips of microvilli. (C and D) Culture of MCDK cells, transfected with genes for a protein (espin) that enhances microvillus formation and a secreted protein of interest (sample courtesy of G.P. Richardson and R. Goodyear). (C) The formation of long microvilli differentiates cells (marked with asterisks), which are expressing the transfected gene from those neighbours that have not. (D) Cells with elongated microvilli show immunogold labelling for secreted protein of interest, whereas the adjacent cell with normal short microvilli, which presumably has not been transfected, is unlabelled. (E and F) Gold particles of two different sizes, 10 nm and 5 nm, label at different ends along the line of the tip links between adjacent stereocilia (arrow). Anaglyph (E) reveals separate localisation of the two sizes of gold particle in the region of the tips of stereocilia. A high power view (F) confirms smaller particles at the bottom end of the link (tip of shorter stereocilium) and larger particle at the top end (shaft of the longer stereocilium) (a chick vestibular hair cell). (G and H) Separate localisation of immunogold-labelled proteins: H shows a higher power view of hair bundle in G, in which gold particle labelling is present in the space between the kinocilium (true cilium) and the longest stereocilia. Black arrows indicate larger (10 nm) particles, white arrows the smaller (5 nm) particles. Scale bars: A, 5 μm; B, 0.25 μm; C, 5 μm; D, 1 μm; E, 100 nm; F, 100 nm; G, 0.5 μm; H, 0.25 μm.

interest that is secreted from the cell. The gold particles that label the protein of interest are present on cells with elongated microvilli, but adjacent cells that have not expressed espin, that is that have short microvilli, do not express the secreted protein (Figure 13.4D). The high resolution capability of the FESEM also allows visualisation of gold particles of different sizes, at least as small as 5 nm (Müller *et al.*, 1989; Walther and Müller, 1986), even 1 nm (Hermann, Schwarz and Müller, 1991), to enable double, or even triple, gold labelling, in order to distinguish closely adjacent molecular components within macromolecular complexes (Goodyear *et al.*, 2010). In the example (Figure 13.4E and F), the two proteins that form the 'tip-link' (Figure 13.1F and G) – which gates the transduction channels of the sensory 'hair' cells in the inner ear and which are located at the distal ends of the 'stereocilia' (modified microvilli) of which the hair bundle on the sensory cell is composed – are labelled separately with gold particle of different sizes. The smaller 5 nm particles localise to the bottom end of the link (at the tip of the stereocilium) and the larger, 10 nm, particle to the top, at the shaft of the adjacent stereocilium in the next (longer) row (Goodyear *et al.*, 2010) (Figure 13.4F), indicating the differential location of the two proteins that compose the link. The same two proteins form the links between the single kinocilium – a true cilium – and the stereocilia in the longest row and again the two gold labels are easily distinguished in the FESEM in backscatter mode (Figure 13.4G and H).

Details of intracellular structural organisation can also be revealed following demembranation prior to tissue processing. This provides a means of examining in particular details of the cytoskeleton. A number of procedures for removing membranes have been described (DesMarais *et al.*, 2004; Shao *et al.*, 2006; Schliwa, van Blerkom and Porter, 1981; Svitkina, Verkhovsky and Borisy, 1995, 1996) but broadly they involve exposing unfixed cells or tissue to a detergent, Triton (sometimes with 0.5% saponin added), in a buffered (PIPES or HEPES) solution (pH 6) of physiologically relevant salts (5 mM KCl, 137 mM NaCl, 4 mM $NaHCO_3$, 0.4 mM KH_2PO_4, 1.1 mM Na_2HPO_4, 2 mM $MgCl_2$) containing EGTA, glucose and protease inhibitors as well as actin stabilisers such as phalloidin or phallocidin to preserve the actin cytoskeleton and microtubule stabilisers such a taxol. After a period of incubation in the demembranating conditions, the samples are washed in the buffered salt solution, then fixed and processed for SEM. Application of these procedures to the inner ear shows a network of cytoskeletal elements in the apical cytoplasm of supporting cells, which surround each hair cell, of the sensory epithelia, that is much denser and more organised than that in the cells of the non-sensory epithelium that surrounds each sensory patch (Figure 13.5A and B). Supporting cells play a crucial role in providing a rigid mechanical framework to support the apical transduction apparatus of the hair cell, and the relationship of the cytoskeletal network with the tight-adherens junctions region around the neck of the cell is exposed. A meshwork of fine fibrils of the size of actin filaments is seen to cover the inner surface of the membrane on the supporting cell side of the junction and to associate with filaments of the size of microtubules that fan out towards the intercellular

Figure 13.5 Imaging intracellular details. (A and B) Differences in complexity of cytoskeleton in different cell types revealed following detergent extraction of membrane. (A) In unspecialised cells in tissue surrounding the sensory epithelium there are relatively few loosely organised elements of the cytoskeleton. (B) In supporting cells (sc) in the sensory epithelium, which provide mechanical support to hair cells (hc), there is an extensive complex cytoskeletal network at the level of the intercellular junctions between cells (newt saccular macula). (C and D) Details of cytoskeletal organisation (in supporting cells of mouse organ of Corti). (C) Filaments of the size of microtubules (arrows) radiate out and associate with a meshwork of fibrils of the size of actin filaments that form a dense network on the inner surface of the plasma membrane (arrowheads). (D) Anaglyph shows the 3D organisation of the cytoskeletal elements that support the apical end of the outer hair cell (ohc). Microtubules (arrows) in a bundle running up towards the apical end of the cell fan out in the head region of the cell to associate with filamentous meshwork (arrowhead) on the inner surface of the supporting cell at the junction with the ohc. (E) Individual filaments resolved in demembranated stereocilia. (F) In demembranated cilium (kinocilium), parallel fibrils consistent with axonemes are resolved. (F′) Image processing (reversed FFT) suggests subunit structure (arrows) along the axoneme-like fibrils. (G) Immunogold labelling for myosin VIIa in partially demembranated sample, which localises protein to tips of the stereocilia (arrows). (H) Sample immunolabelled then processed through the OTOTO procedure. Intense labelling for acetylated α-tubulin along microtubules, but fibrils in meshwork that is seen, in anaglyph, to ensheath the microtubules are unlabelled. Scale bars: A, 1 μm; B, 1 μm; C, 0.5 μm; D, 1 μm; E, 100 nm; F, 100 nm; G, 0.150 nm; H, 100 nm.

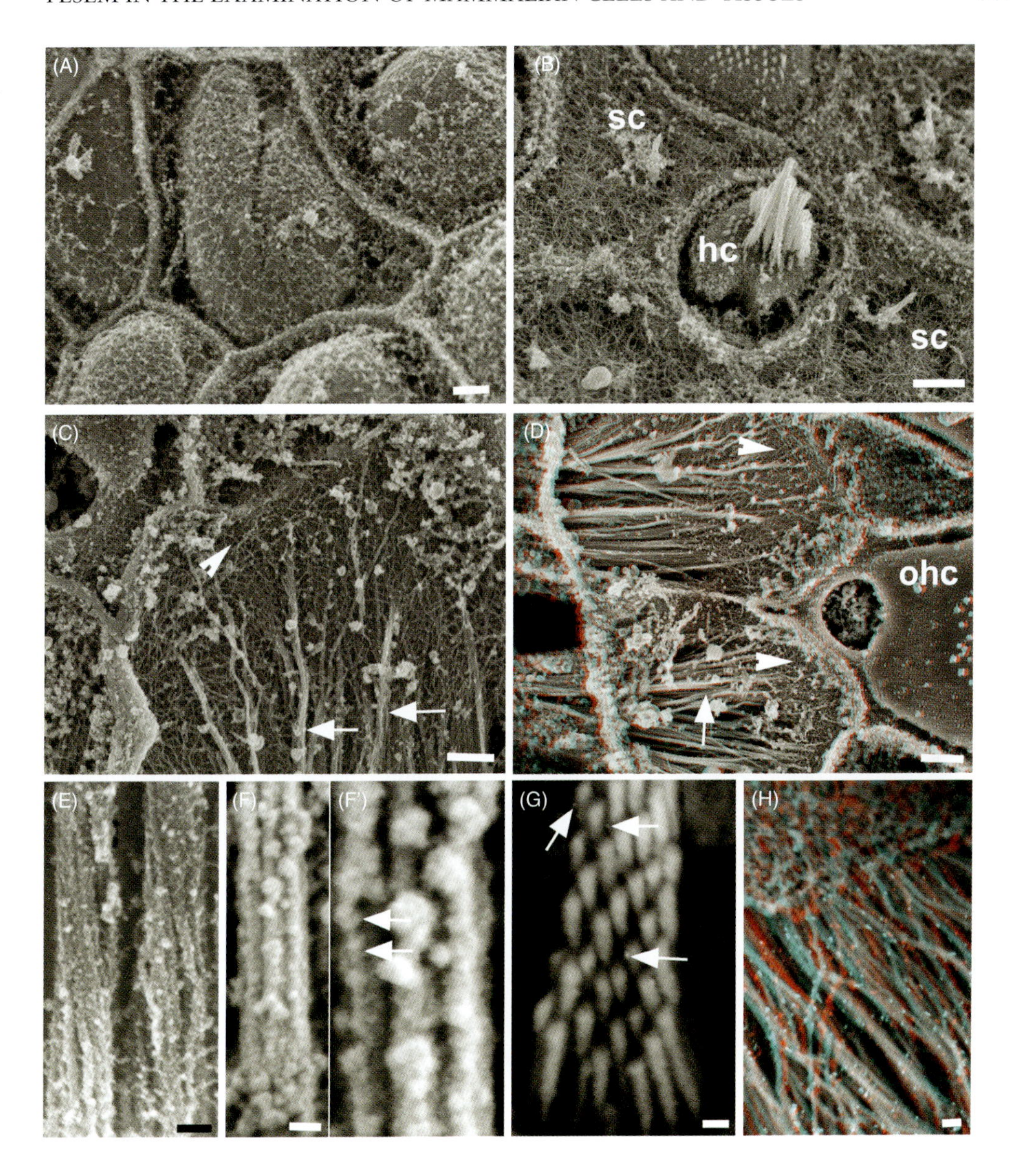

junctions (Figure 13.5C). Stereoimaging shows the 3D organisation of this cytoskeletal assembly (Figure 13.5D). The closely packed network of filaments that forms the cuticular plate at the apical end of the hair cell is also revealed. Removal of the membrane around the stereocilia shows the individual actin filaments (Figure 13.5E). Likewise, the axonemes within the kinocilium are resolved (Figure 13.5F) and image processing suggests that individual tubulin subunits can be resolved (Figure 13.5F′). Application of immunogold labelling procedures following demembranation allows identification of molecular components of, for example, cytoskeletal structures (Figure 13.5G) and distinctions in the molecular compositions with such structures. In the example illustrated in Figure 13.5H, the fibrils labelled with antibodies to alpha-tubulin are enclosed within a meshwork of filaments that are not labelled and have the dimensions of actin filaments. Stereoimaging more clearly reveals the enclosure of the bundle microtubules within the filamentous meshwork.

The ability of FESEM to produce high resolution imaging at low accelerating voltages, increasing the signal to noise ratio of images whilst limiting sample damage, makes it particularly suited to examining the complex and fragile structures present in mammalian tissues. Using a variety of tissue preparation techniques it is possible to perform a comprehensive analysis of the molecular architecture of complex cells and tissues, encompassing three-dimensional reconstruction, low and high resolution imaging of surfaces and internal structures, and examination of macromolecular complexes by high resolution imaging and immunolabelling. These methods combined produce not just important structural information in their own right, but may also be used to place structures in their biological context and to understand the relationships between tissue, cellular and macromolecular architecture.

ACKNOWLEDGEMENTS

Work was supported by grants from the UK MRC, the UK BBSRC, The Rosetrees Trust, Action on Hearing Loss (with Deafness Research UK) and The Dunhill Medical Trust. The authors are grateful to Guy Richardson and Richard Goodyear for the samples presented in Figure 13.4C and D.

REFERENCES

Boyde, A. and Maconnachie, E. (1980) Treatment with lithium salts reduces ethanol dehydration shrinkage of glutaraldehyde fixed tissue. *Histochemistry*, 66, 181–187.

Davies, S. and Forge, A. (1987) Preparation of the mammalian organ of Corti for scanning electron microscopy. *Journal of Microscopy*, 147, 89–101.

Denk, W. and Horstmann, H. (2004) Serial block-face scanning electron microscopy to reconstruct three-dimensional tissue nanostructure. *PLoS Biology*, 2, e329.

DesMarais, V., Macaluso, F., Condeelis, J. and Bailly, M. (2004) Synergistic interaction between the Arp2/3 complex and cofilin drives stimulated lamellipod extension. *J. Cell Sci.*, 117, 3499–3510.

Echlin, P. (1978) Low temperature scanning electron microscopy: A review. *Journal of Microscopy*, 112, 47–61.

Goodyear, R.J., Forge, A., Legan, P.K. and Richardson, G.P. (2010) Asymmetric distribution of cadherin 23 and protocadherin 15 in the kinocilial links of avian sensory hair cells. *J. Comp Neurol.*, 518, 4288–4297.

Hermann, R., Schwarz, H. and Müller, M. (1991) High precision immunoscanning electron microscopy using Fab fragments coupled to ultra-small colloidal gold. *J. Struct Biol.*, 107, 38–47.

Horstmann, H., Korber, C., Satzler, K., Aydin, D. and Kuner, T. (2012) Serial section scanning electron microscopy (S3EM) on silicon wafers for ultra-structural volume imaging of cells and tissues. *PLoS One*, 7, e35172.

Hunter-Duvar, I.M. (1978) A technique for preparation of cochlear specimens for assessment with the scanning electron microscope. *Acta Otolaryngol. Suppl.*, 351, 3–23.

Hunter-Duvar, I.M. and Mount, R.J. (1978) A technique for preparation of large cochlear specimens for assessment with the transmission electron microscope. *Acta Otolaryngol. Suppl.*, 351, 33–44.

Jongebloed, W.L., Stokroos, I., Van der Want, J.J. and Kalicharan, D. (1999) Non-coating fixation techniques or redundancy of conductive coating, low kV FE-SEM operation and combined SEM/TEM of biological tissues. *Journal of Microscopy*, 193, 158–170.

Kaab, M.J., Notzli, H.P. and ap Gwynn, I. (1998) Dimensional changes of articular cartilage during immersion freezing and freeze-substitution for scanning electron microscopy. *Scanning Microscopy*, 12, 465–474.

Malik, L.E. and Wilson, R.B. (1975) Evaluation of a modified technique for SEM examination of vertebrate specimens without evaporated metal layers. *IITRII SEM*, 1975.

May-Simera, H.L., Ross, A., Rix, S., Forge, A., Beales, P.L. and Jagger, D.J. (2009) Patterns of expression of Bardet-Biedl syndrome proteins in the mammalian cochlea suggest noncentrosomal functions. *J. Comp. Neurol.*, 514, 174–188.

Müller, M., Walther, P., Hermann, R. and Schwarb, P. (1989) SEM immunocytochemistry with small (5 to 15 nm) colloidal gold markers, in *Immuno-Gold Labeling in Cell Biology* (eds A.J. Verkleij and J.L.M. Leunissen), CRC Press, Boca Raton, FL, pp. 199–216.

Osumi, M., Konomi, M., Sugawara, T., Takagi, T. and Baba, M. (2006) High-pressure freezing is a powerful tool for visualization of *Schizosaccharomyces pombe* cells: ultra-low temperature and low-voltage scanning electron microscopy and immunoelectron microscopy. *J. Electron Microsc. (Tokyo)*, 55, 75–88.

Read, N.D. and Jeffree, C.E.. (1991) Low-temperature scanning electron microscopy in biology. *Journal of Microscopy*, 161, 59–72.

Richards, R.G., Owen, G.R. and ap Gwynn, I. (1999) Low voltage backscattered electron imaging (<5 kV) using field emission scanning electron microscopy. *Scanning Microscopy*, 13, 55–60.

Schliwa, M., van Blerkom, J. and Porter, K.R. (1981) Stabilization of the cytoplasmic ground substance in detergent-opened cells and a structural and biochemical analysis of its composition. *Proc. Natl Acad. Sci. USA*, 78, 4329–4333.

Schwarz, H. and Humbel, B.M. (2014) Correlative light and electron microscopy using immunolabeled resin sections. *Methods Mol. Biol.*, 1117, 559–592.

Shao, D., Forge, A., Munro, P.M. and Bailly, M. (2006) Arp2/3 complex-mediated actin polymerisation occurs on specific pre-existing networks in cells and requires spatial restriction to sustain functional lamellipod extension. *Cell Motil. Cytoskeleton*, 63, 395–414.

Svitkina, T.M., Verkhovsky, A.B. and Borisy, G.G. (1995) Improved procedures for electron microscopic visualization of the cytoskeleton of cultured cells. *J. Struct. Biol.*, 115, 290–303.

Svitkina, T.M., Verkhovsky, A.B. and Borisy, G.G. (1996) Plectin sidearms mediate interaction of intermediate filaments with microtubules and other components of the cytoskeleton. *J. Cell Biol.*, 135, 991–1007.

Taylor, R., Bullen, A., Johnson, S.L., Grimm-Gunter, E.M., Rivero, F., Marcotti, W., Forge, A. and Daudet, N. (2015a) Absence of plastin 1 causes abnormal maintenance of hair cell stereocilia and a moderate form of hearing loss in mice. *Hum. Mol. Genet.*, 24, 37–49.

Taylor, R.R., Jagger, D.J., Saeed, S.R., Axon, P., Donnelly, N., Tysome, J., Moffatt, D., Irving, R., Monksfield, P., Coulson, C., Freeman, S.R., Lloyd, S.K. and Forge, A. (2015b) Characterizing human vestibular sensory epithelia for experimental studies: New hair bundles on old tissue and implications for therapeutic interventions in ageing. *Neurobiol. Aging*, 36, 2068–2084.

Walther, P. (2008) High-resolution cryo-SEM allows direct identification of F-actin at the inner nuclear membrane of *Xenopus oocytes* by virtue of its structural features. *Journal of Microscopy*, 232, 379–385.

Walther, P. and Müller, M. (1986) Detection of small (5–15 nm) gold-labelled surface antigens using backscattered electrons, in *The Science of Biological Specimen Preparation 1985* (eds M. Müller, R.P. Becker, A. Boyde and J.J. Wolosewick), SEM Inc., AMF O'Hare, IL, pp. 195–201.

Walther, P., Kriz, S., Müller, M., Ariano, B.H., Brodbeck, U., Ott, P. and Schweingruber, M.E. (1984) Detection of protein A gold 15 nm marked surface antigens by backscattered electrons. *Scanning Electron Microsc.*, 1984, 1257–1266.

14

Public Health/Pharmaceutical Research – Pathology and Infectious Disease

Paul A. Gunning[1] **and Bärbel Hauröder**[2]

[1]*Smith & Nephew Advanced Wound Management, Hull, UK*
[2]*Zentrales Institut des Sanitätsdienstes der Bundeswehr, Koblenz, Germany*

14.1 INTRODUCTION

Once more likely to be found in academic research laboratories or national research institutes, FEGSEMs are increasingly common in public health institutions and in commercial research and development (R&D), quality assurance (QA) and regulatory support roles. This is not solely due to decreased capital costs for FEGSEM instruments but to the greatly increased flexibility provided by FEGSEM when handling tricky 'real world' samples. Many pharmaceutical and medical device products are composed of materials that may be extremely beam-sensitive and troublesome with regard to electrostatic charging (soft polymers, plasticised materials, microfibrous, highly lofted polymer textiles, biological tissue/material interfaces, colloids/particulates, etc.). FEGSEM provides better control over these issues and greatly enhanced resolution at the lower accelerating voltages needed to successfully image such samples. Of course, FEGSEM continues to be used for basic academic research related to public health; high-end FEGSEM/FEG-STEM instruments encroach on the resolution performance of TEM, but with greatly reduced energy dosage to fragile specimens via short exposure durations (raster scanned electron beam at low accelerating voltages as opposed to continuous irradiation at high accelerating voltages typical of TEM). New technologies such as aberration correction and energy monochromation continue to push the resolving power of FEGSEM ever closer to that of TEM, helping

Biological Field Emission Scanning Electron Microscopy, First Edition.
Edited by Roland A. Fleck and Bruno M. Humbel.

to address rapidly emerging industrial applications of nanoparticle and nanostructured biomaterials development.

14.2 BIOLOGICAL TISSUE SPECIMEN PREPARATION

14.2.1 Fixation Considerations

Several methods can be used to 'fix' tissue specimens; these methods may involve heating or microwave heating (thermal fixation) or freezing/flash freezing (cryo-fixation/cryo-preservation). However, for the purposes of this chapter, fixation of tissue by chemical means will be considered in the following paragraphs.

Biological tissue specimens can be grouped into three broad types: soft tissue (organs, fat, muscle, vasculature and nervous system), hard tissue (usually mineralised tissues, bone, teeth, mineralised plaques) and dense tissue (tendons, ligaments, skin and some connective tissues). Categorising tissue in this manner is helpful when deciding how best to achieve optimal fixation and dehydration. Somewhat counterintuitively, harder, denser tissue can be more problematic to prepare successfully.

Both hard tissue and dense tissue present substantial barriers to the diffusion of fixative and, as a result, often require extended fixation times of days, or even weeks, if larger blocks of more than just a few millimetres of such tissue are to be prepared. Poor fixation is also likely to cause difficulty if the tissue is to be examined in cross-section, for instance to examine the depth of mineralisation across tendon tissue where it inserts into bone. Poorly fixed specimens will be difficult to cut with a microtome, frequently causing distortion or tearing of the central portions of the sections or block-face.

Fixation of any tissue is often hampered by the rate of diffusion of fixative chemicals through the entire volume of the specimen and, for this reason, the size of the specimen also affects the ultimate quality of the fixation achieved; that is, large blocks (larger than 10 mm) of tissue may fix well enough around exterior surfaces, but inside the fixation is likely to be poor or even completely absent at the centre.

Soft tissue is also prone to poor fixation if large blocks are fixed. In this case the fixation process itself causes its own restriction to diffusion of fresh fix into the tissue, as the exterior surfaces gel or stiffen, causing something akin to 'skinning' around the block that the fixative must diffuse through to reach the tissue at the centre. Delicate tissues may often 'over-fix' at the periphery of the tissue block, resulting in densification, distortion or partial destruction of cells or organelles. This effect can, however, be mitigated by adding fixative in a controlled manner, gradually increasing the concentration of fix over several hours. This allows a more gradual permeation of fix through the block without a sudden heavy fixation of the exterior surfaces with associated penetration problems. A suitable approach is to transfer the specimen block through a series of fixative solutions of slowly increasing concentrations, leaving the specimen to equilibrate for 20 minutes or more between each transfer. Red blood cells and blood clots are a good example of 'soft tissue' that is very sensitive to fixation conditions. A staged approach to fixation usually produces specimens that are relatively free of osmotic shock/over-fixation effects that will deform the red blood cells toward echinocyte ('burr cell') morphologies, but yield well preserved discocyte morphologies (Figure 14.1). Again, optimal duration in each sequential solution depends on the size of the specimen; larger specimens require longer equilibration times.

Figure 14.1 Blood clot, showing good morphological (discocyte) preservation of red blood cells enmeshed in fibrin network. Note the activated platelet at the lower right of the clot.

Choice of fixative chemistry is also important, as most fixatives are somewhat specific/selective in the tissue components that they will fix. Aldehydes, commonly glutaraldehyde and/or formaldehyde and acrolein (acrylic aldehyde) will fix proteins and some lipoproteins and glyco-proteins. Traditionally, formalin solutions tend to be selected for light microscopy and glutaraldehyde selected for electron microscopy, because of the superior retention of ultra-structural detail resulting from glutaraldehyde fixation. However, glutaraldehyde is known to be slow-penetrating through tissue and is more susceptible to poor fixation problems at the centre of even moderately large tissue blocks with dimensions of more than just a few millimetres. Both chemistries can be used at once as a compromise, but a good understanding of the effects of fixatives is valuable before committing to a particular method. Useful information regarding the action of aldehyde fixatives was published in a comprehensive article written by Kiernan (2000).

Fats, lipids and oils will not be fixed by aldehydes, but most can be fixed using osmium tetroxide or ruthenium red. Osmium tetroxide (OsO_4) can solubilise some proteins, but it is generally preferable to use OsO_4 for secondary fixation, after any protein moieties have been stabilised by primary fixation with one of the aldehydes. Sufficient pH buffering capacity is important when using osmium and other fixatives, as many (including most aldehydes) are acidic and corrosive when made up in unbuffered aqueous solution.

Most fixatives are volatile compounds, so fixation can be undertaken by exposing the specimen to a vapour, as opposed to immersion in a liquid solution. Vapour fixation is usually accomplished by placing the samples in a desiccator or similar gas-tight vessel containing a saturated atmosphere of the fixative compound. This method of fixation can be helpful if there are concerns about dissolution or migration of some of the specimen materials during immersion in liquid for extended periods of time (e.g. if spatial distribution or crystallinity of a water-soluble pharmaceutical compound is to be examined). Notably, because of their volatility and propensity to cross-link and bind to biological tissue, significant health and safety issues must be considered when using most chemical fixatives. Suppliers of microscopy consumables therefore provide comprehensive safety data and advice when supplying these compounds.

Appropriate tissue fixation is a significant discipline in its own right. A more comprehensive review of fixation is outside the scope of this book chapter, but a very useful guide written by M.A. Hayat (1981) is still in print.

Appropriate, effective fixation is therefore of great importance for FEGSEM observations, especially if poor preservation of the very ultra-structural detail that FEGSEM is particularly suited to imaging is to be avoided.

14.2.2 Dehydration Considerations

With each successive preparative step undertaken, the risk of introducing preparation artefacts increases. Different dehydration methods may cause different artefacts (most commonly shrinkage) or differ in their severity. Biological specimens are usually dehydrated via critical point drying (CPD), freeze-substitution followed by critical point drying or freeze-drying (sometimes referred to as 'lyophilisation'). Shrinkage varies both by method and specimen type, and there are differing observations as to how much shrinkage occurs. Reports vary for CPD between 10 and 15% shrinkage for central nervous system tissue (Dykstra 1992) and up to 20% shrinkage for liver, lung or kidney tissue (Hayat 1978) or up to 41% reduction of original volume (Boyde and Machonnachie 1979). Freeze-drying and freeze-substitution methods are also known to impart artefacts caused by the formation of ice crystals within the hydrated material/tissue, but these artefacts can be minimised by including very small quantities of 'cryoprotectant' substances (glycerol, methanol, etc.) into the aqueous medium and ensuring as rapid a freezing process as possible. Small specimen sizes are, as always, preferable. Cryogenic liquids at extremely low temperatures such as liquid or 'slushed' nitrogen are used to achieve the most rapid 'vitrification' possible. Sometimes intermediate liquids, propane, ethane, isopentane or Freon, or methods such as slam freezing or high pressure freezing using specialised equipment, are also used to help achieve a freezing rate at conditions that allow for vitrification.

Critical point drying employs supercritical liquid carbon dioxide (CO_2); being an excellent solvent for many transition fluids, this method ensures excellent specimen dehydration, but it does require the use of a specialised temperature-controlled pressure vessel to achieve the conditions necessary for CO_2 to be maintained as a liquid (the vessel is supplied from a liquid CO_2 siphon cylinder). The transition fluid is gradually rinsed from the specimen by means of multiple rinses in liquid CO_2, usually involving some degree of agitation to ensure good mixing of transition fluid and CO_2. When the specimen has been entirely impregnated with liquid, final dehydration is accomplished by increasing the temperature and pressure in the CPD vessel to push the liquid beyond its critical point (31.1 °C at 7.39 MPa pressure) to become gaseous CO_2. The transition across the critical point is very gentle, ensuring minimal surface tension forces are exerted on the specimen.

Hexamethyldisilazane (HMDS) can provide an alternative method that also employs low surface tension to minimise potential disruption of fine microstructures. This organic solvent has a very low surface tension, low viscosity and good penetrative properties. It is miscible with typical transition fluids and has the advantage that the immersed specimens can simply be left to dry inside a fume hood, as opposed to the need for a specialised pressure vessel. Although shrinkage may be somewhat increased compared to CPD, less agitation of the sample is needed and it may yield better results with very delicate brittle or friable specimens.

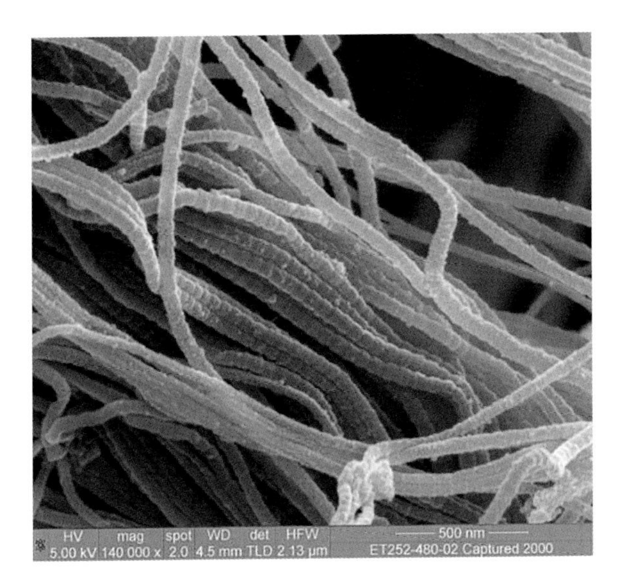

Figure 14.2 Collagen fibril ultra-structure, CPD only, ca. 5 nm gold–palladium coating.

Selecting the most appropriate dehydration method requires careful thought about what information is to be obtained from the specimen; it is not always the case that a more complex dehydration procedure will yield benefits at the highest magnifications. Ultrastructural details of collagen fibrils can be successfully imaged following chemical (glutaraldehyde) fixation and critical point drying (Figure 14.2). However, microstructural details within cartilage tissue may be lost by a simple CPD approach, but will be preserved by including an intermediate freeze-substitution step as described by Iolo ap Gwynn *et al.* (2000). Consulting the literature and discussion with other microscopists is always useful, particularly where an unfamiliar specimen type is to be prepared for FEGSEM. Evaluation of different methods is also valuable, as so often in microscopy multiple perspectives from different preparative methods are likely to yield higher quality information than reliance on a single method.

Methods involving CPD or HMDS can be problematic because the solvents used and the transition liquids used (ethanol, acetone, supercritical liquid CO_2 or HMDS) may cause selective or partial dissolution of some of the specimen components (e.g., extracellular polysaccharide layers on biofilm specimens can be partially removed by CPD).

The two major alternatives to dehydration methods of cryomicroscopy and ESEM ('environmental' SEM) are not discussed here because they are discussed elsewhere in this volume (see Chapter 17 by Hazekamp and van Ruijven, Chapter 18 by Walther and Chapter 20 by Reimer, Eggert and Hohenberg).

14.2.3 Alternative Preparation Methods

Occasionally, traditional fixation and dehydration may be impractical, due to reasons of time or experimental design. Dehydration of the specimen will alter its mechanical properties from its native state, which is problematic where time-course experiments are to be carried out on the same area of an individual specimen (studies examining the effects of mechanical wear on a surface or studies of surface dissolution, for example). Making a

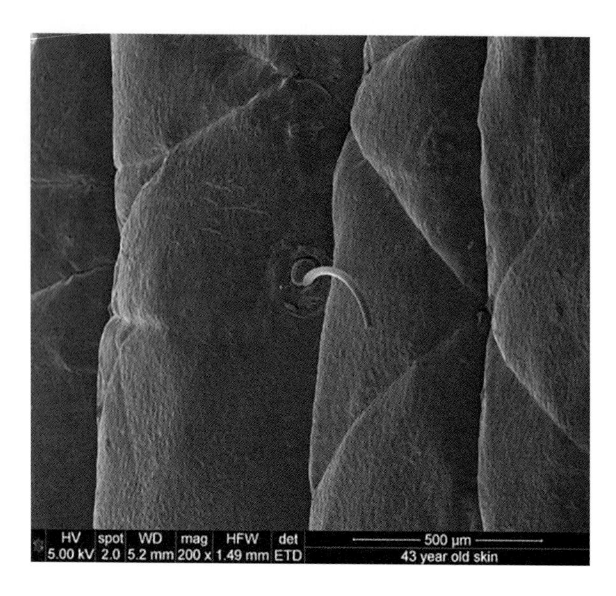

Figure 14.3 Moisturised skin replica, showing hair follicle.

replica of the specimen can prove useful by enabling rapid reproduction of the specimen to a non-perishable facsimile of the tissue surface. Replication is also useful where the specimen cannot be obtained, for example where a skin surface is to be examined at various time-points in a study of skin-cream or topical pharmaceutical applications (Figure 14.3). Replicas can be of surprisingly high fidelity (often conservatively stated as 0.1 μm feature reproducibility; http://kulzer-technik.de/en_kt/kt/maerkte/metallographie/produktbereiche /abdruckverfahren_1/provil_novo_1.aspx) and have great advantages in terms of repeat 'sampling' of the same area and in terms of ethics and welfare. Several replication materials can be used; cellulose acetate sheets have been popular in the past because of their tolerance to water/moist specimens (wet cartilage replication shown in Figure 14.4, micrographs of replicas shown in Figure 14.5). Silicone elastomers have become increasingly available and moisture tolerant dental replication compounds can provide excellent fidelity and great convenience, being easy to apply and have rapid curing times. These compounds impart very low/negligible thermal effects on the sample and the rapid cure helps to minimise specimen dehydration. Shrinkage on curing is 1% or less for polydimethylsiloxane (PDMS) or vinylpolysiloxane (VPS) compounds (Galbany *et al.* 2006), comparing very favourably with shrinkage of at least 5% to 40% for most dehydration processes including critical point drying, freeze-drying or freeze-substitution, or upwards of 20% for simple dehydrations from ethanol or acetone.

Non-destructive progressive snapshots of mechanical wear (cartilage examples are shown in Figure 14.6), dissolution processes or other progressive surface modifications necessitate working uncoated, because the presence of a coating risks modifying the very process under investigation. Even where the biomaterial sample may not require dehydration, low vacuum or low kV high vacuum imaging is likely to provide the best results in such cases. Replication of the surface can also be useful for such studies; maximising sample throughput may be important to allow the test sample to be returned quickly for continuation of progressive wear or surface modification experiments. Replication may be preferable if there are concerns that exposure of the sample material to vacuum or low vacuum conditions may change its mechanical/wear resistance properties or if the specimen is perishable (e.g. animal

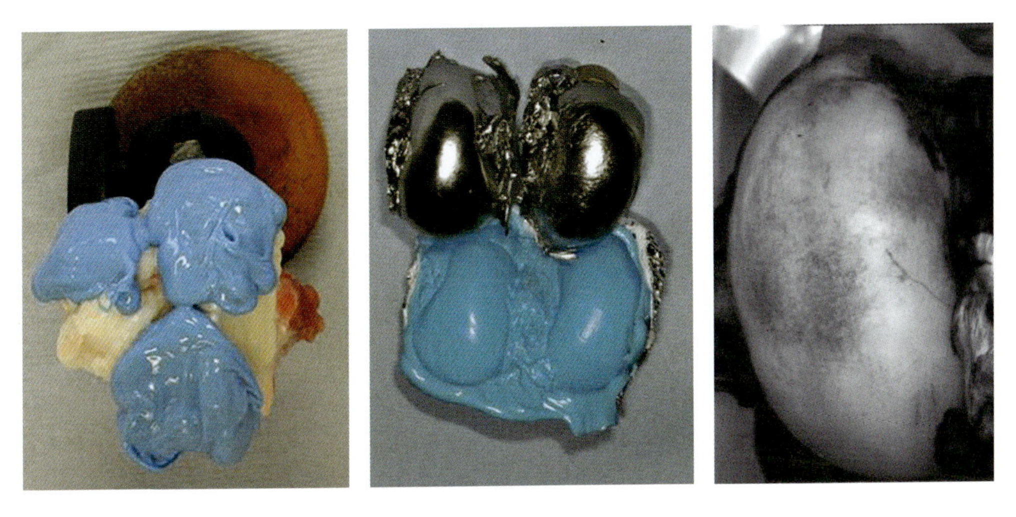

Figure 14.4 VPS (Provil® novo light, ISO 4823, Heraeus Kulzer, Wehrheim, Germany) replication of ovine knee condyles; Indian ink stained scratches on condyle (at right).

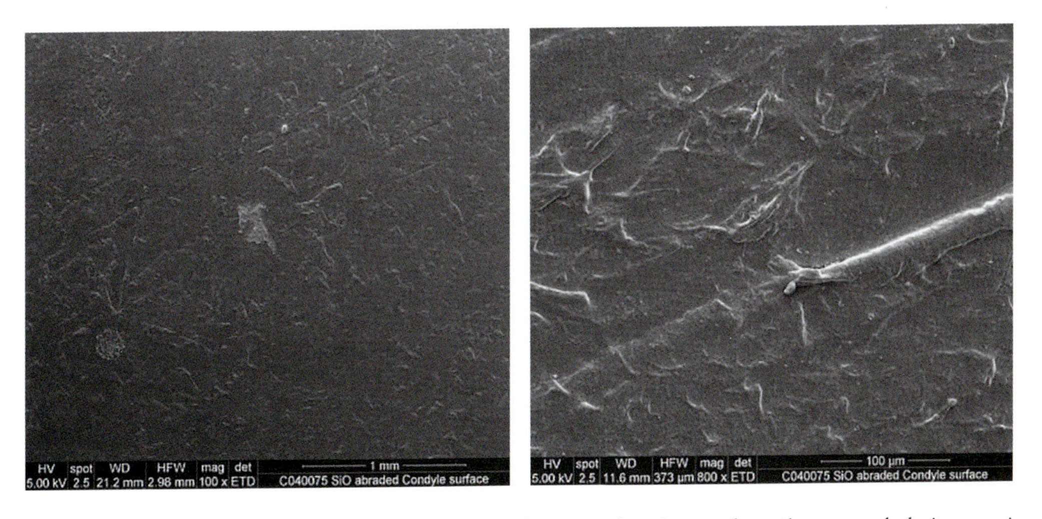

Figure 14.5 Superficial scratches in glucosaminoglycan surface layer of cartilage revealed via examination of replica.

or vegetable tissue). Replicas can also be coated without any concerns about interference to the actual specimen.

14.3 BIOMATERIALS SAMPLE PREPARATION

The term 'biomaterials' is very broad and covers a multitude of different materials, including tissue such as bone or biological substances such as collagen that may be used in tissue engineering or allografts/autografts, as well as synthetic materials. For the purposes of this chapter, synthetic materials that are designed for use in healthcare will be termed biomaterials. This is a somewhat arbitrary classification, because many real world samples

Figure 14.6 Replica of a pristine cartilage surface. Following wear down to the hypercellular layer, where chondrocyte lacunae are revealed.

may include both synthetic biomaterials and biological tissue, but for clarity each sample type is discussed separately in this chapter. Having 'narrowed down' biomaterials to synthetics still leaves an immense variety of materials, covering large areas of materials science and engineering. Many pseudo-synthetic materials of biological origin are also utilised, following processing, such as cotton, chitin, collagen, alginates, etc. The microscopist is faced with very wide-ranging demands with regard to optimal sample preparation methods, as biomaterials include hard and soft polymers (resorbable and non-resorbable), composite materials, glasses/ceramics and metals. The physical form of these materials varies widely from bulk solids to particulates and powders, densely packed or highly lofted fibres (woven or non-woven), films, laminates, cemented/welded/sintered/tempered/annealed structures, crystalline/amorphous or partially crystalline, foams, gels, emulsions/colloids, etc. Many biomaterials are surface-modified to impart greater biocompatibility; typical surface modifications can include plasma deposited layers or engineered surface textures (physical embossing/moulding methods, chemical, electrochemical or electron beam etching). Different bulk material components or surface coatings/treatments can be present at multiple length scales from macroscopic to true nanoscale.

The sheer breadth of biomaterials necessitates great flexibility from the SEM instrumentation used. FEGSEM has significant advantages over conventional tungsten-gun SEM, such as brighter, better resolved low kV and/or low vacuum imaging and greater flexibility in utilising energy selective electron detectors (particularly useful with in-lens electron-optics). The extra flexibility provided by each of these FEGSEM advantages is invaluable when handling such diverse sample types.

14.3.1 Cross-sections; Soft Tissues and Soft/Semi-soft Biomaterials

FEGSEM technology has enabled a TEM-like performance in many respects, but perhaps one of the most striking is the increased image quality achieved when using an FEGSEM with an STEM (scanning transmission electron microscopy) detector or when using low-kV sensitive backscattered electron detectors to examine a well prepared block face. The specimen must be either cross-sectioned to a planar block face (backscatter) or an ultrathin section (some 200 nm or thinner for STEM) must be cut using an ultramicrotome, ion mill or FIB-SEM (focussed ion beam SEM; see Chapter 19 by Fukuda, Leis and Rigort, Chapter 24 by Giannuzzi and Chapter 26 by Kizilyaprak *et al.* for detailed information regarding FIB-SEM). Preparing a suitable block face or section will require embedding of the specimen tissue within a supporting medium, usually a polymer resin (various low to medium viscosity epoxy and acrylic resins are most commonly employed). Different embedding resins can be polymerised in different ways; some are polymerised (cured) by addition of an 'activator' or catalyst and many such resins cure by means of an exothermic polymerisation, so specimen heating can be an issue. Some resins are cured using photochemistry; exposure to light of certain wavelengths (often yellow/white followed by blue) causes the resin to polymerise. Other resins can be cured using heat; this is suitable for heat stable specimens, but less so for heat sensitive specimens. Heat induced sample degradation can manifest itself in less than obvious ways than simply loss of microstructural detail or expansion/shrinkage artefacts: for example, exothermic polymerisation or thermal curing induced specimen heating can denature proteins, causing problems with the antigenicity of tissue if subsequent immuno-labelling methods are to be used.

Essentially the same guidelines apply for cross-sectional preparation as for dehydration, that is small specimen sizes work best, consideration of tissue density and hardness will help achieve optimal fixation and subsequent infiltration with a suitable support medium for sectioning. The processing steps are broadly similar: a fixation step, followed by step-wise transfer into a non-aqueous transition solvent, typically anhydrous acetone. Care must be taken to ensure that the fixation and solvent transfer steps do not solubilise or rinse out/redistribute any of the substances of interest; for illustrative purposes, we will assume that this substance is an active pharmaceutical ingredient (a.p. i.). The same care must be observed when selecting a suitable support medium; some resins/monomers/catalysts may also solvate or redistribute substances of interest in the specimen block. Many embedding resins are purposefully designed to completely infiltrate the specimen material at the molecular level. This provides the best possible mechanical support for delicate biological microstructures when the specimen is mechanically sliced using a microtome or ultramicrotome, but it may not be desirable if the infiltration causes redistribution of the substances of interest. In such cases a compromise may be necessary, whereby a less penetrating resin (examples include Araldite® epoxy resin or viscous metallurgical resin such as Met Prep Epo-Set; Metprep Ltd, Coventry, UK) is used to mitigate any redistribution; such a compromise is acceptable where observation of the spatial distribution of the substance in question is most important. Some loss of biological structure may be acceptable in such cases, as long as enough 'recognisable' structure is still present to allow the 'context' of substance distribution to be established. The use of a cryomicrotome can also be useful in such circumstances; even if ice crystal formation causes some disruption to the specimen microstructures, the preservation of spatial distribution of the a.p. i. may outweigh some loss of microstructure information due to ice artefacts. If a cryomicrotome is to be used, care must be taken to ensure that any condensation on the cold surface of either the section or the block-face does not itself cause surface redistribution of the a.p. i., although many current cryomicrotomes are designed to minimise condensation problems.

14.3.2 Cross-sections; Hard Tissues and Hard Biomaterials

Preparing cross-sections of mineralised hard tissue specimens and hard biomaterials presents a particular challenge. Preparation of resin embedded blocks may be accomplished very similarly to resin-embedding of soft tissue, but the fixation and infiltration time required are far lengthier, typically requiring weeks or months to successfully embed large volume specimens such as orthopaedic implants in bone, where specimens of several centimetres in length/width are to be examined. Vacuum assisted infiltration is often necessary, but most of such specimens are robust enough to withstand such methods. Much harder resins, such as glycol methacrylate (GMA) or Technovit 7200 (Heraeus Kulzer GmbH, Division Technik, Wehrheim, Germany) must also be used in order to try and better match the mechanical properties of the specimen and support it during cutting and polishing to achieve a good quality cross-section. A typical sectioning, grinding and polishing process is shown in the flow chart in Figure 14.7. As described in the final step of the flow chart, it sometimes helps to incorporate some self-adhesive metal tape (copper or aluminium tape work well), placed across any unimportant areas of the block face and extending down on to the sample stub to further enhance electrostatic charge conduction to ground. Figure 14.8 shows a stitched micrograph of a cross-section through a titanium implant in bone, consisting of 14 individual image fields (total length of the implant is 12 mm). Backscatter imaging provides detailed information regarding the extent of mineralisation of the bone tissue, with the brightness of the tissue related directly to the quantity of calcium hydroxyapatite present. Several authors (Roschger *et al.* 1998; Boyde *et al.* 1995; Bloebaum *et al.* 1997) have developed and used an analytical protocol for calibration of a given SEM or FEGSEM and backscatter detector system such that image analysis can be used to determine the quantity of calcium hydroxyapatite in the bone directly from the greyscale values recorded in the

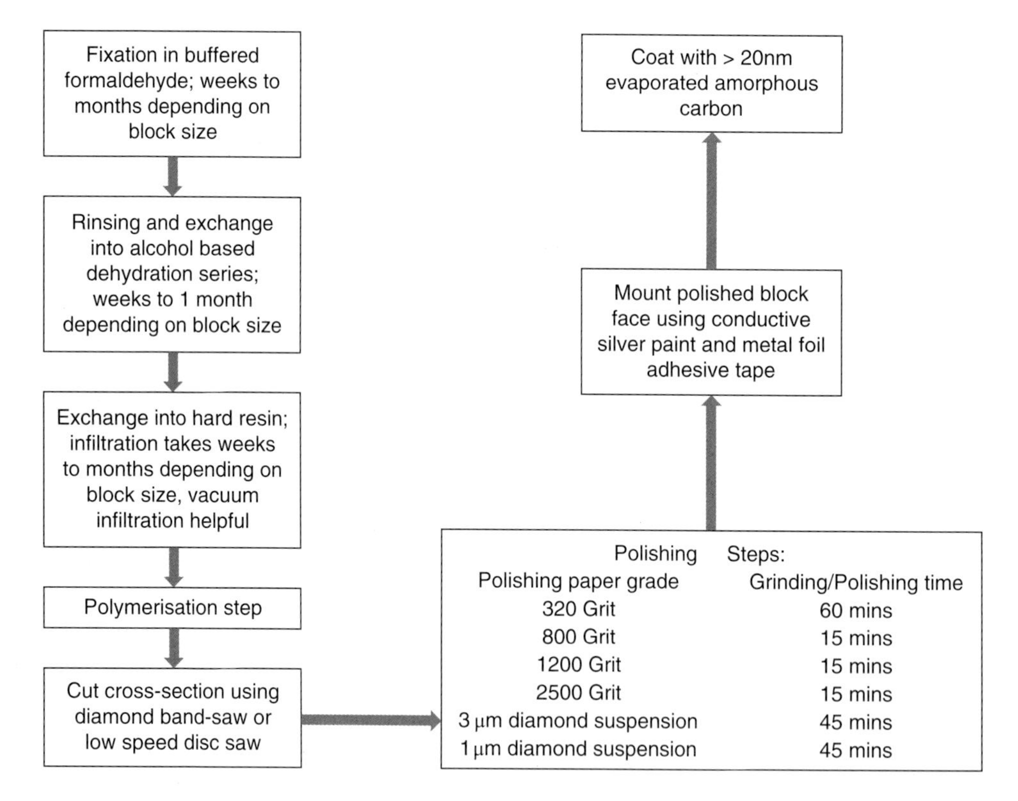

Figure 14.7 A flow chart showing the typical process cycle for hard tissue/hard biomaterial sectioning and polishing.

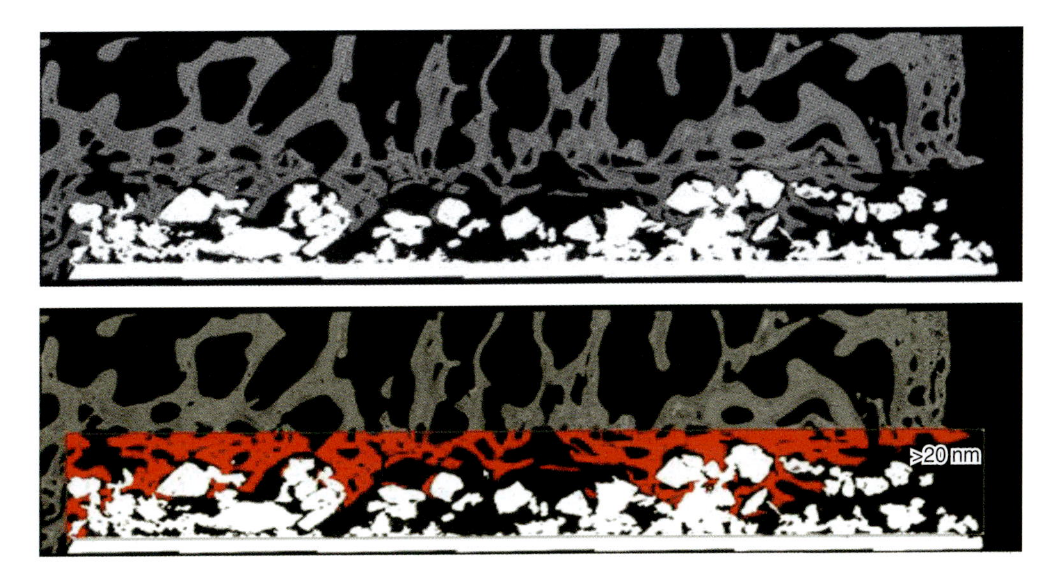

Figure 14.8 Resin-embedded bone/titanium implant interface, polished cross-sectional block-face and backscattered electron image (greyscale segmentation for image analysis – lower).

backscatter image. Even if mineralisation is not to be quantified, the backscatter image provides an excellent contrast range for determining the percentage area of void space in a porous (bone) in-growth surface that has been occupied by new bone tissue, based on simple and robust image analysis from greyscale segmentation/thresholding as shown in Figure 14.8. FEGSEM is particularly well suited to such work, because modern FEG sources provide the stable and reproducible beam currents required for comparative studies.

Wide area ion mills and focussed ion beam (FIB) instruments (Chapter 28 by Warley and Skepper) can be used to provide excellent final polishing of resin-embedded hard tissue/metal implant specimens, but many hours of milling are required and these methods do not obviate the need for very time consuming fixation, embedding and section cutting. Rather they merely replace or improve the final diamond polishing steps in a typical preparative protocol.

14.4 DETERMINATION OF AN OPTIMAL INSTRUMENTAL APPROACH

Any microscopist faced with a request to prepare a Standard Operating Procedure (SOP) for an FEGSEM instrument used in a Research or Investigative environment will attest to the fact that covering all of the possible instrumental approaches in such a prescriptive document is likely to yield a weighty tome. Condensing such information into a pragmatic set of broad guidelines is more practical than producing a detailed list of steps. The flow chart presented in Figure 14.9 can be used as a starting point when determining how best to approach the FEGSEM of different biomaterials.

The information presented is not by any means comprehensive; the flow chart merely provides guidelines for basic secondary electron imaging of different biomaterials. This would differ for backscattered electron imaging or X-ray microanalysis applications. The choice of the instrumental approach is typically easier for electrically conductive dry, solid materials;

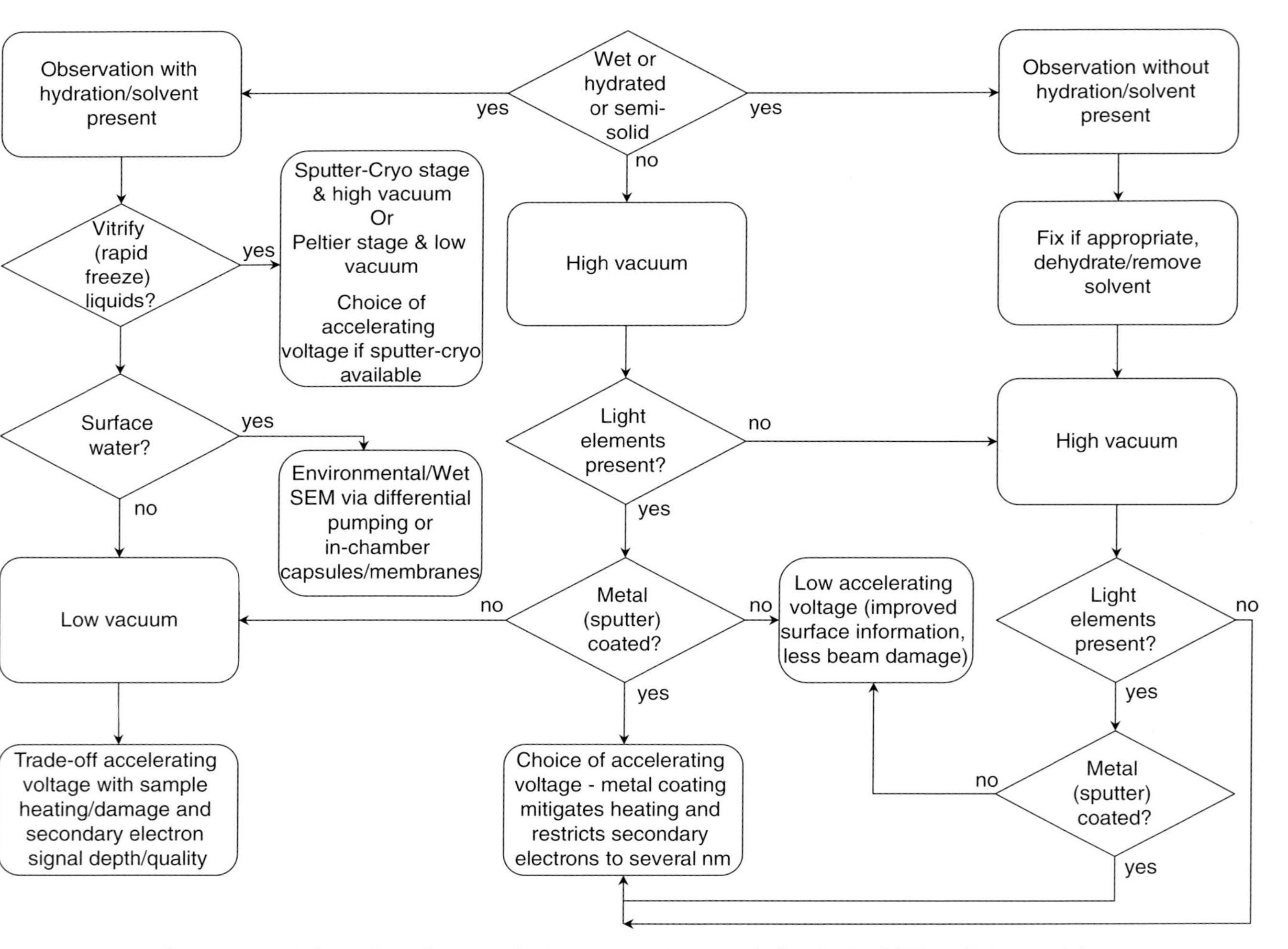

Figure 14.9 A flow chart showing the instrumental approach for the FEGSEM of biomaterials.

high vacuum is unlikely to be troublesome and the microscopist will have flexibility in the choice of the accelerating voltage used.

Several examples of different instrumental approaches are provided below to help illustrate the flexibility provided by FEGSEM, where the sample is more representative of those covered in the flow chart in Figure 14.9.

14.4.1 Example 1. Nanocrystalline Silver Coated Wound Dressing

Many biomaterials are surface-treated to impart desirable mechanical or biological properties, for example improved infection control, biocompatibility or bioactive properties. The example discussed here is a polymer-based anti-bacterial wound dressing whose infection control properties are provided by a plasma vapour deposited nanocrystalline silver surface. Taken at face value, it may be expected that this biomaterial would be relatively easy to handle for SEM; silver is electrically conductive and the dressing should behave well under a scanning electron beam. This expectation is to some extent correct, but if the nanoscale morphology of the deposited layer is to be examined at high magnification, the tiny area covered by the electron beam scan raster results in significant energy being absorbed by the silver layer. Dwelling on a single area for more than a short time results in deformation of the nanoscale silver, as revealed by subsequently imaging the same area at lower magnification (Figure 14.10, left and centre). The brightness of the FEG electron source and the improved spatial resolution at low kV enable imaging of delicate surface features with minimal beam damage. Minimal beam damage was achieved for the nanocrystalline silver coated dressing by decreasing the accelerating voltage, probe current and electron beam dwell time, thereby minimising the energy deposited at the sample surface, although a small amount of beam-induced carbon contamination can still be seen in the previously scanned area (Figure 14.10, right). The electrical conductivity provided by the silver layer of this biomaterial is good enough to avoid the need for sputter coating, enabling the sample to be imaged uncoated and at high vacuum conditions. Imaging the sample without a coating was important for such a finely structured surface, because the addition of even 5 nm or more of sputter coating would modify the surface morphology.

14.4.2 Example 2. Biofilms on Fibrous Gauze Wound Dressing

Microfibrous materials present challenging samples if high quality images are to be obtained. The difficulty encountered arises from the very poor electrical conductivity of such highly lofted materials; the physical contact areas between each microfibre are tiny, not to mention

Figure 14.10 Nanocrystalline silver surface, left and centre, show the effects of beam damage, with negligible beam damage at right, but some carbon contamination can be seen in the region of a previous higher magnification scan.

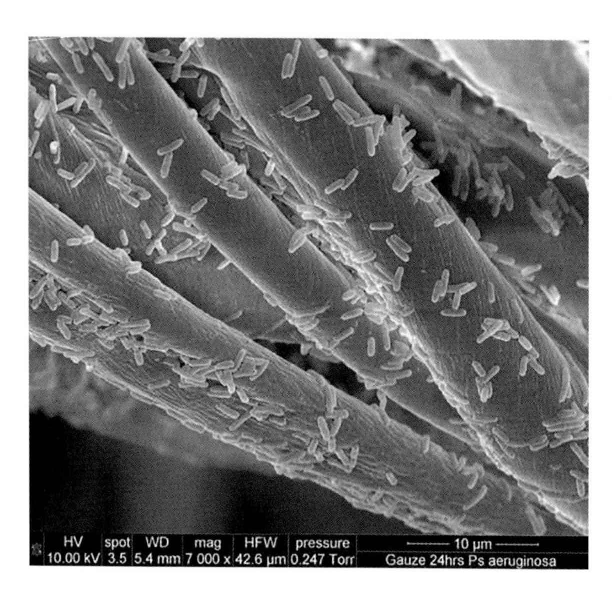

Figure 14.11 *Pseudomonas aeruginosa* bacterial biofilm on gauze fibres.

line-of-sight issues when trying to sputter-coat such samples effectively. Electrostatic charging under the scanning beam often causes sample drift with microfibre materials, hampering attempts to record images, particularly at the magnifications needed to visualise any bacteria or biofilms on the dressings in any detail. The sample shown in Figure 14.11 had been incubated with a suspension of bacteria in order to study biofilm formation. Preservation of the bacteria was achieved by a glutaraldehyde fixation followed by rinsing and 10% incremental ethanol dehydration and critical point drying via supercritical liquid carbon dioxide. Low voltage imaging can help when imaging such samples, but not always; lower accelerating voltages can result in more electrons remaining on the surface of the fibres as opposed to penetrating the fibre surface, so although the energy deposited is less at low kV and the likelihood of beam damage is reduced, electrostatic charge build-up on the fibres can actually worsen at lower accelerating voltages. Decreasing the probe current will help, but imaging at low kV (e.g. 5 kV or lower) and a low probe current (e.g. less than 100 pA) may yield increasingly noisy results, necessitating slower scan rates where even small amounts of sample drift will cause image degradation if the microscope is not equipped with drift correction facilities. The image shown in Figure 14.11 was obtained by using low vacuum conditions (0.25 Torr, water vapour at room temperature) and moderate accelerating voltage of 10 kV to provide good charge neutralisation of the fibres; no drift correction was necessary to achieve stable imaging conditions. Low vacuum conditions can be achieved using different gases; water vapour is often used because of the low ionisation potential of water molecules, producing good signal generation and amplification across the distance from the sample surface to the low vacuum detector, but dry nitrogen or air can also be used if the sample is water sensitive. Note that there is little evidence of extracellular polysaccharide (EPS) in Figure 14.11; the sample was dehydrated using an ethanol series followed by supercritical liquid carbon dioxide, resulting in dissolution and removal of much of the EPS from the sample surface.

14.4.3 Example 3. Resorbable Polymer Microspheres

Extremely beam-sensitive biomaterials such as soft polymers present even greater challenges for artefact-free imaging. Figure 14.12 shows several resorbable polymer microspheres

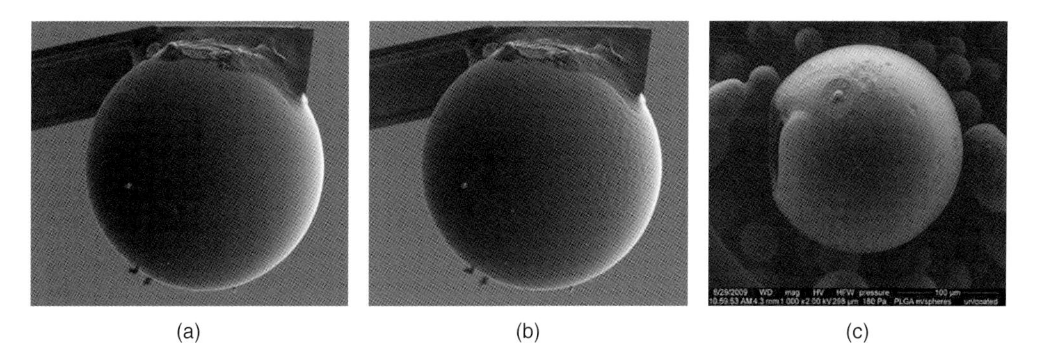

(a) (b) (c)

Figure 14.12 Resorbable polymer microspheres, sputter coated, high vacuum (a), but note beam damage when re-scanned (b); no beam damage even when imaged uncoated at low kV in low vacuum (c).

typical of formulations intended to deliver active pharmaceutical ingredients. The first two microspheres (a and b) are shown attached to the cantilever of an atomic force microscope, having originally been attached to facilitate mechanical testing of the microspheres. These microspheres were imaged following a 10 nm sputter coating of gold–palladium and using a modest accelerating voltage of 5 kV with a small probe current, yet just a single slow scan (a) proved enough to induce wrinkling and dimpling of the polymer surface (b), as seen in the scan acquired immediately after the first slow scan was acquired. Just as with the silver-coated wound dressing surface, this example serves to illustrate the importance of occasionally re-scanning a field of view to check for any beam-induced artefacts where the sample may be considered to be beam-sensitive; zooming out in magnification sometimes helps with visualising even quite subtle differences between previously scanned areas and those freshly scanned. The polymer microsphere shown in (c) was imaged at low vacuum, but using a much lower accelerating voltage of 2 kV, allowing uncoated imaging without causing any shrinkage or wrinkling artefacts during a single slow scan.

14.4.4 Example 4. Gel Networks

Hydrogels and protein gels can be difficult biomaterials to handle for FEGSEM. Cryo techniques can be used, with extreme care taken to avoid the generation of ice-crystal artefacts, but they can also be successfully dehydrated with some thought given to protecting such delicate, highly lofted and friable biomaterials whilst they undergo any fixation and dehydration steps. Dehydrating the gels as very small pieces in small, porous polyethylene baskets ensures that dehydration media such as alcohols/acetone or supercritical liquid CO_2 can perfuse through the gel completely over reasonable timescales, whilst protecting it from any shear forces when the solutions are mechanically agitated to ensure mixing. Mounting the dehydrated gels in a custom-made sample mount can provide a relatively inexpensive and practical solution. Figure 14.13 shows a metal disc mount, into which several recesses have been drilled. After moistening the edges of the recesses with conductive mounting paint, the gels depicted were placed within each little recess. Ensuring that the recesses are shallow and contain only limited volumes of dehydrated gel produces the best results in terms of protection during introduction to the vacuum chamber in the FEGSEM during evacuation or venting steps and also in providing good electrostatic contact between the (subsequently) sputter-coated gel network and the sample stage in the FEGSEM, countering any tendency to charging up under the scanning electron beam. Figure 14.14 illustrates that high quality images can be obtained from such materials when low accelerating voltages are used, even

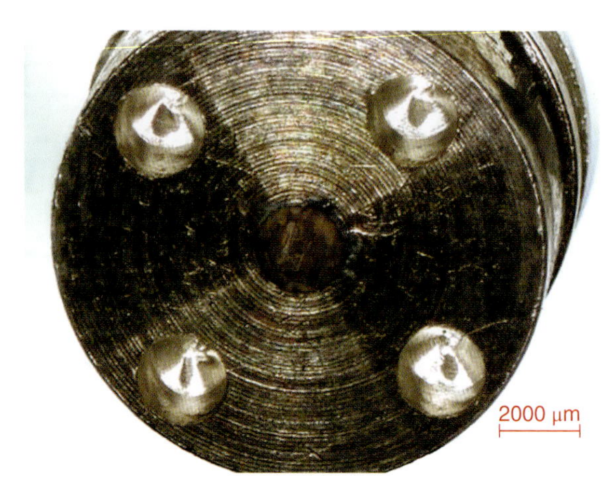

Figure 14.13 Sample holder for dehydration and imaging of hydrogels (the central receptacle contains gel).

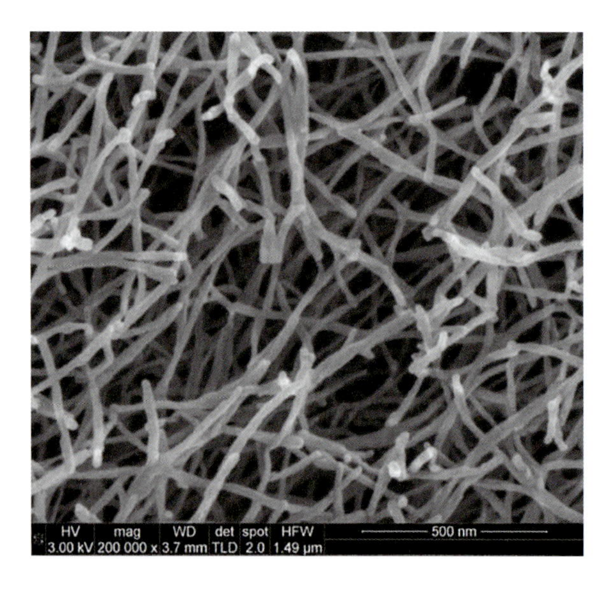

Figure 14.14 Proteinaceous gel network, dehydrated using graded alcohols and CPD, sputter-coated with ca. 5 nm gold–palladium and imaged at low kV in high vacuum.

at very high magnifications, without any charging artefacts or movement of highly lofted nanoscale fibres degrading the image.

14.5 ENERGY DISPERSIVE X-RAY MICROANALYSIS

X-ray microanalysis provides a powerful extra investigative method to the electron microscopist working in Public Health or Pharmaceutical/Medical Device Research. Energy dispersive X-ray microanalysis (commonly abbreviated to EDS or EDX, but sometimes termed XEDS or EDXMA) tends to be more commonly used, owing to somewhat less stringent sample presentation requirements than its more quantitative cousin wavelength dispersive X-ray microanalysis (WDX). A detailed description of the principles behind X-ray microanalysis methods can be found in Volume I, Chapter 1 by Joy and Volume II, Chapter 28 by Warley and Skepper. This section aims to provide several examples of EDS applications in both public health monitoring and medical device research and troubleshooting.

For most samples examined at higher accelerating voltages, it is advisable to provide a thin electrically conductive coating of amorphous carbon across the sample surface; even 'conductive' samples such as metals may have surface features composed of insulating compounds or oxidised material that can be the cause of charged areas when examined in the FEGSEM. However, a major advantage for the application of a FEGSEM in a medical diagnostic setting is the ability to look at biological specimens without any coating and to perform X-ray microanalysis even at low kV, for example below 5 kV.

Consideration of appropriate sample preparation and avoidance of unnecessary preparative steps are critical to obtaining high quality, accurate data. Charging is usually more likely when working at EDS analytical electron-optical conditions of higher probe current (e.g. in the nA range) and higher accelerating voltages. Electrostatic charge on such features can in turn have the effect of decelerating the electron beam energy, hampering X-ray excitation efficiency and causing the EDS system software algorithms to potentially 'misinterpret' the data collected; in particular this may cause increased inaccuracy for any semi-quantitative/quantitative analysis.

EDS is frequently used in change control or in medical diagnostic applications and in pseudo-forensic troubleshooting work related to medical devices or pharmaceuticals, but 'change control' is also a critical activity in these heavily regulated industries. Whenever a significant change is made to the manufacture of a device or pharmaceutical product, whether change to the production process itself or to the raw materials or their source, any resultant change to the packaged product must be characterised in order to demonstrate physicochemical equivalence to the old product. This includes any potential changes to its packaging.

14.5.1 EDS Example 1. Change Control

A change-control example is illustrated by Figure 14.15, which shows a particularly challenging microfibrous polyester wound dressing material coated with a hydrophobic surface modifier. The supplier of the original material provided polyester fibres modified by a submicrometre PTFE coating. The new material sourced from a different supplier was indeed hydrophobic (see the macrophotograph inset on the EDS spectra), but EDS was needed to try to determine whether a fluorinated hydrophobic agent was present on the fibre surfaces. This particular example helps to demonstrate some of the limitations of SEM/EDS and also emphasizes the importance of analysing the sample at more than one accelerating voltage when undertaking EDS on softer 'light element' materials such as polymers or other organics and biologics. Chapters 1 and 12 in Volume I describe the practical effects of the electron interaction volume, with regard to collecting X-rays from below the surface of the sample under investigation. EDS spectra were collected at three different accelerating voltages, 20 kV, 5 kV and 2.5 kV, in order to manipulate the electron interaction volume from deeper subsurface at 20 kV to a shallow one at 2.5 kV (see Monte Carlo simulations inset, Figure 14.15). The elements of interest, namely oxygen (the polyester is composed of

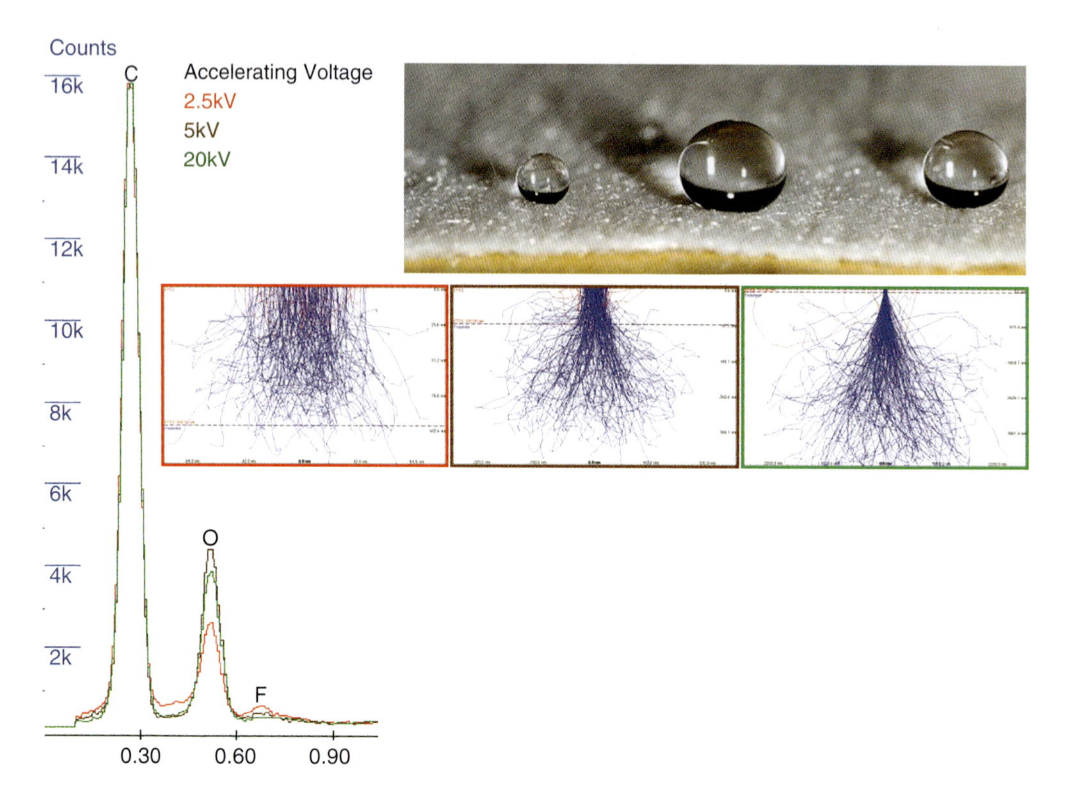

Figure 14.15 EDS spectra obtained at progressively lower accelerating voltages. Note the change in oxygen and fluorine peaks; the fluorine peak is highest at the lowest accelerating voltage because the fluorinated coating contributes a greater proportion of the X-ray emission detected from a shallower electron beam interaction volume. Electron flight trajectory Monte Carlo theoretical simulation graphics are shown inset at each kV from low to high, left to right. The boundary between PTFE and PET is demarked by a dotted horizontal line on each simulation graphic (software generated data, CASINO v2.84, Université de Sherbrooke, Québec, Canada).

carbon, hydrogen and oxygen) and fluorine (PTFE is composed of carbon and fluorine), are light elements with X-ray emission Kα peaks in the low energy range (all less than 1 keV), so all will excite efficiently at lower primary electron beam energies of 2 kV or more. For lighter elements typical of biological tissue or polymers/organic compounds, even very modest accelerating voltages in the range of approximately 2–7 kV usually provide sufficient 'overvoltage' to efficiently excite X-rays from the sample for elements up to calcium. Overvoltage is a generally accepted term referring to accelerating voltages of between 1.5 and 3 times higher than the X-ray emission line to be excited; for example, for the calcium Kα line, the emission energy is 3.692 keV, so an overvoltage of 7 kV will excite X-rays from this element efficiently and allow good detection sensitivity. Although electron beam energy at 2 keV provides sufficient overvoltage for fluorine, the probe current may need to be increased to excite sufficient X-ray flux to avoid noisy spectra. With future developments and refinements to both X-ray detector technology and the processing of spectral data, the practicality of such approaches for thin coatings of light element materials is likely to increase still further. For samples similar to the hydrophobic fibre dressing, whilst EDS can provide an indication of the presence of a fluorinated compound, unambiguous confirmation of the presence of PTFE necessitates the use of ToF-SIMS (time-of-flight secondary ion mass spectrometry), a highly surface-specific and sensitive chemical mapping technique providing directly sampled, spatially resolved chemical speciation. Auger electron spectroscopy (AES) and X-ray photoelectron spectroscopy (XPS) both provide great surface specificity; for AES the interaction volume is only 5–100 Å, with most of the Auger signal originating from the first 5–20 Å of the surface, XPS recovers information from the top few nm, but AES is also an elemental speciation technique capable of definitive elemental identification and chemical bond identification for simple molecular compounds. XPS can also only provide basic chemical bond information as opposed to fuller chemical fingerprint spectra from complex molecules. If more information about truly surface-specific analyses such as ToF-SIMS, XPS and Auger are required, further information can be found in an excellent textbook primer edited by Vickerman and Gilmore (2009).

Other spectroscopic methods such as Fourier transform infra-red (FT-IR) and Raman spectroscopy are capable of detecting and identifying a fuller chemical compound 'fingerprint spectrum' of PTFE and other compounds, but if the surface coatings are significantly less than approximately 0.3 μm thick neither method has sufficient surface specificity to provide any practicable sensitivity advantage over EDS carried out at a low accelerating voltage. The differences observed in elemental spectra obtained at different accelerating voltages (different interaction volumes/depths) can be seen as a limitation, but another way to consider this 'limitation' is that EDS can provide a quick and convenient, albeit relatively qualitative, assessment of the nature of a surface coating, bloom or otherwise surface modified material.

The high source brightness and superior spatial resolution provided by FEGSEM offers further advantages for EDS if used in conjunction with a STEM (scanning transmission electron microscopy) detector, enabling observations and analysis of ultrathin sectioned samples. In such cases the electron beam is passed through sample sections as thin as perhaps 60–100 nm, thereby limiting the beam interaction volume through which electrons are scattered and improving the spatial resolution of EDS to better than 100 nm. It can be argued that the FEGSEM-STEM approach is better than pure TEM-based EDS, because exposure of the sectioned sample to the beam is less damaging, owing to the fact that the FEGSEM electron beam is raster scanned with very short dwell times across the sample, as opposed to continuous beam exposure from conventional TEM.

14.5.2 EDS Example 2. Contaminant Identification

Figure 14.16 shows a backscattered electron micrograph of the edge of a contaminant particle and a corresponding EDS elemental dot map. Backscatter is a quick and convenient method for locating any inorganic contaminant flecks or particles within a polymer/organic-based product, with any contaminants represented by much brighter Z-contrast. This helps to 'find the needle in the haystack' quickly, even where it may be a tiny submicrometre particle (assuming it is close to the surface of the sample material). EDS can provide particularly useful information regarding the possible origin of the contamination; the example shown is composed of elements consistent with steel, but there is also some abraded surface coating of zinc, suggesting that the fleck originated from galvanized steel. The small patches of silicon-rich material originated from a polydimethylsiloxane (PDMS)-type compound from which the contaminant was isolated.

14.5.3 EDS Example 3. Organic Materials

Figure 14.17 shows a backscatter image and corresponding EDS elemental dot map of a medical device containing a layer of sodium polyacrylate superabsorber material

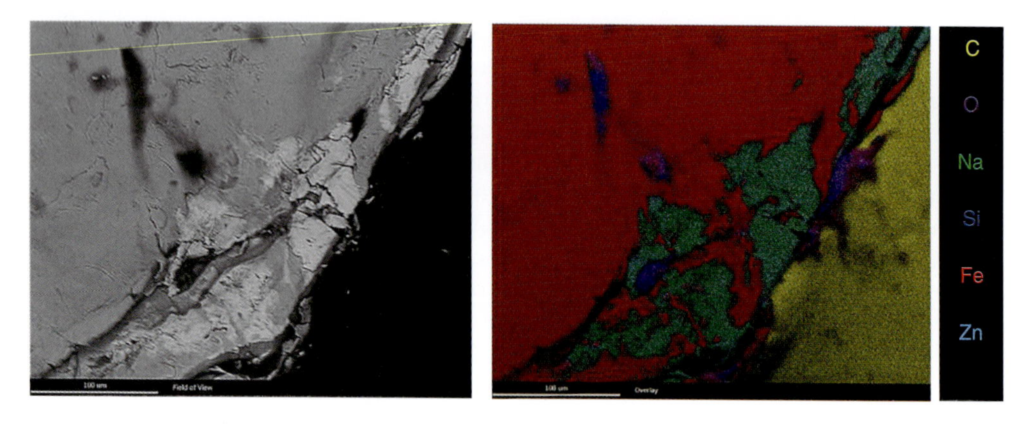

Figure 14.16 Backscatter image and corresponding EDS map of an inorganic contaminant fleck.

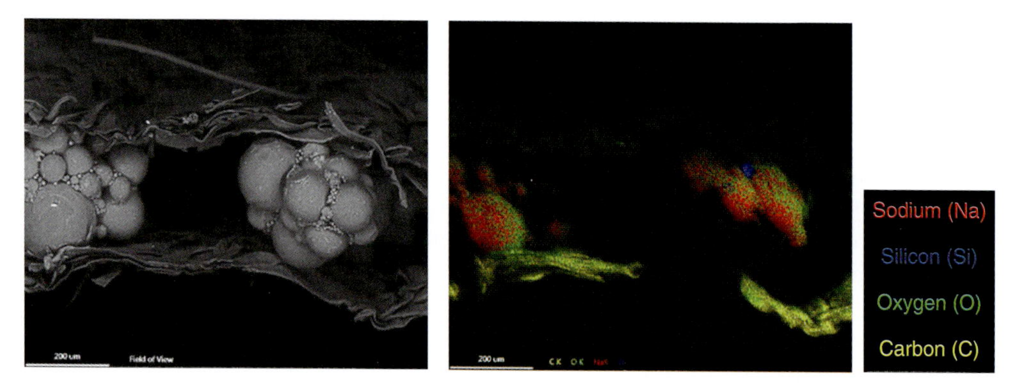

Figure 14.17 Backscattered electron image and corresponding EDS elemental dot map of superabsorber material. Note the high sodium and oxygen content in spray-dried superabsorber granules.

sandwiched between two woven sheets of cellulose-based material. SEM and EDS can provide enough information to determine the form and location of the superabsorber, which can be visualised because of the sodium content. Sodium polyacrylate is composed of hydrogen (not detectable using EDS), carbon, oxygen and sodium, with its high oxygen and sodium content reflected in the resulting EDS map, whilst cellulose is chiefly hydrogen, carbon and oxygen (the carbon and oxygen are detectable and shown clearly in the EDS map). Note the slight difference in appearance of the spray-dried polyacrylate granules in the EDS map, which look less substantial than shown in the backscatter image. This difference is caused by topography and EDS detector geometry, with larger granules 'shielding' off the X-ray signal from reaching the EDS detector that was mounted at top right with respect to the map image (areas of lower topography shown at the lower left are therefore shielded from the detector). For simple organic molecules containing other inorganic elements, EDS can provide good information regarding a likely chemical identity, but where the organic substances are just a polyolefin, polyester or protein (with a low inorganic content), any robust chemical identification is challenging.

Whilst EDS provides only chemical element information as opposed to fuller compound information from a 'chemical fingerprint' when investigating organic or biological materials, it may seem like an inferior choice of technique. However, compared to a high-end expensive technique such as ToF-SIMS, MALDI or XPS, EDS can provide a relatively convenient 'first

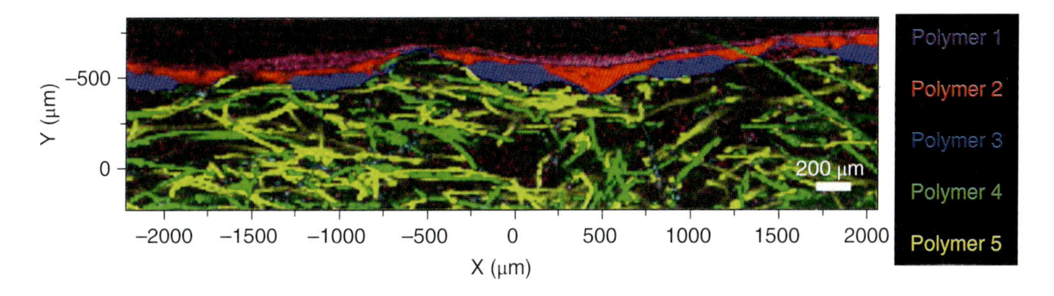

Figure 14.18 Raman chemical map; spatially resolved chemical identification of a complex mixed polymer sample.

look' and is a powerful analytical tool when FEGSEM is used alongside modern low kV backscatter detectors and vibrational spectroscopy techniques such as FT-IR or Raman. FTIR and Raman spectroscopy are relatively convenient in terms of sample presentation. With chemical mapping functionality, each possesses similar spatial resolution to SEM-EDS, making these methods excellent complementary partners for analysis of organic substances. Raman spectroscopy has the added advantage of being capable of analysing wet samples (water absorbs IR, interfering with analysis).

Another medical device with a complex microstructure is shown in Figure 14.18. This sample is composed of a mixture of different polyolefins and polyesters, which are difficult to identify with confidence from EDS analysis alone. Using Raman spectroscopy to produce a chemical map based on the spectral chemical fingerprint of the compounds present shows that many different polymers are present in this sample, with good spatially resolved information provided by the Raman map. Once the polymers are identified by Raman spectroscopy, the information can be used to inform a higher resolution examination of such materials using backscatter imaging to help visualise subtle differences in oxygen and/or nitrogen content of the compounds present.

14.5.4 EDS Example 4. Detection of Asbestos and Mineral Fibres in Lung Tissue

For centuries asbestos had fascinated man for its unique physicochemical properties. Inextinguishable and fire-resistant asbestos fibres had been used to fortify pottery, to make wicks for oil lamps and had been woven into cloth that was durable and heat resistant (Dobbertin 1980; Cugell and Kamp 2004).

During the industrial period the demand for asbestos increased constantly and exploded after World War II. On the other hand, first hints regarding the potential health risks of asbestos were given by Monatque Murray in 1900 and reported to the 'Departmental Committee on Industrial Disease' (Cugell and Kamp 2004; Murray 1907). Machand was the first to describe strange pigmented crystals in lung tissue (Marchand 1906) that were later correlated with asbestos dust (Cook 1929).

It took another 29 years before asbestosis became acknowledged (1936) as an occupational disease in Germany (Dobbertin 1980) followed by lung cancer (Dobbertin 1980; Enterline 1991) related to an extended exposure to asbestos fibres (1943) (Dobbertin 1980; Enterline 1991). Many more years went by before asbestos was banned in Germany in 1993 (UK, 1999, and EU, 2005). The development of diagnostic criteria led to a better understanding of asbestos-imposed disorders, improved preventive regulations and appropriate compensation.

Today in most industrial countries asbestos-related diseases fall into four categories and the Helsinki criteria, a consensus report published in 1997 by a group of international experts (Henderson *et al.* 1997), are widely accepted.

One of the problems is very long latent periods between exposure and the possible onset of the illness, which may range from 20 up to 40 years or longer. Right up to the present day we receive samples from patients with questionable exposure-related lung cancers.

Electron microscopy in combination with X-ray microanalysis can be helpful to analyse asbestos fibre contents in lung tissue (Henderson *et al.* 1997; Riediger 1985; Roggli 1991). There are basically two different procedures to extract fibres out of lung tissue: digestion with sodium hypochlorite and cold plasma ashing with oxygen. Various protocols have been published (Riediger 1985; Roggli 1991). The protocol flow chart shown in Figure 14.19 works best in our opinion. Make sure all instruments and vessels are carefully cleaned and fibre-free and all liquids are filtered through 0.25 μm filters to avoid contamination. Cubes of 1 cm × 1 cm × 1 cm are cut from formalin fixed wet material and freeze-dried. If possible we prefer to analyse at least four samples from different areas of the lung since fibres are not evenly distributed within the lung and accumulate in the lower parts rather than the upper lobule. Organic matter is removed by plasma ashing with oxygen. Inorganic residues are re-suspended in 40 ml 0.5 N HCl. Then 2 ml of the suspension are filtered on to cellulose nitrate membrane filters for light-microscopy and 10 ml are filtered on to polycarbonate filters for electron microscopy.

Figure 14.20 illustrates typical morphologies and an accompanying EDS spectrum of crocidolite fibres that were isolated for analysis as described above (see Figure 14.21).

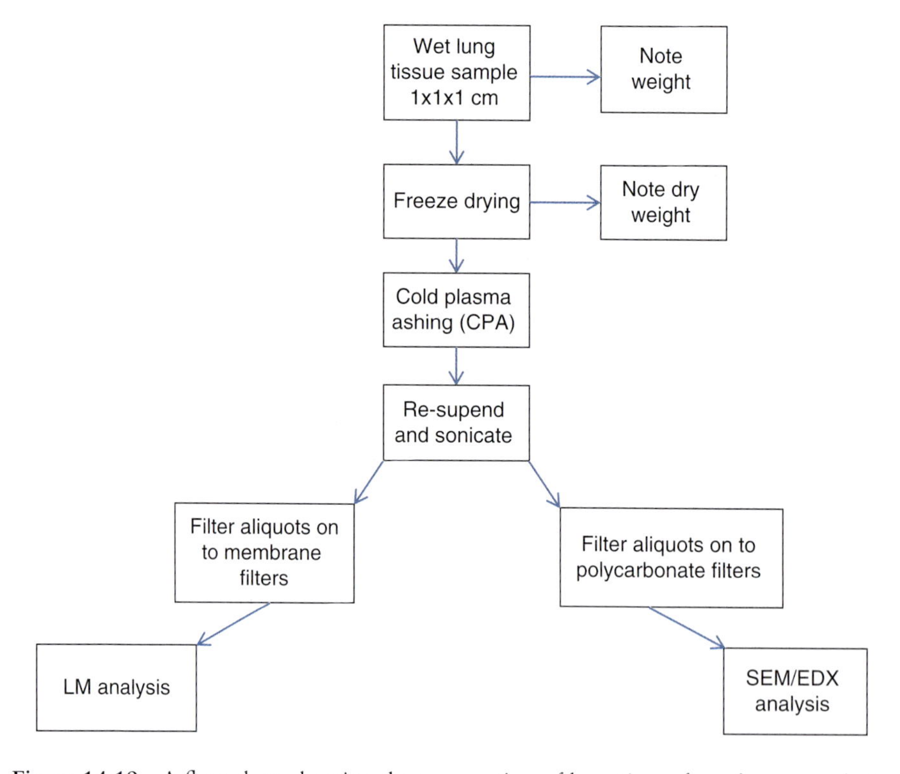

Figure 14.19 A flow chart showing the preparation of lung tissue for asbestos analysis.

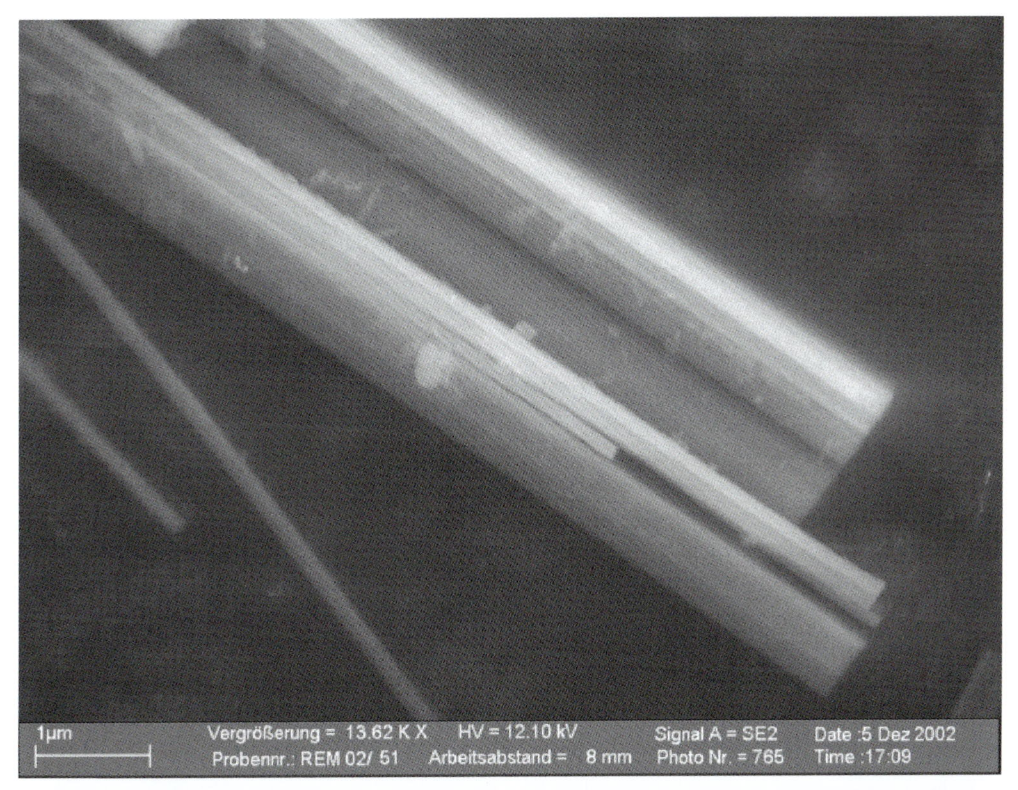

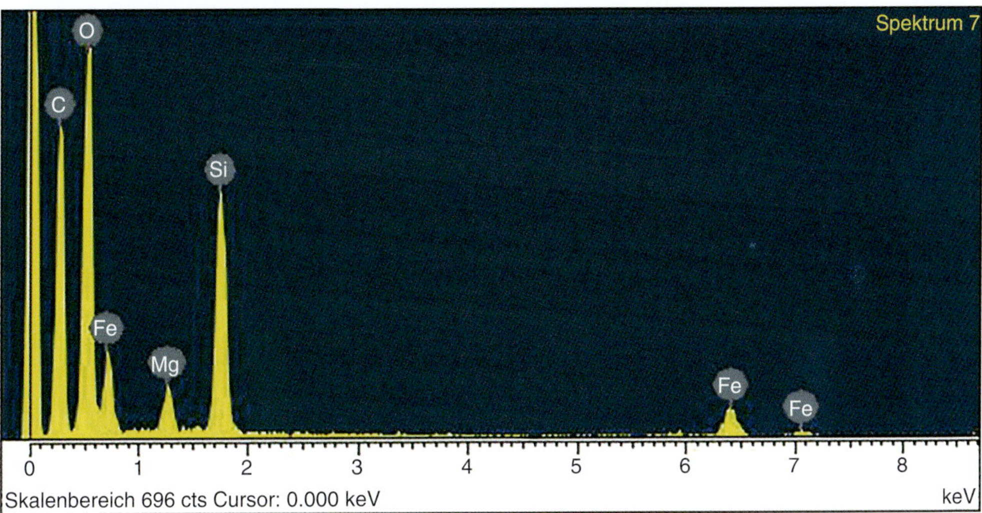

Figure 14.20 Image and spectrum of a typical crocidolite fibre.

14.5.5 EDS Example 5. Detection and Analysis of Foreign Bodies and Inclusions in Tissue

Foreign bodies or inclusions in tissue can be easily identified and analysed in thick (5–10 μm) sections. We use histological paraffin sections. Sections are mounted on polycarbonate cover

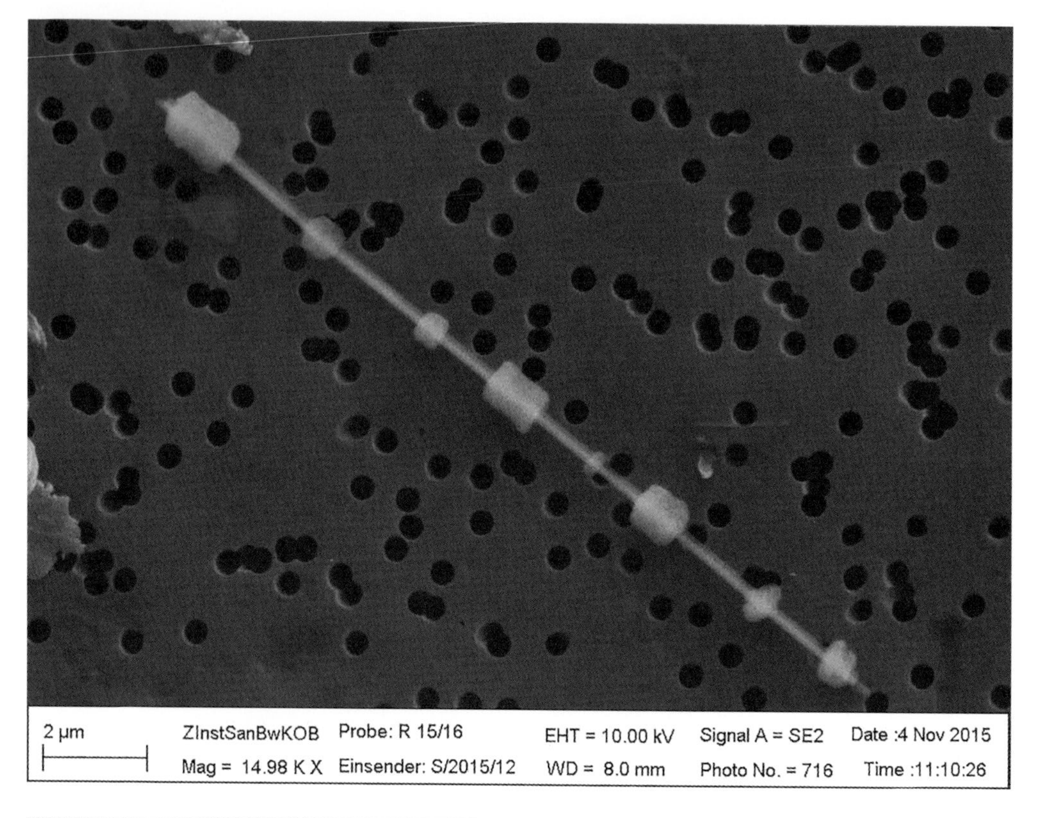

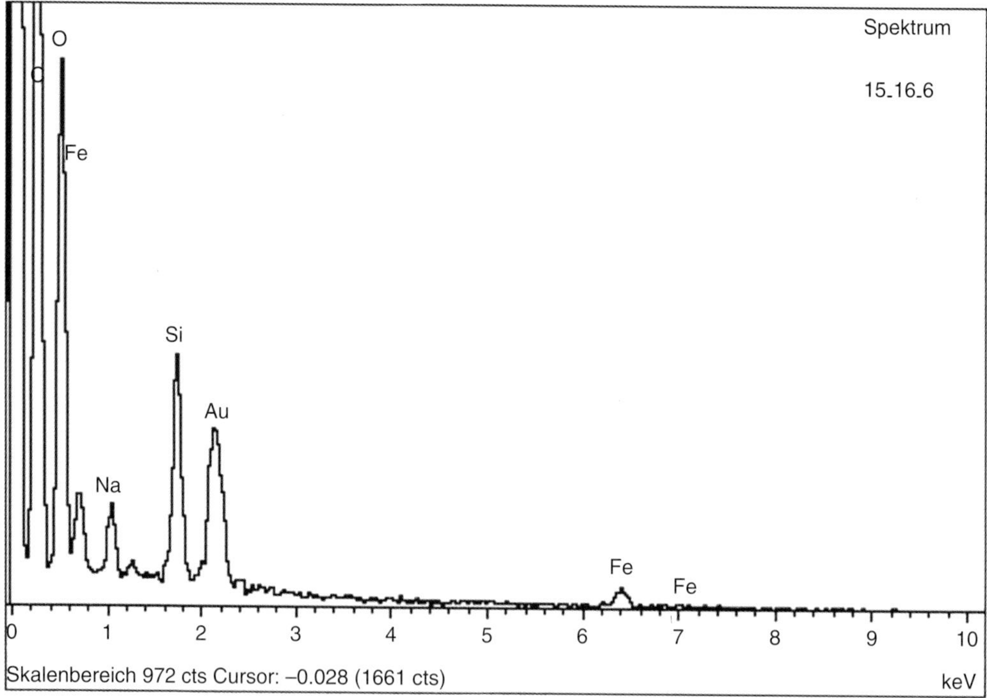

Figure 14.21 Image and spectrum of an asbestos body formed from a crocidolite fibre from the lung of a patient who had been exposed to asbestos as a welder in a ship wharf.

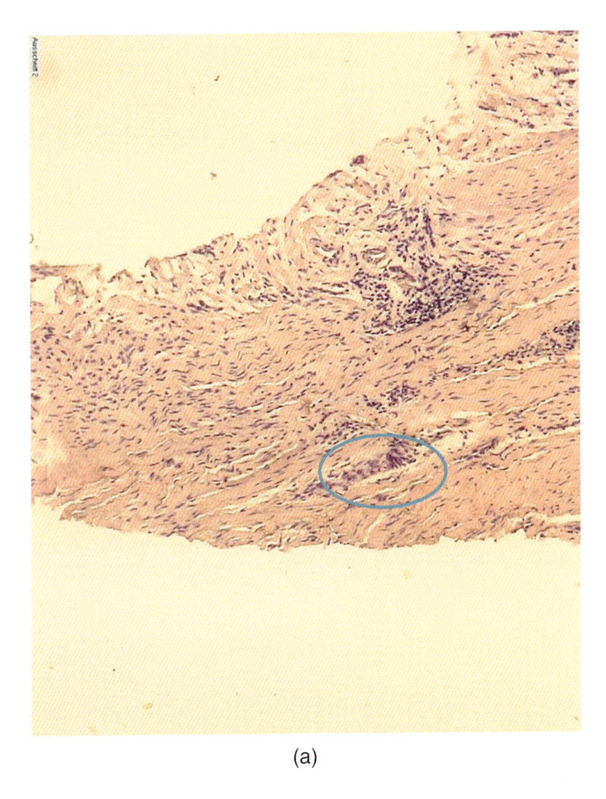

(a)

Figure 14.22 (a) H&E picture of a paraffin section with the analysed area encircled.

(b)

Figure 14.22 (*continued*) (b) Deposits in connective tissue of gingiva: the backscattered electron image is mixed with a secondary electron image. EDS analysis of bright white appearing inclusions (circle) revealed titanium that had been leached out from the dental implant.

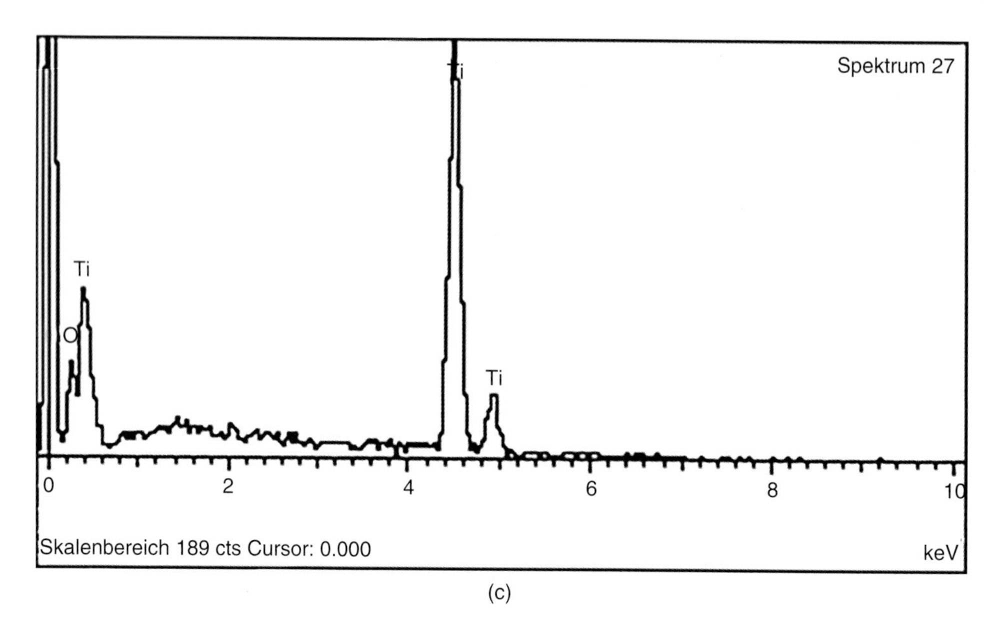

Figure 14.22 (*continued*) (c) EDS spectrum of gingival deposits.

slides, deparaffinised in xylene and dried. In this way SEM images can be directly compared with a section stained for LM (e.g. H&E). It can be of great advantage if the SEM is equipped with a backscattered electron detector (BSE). The brightness of a BSE image correlates with the atomic number (Z) of an object (Goldstein *et al.* 1984). The higher the Z value, the brighter the image. Inclusions with a different composition than the surrounding tissue can therefore be detected due to the intensity difference in the image. If BSE and SE images can be mixed the composition and topography can be combined.

An example is a biopsy from the gingiva of a patient who developed frequent inflammation with or without pus. The H&E section showed lymphocytic infiltration around black inclusions within the connective tissue. Combined BSE/SE images showed inclusions of high intensity that could be identified as titanium by EDS X-ray analysis. The titanium had been leached out from a dental implant (see Figure 14.22).

14.6 QUALITY CONTROL OF BIOMEDICAL PRODUCTS

Patients complained about the quality of razors that were used to prepare them for surgery. Depending on the batch, they said that shaving was painful and their skin was irritated. FEGSEM at low kV revealed that the brand of razors that caused complaints showed insufficient coating with lubricant (Figure 14.23).

Another complaint of patients addressed painful immunisation procedures. SEM could demonstrate that the bifurcation needles of one brand showed a very big deviation in terms of product quality, with needles having broken, uneven or dull tips, leading to painful incisions (Figure 14.24).

14.6.1 EDS Example 6. 'White Powder' as a Potential Biohazards Material

Bacterial spores, especially spores of *Bacillus anthracis*, have long been a putative biological weapon and were used in the Washington, DC area in 2001, leading to death and infection.

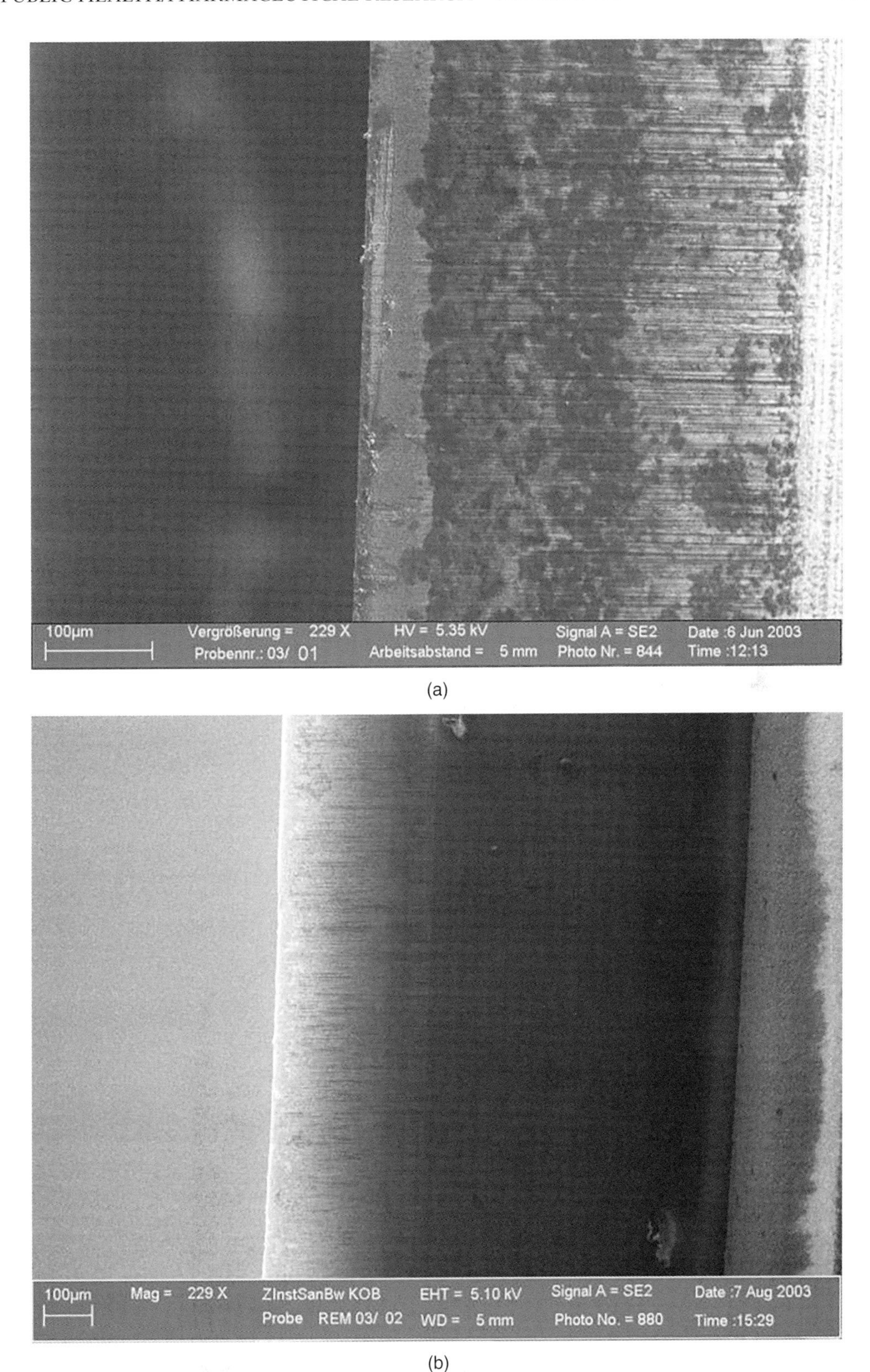

Figure 14.23 (a) Razor with insufficient coating, combined SE and in-lens detector. (b) Razor with good coating, combined SE and in-lens detector.

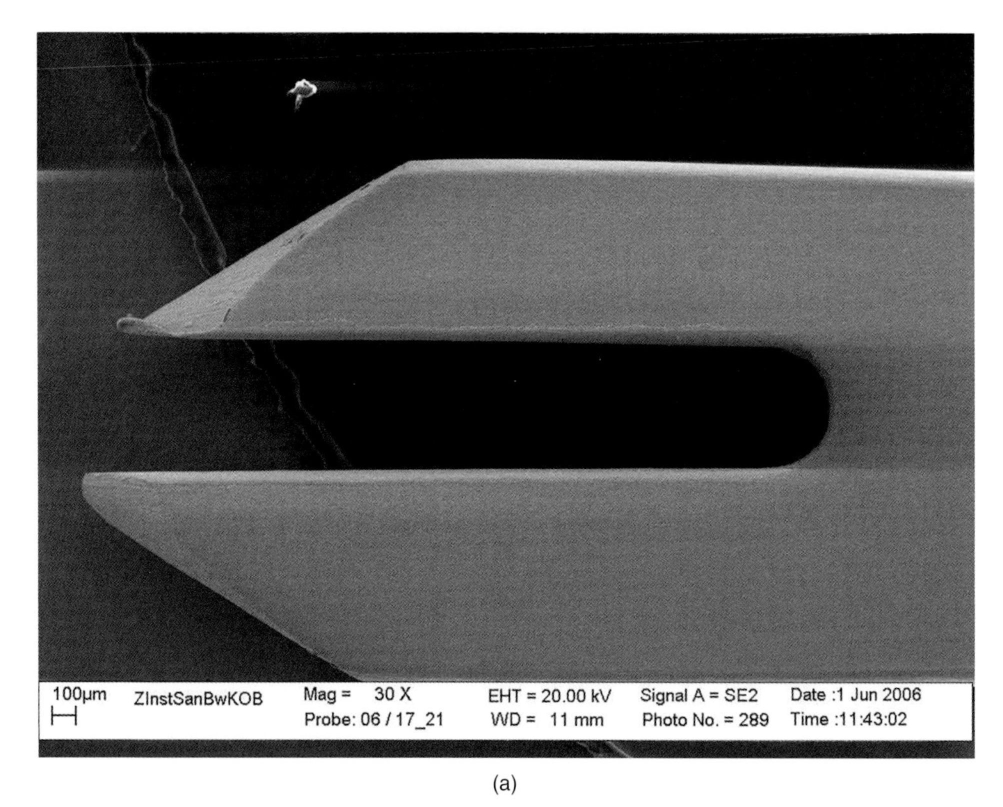

(a)

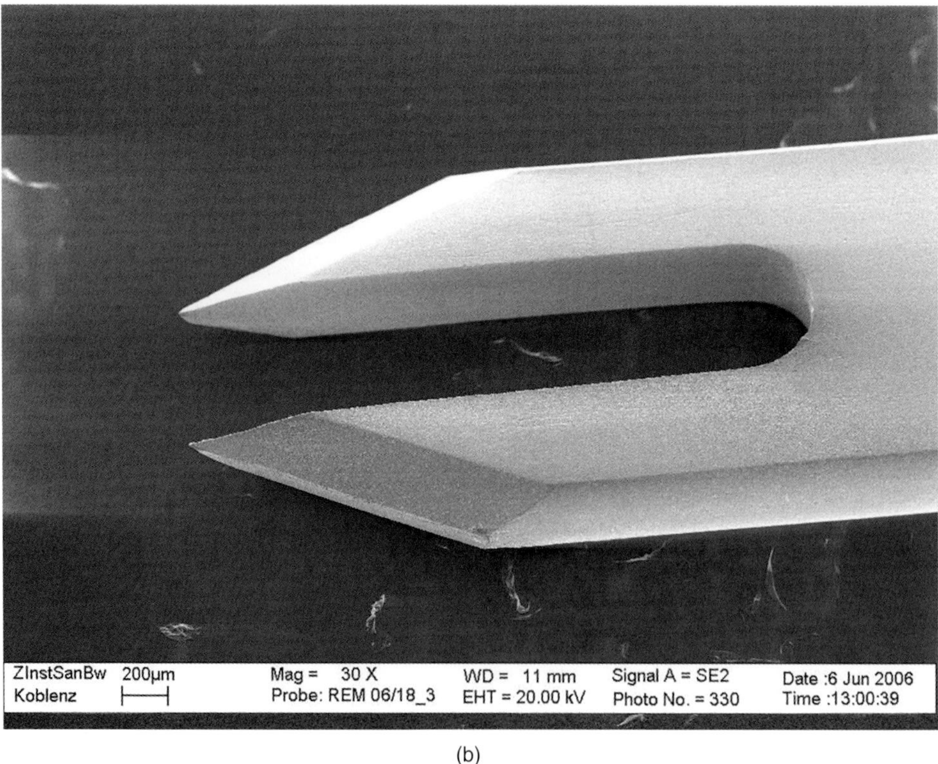

(b)

Figure 14.24 Quality control: (a) image of a bifurcation needle with an uneven and blunt tip and (b) image of a bifurcation needle with a good tip.

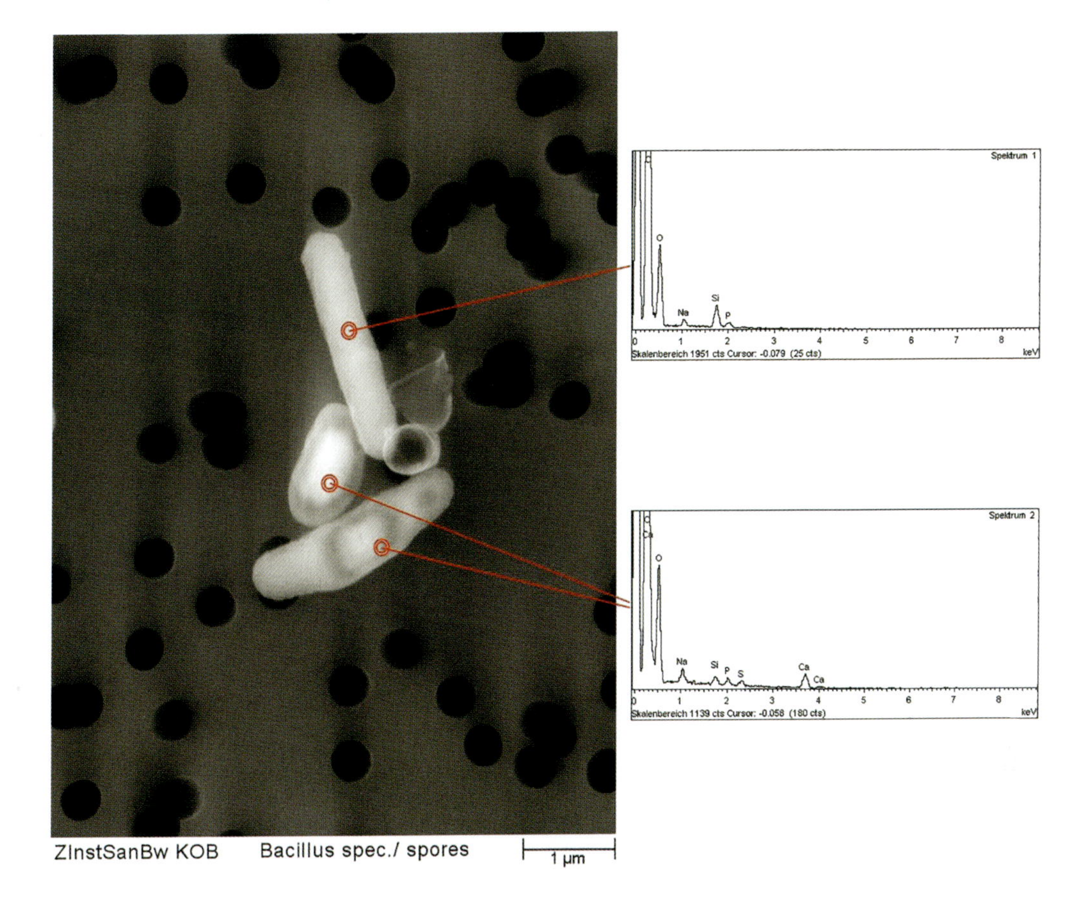

Figure 14.25 Bacterial spores and vegetative cells.

Ever since, envelopes with 'white powder' were considered as a potential biohazard material and emerge from time to time.

FEGSEM in combination with X-ray microanalysis can provide a reliable and quick tool for a first-line diagnosis to detect bacterial spores in environmental samples (Figure 14.25). While EDS microanalysis of vegetative cells shows a spectrum of sodium, silicon and phosphorus besides carbon and oxygen, the typical spectrum of spores includes in addition calcium and sometimes small amounts of sulphur.

An intentional mixture of various white, powdery materials such as talcum powder or flour and freeze-dried suspensions of spore-forming bacteria demonstrate that bacterial cells and their spores can be well distinguished by morphology and composition (Figure 14.26). While flour shows only peaks from carbon and oxygen, the flaky structures of talcum powder present themselves with carbon, oxygen, magnesium and silicon peaks; a bacterial cell with endospores can be identified by their characteristic calcium peak. The accumulation of calcium in bacterial spores had been described before and was related to their heat resistance (Scherrer and Gerhardt 1972). Since this is a common feature of all thermoresistant endospores it is not possible to discriminate different spore-forming bacterial species (Laue and Fulda 2013). However, screening environmental probes for potential biohazard contamination in an FEGSEM provides a very quick and easy way to direct further investigations.

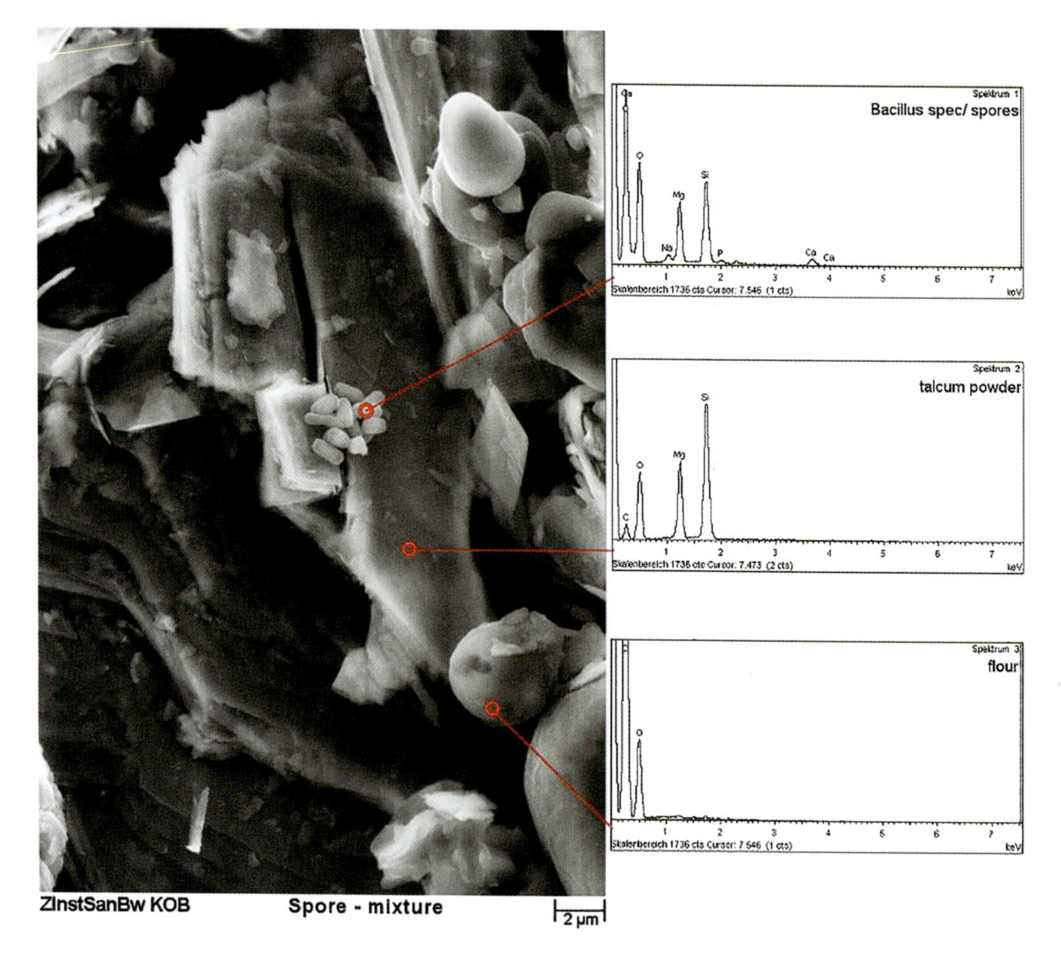

Figure 14.26 Localisation of spore-forming bacteria in a mixture of flour and talcum powder.

14.7 SUMMARY

This chapter has considered the usefulness of FEGSEM in public health, pharmaceutical and medical device research, pathology and infectious disease applications. Whilst many applications can be approached using conventional SEM, the advantages of FEGSEM lie in its great flexibility with regard to sample handling. The innate ability of FEGSEM to obtain high quality observations and data from electron beam sensitive samples gives these instruments a significant advantage over SEM. These advantages are principally derived from the greatly enhanced low accelerating voltage performance of FEGSEM. Having established the value of using FEGSEM, the chapter has provided some perspective on its limitations, aiming to illustrate with many diverse examples how specimen appropriate preparation and instrumental technique selection are critical to maximising the advantages of FEGSEM over conventional SEM, with some hints and tips provided to help the user achieve the best results from trickier specimens/materials. Lastly, the chapter has provided several examples of the value of backscatter and EDS microanalysis detectors, highlighting their usefulness in industrial troubleshooting and public health applications and illustrating the complementary nature and power of these techniques when used alongside other spectroscopic and chemical fingerprinting techniques.

ACKNOWLEDGEMENTS

The authors wish to thank the following individuals for their help and contributions to the chapter: Mr David McCarthy for his micrograph of a resorbable polymer microsphere (Figure 14.12c) and Dr Angela Harrison for her Raman spectral map of a polymer-based medical device (Figure 14.18).

REFERENCES

Bloebaum, R.D., Skedros, J.G., Vajda, E.G., Bachus, K.N. and Constantz, B.R. Determining mineral content variations in bone using backscattered electron imaging, *Bone*, 20 (5), 485–490, May 1997.

Boyde, A. and Machonnachie, E. *Volume Changes During Preparation of Mouse Embryonic Tissue for Scanning Electron Microscopy, Scanning*, vol. 2, pp. 149 -163, G. Witzstrock Publishing House, 1979.

Boyde, A., Jones, S.J., Aerssens, J. and Dequeke, J. Mineral density quantitation of the human cortical iliac crest by backscattered electron image analysis: Variations with age, sex, and degree of osteoarthritis, *Bone*, 16 (6), 619-627, June 1995.

Cook, W.E. Asbestos dust and curious bodies found in pulmonary asbestosis, *Brit. Med. J.*, 2, 578 -580, 1929.

Cugell, D.W. and Kamp, D.W. Asbestos and the plasma: A review, *Chest*, 125,1103 -1117, 2004.

Dobbertin, S. in Berichte 7/80 Luftqualitätskriterien -Umweltbelastung durch Asbest und Andere Faserige Feinstäube, *eds* Umweltbundesamt, *pp. 9 -11*, Erich Schmidt, Berlin, 1980.

Dykstra, M.J. *Biological Electron Microscopy: Theory, Techniques and Troubleshooting*, Chapter 9, p. 239, Plenum Press, New York, 1992, ISBN-13 978-1-4684-0012-0.

Enterline, P.E. Changing attitudes and opinions regarding asbestos and cancer 1934 -1965, *Am. J. Ind. Med.*, 20, 685 -700, 1991.

Galbany, J., Estebaranz, F., Martínez, L.M., Romero, A., De Juan, J., Turbón, D. and Pérez-Pérez, A. Comparative analysis of dental enamel polyvinylsiloxane impression and polyurethane casting methods for SEM research, *Microsc. Res. Tech.*, 69 (4), 246 -52, April 2006.

Goldstein, J.I., Newbury, D.E., Echlin, P., Joy, D.C., Fiori, C. and Lifshin, E. *Scanning Electron Microscopy and X-Ray Microanalysis - A Text for Biologists,Materials Scientists, and Geologists*, Plenum Press, New York, London, 1984.

ap Gwynn, I., Wade, S., Kääb, M.J., Owen, G.R. and Richards, R.G. Freeze-substitution of rabbit tibial articular cartilage reveals that radial zone collagen fibres are tubules, *Journal ofMicroscopy*, 197 (Pt 2), 159 -172, February 2000.

Hayat, M.A. *Fixation for Electron Microscopy*, Academic Press, London, 1981, ISBN-10: 012414411X, ISBN-13: 978-0124144118.

Hayat, M.A. *Introduction to Biological Scanning Electron Microscopy*, University Park Press, Baltimore, 1978, ISBN 0839111738.

Henderson, D.W. *et al.*, Asbestos, asbestosis, and cancer: The Helsinki criteria for diagnosis and attribution. A consensus report, *Scand. J. Work Environ. Health*, 23, 311 -316, 1997.

Kiernan, J.A. Formaldehyde, formalin, paraformaldehyde and glutaraldehyde: What they are and what they do, *Microscopy Today*, 8 (00-1), 8 -12, January 2000.

Laue, M. and Fulda, G. Rapid and reliable detection of bacterial endospores in environmental samples by diagnostic electron microscopy combined with X-ray microanalysis, *J. Microbiol. Methods*, 94, 13 -21, 2013.

Marchand, F. Über eigentümliche Pigmentkristalle in den Lungen, *Verh. Deutsch Ges.*, 17, 223 -228, 1906.

Murray, H.M. Departmental Committee on Compensation for Industrial Diseases, Minutes of Evidence, p. 127, HMSO, London, 1907.

Riediger, G. Bestimmungen von anorganischen Fasern im menschlichen Lungengewebe (Methode unter Verwendung eines Feldemissions-Rasterelektronenmikroskops - FE-REM-Methode), 1985; https://www.ifa-arbeitsmappedigital.de/IFA-AM_7489-1-1.

Roggli, V.L. Scanning electron microscopic analysis of mineral fiber content of lung tissue in the evaluation of diffuse pulmonary fibrosis, *Scanning Microsc.*, 5, 71 -83, 1991.

Roschger, P., Fratzl, P., Eschberger, J. and Klaushofer, K. Validation of quantitative backscattered electron imaging for the measurement of mineral density distribution in human bone biopsies, *Bone*, 23 (4), 319 -326, October 1998.

Scherrer, R. and Gerhardt, P. Location of calcium within Bacillus spores by electron probe X-ray microanalysis, *J.Bacteriol.*, 112, 559 -568, 1972.

Vickerman, J.C. and Gilmore, I.S. (eds), *Surface Analysis: The Principal Techniques*, 2nd edn, John Wiley & Sons, March 2009, ISBN: 978-0-470-01764-7.

15

Field Emission Scanning Electron Microscopy in Cell Biology Featuring the Plant Cell Wall and Nuclear Envelope

Martin W. Goldberg
Science Laboratories, School of Biological and Biomedical Sciences, Durham University, Durham, UK

15.1 INTRODUCTION

Scanning electron microscopy (SEM) produces beautiful images. They have 3D perspective (topography), they are intuitive to interpret and have a 'real world' feel. SEM also produces quantitative 3D data and can resolve small clusters of atoms (Figure 15.1) and single biological molecules. SEMs can image at low magnifications with greater depth of field than light microscopes, making them the preferred technique for imaging rough and topographically complex surfaces. This also results in excellent contextual information for the high resolution data obtained at magnifications up to 2 million times and 0.4 nm resolution. However, the depth of field is compromised when the SEM is optimised for high resolution imaging, which requires very short working distances and the finest possible beam diameter. This emphasises the need to be flexible with microscope settings (keV, probe diameter, current, apertures, working distance) when imaging different objects, or even different aspects of the same object. There has been 50 years of development of instrumentation and sample preparation. Despite this, SEM has not had the general impact on cell biology as that of other imaging methods such as confocal light microscopy and transmission electron microscopy (TEM). This may be explained because until recently SEM was perceived as a comparatively

Biological Field Emission Scanning Electron Microscopy, First Edition.
Edited by Roland A. Fleck and Bruno M. Humbel.

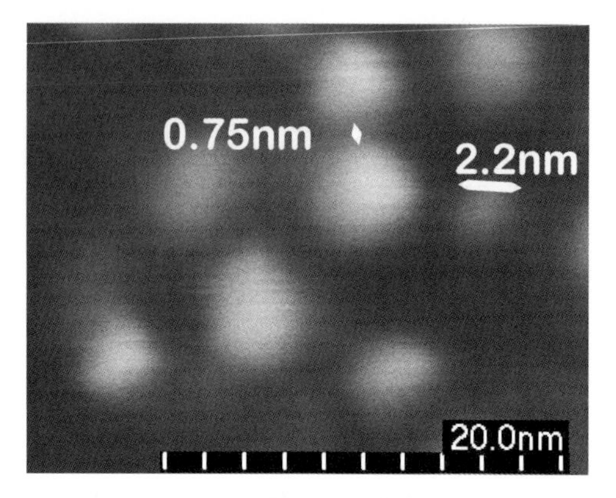

Figure 15.1 Platinum nanoparticles, prepared via a polyol synthesis method (Pushkarev *et al.*, 2012).

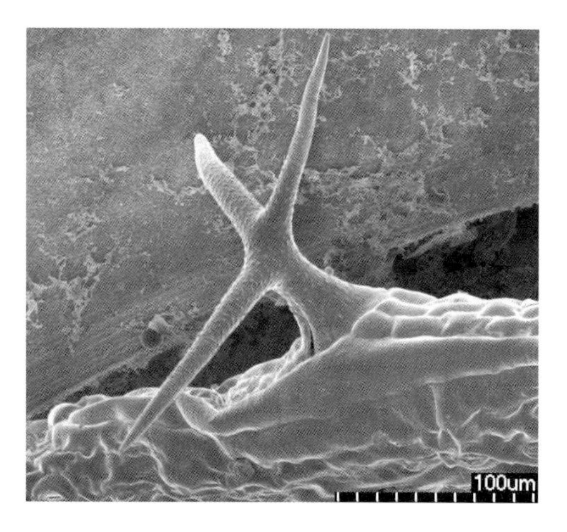

Figure 15.2 *Arabidopsis* trichome using cryo-FEGSEM.

low resolution technique whose sole advantage was the great depth of field, useful for imaging relatively large objects such as phenotypes of homeobox gene mutations in *Drosophila* (Halder, Callaerts and Gehring, 1995), trichomes (hairs) of plant leaves (Figure 15.2) or pollen (Figure 15.3).

High resolution SEM has been available since the beginning of the 1970s with the introduction of the first commercial field emission gun SEM (FEGSEM, Coates and Welter Instrument Cooperation Inc.; Coates and Welter, 1972), the later addition of field emission in-lens instruments in the 1980s (Hitachi) and moderate resolution instruments with tungsten filaments and in-lens or semi-in-lens optics. Subnanometre resolution instruments are now available to cell biologists and instruments with resolution in the 1–4 nm range are relatively common. This level of detail is sufficient to study cellular components and tissue organisation. The resolving power of the instruments is shown by imaging platinum particles on a silicon chip, where we observe particle–particle distances of less than a nanometre (Figure 15.1). Biological molecules are more difficult to image at high resolution, but a ~200 kDa nucleosome, for instance, is ~10 nm in diameter with a ~2 nm spacing between the two strands of DNA wrapped around the histone octamer, all of which can be readily

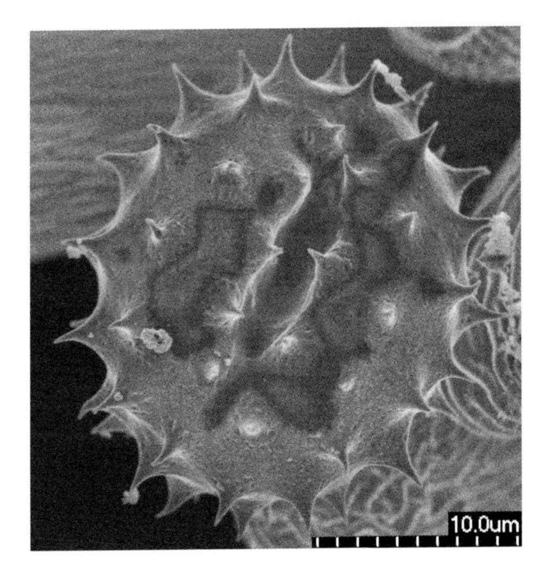

Figure 15.3 Daisy pollen using cryo-FEGSEM.

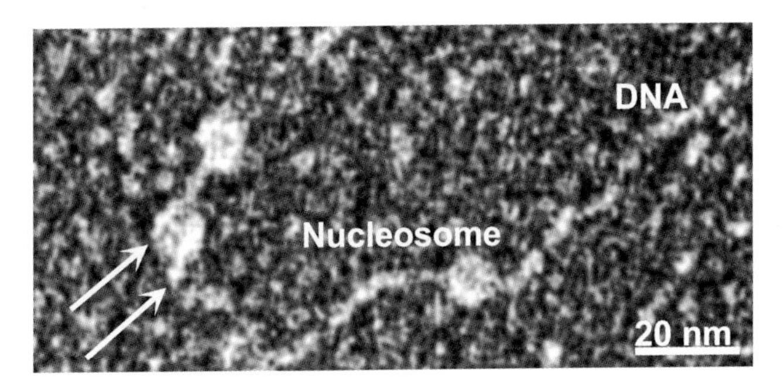

Figure 15.4 Isolated chromatin on carbon support film, fixed with glutaraldehyde, stained with uranyl acetate and coated with 1 nm chromium. Arrows indicate the two strands of DNA wrapped around a histone octamer.

resolved by FEGSEM (Figure 15.4). Modest sized proteins, in the range of several tens of kDa in size are therefore possible to image. However, this is not where the primary value of SEM lies. A nucleosome can be crystallized and its atomic structure determined by X-ray crystallography (Richmond *et al.*, 1984) or imaged easily by TEM. The value of FEGSEM lies with the imaging of large structures or structures within a cellular or subcellular context. It is also of particular value when the structure of interest is highly variable in its morphology, so that ensemble methods, like X-ray crystallography or most high resolution TEM methods, which essentially average many particles to obtain noise-free high resolution 3D data, are not suitable. Under suitable conditions FEGSEM can be used to obtain noise-free, medium resolution surface images of individual structures. It is therefore useful for observing rare structures or structures that vary from individual to individual and therefore preclude methods involving averaging. For example, although the core scaffold of the nuclear pore complex (NPC) is rigid and amenable to averaging methods (Bui *et al.*, 2013; Maimon *et al.*, 2012) its peripheral components are not. Such structures are usually missed by these methods or lack detail.

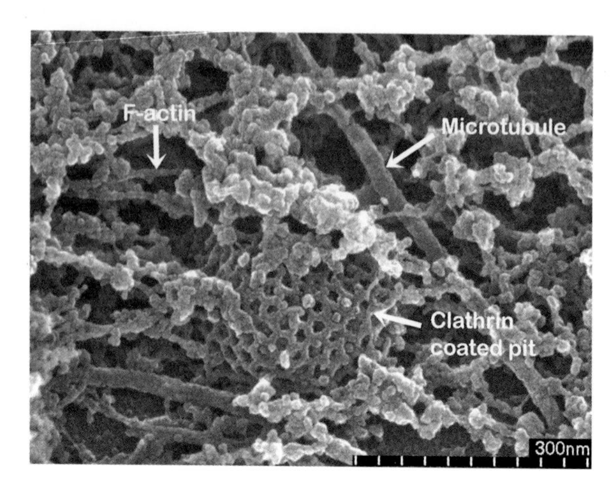

Figure 15.5 HaCaT keratinocyte grown on a silicon chip, detergent extracted, fixed, dried then fractured and sputter coated with 1.5 nm chromium.

SEM is a surface imaging technique and is therefore most obviously useful for looking at surfaces and interfaces. These can be surfaces that are readily accessible such as the surface of a *Caenorhabditis elegans* worm (Costa, Draper and Priess, 1997). They may be internal surfaces from structures, such as the nuclear envelope, that can be exposed by fracturing (Allen *et al.*, 2007) or mitochondria in apoptotic cells exposed by isolation (Kluck *et al.*, 1999).

Here, two cellular interfaces, the plant cell wall and the nuclear envelope, where FEGSEM has been used extensively and fruitfully, are used to illustrate what can be achieved. It is hoped that the success in these fields may stimulate the use of FEGSEM in other areas of cell biology interested in surfaces, interfaces and large structures. These could include fields such as membrane trafficking and the many roles of the cytoskeleton (Figure 15.5).

15.2 PLANT CELL WALL

Due to its fundamental importance to plant cell biology and economical significance, the structure and synthesis of the plant cell wall has been an intense area of research. Due to the position at the external surface and the role of other interfaces in its development at different layers, FEGSEM has played a significant role in this research. Modern FEGSEMs have the ability to resolve individual fibres within the structure, enabling detailed questions to be answered concerning cell wall structure and synthesis.

All plant cells are enclosed in a rigid cell wall that gives strength to plant tissue, protecting it from challenges such as the forces of gravity, environmental factors and pathogen attack. There are two distinct types of cell wall: primary and secondary (Endler and Persson, 2011). Primary cell walls are laid down first, are associated with growing tissues and are composed of cellulose, hemicelluloses, pectins and glycosylated proteins, giving strength and plasticity. The secondary cell wall in some tissue is laid down subsequently, is more rigid and is composed of cellulose, hemicelluloses and lignin. Because of the way these matrices are assembled, the secondary cell wall, when present, is next to the plasma membrane. Outside this is the primary cell wall, then the middle lamella, so called because in a plant tissue it is the layer of material in between the cell walls of adjacent cells. Very early TEM studies using metal shadowing showed the highly ordered parallel fibres of the secondary wall of flax fibres and the less ordered organisation of the primary wall (Mühlethaler, 1950).

Although this predates the first commercial SEM and was at the early stages of SEM development, it demonstrated the value of surface imaging to study complex networks of fibres such as cell wall components. It showed that, unlike light microscopy, individual fibres could be imaged and, unlike TEM cross-sections, individual fibres could be traced for some distance. This is the advantage of surface imaging. SEM, however, has one further advantage, which is the depth of imaging. Techniques such as replicas imaged in the TEM, or atomic force microscopies (AFMs), are in practice restricted to a narrow depth from the surface. Replicas and AFM are essentially like looking into a forest and only seeing the trees around the periphery. For SEM, however, samples can be treated with heavy metals (e.g. osmium, uranium), which stain throughout the sample, enabling the generation of signal from depth. Such bulk staining can be enhanced using mordants like thiocarbohydrazide or tannic acid (Seligman, Wasserkrug and Hanker, 1966).

In this way we can see the trees beyond the periphery. Of course, as with the forest analogy, the deeper in we try to image, the less detail can be obtained, because less primary beam electrons will reach the deeper surfaces, less secondary electrons will escape the sample to be detected and more and more structures will be occluded. If we rely on metal coating for signal generation this problem is exacerbated because of the limited penetration of the coating metal, so that the deeper you go, the thinner the coating will be. However, SEM remains one of the best methods for tracing the organisation of complex networks at high resolution.

In the mid-1970s a TEM fitted with scanning coils was used to image the surface of protoplasts (cells with the wall removed) from tobacco leaves in order to study the early stages of cell wall reformation (Bugess and Linstead, 1976), showing short, straight fibres emanating from small protrusions, which developed into a two-dimensional fibrous mat where great detail of the structural organisation could be seen. In a similar way, Osumi and colleagues studied the regeneration of yeast cell walls using the iconic Hitachi S-900 in-lens FEGSEM (Osumi *et al.*, 1989). Importantly, this group correlated their FEGSEM results with several other electron microscopy techniques, including thin section TEM, negative staining medium voltage TEM and freeze-fracture replicas, all giving a unique insight and emphasising the need not to focus on a single technique (Osumi, 1998) (see Chapter 16, Osumi).

This group also chose to operate the microscope using low accelerating voltages. Because of the limited beam penetration and reduced production of SEII secondary electrons, low voltages can give a better contrast and, in specific samples, also reduce charging effects. However, this is at the cost of instrument resolution. Having said this, they were able to image unidentified particles as small as 1.4 nm in diameter. Although such instruments can resolve into the subnanometre range using high keVs (see Figure 15.1), this demonstrates how flexible they can be when it comes to imaging highly varied samples from small nanoparticles to macroorganisms, retaining much of the imaging power regardless.

Using a simple freeze-fracturing technique, Fowke *et al.* (1999) imaged the cell wall from the 'inside out' in the vacuolated suspensor cells in embryos, showing the highly organised arrays of parallel microfibrils. These were seen through fractures in the plasma membrane, which was also imaged with its associated cortical microtubules, which are thought to play a role in determining the organisation of the cell wall. Again these results were compared to other imaging methods, using TEM and immunofluorescence to identify components such as clathrin coated pits and microtubules. This does highlight one of the challenges in FEGSEM cell biology: the identification of structures. When looking at images of the cell wall you basically see a lot of fibres. You can detect and even quantify their organisation

(Jacques *et al.*, 2013), but with the exception of a few very distinctive structures such as plasmodesmata (Bell and Oparka, 2011) or nuclear pore complexes (Goldberg and Allen, 1992) it may be difficult or impossible to identify a structure based purely on its morphology. To solve this it is necessary to use immunogold labelling. Colloidal gold particles can be coated with antibodies or covalently linked to small gold nanoparticles. Because the surface of interest is usually exposed during the preparation of SEM samples, it is in principle straightforward to immunogold label FEGSEM samples. Although glucomannan was shown to be present in the developing cell wall only during the night using indirect immunogold labelling with 15 nm gold particles attached to secondary antibodies (Hosoo, Imai and Yoshida, 2006), this approach does not appear to have been used extensively in the cell wall field, and only to show the presence/absence of an epitope, rather than the details of how it is organised.

FEGSEM has been used more to determine the degree of organisation of fibres as a whole with different plants/tissues, under different conditions or in the presence of mutations that affect cell wall synthesis and construction.

The main component of the cell wall is cellulose, which is composed of β-1,4-linked glucan chains. A temperature sensitive mutant was isolated (Sato *et al.*, 2001) that displayed impaired tissue elongation at the non-permissive temperature, and showed reduced cellulose, but increased pectin, content. Although the precise role of the identified endo-1,4-β-glucanase, KORRIGAN, in cellulose synthesis is uncertain, FEGSEM imaging showed a clear alteration in the organisation of fibrils of the cell wall in the mutant, providing crucial information on the biosynthetic pathways of cell wall synthesis. Correlative TEM imaging did not show the same level of detail, but confirmed a denser fibril packing, consistent with an increased pectin content. This is important because, as is often the case for this surface imaging method, the surface of interest had to be exposed, often involving extraction procedures that could alter the structures of interest. In this case non-crystalline polysaccharides were extracted with 4 M NaOH, clearly a harsh procedure. To confirm the crystalline nature of the cellulose filaments, X-ray scattering studies were performed and showed that the crystallinity of cellulose from the mutant was indeed reduced, as suggested by the FEGSEM images.

It has been known for a long time that cell wall organisation is dependent on cortical microtubule organisation, although the mechanism is still debated. It is also thought that the anisotropic growth of plant cells is related to the organisation of the cell wall. This was tested using mutants of *Arabidopsis thaliana*, whose tissues swelled radially rather than in the normal directional growth. Immunofluorescence was used to demonstrate the normal transverse orientation of the microtubules in these mutants. However, because of the tight and dense packing of the cell wall microfibrils, FEGSEM proved to be essential to show that these also had the normal transverse orientation, showing that microfibril orientation is not essential for determining anisotropy (Wiedermeier *et al.*, 2002).

Again, the inner cell wall that abuts the plasma membrane was imaged after extraction of overlying material and the resolution of individual filaments was sufficient to quantify their orientation. In a similar way the contribution to wall ultrastructure of lignin and xylan in the inner secondary wall of woody tissue from Japanese beech, was determined using FEGSEM (Awano, Takabe and Fujita, 2002). On the other hand, mutations in a protein involved in cellulose synthesis KOB1, had disorganised microfilaments in the elongation zone, but an apparent wild type organisation in the division zone (Pagant *et al.*, 2002).

FEGSEM was further used to study the relationship between microtubule organisation and cellulose microfilament orientation (Himmelspach, Williamson and Wasteneys, 2003). As immunofluorescence is an excellent way to determine microtubule organisation, but with insufficient resolution to image the tightly packed cell wall microfibrils, a correlative approach was again taken. It was shown that when microtubules are disrupted using a temperature sensitive mutation (mor1-1), newly synthesised microfilaments are still oriented correctly, in a parallel transverse organisation. This provided evidence against the hypothesis that microtubules determine microfibril organisation. Conversely, it was shown, by FEGSEM/immunofluorescence, that treatment of *Arabidopsis* roots with moderate levels of a microtubule disrupting drug, oryzalin, caused a more random orientation of the microfibrils, although they were still well organised into parallel arrays (Baskin *et al.*, 2004).

Similar studies have shown that the innermost wall is organised into parallel transverse arrays regardless of the stage of elongation or subsequent organisation during tissue development (Refrégier *et al.*, 2004). In other words the microfibrils are laid down with this organisation and must be reorganised later. Importantly, from a technical point of view, inner cell walls from unextracted tissue were compared using AFM and FEGSEM (Marga *et al.*, 2005).

Although the information and level of detail are different in the AFM than FEGSEM, the results and conclusions were entirely consistent between the two techniques. This is an important point because SEM sample preparation often involves strong fixation, dehydration and critical point drying, which are all treatments that could potentially introduce artefacts. This study also developed a procedure to quantify the order within the microfibril array using fast Fourier transforms, which were able to detect regularities that were smaller than the easily observable filaments. This suggests the possibility of using such quantification to determine information that is not obvious from simply looking at the images.

Another regulator of cellulose biosynthesis, COBRA, was also shown by a combination of FEGSEM and polarized light microscopy to be involved in the organisation of microfibrils at the inner cell wall into parallel arrays, essential for anisotropic growth in Arabidopsis roots (Roudier *et al.*, 2005). Here, roots were 'cryo-planed' and fractured to reveal the inner surface of the cell wall (Himmelspach *et al.*, 2003). Polarized light microscopy was used to quantify the amount of cellulose within the wall as well as its orientation. Although the information on orientation provided by FEGSEM has more detail, it is restricted to small random patches and is only of the exposed surface. Therefore the complementary light microscopy was necessary to give a wider view in both area and depth. These results were also complemented with immunogold labelling of TEM sections of high pressure frozen roots, showing transport of COBRA through the secretory pathway from the Golgi apparatus to the cell wall. Although immuno-FEGSEM could have been used to show the presence in the cell wall, immuno-TEM is probably more convenient for localisation to intracellular organelles.

In order to study cell wall microfibril organisation in a variety of plant tissues, a combination of fixation, extraction and cryo-planing (Sugimoto, Williamson and Wasteeneys, 2000) was used to preserve and expose the relevant surfaces (Fujita and Wasteneys, 2014) (Figure 15.6A to C). In particular, this study provided new information about the organisation of microfibrils in newly synthesised cell walls of dividing cells, showing that during cytokinesis, short disordered filaments were present instead of the usual parallel arrays, and even the longer filaments, although more ordered, were in a variety of orientations. The same was observed in newly divided stomatal guard cells, which had randomly oriented microfibrils, whereas microfibrils of mature guard cells had a circumferential organisation. It was also found that microfibrils took on a circular organisation around the plasmodesmata (see the next section), with a spoke-like arrangement of filaments. It was further found that microfibrils are disorganised in epidermal inflorescence stem cells in a CseA1 mutant (called *anisotropy 1*, *any1*) (Figure 15.6D), potentially leading to mechanical weakening (Fujita *et al.*, 2013).

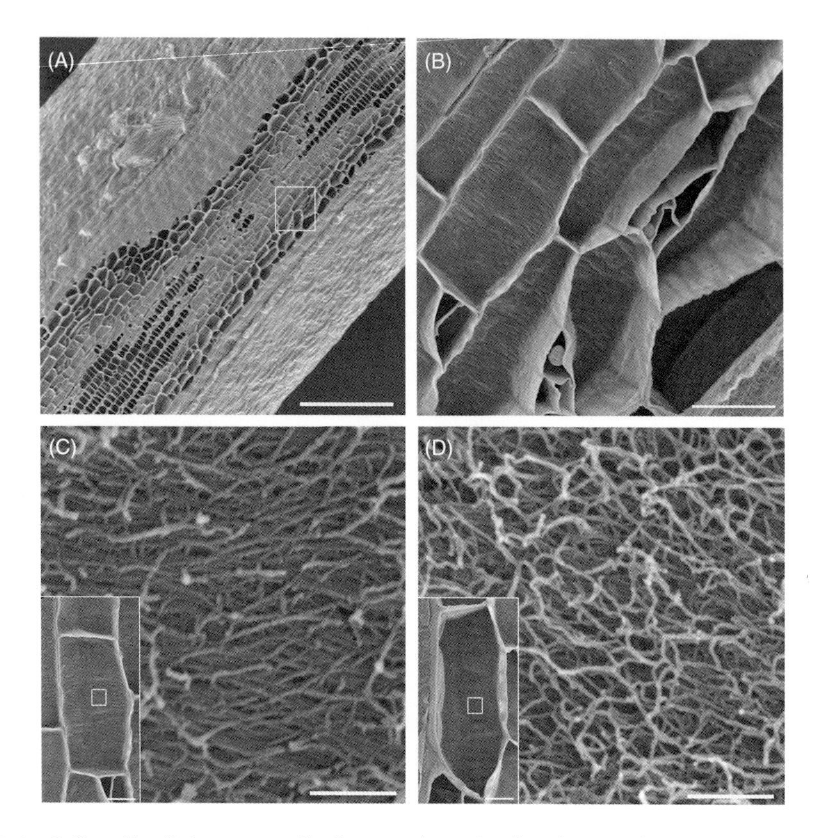

Figure 15.6 Cell wall cellulose microfibrils in epidermal cells of an *Arabidopsis* inflorescence stem. (A) A longitudinal section of cryo-planed inflorescence stem shows both epidermal and cortex cells. The square in (A) indicates the area shown in (B), showing a higher magnification of the epidermal cell. (C) Cellulose microfibrils at the inner periclinal surface of the wild-type cell. The inset shows the elongating epidermal cell and the square indicates the inner periclinal wall from which the higher magnification image was taken. (D) Cellulose microfibrils in the *any1* mutant, which has a mutation in CesA1, an enzyme subunit of cellulose-synthase complexes, have no predominant orientations compared to the ones in the wild type shown in (C). Scale bar = 100 μm (A), 1 μm (B), 200 nm (C, D) and 500 nm (inset for C, D).

15.3 PLASMODESMATA: CONNECTING PLANT CELLS

Plasmodesmata are small connections between plant cells where the plasma membrane of one cell joins to the plasma membrane of an adjacent cell via a membrane tube, of about 100 nm diameter. These may be simple or branched tubes and are thought to contain an extension of the endoplasmic reticulum (ER), known as the desmotubule, connecting the ER of the two cells (Oparka, 2004). Plasmodesmata were first imaged by SEM and TEM in the mid-1970s (Olesen, 1975). In order to study plasmodesmata formation (Faulkner *et al.*, 2008), advantage was taken of the fact that trichomes (leaf hair cells) can be easily fractured after freezing to reveal many plasmodesmata connections with the underlying epidermis. During trichome development many new plasmodesmata are assembled and FEGSEM data suggested a model where new plasmodesmata arise by fission of existing ones.

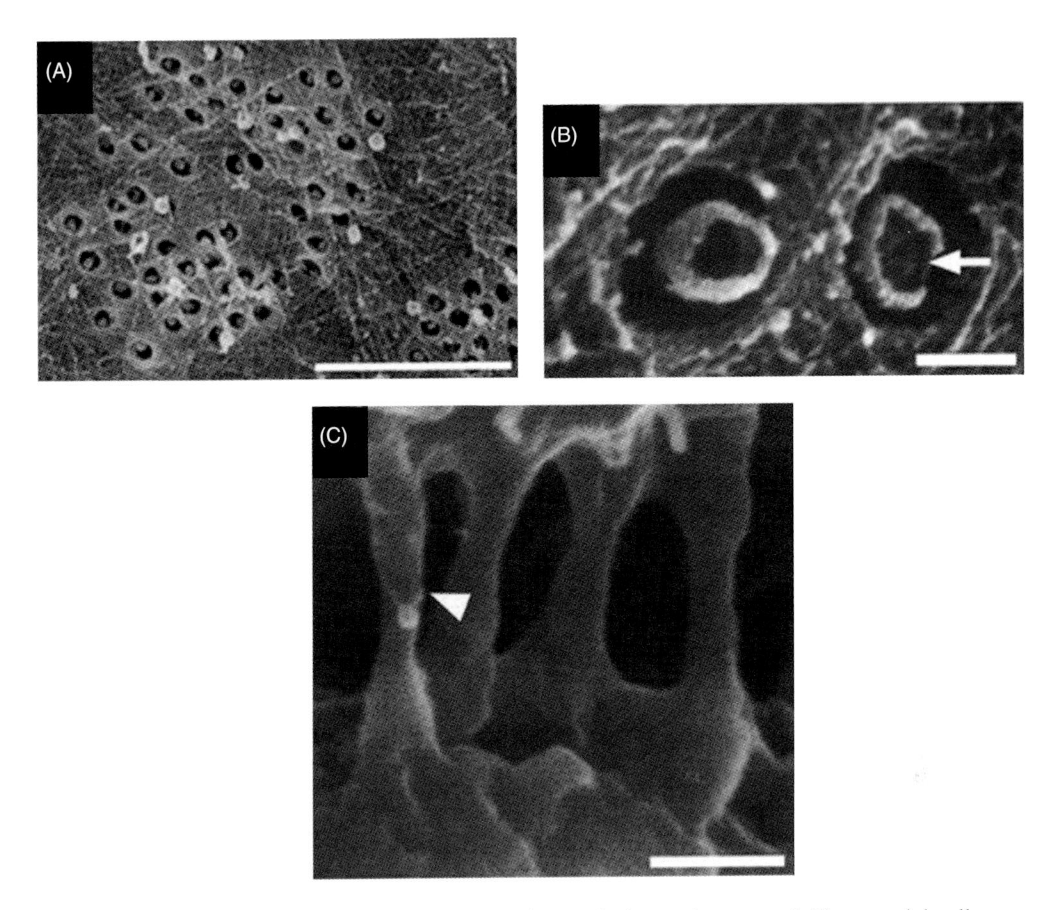

Figure 15.7 FEGSEM of plasmodesmata. (A) Clustered plasmodesmata of *Chara* nodal cells, permeabilised to remove the cytoplasm, then fixed, frozen, fractured, thawed, dehydrated and critical point dried. (B) Freeze-fractured *Chara* plasmodesmata showing internal structure (arrow). (C) External surface, side view, of *Azolla* root plasmodesmata after cryo-sectioning. Scale bars = 1 μm (A) and 100 nm (B&C). Adapted from Brecknock *et al.* (2011).

Consistent with this, using cryo-fracturing, plasmodesmata distribution was shown to be apparently random when at high density or in clusters at lower density (Figure 15.7A), with high levels of variation between different cells (Brecknock *et al.*, 2011). Because freeze-fractures travel through different membranes and other planes (representing planes of least resistance), different levels of plasmodesmata structure and the surrounding cell wall microfibrils could be studied. Most notable are images of the branched and simple plasmodesmata as they pass through the cell wall, giving pseudo-3D images of the membrane structures from both the top and side views (Figure 15.7B and C). Fractures through the plasmodesmata revealed the internal structure, including a central nodule surrounded by particles or a central aperture (Figure 15.7B). Due to the dimensions of the plasmodesmata (~100 nm diameter) it should be possible, with developments in sample preparation, to obtain further information on its controversial structural organisation.

15.4 NUCLEAR ENVELOPE IN PLANTS, ANIMALS AND FUNGI

Another cellular component of similar size to the plasmodesmata is the nuclear pore complex (NPC), which has been extensively studied by FEGSEM (Ris, 1991; Ris and Malecki, 1993; Goldberg and Allen, 1996). The nuclear envelope (NE) is the double membrane barrier surrounding the nuclear contents. Its evolution was a pivotal point in the evolution of complex eukaryotes and was probably a necessary development for the existence of multicellular organisms. The NPCs are selective channels that occur at points of fusion between the inner and outer membrane and control the import of proteins such as transcription factors and export of mRNA and other macromolecular complexes. They are therefore pivotal in the complex control of gene expression.

Like the plasmodesmata (Brecknock *et al.*, 2011), NPCs in rapidly dividing plant cells were shown by FEGSEM to be non-randomly distributed when at low density, but more evenly distributed in mature cells (Fiserova *et al.*, 2009), indicating that NPCs in plants are assembled via a co-operative mechanism or that there is some underlying organising structure such as a plant nuclear lamina (or 'plamina'). The nuclear lamina of animal cells is a well-established intermediate filament containing structure lining the inner nuclear membrane, extensively studied by FEGSEM (Goldberg and Allen, 1992, 1993, 1996; Goldberg *et al.*, 2008) but was not known to exist in plants. Although evidence existed for intermediate filament-like proteins in plants (Ciska and Moreno Díaz de la Espina, 2014) that may locate to the nuclear periphery, FEGSEM provided direct evidence for an equivalent structure (Fiserova *et al.*, 2009). This was achieved by fracturing dried nuclei (Allen *et al.*, 2007). Because the plant lamina is a simple single molecule thick mat of filaments on the inner nuclear membrane, only a high resolution surface imaging method is suitable to image it in any detail. However, freeze-etch-metal shadow TEM methods (Heuser, 2011) or AFM (Kramer *et al.*, 2008) are only suitable for relatively flat surfaces, whereas FEGSEM could image down through the complex surrounding material to see the small patches of relevant surfaces.

Because it is a thin flat sheet of filaments, surface imaging is essential to study the details of the nuclear lamina. Vertebrate nuclear lamina organisation was first imaged by metal shadow TEM (Aebi *et al.*, 1986; Whytock, Moir and Stewart, 1990), but in order to achieve sufficient discrimination between filaments, the attached membrane had to be removed by detergent extraction. This revealed a remarkable orthogonal array of relatively thick filaments with ~50 nm spacing. However, later FEGSEM studies (Goldberg and Allen, 1996; Goldberg *et al.*, 2008) suggested that this organisation could have resulted from lateral aggregation of filaments after membrane removal and, in fact, with the membrane still intact, thin (<10 nm) filaments can be resolved to lie in parallel arrays on the membrane surface separated by 5 nm 'spacers' (Figure 15.8). Interestingly, this simple parallel organisation becomes more complex around NPCs, opening up new questions concerning the interactions between the lamina and NPCs (Goldberg *et al.*, 2008).

Although the 5 nm spacing between the filaments should not be difficult to resolve by FEGSEM, two aspects of specimen preparation were crucial for achieving this. Firstly, a brief wash of the isolated nuclear envelope in a low salt buffer 'cleans out' material from between the filaments. This must be a brief wash, otherwise structural changes occur in the NPC. This emphasises one of the problems with FEGSEM in cell biology. The signal comes from the surface of the material and the signal is the same, almost regardless of what the material is, particularly in the case of biological samples, which are mostly similar light atoms (C, N, O). Heavy metal fixes, such as osmium tetroxide, will change this but, depending on the fixation protocol, may be fairly evenly distributed in the sample.

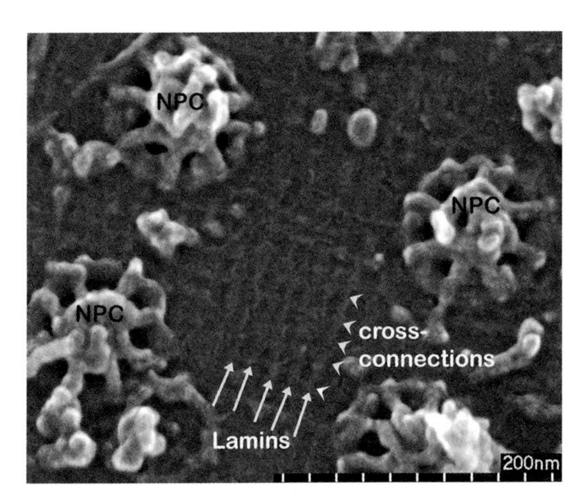

Figure 15.8 Isolated *Xenopus* oocyte nuclear envelope showing nuclear pore complexes and nuclear lamina.

Furthermore, samples are usually coated with a thin metal film, so that most of the signal is actually from the coating. Additionally, the use of low electron beam accelerating voltages ($<\sim 3$ keV) ensures that the image is generated almost exclusively from the surface. Therefore any material that exists between two objects may fill in the gap and prevent resolution of structures such as adjacent lamin filaments. Also, an effective metal coating must have a grain size of less than the resolution required, and will be "continuous" in order to achieve electrical conductivity. From this point of view, magnetron sputter coated chromium has proved effective (Apkarian, Gutekunst and Joy, 1990; Goldberg and Allen, 1992).

As a refractory metal, chromium has a low surface mobility and therefore produces a small grain size (too small to be resolved by FEGSEM), as there is limited aggregation of chromium atoms on the specimen surface, resulting in good conductivity with very thin coats. A disadvantage of chromium is that it oxidises rapidly, making it essential to initially pump to a high vacuum to remove all traces of oxygen and water before introducing very pure argon for sputtering (Apkarian, 1994). Samples also have to be imaged straightaway and long term archiving of samples is not possible. An additional advantage to chromium, however, is that it has a relatively low atomic number compared to gold. This is important because gold nanoparticles are used to tag antibodies for immuno-SEM in order to locate a specific protein within a sample. Gold nanoparticles are imaged by FEGSEM using a backscatter detector (Walther *et al.*, 1991), which can distinguish material based on atomic number differences. Therefore coating samples with metals such as gold or platinum can make it difficult or impossible to distinguish the immunogold particles from surrounding material. On the other hand, gold particles on chromium-coated material are easily distinguished.

Chromium was also deposited by electron beam evaporation using a double axis rotary shadow technique after freeze-drying (Hermann, Schwarz and Müller, 1991). Unlike magnetron sputter coating, which is done under low vacuum conditions in an atmosphere of argon, electron beam evaporation is done at high vacuum. In magnetron sputter coating, chromium atoms are deflected off the argon gas and will, as a result, coat the sample from multiple directions. Electron beam evaporation conversely is highly directional, as there is little gas to interact with the evaporated chromium. This means a more complex setup is needed to ensure multidirectional even coating on the sample. However, very detailed imaging of bacteriophage substructure was achieved, with extraordinary simultaneous backscatter images of 0.8 nm immunogold particles. Other metals such as tungsten or tantalum can also give good resolution samples, as well as samples that can be stored more long term, but because their atomic numbers are similar to gold, they have not been used for immunogold labelling experiments. Electron beam evaporated tungsten was used to identify actin filaments, purely from their substructural morphology, on the surface of the inner nuclear membrane (Walther, 2008). This was possible because the very thin, even, continuous layer of metal provided enough signal and conductivity without filling in the space between individual actin proteins within the filament structure. Interestingly, despite the level of detail available for the actin filaments, little detail of lamina organisation could be seen in the published images, showing how important it is to experiment with different isolation, fixation, drying and coating methods to show the features of interest.

Overall, the choice of metal is crucial for successful FEGSEM imaging of biological material. Although different deposition methods can give equivalent results, there may be stringent requirements for coating conditions to enable 'grain-free' high resolution coatings, such as the elimination of oxygen from the vacuum during sputter coating of chromium. An ideal coating will: (1) be continuous, to allow the easy flow of electrons from the point of entry to earth (and hence prevent charge build-up); (2) have little intrinsic structure (i.e. it will be smooth) so that it evenly follows the contours on the coated surface and reflects the structure of the underlying material; (3) large numbers of secondary electrons emitted from its surface (therefore a high atomic number is desirable); (4) limit the penetration of primary beam electrons, in order to minimise the generation of SEIIs (sample-derived noise) from primary electrons backscattered by subsurface structures; (5) have all these properties when deposited as a very thin film (around 1–2 nm) to avoid obscuring fine detail; and (6) for immunogold labelling, must have a lower atomic number than gold and/or it be extremely thin. Although chromium fulfils this last requirement it falls short on others, as well as oxidising very quickly. Metals such as iridium, despite the high atomic number, may be a good alternative because very thin, smooth coats have been obtained (Fitchman, Shaulov and Harel, 2014). In general, metals such as gold and platinum, which have a high surface mobility and aggregate into discrete particles, are not suitable for high resolution imaging. However, due to their high electrical conductivity, thicker coatings (5–10 nm) are useful for coating large, bulky, non-conductive samples. These metals also generate a very high SEI signal.

Due to its intermediate atomic number (49), indium may be a useful coating material, although, to my knowledge, it has not been explored for biological samples. On the other hand, indium-tin-oxide (ITO) coating is commonly used for creating light (photon) transparent and electrically conductive coatings for glass slides in order to carry out correlative light microscopy with SEM (Pluk *et al.*, 2009). Mammalian culture cells are grown on this highly conductive surface. Using fluorescence microscopy, cellular components are identified (e.g. MitoTracker® to label mitochondria). Samples are then processed and imaged, uncoated, in the FEGSEM and the ultrastructural origin of the fluorescence is identified. Because mammalian culture cells often spread out into a very thin structure, which in this case is growing on a conductive substrate, charging is not a problem. Using low primary beam acceleration, good surface images can be obtained. By increasing the accelerating voltage on the same area, because of the lack of coating, a very high contrast backscatter image of immunogold particles can be obtained and superimposed on the low keV SE image.

The use of ITO to coat the actual biological material has also been tried successfully, although rather thick coatings (20 nm) seemed to be necessary to prevent charging (Rodighiero *et al.*, 2015). The 12 nm immunogold particles are visible even with this thick coating, but whether the coating enables the imaging of fine features (in the 1–5 nm range) remains to be shown.

The NPC, which is attached to the nuclear lamina, traverses the two nuclear membranes and mediates nucleocytoplasmic transport. It has a repetitive structure with rotational symmetry and symmetry about its central plane (Unwin and Milligan, 1982). It therefore lends itself to structural analysis by TEM methods employing particle averaging (Hinshaw, Carragher and Milligan, 1992; Akey and Radermacher, 1993) or cryo-electron tomography with averaging (Bui *et al.*, 2013; Maimon *et al.*, 2012). Such techniques have indeed produced impressive 3D structural models of the NPC and have recently been augmented with X-ray crystallography structures of individual nucleoporin components into the low resolution cryo-electron-tomography structures (Bui *et al.*, 2013) as well as super-resolution light microscopy (Löschberger *et al.*, 2012; Szymborska *et al.*, 2013) to provide information on protein complex orientation and position. However, these methods have not proved particularly successful for imaging structures at the periphery of the NPC, which are fragile, complex and less ordered than the core of the structure, and although there may be some loose rotational symmetry, the cytoplasmic and nucleoplasmic components appear to be completely different. The main reason we know this is because of studies using FEGSEM. The first FEGSEM of NPCs (Kirschner, Rusli and Martin, 1977) used isolated mouse liver nuclei and imaged the cytoplasmic surface of the NE before and after detergent extraction to remove the membranes. NPCs appeared as doughnut-shaped structures standing proud of the outer nuclear membrane, with a subunit composition. Although detailed analysis of the structure from this and other early studies (Schatten and Thoman, 1978) was limited, it did prove ideal for identifying some peripheral and interconnecting structures, which were not considered in later TEM studies. Importantly, FEGSEM was an ideal method to quantify the numbers and distribution of NPCs over the surface of the nucleus. Although certain assumptions have to be made because only about half the nuclear surface is visible in the SEM, many nuclei can be analysed (100s to 1000s), making this much more convenient than techniques such as serial section reconstruction (Winey *et al.*, 1997) and stereology for determining 3D quantitative data such as this. Recent FEGSEM analysis of isolated nuclei (Shaulov and Harel, 2012) showed further details of mammalian NPCs from the cytoplasmic side, which were consistent with, but not identical to, similar images of *Xenopus* oocyte NPCs (see below).

The first major new component of the NPC was discovered by Hans Ris (1991, 1997) when he isolated *Xenopus* oocyte nuclear envelopes and, crucially, fixed them with a mixture of tannic acid with glutaraldehyde, before stabilising membranes with osmium tetroxide, critical point drying, coating with platinum and imaging using low keV FEGSEM. Tannic acid appears to be necessary to stabilise fragile filamentous structures during dehydration/critical point drying and hence preserved a cage-like structure on the nucleoplasmic face of the NPC, which was termed the fishtrap but is now more commonly referred to as the nucleoplasmic basket (Goldberg and Allen, 1992). Without tannic acid, baskets are present but are flattened and badly preserved. Using chromium coating and high accelerating voltages (30 keV), to take full advantage of the resolving power of the instrument, later studies (Goldberg and Allen, 1992, 1993, 1996) showed how the basket consisted of the eight filaments attached to the outer periphery of the NPC nucleoplasmic ring, then branching at the distal end to be woven together to form a ring. Interestingly, this ring was shown by FEGSEM to be dynamic, existing in open and closed conformations, depending on the presence or absence of transport substrates (Kiseleva *et al.*, 1996).

The use of high accelerating voltages can be a contentious issue. Clearly most FEGSEMs have higher instrument resolution at high keVs. However, potentially, charging artefacts and generation of SEII noise (backscatter generated secondary electrons) may also be higher at higher keVs, limiting the detail that can be detected from the sample. However, this is highly sample-dependent. For instance, a very thin sample such as an isolated nuclear envelope will not charge when scanned with a high keV beam because the majority of primary electrons will pass through the sample into the conductive mount (e.g. silicon chip). Similarly, because there is little underlying material to backscatter primary electrons, there is little generation of SEIIs. Therefore the best detail can be achieved with high voltages. For more bulky samples, however, it may be necessary to compromise on instrument resolution, by using low accelerating voltages (<3 keV), to limit these undesirable effects. It may also be useful to use thicker more conductive metal coating, such as platinum, rather than chromium, because chromium has less electron 'stopping power', as relatively light metals will allow more electrons to penetrate the surface and build up inside a bulky specimen: in a bulky, non-conductive sample the only electrically conductive part of the specimen may be the metal coat.

One of the major limitations of FEGSEM is that it is not truly three-dimensional in the sense that, for instance, transmission electron tomography is, because it does not give 3D information throughout the volume of the structure. The information is restricted (mainly) to the surface and, furthermore, only to the surface that is not attached to the support. Clearly FEGSEM gives a 3D perspective, especially with the use of tilting (Figure 15.9) and a real 3D image can be obtained by taking two images of the same object tilted at 6° to each other, thus creating a stereo pair. Height calculations can be performed from stereo pairs and recently novel methods were devised to calculate quantitative 3D surface structures of helical macromolecules (Woodward, Wepf and Sewell, 2009) and non-helical bulk specimens (Woodward and Sewell, 2010). Similarly, Luck *et al.* (2010) developed an algorithm to extract graph structure of intermediate filament networks from FEGSEM tomographic reconstructions from tilt series.

However, information is still restricted to the surface. In order to get below the surface it is necessary to expose or create new surfaces. Cross-sectional 'side' images of NPC baskets were obtained by embedding nuclei in resin, cutting semi-thin sections and then removing the resin (Ris and Malecki, 1993).

Alternatively, nuclei can be fixed, frozen, cryo-sectioned or cryo-planed, thawed and processed for FEGSEM imaging (Figure 15.10). However, these methods did not give much detail concerning the internal structure of the NPC. As a structure that is partially embedded in a membrane, detergent extraction uncovers some new details, such the attachment of the 'spoke ring complex' to the nuclear lamina. Limited proteolysis also revealed considerable

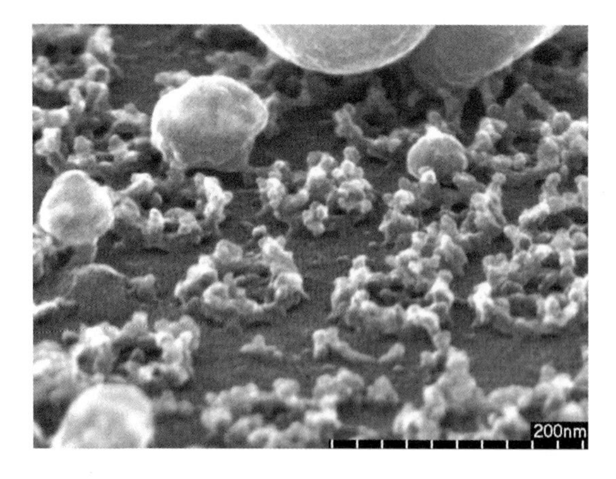

Figure 15.9 Cytoplasmic face of an isolated *Xenopus* oocyte nuclear envelope tilted to 35°.

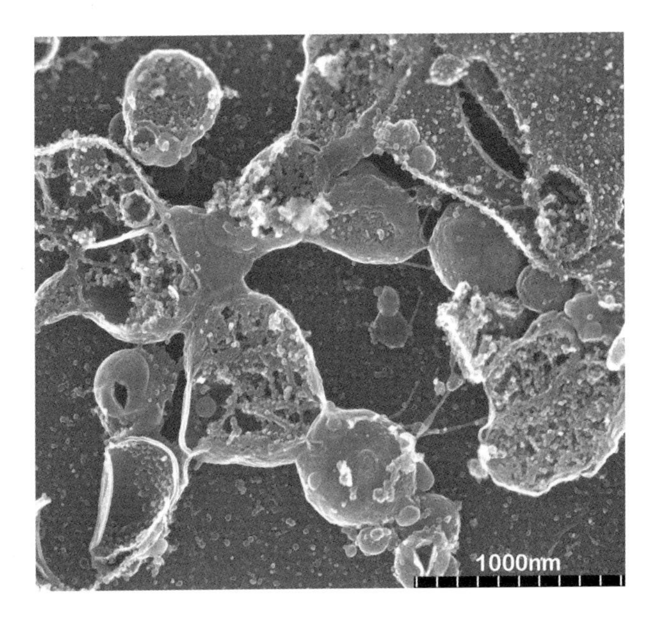

Figure 15.10 Cryo-planed membrane vesicles from *Xenopus* egg extracts.

internal organisation by stripping off more peripheral structures. When combining detergent extraction with proteolysis (Goldberg and Allen, 1993, 1996), a spoke ring complex whose structure was entirely consistent with 3D cryo-TEM studies (e.g. Akey and Radermacher, 1993) was observed. Fracturing methods can also be employed. Firstly, Unwin and Milligan (1982) showed that simply 'rolling' a *Xenopus* oocyte nucleus results in detachment of the cytoplasmic ring of the NPC because it adheres strongly to the EM support. This enabled details of the cytoplasmic ring to be studied by TEM (Unwin and Milligan, 1982), with information on its molecular weight determined by scanning TEM (Reichelt *et al.*, 1990).

However, using FEGSEM to image the membrane from which the cytoplasmic ring had been removed revealed a further underlying ring structure, the 'star ring', which appears to anchor into the membrane (Goldberg and Allen, 1996). Although this could have been an artefact of the fracturing process, the same structure was later observed during NPC assembly in a cell-free system (Goldberg *et al.*, 1997), as well as during disassembly (Cotter *et al.*, 2007) in *Drosophila* embryos (Kiseleva *et al.*, 2001) and rapidly dividing plant cells (Fiserova *et al.*, 2009). Therefore, using natural biological processes, to assemble large structures in a stepwise fashion for FEGSEM analysis is another approach to image internal structures.

One of the great advantages of FEGSEM is the ease with which samples can be routinely prepared, once the methods have been developed. Large areas can be easily surveyed. For macromolecular complexes such as the NPC, this means that complex experiments with many variables can be performed and large numbers of individual structures can be studied. In contrast, this may not be feasible for more involved methods like electron tomography, which requires collection and averaging of thousands of images (Bui *et al.*, 2013), although categorisation does allow study of differences between categories of NPCs (Beck *et al.*, 2004).

Using a combination of atomic force microscopy and FEGSEM, Perez-Terzic *et al.* (1996) showed a role for calcium in determining the conformation of the NPC. It was also shown that high concentrations of RanGTP (a small GTPase that terminates nuclear import, stimulates nuclear export and binds to the NPC) also affect the conformation of the NPC cytoplasmic filaments (Goldberg *et al.*, 2000), stimulating them to form a basket-like structure. Interestingly, a similar cytoplasmic basket-like structure was also observed in nuclei isolated from budding yeast cells (Fiserova *et al.*, 2014). Successful imaging of yeast NPCs, where many mutants are available, will allow large scale experiments into NPC structure and function in the future.

HIV capsid shells were imaged by FEGSEM at the NPC, helping to show the mechanism of uncoating (Arhel *et al.*, 2007). A combination of freeze-fracture and dry-fracture FEGSEM was also used to quantify the number and distribution of NPCs after Herpes Simplex Virus-1 infection, showing a dramatic reduction in the number of NPCs and a large dilation of the NPC itself (Wild *et al.*, 2009). A novel *in vitro* system for studying viral capsid binding to the NPC was developed using FEGSEM as part of the assay for determining the role of specific proteins, both in the virus and NPC, in nuclear envelope targeting (Anderson *et al.*, 2014).

These latter experiments show the value of FEGSEM as a simple tool that can be incorporated into complex experiments to give straightforward answers to specific questions (such as, is a specific protein involved in a specific interaction?). Together with its ability to provide new details of complex structures at medium resolution should establish FEGSEM as an invaluable tool for many cell biologists.

15.5 CONCLUSION

The introduction of the field emission gun, as well as subsequent developments over the last 40 years, have transformed the SEM into a powerful tool for studying subcellular and molecular structures and mechanisms of biological systems. Together with its ability to also image large highly complex and topographical surfaces at a wide range of magnifications, makes this a versatile technique that can add a unique perspective to questions regarding structure and particularly organisation of molecular components over large areas, as exemplified by work on the cell wall. With resolution capabilities approaching that of the TEM and the ability to image individual biomolecular structures such as the NPC or plasmodesmata, at that resolution, FEGSEM can provide structural information that perfectly complements other methods, such as cryo-electron tomography, where certain components, such as the NPC basket/fishtrap, may not be detected

ACKNOWLEDGEMENTS

This work was supported by grants from the Biotechnology and Biological Sciences Research Council, UK (Grant Numbers BB/E015735/1 and BB/G011818/1). Thanks to Laura Bingham, Durham University, for the material imaged in Figure 15.1. Thanks also to Miki Fujita and Geoffrey Wasteneys, University of British Columbia, for providing Figure 15.6 and Robyn Overall, University of Sydney, and Springer for permission to use the images in Figure 15.7.

REFERENCES

Aebi, U., Cohn, J., Buhle, L. and Gerace, L. (1986) The nuclear lamina is a meshwork of intermediate-type filaments. *Nature.*, 323, 560–564.

Akey, C.W. and Radermacher, M. (1993) Architecture of the *Xenopus* nuclear pore complex revealed by three-dimensional cryo-electron microscopy. *J. Cell Biol.*, 122, 1–19.

Allen, T.D., Rutherford, S.A., Murray, S., Gardiner, F., Kiseleva, E., Goldberg, M.W. and Drummond, S.P. (2007) Visualization of the nucleus and nuclear envelope in situ by SEM in tissue culture cells. *Nat. Protoc.*, 2, 1180–1184.

Anderson, F., Savulescu, A.F., Rudolph, K., Schipke, J., Cohen, I., Ibiricu, I., Rotem, A., Grünewald, K., Sodeik, B. and Harel, A. (2014) Targeting of viral capsids to nuclear pores in a cell-free reconstitution system. *Traffic*, 15, 1266–1281.

Apkarian, R.P. (1994) Analysis of high quality monatomic chromium films used in biological high resolution scanning electron microscopy. *Scanning Microsc.*, 8, 289–299.

Apkarian, R.P., Gutekunst, M.D. and Joy, D.C. (1990) High resolution SE-I SEM study of enamel crystal morphology. *J. Electron Microsc. Tech.*, 14, 70–78.

Arhel, N.J., Souquere-Besse, S., Munier, S., Souque, P., Guadagnini, S., Rutherford, S., Prévost, M.C., Allen, T.D. and Charneau, P. (2007) HIV-1 DNA flap formation promotes uncoating of the pre-integration complex at the nuclear pore. *EMBO J.*, 26, 3025–3037.

Awano, T., Takabe, K. and Fujita, M. (2002) Xylan deposition on secondary wall of *Fagus crenata* fiber. *Protoplasma*, 219, 106–115.

Baskin, T.I., Beemster, G.T., Judy-March, J.E. and Marga, F. (2004) Disorganization of cortical microtubules stimulates tangential expansion and reduces the uniformity of cellulose microfibril alignment among cells in the root of *Arabidopsis*. *Plant Physiol.*, 135, 2279–2290.

Beck, M., Förster, F., Ecke, M., Plitzko, J.M., Melchior, F., Gerisch, G., Baumeister, W. and Medalia, O. (2004) Nuclear pore complex structure and dynamics revealed by cryoelectron tomography. *Science*, 306, 1387–1390.

Bell, K. and Oparka, K. (2011) Imaging plasmodesmata. *Protoplasma*, 248, 9–25.

Brecknock, S., Dibbayawan, T.P., Vesk, M., Vesk, P.A., Faulkner, C., Barton, D.A., Overall, R.L., Verbelen, J. and Vissenberg, K. (2011) High resolution scanning electron microscopy of plasmodesmata. *Planta*, 234, 749–758.

Bugess, J. and Linstead, P.J. (1976). Scanning electron microscopy of cell wall formation around isolated plant protoplasts. *Planta*, 131, 173–178.

Bui, K.H., von Appen, A., DiGuilio, A.L., Ori, A., Sparks, L., Mackmull, M.T., Bock, T., Hagen, W., Andrés-Pons, A., Glavy, J.S. and Beck, M. (2013) Integrated structural analysis of the human nuclear pore complex scaffold. *Cell*, 155, 1233–1243.

Ciska, M. and Moreno Díaz de la Espina, S. (2014) The intriguing plant nuclear lamina. *Front Plant Sci.*, 5, 166.

Coates, V.J. and Welter, L. (1972) See it! Analyze it! *Anal. Chem.*, 44, 60A.

Costa, M., Draper, B.W. and Priess, J.R. (1997) The role of actin filaments in patterning the *Caenorhabditis elegans* cuticle. *Dev. Biol.*, 184, 373–384.

Cotter, L., Allen, T.D., Kiseleva, E. and Goldberg, M.W. (2007) Nuclear membrane disassembly and rupture. *J. Mol. Biol.*, 369, 683–695.

Endler, A. and Persson, S. (2011) Cellulose synthases and synthesis in *Arabidopsis*. *Mol. Plant*, 4, 199–211.

Faulkner, C., Akman, O.E., Bell, K., Jeffree, C. and Oparka, K. (2008) Peeking into pit fields: a multiple twinning model of secondary plasmodesmata formation in tobacco. *Plant Cell*, 20, 1504–1518.

Fiserova, J., Kiseleva, E. and Goldberg, M.W. (2009) Nuclear envelope and nuclear pore complex structure and organization in tobacco BY-2 cells. *Plant J.*, 59, 243–255.

Fiserova, J., Spink, M., Richards, S.A., Saunter, C. and Goldberg, M.W. (2014). Entry into the nuclear pore complex is controlled by a cytoplasmic exclusion zone containing dynamic GLFG-repeat nucleoporin domains. *J. Cell Sci.*, 127, 124–136.

Fichtman, B., Shaulov, L. and Harel, A. (2014) Imaging metazoan nuclear pore complexes by field emission scanning electron microscopy. *Methods Cell Biol.*, 122, 41–58.

Fowke, L., Dibbayawan, T., Schwartz, O., Harper, J. and Overall, R. (1999) Combined immunofluorescence and field emission scanning electron microscope study of plasma membrane-associated organelles in highly vacuolated suspensor cells of white spruce somatic embryos. *Cell Biol. Int.*, 23, 389–397.

Fujita, M. and Wasteneys, G.O. (2014) A survey of cellulose microfibril patterns in dividing, expanding and differentiating cells of *Arabidopsis thaliana*. *Protoplasma*, 251, 687–698.

Fujita, M., Himmelspach, R., Ward, J., Whittington, A., Hasenbein, N., Liu, C., Truong, T.T., Galway, M.E., Mansfield, S.D., Hocart, C.H. and Wasteneys, G.O. (2013) The anisotropy 1 D604N mutation in the *Arabidopsis* cellulose synthase 1 catalytic domain reduces cell wall crystallinity and the velocity of cellulose synthase complexes. *Plant Physiology*, 162, 74–85.

Goldberg, M.W. and Allen, T.D. (1992). High resolution scanning electron microscopy of the nuclear envelope: demonstration of a new, regular, fibrous lattice attached to the baskets of the nucleoplasmic face of the nuclear pores. *J. Cell Biol.*, 119, 1429–1440.

Goldberg, M.W. and Allen, T.D. (1993) The nuclear pore complex: Three-dimensional surface structure revealed by field emission, in-lens scanning electron microscopy, with underlying structure uncovered by proteolysis. *J. Cell Sci.*, 106, 261–274.

Goldberg, M.W. and Allen, T.D. (1996) The nuclear pore complex and lamina: three-dimensional structures and interactions determined by field emission in-lens scanning electron microscopy. *J. Mol. Biol.*, 257, 848–865.

Goldberg, M.W., Wiese, C., Allen, T.D. and Wilson, K.L. (1997) Dimples, pores, star-rings, and thin rings on growing nuclear envelopes: Evidence for structural intermediates in nuclear pore complex assembly. *J. Cell Sci.*, 110, 409–420.

Goldberg, M.W., Rutherford, S.A., Hughes, M., Cotter, L.A., Bagley, S., Kiseleva, E., Allen, T.D. and Clarke P.R. (2000). Ran alters nuclear pore complex conformation. *J. Mol. Biol.*, 300, 519–529.

Goldberg, M.W., Huttenlauch, I., Hutchison, C.J. and Stick, R. (2008) Filaments made from A- and B-type lamins differ in structure and organization. *J. Cell Sci.*, 121, 215–225.

Halder, G., Callaerts, P. and Gehring, W.J. (1995) Induction of ectopic eyes by targeted expression of the eyeless gene in *Drosophila*. *Science*, 267, 1788–1792.

Hermann, R., Schwarz, H. and Müller, M. (1991) High precision immunoscanning electron microscopy using fab fragments coupled to ultra-small colloidal gold. *J. Struct. Biol.*, 107, 38–47.

Heuser, J.E. (2011) The origins and evolution of freeze-etch electron microscopy. *J. Electron Microsc.*, 60 (Suppl. 1), S3–29.

Himmelspach, R., Williamson, R.E. and Wasteneys, G.O. (2003) Cellulose microfibril alignment recovers from DCB-induced disruption despite microtubule disorganization. *Plant J.*, 36, 565–575.

Hinshaw, J.E., Carragher, B.O. and Milligan, R.A. (1992) Architecture and design of the nuclear pore complex. *Cell*, 69, 1133–1141.

Hosoo, Y., Imai, T. and Yoshida, M. (2006) Diurnal differences in the supply of glucomannans and xylans to innermost surface of cell walls at various developmental stages from cambium to mature xylem in *Cryptomeria japonica*. *Protoplasma*, 229, 11–19.

Jacques, E., Buytaert, J., Wells, D.M., Lewandowski, M., Bennett, M.J., Dirckx, J., Verbelen, J.P. and Vissenberg, K. (2013) MicroFilament Analyzer, an image analysis tool for quantifying fibrillar orientation, reveals changes in microtubule organization during gravitropism. *Plant J.*, 74, 1045–1058.

Kirschner, H., Rusli, M. and Martin, T.E. (1977) Characterisation of the nuclear envelope, pore complexes, and dense lamina of mouse liver nuclei by high resolution scanning electron microscopy. *J. Cell Biol.*, 72, 118–132.

Kiseleva, E., Goldberg, M.W., Daneholt, B. and Allen, T.D. (1996) RNP export is mediated by structural reorganization of the nuclear pore basket. *J. Mol. Biol.*, 260, 304–311.

Kiseleva, E., Rutherford, S., Cotter, L.M., Allen, T.D. and Goldberg, M.W. (2001) Steps of nuclear pore complex disassembly and reassembly during mitosis in early *Drosophila* embryos. *J. Cell Sci.*, 114, 3607–3618.

Kluck, R.M., Esposti, M.D., Perkins, G., Renken, C., Kuwana, T., Bossy-Wetzel, E., Goldberg, M., Allen, T., Barber, M.J., Green, D.R. and Newmeyer, D.D. (1999) The pro-apoptotic proteins, Bid and Bax, cause a limited permeabilization of the mitochondrial outer membrane that is enhanced by cytosol. *J. Cell Biol.*, 147, 809–822.

Kramer, A., Liashkovich, I., Oberleithner, H., Ludwig, S., Mazur, I. and Shahin, V. (2008) Apoptosis leads to a degradation of vital components of active nuclear transport and a dissociation of the nuclear lamina. *Proc. Natl Acad. Sci. USA*, 105, 11236–11241.

Löschberger, A., van de Linde, S., Dabauvalle, M.C., Rieger, B., Heilemann, M., Krohne, G. and Sauer, M. (2012) Super-resolution imaging visualizes the eightfold symmetry of gp210 proteins around the nuclear pore complex and resolves the central channel with nanometer resolution. *J. Cell Sci.*, 125, 570–575.

Luck, S., Sailer, M., Schmidt, V. and Walther, P. (2009) Three-dimensional analysis of intermediate filament networks using SEM tomography. *J. Microsc.*, 239, 1–16.

Maimon, T., Elad, N., Dahan, I. and Medalia, O. (2012) The human nuclear pore complex as revealed by cryo-electron tomography. *Structure*, 20, 998–1006.

Marga, F., Grandbois, M., Cosgrove, D.J. and Baskin, T.I. (2005) Cell wall extension results in the coordinate separation of parallel microfibrils: Evidence from scanning electron microscopy and atomic force microscopy. *Plant J.*, 43, 181–190.

Mühlethaler, K. (1950) Electron microscopy of developing plant cell walls. *Biochem. Biophys. Acta*, 5, 15–25.

Olesen, P. (1975) Plasmodesmata between mesophyll and bundle sheath cells in relation to the exchange of C4-acids. *Planta*, 123, 199–202.

Oparka, K. (2004) Getting the message across: How do plant cells exchange macromolecular complexes. *Trends Plant Sci.*, 9, 33–40.

Osumi, M. (1998) The ultrastructure of yeast: Cell wall structure and formation. *Micron.*, 29, 207–233.

Osumi, M., Yamada, N., Kobori, H., Taki, A., Naito, N., Baba, M. and Nagatani, T. (1989) Cell wall formation in regenerating protoplasts of *Schizosaccharomyces pombe*: Study by high resolution, low voltage scanning electron microscopy. *J. Electron Microsc.*, 38, 457–468.

Pagant, S., Bichet, A., Sugimoto, K., Lerouxel, O., Desprez, T., McCann, M., Lerouge, P., Vernhettes, S. and Höfte, H. (2002) KOBITO1 encodes a novel plasma membrane protein necessary for normal synthesis of cellulose during cell expansion in *Arabidopsis*. *Plant Cell*, 14, 2001–2013.

Perez-Terzic, C., Pyle, J., Jaconi, M., Stehno-Bittel, L., Clapham, D.E. (1996) Conformational states of the nuclear pore complex induced by depletion of nuclear Ca^{2+} stores. *Science*, 273, 1875–1877.

Pluk, H., Stokes, D.J., Lich, B., Wieringa, B. and Fransen, J. (2009) Advantages of indium-tin oxide-coated glass slides in correlative scanning electron microscopy applications of uncoated cultured cells. *J. Microsc.*, 233, 353–363.

Pushkarev, V.V., An, K., Alayoglu, S., Beaumont, S.K. and Somorjai, G.A. (2012) Hydrogenation of benzene and toluene over size controlled Pt/SBA-15 catalysts: Elucidation of the Pt particle size effect on reaction kinetics. *J. Catal.*, 292, 64–72.

Refrégier, G., Pelletier, S., Jaillard, D. and Höfte, H. (2004) Interaction between wall deposition and cell elongation in dark-grown hypocotyl cells in *Arabidopsis*. *Plant Physiol.*, 135, 959–968.

Reichelt, R., Holzenburg, A., Buhle, E.L, Jr, Jarnik, M., Engel, A. and Aebi, U. (1990) Correlation between structure and mass distribution of the nuclear pore complex and of distinct pore complex components. *J. Cell Biol.*, 110, 883–894.

Richmond, T.J., Finch, J.T., Rushton, B., Rhodes, D. and Klug, A. (1984) Structure of the nucleosome core particle at 7 A resolution. *Nature*, 311, 532–537.

Ris, H. (1991) The three-dimensional structure of the nuclear pore complex as seen by high voltage electron microscopy and high resolution low voltage scanning electron microscopy. *Electron Microsc. Soc. Am. Bull.*, 21, 54–56.

Ris, H. (1997) High-resolution field-emission scanning electron microscopy of nuclear pore complex. *Scanning*, 19, 368–375.

Ris, H. and Malecki, M. (1993) High-resolution field emission scanning electron microscope imaging of internal cell structures after Epon extraction from sections: A new approach to correlative ultrastructural and immunocytochemical studies. *J. Struct. Biol.*, 111, 148–157.

Rodighiero, S., Torre, B., Sogne, E., Ruffilli, R., Cagnoli, C., Francolini, M., Di Fabrizio, E. and Falqui, A. (2015) Correlative scanning electron and confocal microscopy imaging of labeled cells coated by indium-tin-oxide. *Microsc. Res. Tech.*, 78, 433–43.

Roudier, F., Fernandez, A.G., Fujita, M., Himmelspach, R., Borner, G.H., Schindelman, G., Song, S., Baskin, T.I., Dupree, P., Wasteneys, G.O. and Benfey, P.N. (2005) COBRA, an *Arabidopsis* extracellular glycosyl-phosphatidyl inositol-anchored protein, specifically controls highly anisotropic expansion through its involvement in cellulose microfibril orientation. *Plant Cell*, 17, 1749–1763.

Sato, S., Kato, T., Kakegawa, K., Ishii, T., Liu, Y.G., Awano, T., Takabe, K., Nishiyama, Y., Kuga, S., Sato, S., Nakamura, Y., Tabata, S. and Shibata, D. (2001) Role of the putative membrane-bound endo-1,4-beta-glucanase KORRIGAN in cell elongation and cellulose synthesis in *Arabidopsis thaliana*. *Plant Cell Physiol.*, 42, 251–263.

Schatten, G. and Thoman, M. (1978) Nuclear surface complex as observed with the high resolution scanning electron microscope. *J. Cell Biol.*, 77, 517–535.

Seligman, A.M., Wasserkrug, H.L. and Hanker, J.S. (1966) A new staining method (OTO) for enhancing contrast of lipid-containing membranes and droplets in osmium tetroxide-fixed tissue with osmiophilic thiocarbohydrazide (TCH). *J. Cell Biol.*, 30, 424–432.

Shaulov, L. and Harel, A. (2012) Improved visualization of vertebrate nuclear pore complexes by field emission scanning electron microscopy. *Structure*, 20, 407–413.

Sugimoto, K., Williamson, R.E. and Wasteneys, G.O. (2000) New techniques enable comparative analysis of microtubule orientation, wall texture, and growth rate in intact roots of Arabidopsis. *Plant Phys.*, 124, 1493–1506.

Szymborska, A., de Marco, A., Daigle, N., Cordes, V.C., Briggs, J.A. and Ellenberg, J. (2013) Nuclear pore scaffold structure analyzed by super-resolution microscopy and particle averaging. *Science*, 341, 655–658.

Unwin, P.N. and Milligan, R.A. (1982) A large particle associated with the perimeter of the nuclear pore complex. *J. Cell Biol.*, 93, 63–75.

Walther, P. (2008) High-resolution cryo-SEM allows direct identification of F-actin at the inner nuclear membrane of *Xenopus oocytes* by virtue of its structural features. *J. Microsc.*, 232, 379–385.

Walther, P., Autrata, R., Chen, Y. and Pawley, J.B. (1991) Backscattered electron imaging for high resolution surface scanning electron microscopy with a new type YAG-detector. *Scanning Microsc.*, 5, 301–309.

Whytock, S., Moir, R.D. and Stewart, M. (1990) Selective digestion of nuclear envelopes from *Xenopus oocyte* germinal vesicles: Possible structural role for the nuclear lamina. *J. Cell Sci.*, 97, 571–580.

Wiedemeier, A.M., Judy-March, J.E., Hocart, C.H., Wasteneys, G.O., Williamson, R.E. and Baskin, T.I. (2002) Mutant alleles of *Arabidopsis* RADIALLY SWOLLEN 4 and 7 reduce growth anisotropy without altering the transverse orientation of cortical microtubules or cellulose microfibrils. *Development*, 129, 4821–4830.

Wild, P., Senn, C., Manera, C.L., Sutter, E., Schraner, E.M., Tobler, K., Ackermann, M., Ziegler, U., Lucas, M.S. and Kaech, A. (2009) Exploring the nuclear envelope of herpes simplex virus 1-infected cells by high-resolution microscopy. *J. Virol.*, 83, 408–419.

Winey, M., Yarar, D., Giddings, T.H., Jr and Mastronarde, D.N. (1997) Nuclear pore complex number and distribution throughout the *Saccharomyces cerevisiae* cell cycle by three-dimensional reconstruction from electron micrographs of nuclear envelopes. *Mol. Bio. Cell*, 8, 2119–2132.

Woodward, J.D. and Sewell, B.T. (2010) Tomography of asymmetric bulk specimens imaged by scanning. *Ultramicroscopy*, 110, 170–175.

Woodward, J.D., Wepf, R. and Sewell, B.T. (2009) Three-dimensional reconstruction of biological macromolecular complexes from in-lens scanning electron micrographs. *J. Microsc.*, 234, 287–292.

16

Low-Voltage Scanning Electron Microscopy in Yeast Cells

Masako Osumi

Japan Women's University, Tokyo, Japan and NPO Integrated Imaging Research Support, Tokyo, Japan

16.1 INTRODUCTION

The use of lower voltages in a scanning electron microscope (SEM) increases the topographic contrast and reduces the electronic charging phenomena and beam damage to the biological specimen. However, before the 1980s, microscopists usually did not operate conventional SEMs at low voltage (LV) because of poor resolution in the image of specimens. Even after Pawley (1984), Joy (1984) and Boyes (1984) proposed the LVSEM in 1984, satisfactory studies in the biological application of the method remained limited. The development of the 'in-lens' field emission SEM (FESEM) has greatly improved the resolution (Nagatani and Saito 1986; Nagatani *et al.* 1987b; Pawley 1986), especially in the range of 1–3 keV. The probe size has been calculated to be 0.7 nm in diameter at 30 keV and about 3 nm at 1 keV. Within our group we have been developing techniques to use the in-lens FESEM at LV for direct observation of uncoated biological specimens, and have developed the LVSEM method for its application to the biological field (Osumi *et al.* 1985, 1986, 1987, 1988a,b,c; Yamada *et al.* 1986; Osumi and Nagatani 1987).

Here, we introduce some results of LVSEM in the biological research field, specifically an investigation of the dynamics of the ultrastructure during cell wall formation in *Schizosaccharomyces pombe* using this improved microscope and further application of LVSEM to significance of the high-pressure freezing (HPF) technique in cell biology with reference to a study of septum formation in fission yeast.

Biological Field Emission Scanning Electron Microscopy, First Edition.
Edited by Roland A. Fleck and Bruno M. Humbel.

16.2 DEVELOPMENT AND OUTLINE OF ULTRAHIGH RESOLUTION LOW-VOLTAGE SCANNING ELECTRON MICROSCOPY (UHR LVSEM)

Resolution of the SEM had been markedly improved by means of a field emission (FE) gun with an "in-lens" type design (Nagatani *et al.* 1983; Pawley 1988), which has an 0.7 nm probe size at 30 keV (Nagatani and Saito 1986). However, when these instruments were first conceived and made commercially available the performance at LV, where greater resolving power is needed to visualize details of the surface of a biological specimen, remained insufficient (Osumi *et al.* 1989b). Thus, a commercial model of SEM, S-900 (Hitachi Ltd., Japan), was modified to achieve better performance at low beam energy (about 5 keV or below) with small aberrations of the objective lens and a shorter working distance (Sato *et al.* 1989; Nagatani *et al.* 1990).

This modified in-lens FESEM S-900, referred to here as S-900LV (commercial model SEM S-900H), is a type of UHR LVSEM, which enabled us to observe directly the surface topography of uncoated biological samples with maximum fidelity (Osumi *et al.* 1995).

16.2.1 Resolution of UHR LVSEM

As discussed earlier (Joy 1989; Crew 1985; Nagatani *et al.* 1987a), the spot size of the beam is limited primarily by spherical aberration of the objective lens and diffraction at high voltage (about 10 keV and above), while chromatic aberration and diffraction are the dominant factors at low voltage (about 5 keV or below). The source size of cold field emission is so small that it can be neglected for simplicity. Chromatic aberration can be smaller at higher excitation of a narrow gap objective pole-piece, which also makes the working distance (WD) short. Therefore, some compromise is necessary among minimized aberrations, required specimen size, stage traverse, tilting angle, etc. In practice, tolerable distortion of the image at low magnification and collection efficiency of the secondary electrons are other factors to be considered in designing the instrument.

The UHR SEM, S-900 LV (Sato *et al.* 1989; Nagatani *et al.* 1990) used in this study was designed with an optimized objective lens as shown in Table 16.1. While WD = 2.5 mm (UHR position) can be used below 5 keV, WD = 4.5 mm (standard, STD position) is selected for high voltage from 5 up to 30 keV to avoid saturation of the lens. Figure 16.1a shows a comparison of calculated probe size at different working distances for the S-900 LV and those of S-900 at 1 keV, which indicates that the theoretical resolving power was 2 nm at 1 keV. Actually, the S-900 LV allows visualization of an object of about 1 nm at 2.5 keV and ×300 000 magnification, using gold-evaporated magnetic tape as a resolution test specimen (Figure 16.1b, →) (Osumi *et al.* 1995).

Table 16.1 Specifications of objective lenses used for S-900LV and S-900 (Osumi *et al.* 1995)

	Objective lens	S-900LV		S-900
Geometry	Upper bore diameter	4 ϕ		10 ϕ
	Gap distance	10.5		11
	Lower bore diameter	10 ϕ		2 ϕ
Specifications	Working distance (WD)	2.5(UHR)	4.5 (STD)	5.5
	Focal length (f)	1.7	2.8	3.6
	Spherical aberration coefficient (Cs)	1.0	2.2	1.9
	Chromatic aberration coefficient (Cc)	1.2	2.1	2.5

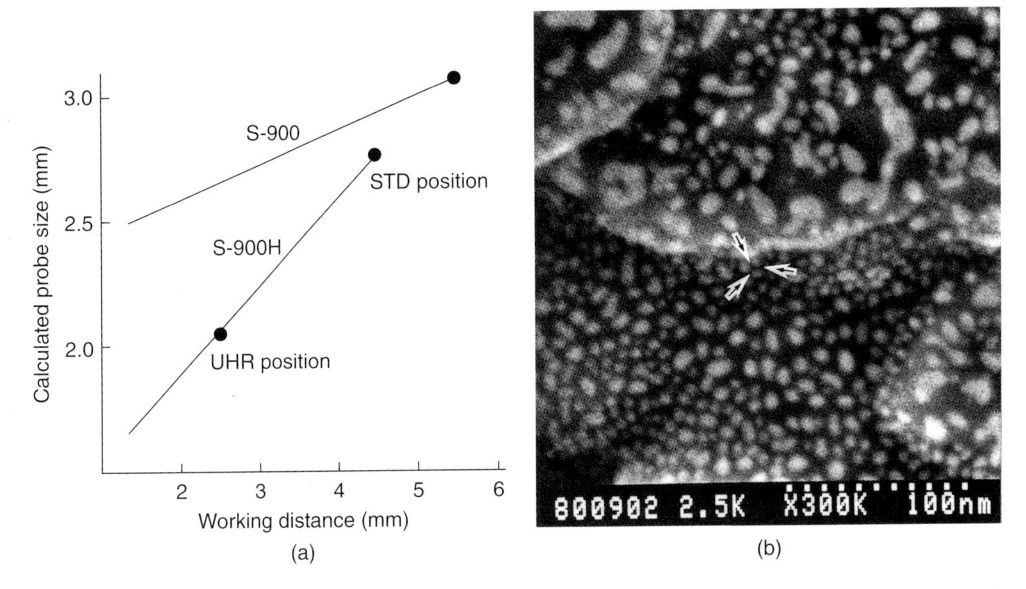

Figure 16.1 Resolution of S-900 LV. (a) Probe size calculated at 1 keV and different working distances (WD). 5 keV or below is used for UHR position at 2.5 mm WD. (b) Arrows indicate the narrowest gaps between gold particles (bright spots) settled on vacuum-evaporated magnetic tape, at 2.5 keV and an original magnification of ×300 000 (Osumi *et al.* 1995; Osumi 2012).

The surface of the uncoated protoplast of *S. pombe*, which appears smooth at low magnification (Figure 16.2a), was seen to be covered with granular materials, when it was observed at ×120 000. The finest detectable particle is 1.4 nm in diameter (Figure 16.2b, arrowhead) (Osumi *et al.* 1995). This is the first image of the protoplast surface and shows the integral membrane proteins of cell membrane, that is, the surface of the cell membrane in yeast cells, which had been shown previously by only the freeze-etching technique.

Furthermore, we tried to take high magnification images of reverting protoplasts at 1 and 2.5 keV taken at ×100 000 with the S-900 LV and found that the resolution of the 2.5 keV image taken at higher magnification was adequate to detect the fine structure of the glucan fibrils (Osumi *et al.* 1995).

16.2.2 Dynamics of the Ultrastructure During Cell Wall Formation in *Schizosaccharomyces pombe*

Using a UHR LVSEM S-900 LV and S-900, we can study the ultrastructure of regenerating cell walls of *S. pombe* protoplasts with uncoated or slightly coated surfaces, respectively (Osumi 2011). Just after digesting the cell wall of the cylindrical *S. pombe* cells with Novozyme 234, no cell wall material was seen on the surface of the round shaped protoplast (Figure 16.2a). The absence of staining with calcofluor confirmed that the surface of freshly prepared protoplasts was free from cell wall materials, glucans and α-galactomannan (Kobori *et al.* 1989). TEM images of thin sections also confirmed the absence of the remnants of old wall materials on the surface (Osumi *et al.* 1998). Many invaginations of the cell membrane, which is unique to *S. pombe*, were found on the surface (Figure 16.2a, →).

After 10 min of incubation of a portion of the surface accumulated invaginations (Figure 16.3a, →), fibrous materials appeared in the area (Figure 16.3b, → and c) at 1.5 h and 3 h. After 5 h, fibrous network covered the whole surface of the protoplast and the

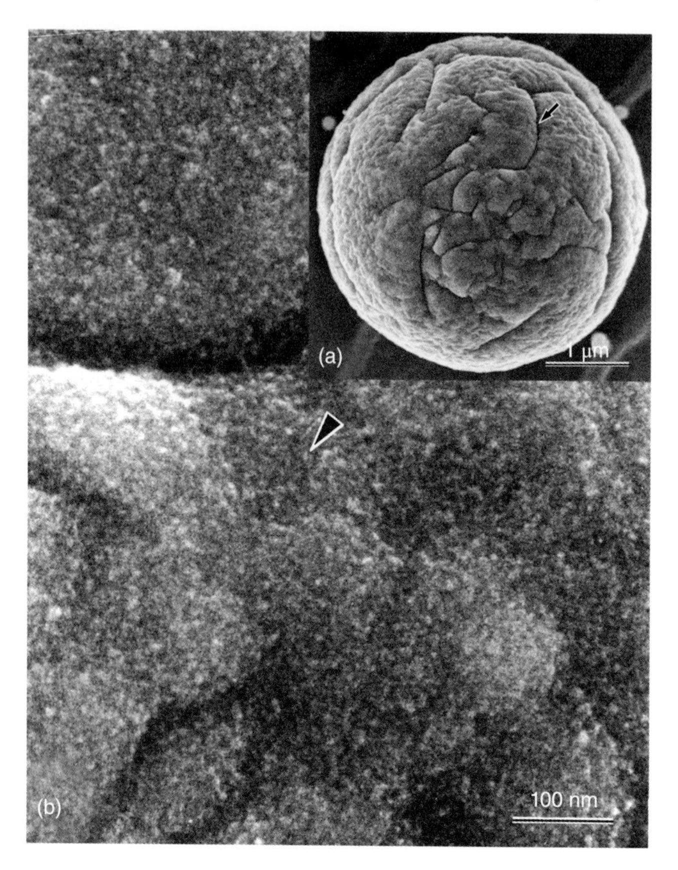

Figure 16.2 An uncoated protoplast of S. *pombe* showing the granular surface and invagination of protoplast membrane. The finest detectable particle is 1.4 nm in diameter (b, arrowhead) (Osumi *et al.* 1995).

shape of the protoplast changed to the elongated form (Figure 16.3e). The microfibrils twisted around each other and developed into thick fibrils, forming flat bundles.

Intrafibrillar spaces were gradually filled with particles (Figure 16.3d, arrowhead). After 5 h of incubation, the whole protoplast was covered with a fibrillar network (Figure 16.3e). After 10 h, the granular materials filled intrafibrillar spaces on the surface (Figure 16.3f). Finally, after 12 h the complete cell wall was regenerated (Figure 16.3g and h) (Osumi *et al.* 1989b, 1995; Osumi 1998).

In the cell-free system using isolated cell membrane of *Candida albicans* and the negative staining method, secreted particles (2 nm thick) converted subsequently to microfibrils of 4 nm thick (Osumi *et al.* 1984; Yamaguchi *et al.* 1985, 1987).

Uncoated UHR LVSEM images of the regenerating protoplast showed scattered particles of 3 nm (Figure 16.4, arrowhead) and 2 nm thick (① →←) microfibrils on the cell surface (Figure 16.4). These microfibrils may be joined end to end, twisted around each other, attached side-by-side to 8 nm thick (② →←) fibrils, and then developed to a ribbon-shaped network of less than 16 nm thick (③ ⇉⇇). The ribbon-shaped structures further contacted side by side and finally formed diversified structures with various widths and lengths, up to 200 nm wide and 1 μm long (→) (Osumi *et al.* 1995). SEM images of microfibrils and ribbon-shaped structures shown here were similar to TEM images of a negatively stained specimen of the reverting protoplast for 3 h (Osumi *et al.* 1989b; Osumi 1998).

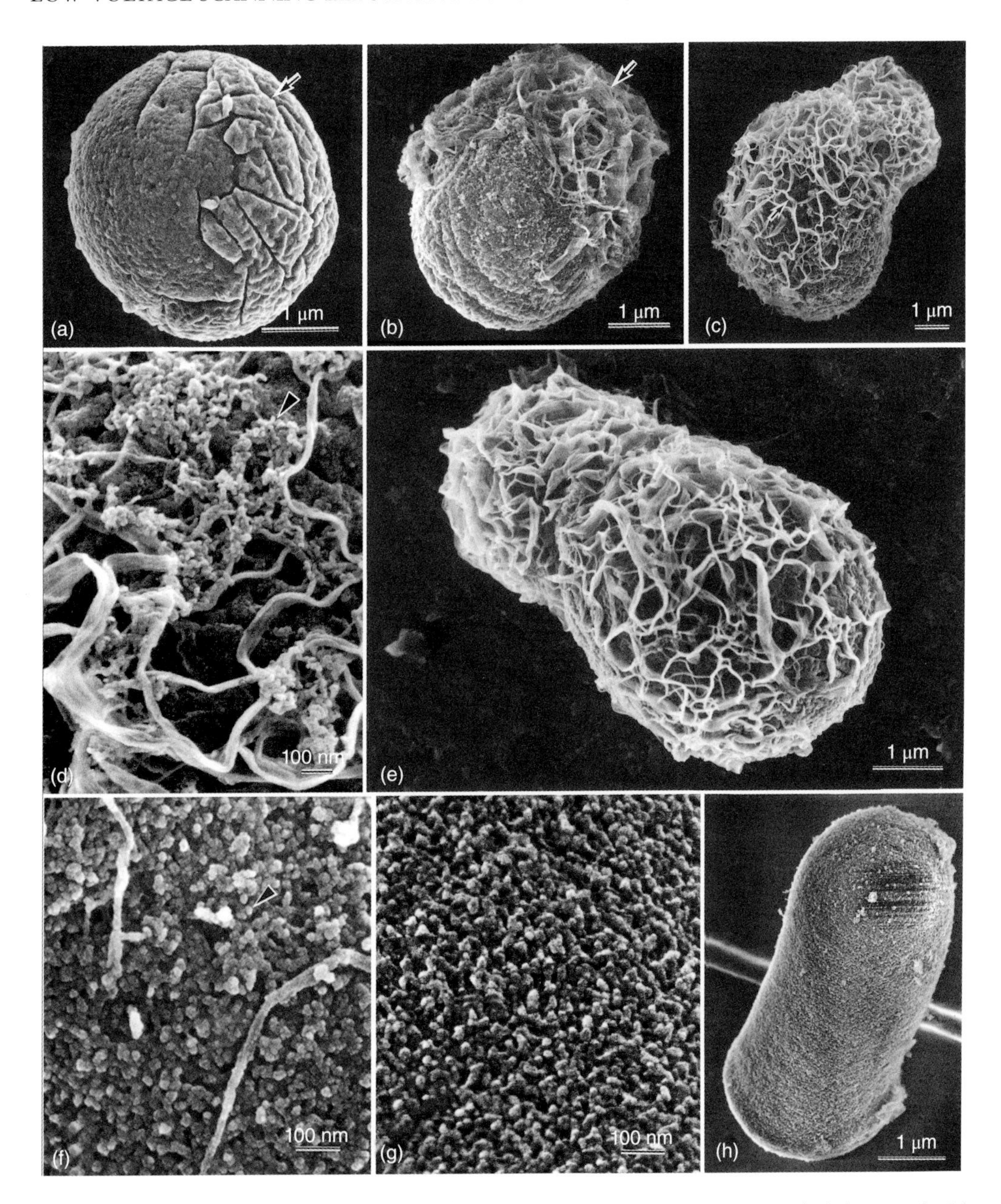

Figure 16.3 Regeneration processes of reverting protoplast. The specimens were slightly coated with a <2 nm layer of platinum–carbon at 2×10^{-7} Pa and 10 °C in a Balzers 500 K freeze-fracture device with the electron gun. Cell wall regeneration process after (a) 10 min, (b) 1.5 h, (c) 3 h, (d) 3 h, (e) 5 h, (f) 10 h, and (g and h) 12 h (Osumi *et al.* 1989b; Osumi 2012).

16.2.3 Identification of Cell Wall Components

Bundles formed on the surface of reverting protoplasts were cleaved into small pieces when the protoplasts were treated with β-1,3-glucanase for 2 h (Osumi *et al.* 1984, 1995) and completely digested with longer treatment. Furthermore, fibrous networks are not formed on the protoplasts in the presence of aculeacin A, an antifungal antibiotic and a potent inhibitor of β-1,3-glucan synthesis (Osumi *et al.* 1984; Yamaguchi *et al.*

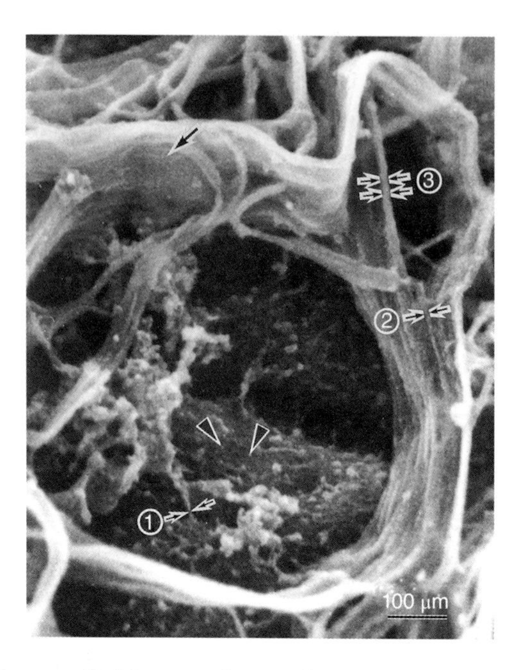

Figure 16.4 More highly magnified images of parts of a reverting protoplast taken at original magnifications of ×100 000. Network (→) was originally formed as fine particles (arrowhead) on the protoplast surface and these were subsequently enlarged and stretched to microfibrils (① →←, 2 nm thick) which twisted around each other and developed into 8 nm fibrils (② →←) forming flat bundles (③ ⇉⇇), 16 nm thick) (Osumi *et al.* 1995; Osumi 2012).

1985, 1987). Immunogold signals for β-1,3-glucan were detected on immunoelectron microscopy (IEM) (Figure 16.5d and e) (Osumi 1998, 2012). Thus, fibrous structures are composed of microfibrils of β-1,3-glucan as in a budding yeast cell wall (Osumi *et al.* 1984; Yamaguchi *et al.* 1987). By using gold-labeled lectin (Osumi *et al.* 1992), α-galactomannan was detected on particles locating in the intrafibrillar space of the reverting protoplast for 3 and 10 h (arrowhead in Figure 16.3d and f) (Osumi *et al.* 1989b). The intrafibrillar spaces were filled with the particles after 12 h (Figure 16.3g), indicating α-galactomannan, and it was also an ingredient of the cell wall (Osumi *et al.* 1992).

In addition to β-1,3-glucan, it was found by NMR spectroscopy (Sugawara *et al.* 2004) and IEM, using an antibody against α-1,3-glucan, that the *S. pombe* cell wall included α-1,3-glucan. Most of all, the α-1,3-glucan distributed just outside the cell membrane, supporting various reports that the α-glucan is synthesized on the cell membrane (Sugawara *et al.* 2003). We also detected Mok 1p, the protein of α-1,3-glucan synthetase, locating on the cell membrane (Figure 16.6) (Konomi *et al.* 2003). Mok 1p moves from the cell tip to the medial region during the cell cycle (Konomi *et al.* 2003). These results revealed that α-1,3-glucan is required for the primary step of glucan bundle formation in new cell wall synthesis during vegetative growth (Osumi 2012; Sugawara *et al.* 2003).

Based on the experiment using anti-β-1,6-glucan antibodies, we proposed that β-1,6-glucan in a highly branched form is also an ingredient of the cell wall. However, β-1,6-glucan was not recognized on the Golgi apparatus (Osumi 2002a; Sugawara *et al.* 2001); rather, they existed in the cytoplasm and beneath the cell membrane. From these results and the fact that the low molecular α-1,3-glucan was transported through the secretory pathway to the cell membrane where the synthase is located and then synthesized into higher molecules, it was speculated that linear β-1,6-glucan was synthesized in the endoplasmic reticulum-Golgi system and that highly branched structures were formed on

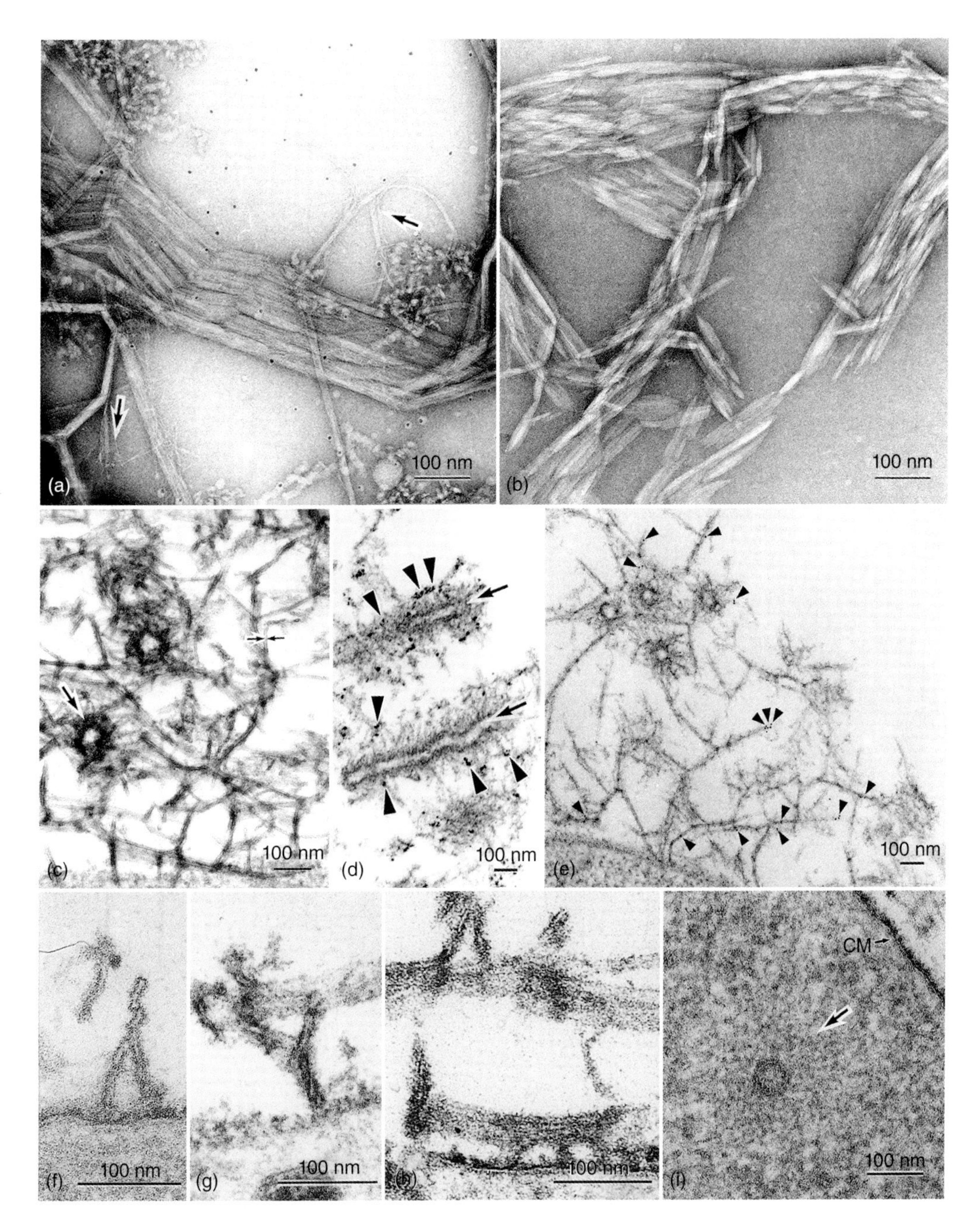

Figure 16.5 β-1,3-Glucan fibrils in the cell wall and filasome. Surface materials forming reverting protoplast for 3 h were negatively stained after treatment with β-1,3-glucanase for (a) 0 h and (b) 2 h. (c) The section images of the network of glucan fibrils. (d and e) IEM images showing the *in situ* localization of β-1,3-glucan (arrowhead). One nm colloidal gold was visualized by the enhanced method. Thin fibril of 1–1.5 nm in diameter (arrows in a and →← in c) and ring-shaped (arrow in c) or canal-shaped forms (arrow in d). (f–h) Development of glucan fibrils. (i) Filasome (arrow) appearing beneath the cell membrane. Specimens were stained with RuO_4 (c–i) (Osumi 1998, 2012). Reproduced with permission from Elsevier.

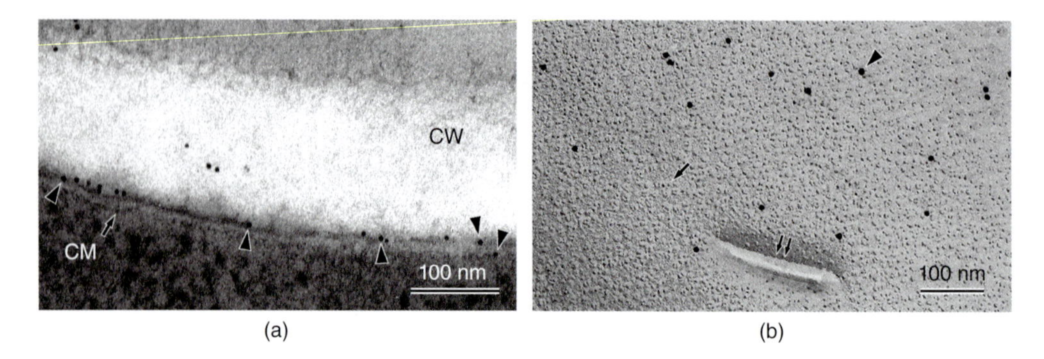

Figure 16.6 Imunoelectron microscopic images of high-pressure frozen *S. pombe* cells labeled with an anti-Mok1p antibody. (a) Ultrathin sectioned image of the freeze-substituted cells by 0.01% OsO_4-aceton and (b) SDS-digested freeze-fracture replica labeling (SDS-FRL9 image) indicate Mok1p localized on the cell membrane. Arrowhead, (a) 5 nm and (b) 10 nm gold particles showing the Mok1p; arrow, integral particles; double arrows, invagination (Konomi *et al.* 2003). CM; cell membrane, CW; cell wall.

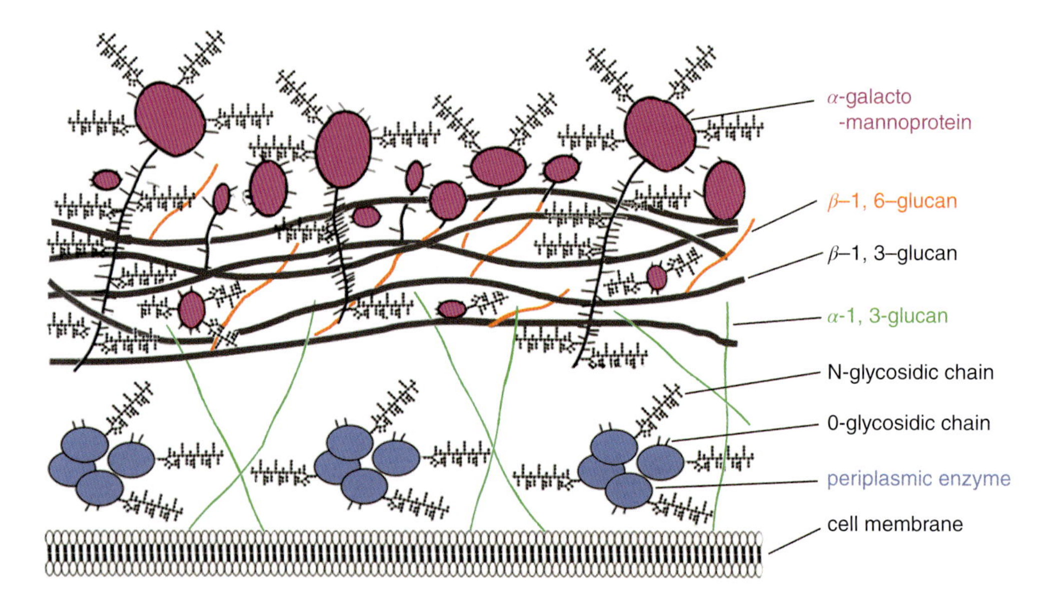

Figure 16.7 Model of molecular composition and structure of cell wall of *S. pombe*. The cell wall consists of two layers. The inner layer provides cell wall strength and is made from β-1,3- and β-1,6-glucan, and α-1,3-glucan. The outer layer consists of α-galactomannoproteins and determines most of the surface properties of the cell. Reproduced by courtesy of Professor F.M. Klis (Osumi 2009, 2012; Klis 1994). Reproduced with permission from Elsevier.

the cell membrane (Osumi 2002a,b; Sugawara *et al.* 2001; Humbel *et al.* 2001). Thus β-1,6-glucan, as well as α-1,3-glucan and β-1,3-glucan, is an ingredient of the cell wall of *S. pombe* (Figure 16.7) (Osumi 2012). Those glucans and α-galactomannan are characteristic components of this yeast (Manners *et al.* 1973). Throughout these experiments, we used antibodies against α-1,3-glucan and highly branched β-1,6-glucan that we had prepared (Sugawara *et al.* 2004).

Treatment with ruthenium tetroxide (RuO_4) (Ranvier 1887; Carpenter and Nebel 1931; Bahr 1954) after conventional fixation provides clear TEM images of glucan fibrils and the outer surface of the cell membrane with high electron density (Naito *et al.* 1991). In addition, filasomes (Figure 16.5i →), Golgi apparatus and secretory vesicles are clearly visible in the cytoplasm. The regenerated fibrils stained with RuO_4 show the twisted, separated, anchored and ring- or canal-shaped forms (Figure 16.5 a–e) (Osumi 1998). The finest regenerating glucan fibril is 1.0–1.5 nm in diameter (Naito *et al.* 1991), which is slightly less than that of the SEM image.

16.3 DEVELOPMENT AND OUTLINE OF ULTRALOW-TEMPERATURE LOW-VOLTAGE SCANNING ELECTRON MICROSCOPY (ULT LVSEM)

16.3.1 High-Pressure Freezing Method for ULT LVSEM

High-pressure freezing (HPF) was considered an alternative method of cryo-fixation in the 1960s and has advanced considerably since then. The commercially available units are now highly reliable with respect to their performance in yeast cells as well as plant tissues (Studer and Muller 1989). In our laboratory, cells of *S. pombe* were cryo-fixed by HPF using the HPM 010 (BAL-TEC AG, Liechtenstein) under 210 MPa. They were observed at −140 °C by Philips XL30 FEG and Hitachi S-4700 SEM using the cryo-system Alto 2500 (Oxford, UK). The Gatan Alto 2500 Cryo Transfer System consisted of a gas-cooled SEM stage module, a vacuum-pumped cryo-operation chamber, a vacuum transfer device (VTD), and a specimen-holder. The samples were cryo-cut (fractured) at −175 to −185 °C and then magnetron sputter coated with platinum (2 nm) in the cryo-operation chamber prior to viewing in the SEM.

S. pombe cells were cryo-fixed using a hat-shaped gold specimen carrier (BTBU 012 130-T; BAL-TEC AG, Liechtenstein) and covered with a dome-shaped gold specimen carrier (BTBU 012 129-T; BAL-TEC AG, Liechtenstein) that had been coated with lecithin (1,2-dipalmitoyl-sn-glycero-3-phosphocholine, Fluka) to allow easy removal of the covers later. The cells were frozen by HPM 010.

The cover (the dome-shaped specimen carrier) was carefully split off and the hat-shaped specimen carrier was mounted on the specimen-holder in liquid nitrogen and then sealed into the VTD. The cryo-preparation chamber attached to either the Philips XL30 S FEG SEM or the Hitachi S-4700 SEM had been pre-cooled with liquid nitrogen and the specimen-holder with the cryo-fixed specimens was moved under vacuum from the VTD to the cryo-preparation chamber via an airlock for fracturing (using the knife assembly within the cryo-operation chamber). The fractured specimens were then subjected to platinum magnetron sputter coating (2 nm), after which the specimens were transferred (via the second airlock) from the cryo-preparation chamber to the gas-cooled SEM stage for observation. The gas-cooled SEM stage module, controllable between −185 and +100 °C, was equipped with an anticontaminator. The specimens were observed with the Philips XL30 S FEG SEM or the Hitachi S-4700 at 1–5 keV at −140 °C (Osumi *et al.* 2006b).

16.3.2 Freeze-fracture Method of Sample Preparation for TEM

Highly concentrated cell suspensions of *S. cerevisiae* grown to mid-log phase were sandwiched between two copper disks and frozen in liquid Freon 23 at −160 °C as quickly as possible (Baba and Osumi 1987; Elder *et al.* 1982) and then transferred to liquid nitrogen, where the copper disks were detached. Freeze-etching was performed in an Ultra High

Vacuum Freeze-Etch Unit BAF 500 K (BAL-TEC AG, Liechtenstein) at −260 °C under 8 $\times 10^{-9}$ Pa. Platinum–carbon shadow coating was carried out at a 45° angle and carbon backing was performed at a 90° angle for high-quality shadowing. The platinum and carbon coats were 1.0 and 2.0 nm thick, respectively. After the coating was completed, the replicas were floated on 70% sulfuric acid overnight, rinsed in distilled water three times, and then mounted on naked grids and examined with a Hitachi H-800 TEM at 200 keV (Baba and Osumi 1987).

16.3.3 Resolution of ULT LVSEM

More highly magnified ULT LVSEM images of the cell membrane of *S. pombe* cells cryo-fixed by HPF were taken at ×120 000 (Figure 16.8a). These images closely resembled high-resolution TEM images of the cell membrane of freeze-fractured cells (Figure 16.8b) that had been cryo-fixed using Freon 23. Integral membrane proteins (arrow) of the cell membrane that are ~15 nm in diameter are also recognizable on the PF face of the cell membrane (Osumi *et al.* 2006b).

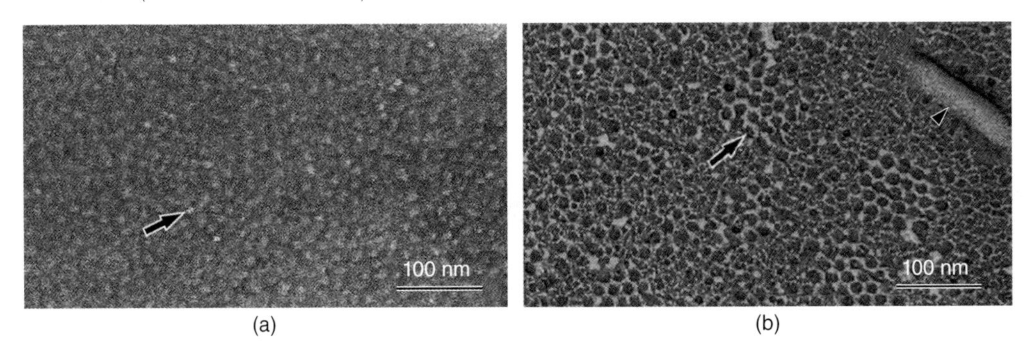

Figure 16.8 (a) ULT-LVSEM images of a part of the PF face of *S. pombe* cell membrane cryo-fixed by HPF taken at higher magnification (×120 000). Arrow, integral particle. (b) TEM image taken at higher magnification of *Saccharomyces cerevisiae* subjected to freeze-fracturing. The PF face showing the cell membrane. The integral membrane particles (arrow) are clearly distinguishable. Some are arranged hexagonally, which is a typical arrangement of a cell membrane component of *S. cerevisiae* (Osumi *et al.* 2006b).

16.3.4 *In situ* Observation of High-Pressure Frozen *S. pombe* Cells by ULT LVSEM

Figure 16.9a and b show fracture images of outer and inside of a cell, respectively. A fracture image of a central plane along the longitudinal axis of a cell (Figure 16.9c) closely resembles the thin-section image of the same HPF specimens (Figure 16.9d). On the PF face of the cell membrane (Figure 16.9a), invaginations found that surface are also observed on the protoplast (Figure 16.2a). Furthermore, images of *S. pombe* cells forming septum during cytokinesis are shown in Figure 16.10 (Osumi *et al.* 2006b).

16.3.5 3D ULT LVSEM Images of Septum Formation

Images of *S. pombe* cells undergoing septum formation during cytokinesis were obtained by HPF and ULT LVSEM. Images of the septum from several different angles during its

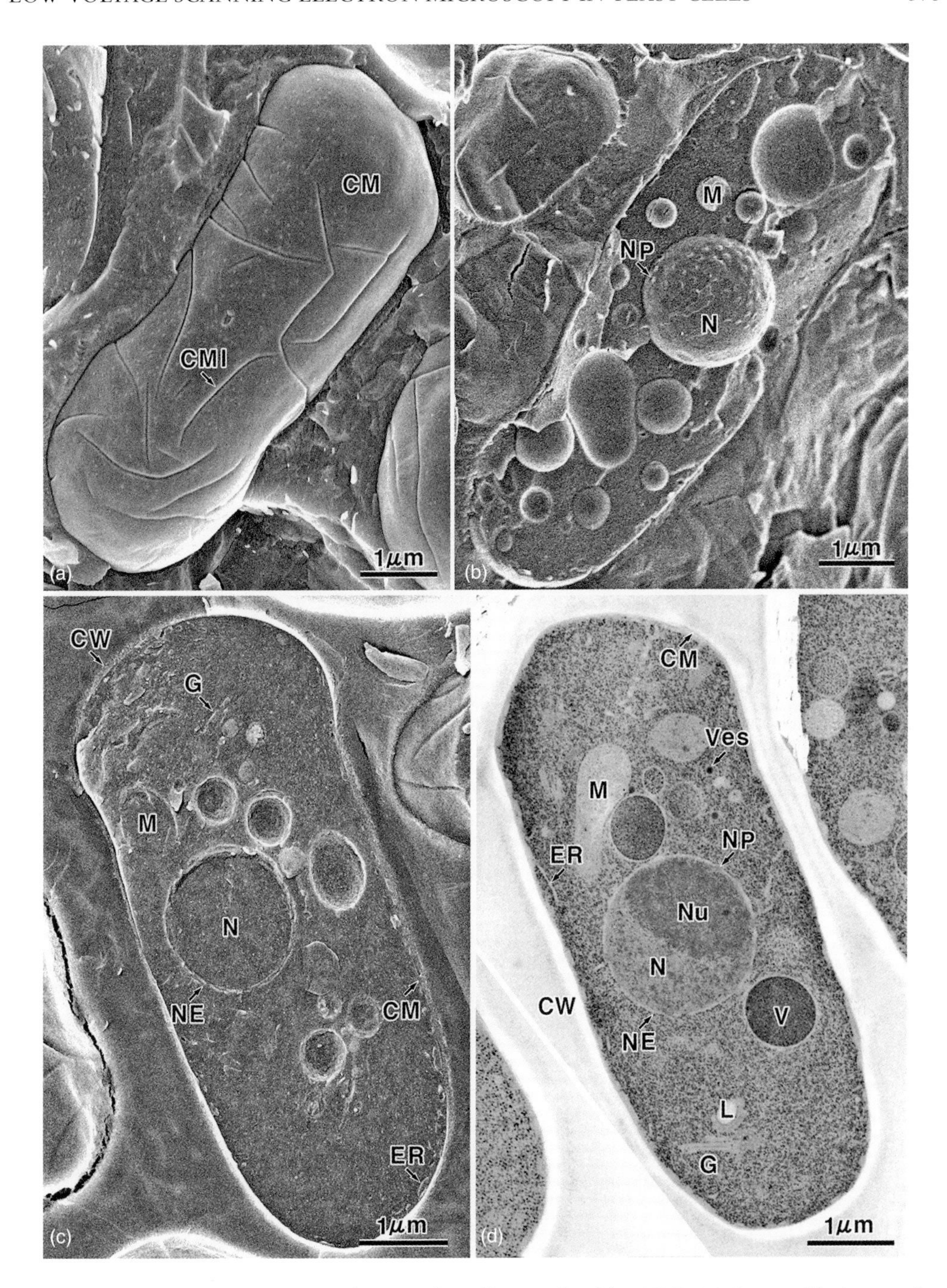

Figure 16.9 ULT LVSEM images of *S. pombe* cells cryo-fixed by HPF. (a) An outside image of a single cell showing the PF face of the cell membrane. (b) An inside image of a cell. (c) An image of a cell cut in the central plane along with the longitudinal axis. (d) TEM image of a thin section (Osumi 2012; Osumi *et al.* 2006b).

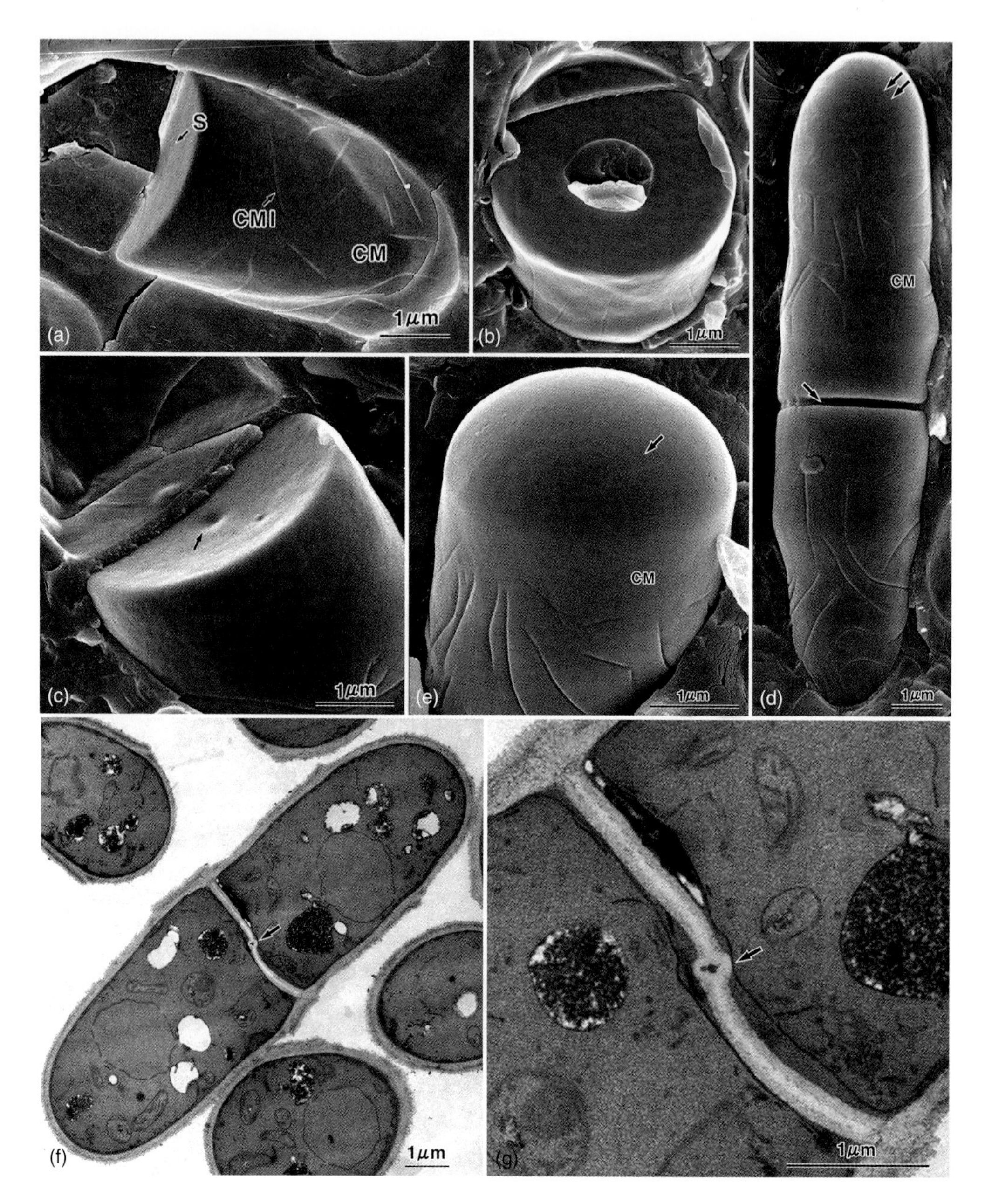

Figure 16.10 ULT LVSEM images of the septum formation in *S. pombe* cells. (a) The incomplete septum (S). (b) The PF face of the incomplete septum. (c) The PF face of the almost completed septum. Both sides of the PF face of the septum can be seen. (d) Bipolar growing cells (double arrows). There is no invagination on either side of the medial region (arrow) of the cell. (e) After the two daughter cells separate, the new ends of the cells expand. There is also no invagination of the cell membrane (CM) in the cells (Osumi 2012; Osumi *et al.* 2006b). (f) TEM image of dividing cells showing completion of the septum. (g) Higher magnification showing corresponding image of the center of a septum in (c) (arrow). Fixed with glutaraldehyde–potassium permanganate.

formation are shown in Figure 16.10. Figure 16.10a shows the incomplete septum plane (S) and the inside of the cell membrane on the EF face showing invaginations of cell membrane (CMIs). Figure 16.10b shows the incomplete septum from another angle. The completed septum is seen in Figure 16.10c and this septum now serves as the "new end" of the daughter cell (Osumi *et al.* 2006b). Figure 16.10d shows the PF faces of the bipolar growing daughter cells during and after their separation. The PF face of the cell membrane showing the divided cell is the location of apical growth at the old end of the cell and no invagination is seen at the cell tip (double arrows). Invagination is also absent from either side of the medial region of the cell (arrow). After separation from the medial region of the elongated cell, the new ends of the daughter cells swelled immediately (Figure 16.10e) (Osumi *et al.* 2006b). Figure 16.10f and g indicates TEM image of a dividing cell.

16.4 *IN SITU* LOCALIZATION OF THE CELL WALL COMPONENT α-1,3-GLUCAN AND ITS SYNTHASE DURING SEPTUM FORMATION

16.4.1 Immunoelectron Microscopy (IEM)

For conventional electron microscopy (EM) and IEM, the specimens were cryo-fixed using an aluminum specimen carrier (BAL-TEC AG, Liechtenstein). For EM, the specimens were then subjected to substitution in 2% OsO_4-acetone for 4 days, while 0.01% OsO_4-acetone was used for IEM, as this created a high contrast in biological membranes (Konomi *et al.* 2000, 2003).

The specimens were then dehydrated and embedded in Quetol 812 (Nissin EM, Japan) (for conventional EM) or LR white resin (medium grade; London Resin Co. Ltd) (for IEM). Ultrathin sections were prepared and blocked for 30 min with normal goat IgG (chromatographically purified; Zymed Laboratories, Inc.) that was diluted 1:30 in 50 mM tris-buffered saline (TBS) containing 0.1% BSA. This was followed by a 1.5 h incubation at room temperature with the primary antibody, an antibody specific for α-1,3-glucan (Konomi *et al.* 2003; Fujimoto 1995) that was diluted 1:250 in 50 mM TBS containing 0.1% BSA and α-1,3-glucan synthase (Mok1p) (Konomi *et al.* 2003). The samples were then incubated for 1 h with the secondary antibodies, 10 and 5 nm colloidal gold-labeled goat anti-rabbit IgG antibodies (British BioCell International, Ltd., UK), which was diluted 1:40 in 50 mM TBS containing 0.1% BSA. The sections were post-fixed with 1% glutaraldehyde in phosphate buffer, stained with 5% uranyl acetate and 0.4% lead citrate, and observed using a JEOL JEM 1200 EXS at 80–100 keV and a Hitachi H-800 TEM at 125 keV (Osumi *et al.* 2006b; Konomi *et al.* 2000).

16.4.2 Freeze-fracture Replica Labeling Method for TEM

The behavior of Mok1p, the α-1,3-glucan synthase, on the cell membrane during cytokinesis was analyzed by the sodium dodecyl sulfate (SDS)-digested freeze-fracture replica labeling (FRL) technique (Fujimoto 1995). Briefly, highly concentrated cell suspensions undergoing cell division were put into a hat-shaped gold carrier and covered with a dome-shaped gold carrier that was coated with lecithin. The cells were frozen under high pressure in the HPM 010 instrument. The specimens were freeze-fractured, labeled by the previously described procedure (Konomi *et al.* 2003) and observed with a Hitachi H-800 TEM at 125 keV.

16.4.3 Analysis of Localization of the Cell Wall Components and its Syntase by IEM

The α-1,3-glucan synthase Mok1p is an essential enzyme in the synthesis of α-1,3-glucan, which is a cell wall component of *S. pombe*. To analyze the behavior of Mok1p during septum formation, high-pressure frozen cells were prepared by SDS-FRL and analyzed by IEM. The immunogold-labeled Mok1p molecules (Figure 16.11a, arrowhead) were seen on the cell membrane in the region of the septum during its formation (Figure 16.11a, arrow). Mok1p also appeared to localize on the membrane in the region of the septum after the septum had been completed (Figure 16.11b, arrowhead). After the daughter cells were separated, Mok1p was observed on the swollen tip of the cell (Figure 16.11c, arrowhead); α-1,3-glucan was detected using thin-sectioning and IEM. During a septum formation, the immunogold-labeled α-1,3-glucan molecules were observed on the adjacent side of the cell membrane, namely, the secondary septum region from initiation (Figure 16.12a, arrowheads) through to completion (new end wall completion; Figure 16.12b, arrowheads). However, the central area of the septum was not labeled (Osumi *et al.* 2006b; Konomi *et al.* 2000).

When we examined the images of the *S. pombe* cell membrane by use of samples frozen by the high-pressure method and viewed with ULT LVSEM, we observed that

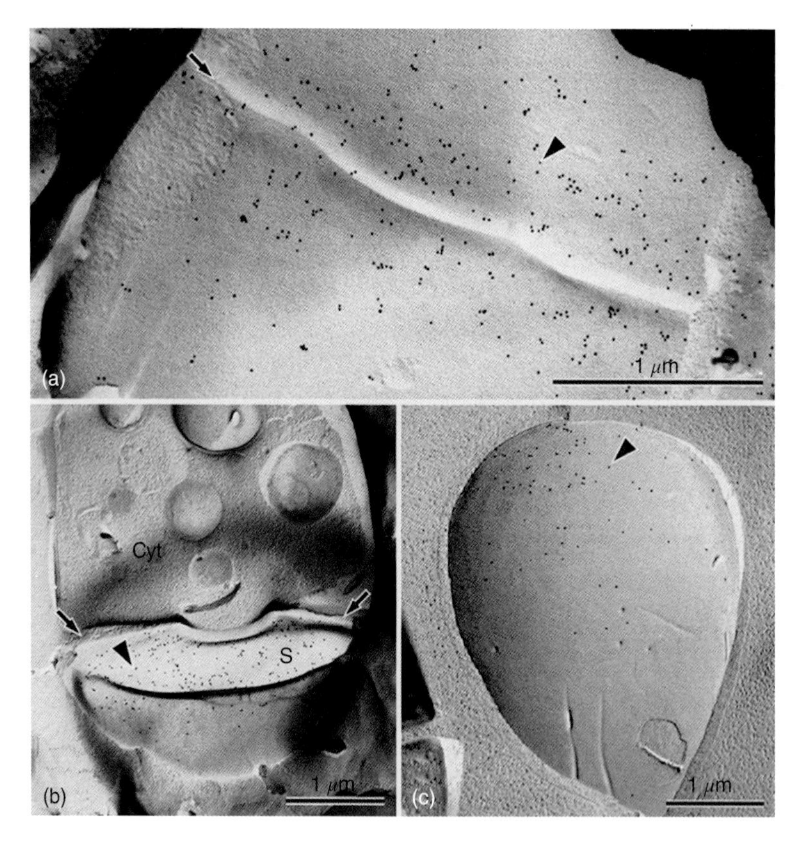

Figure 16.11 SDS-FRL images of HPF *S. pombe* cells labeled with an anti-Mok1p antibody. (a) Mok1p (arrowhead) appears on the PF face of the cell membrane at the septum (arrow). (b) Mok1p also appears on the septum plane (S) when its formation is complete. (c) Mok1p is localized on the tip of the swollen daughter cell (the new end) (Osumi 2012; Osumi *et al.* 2006b).

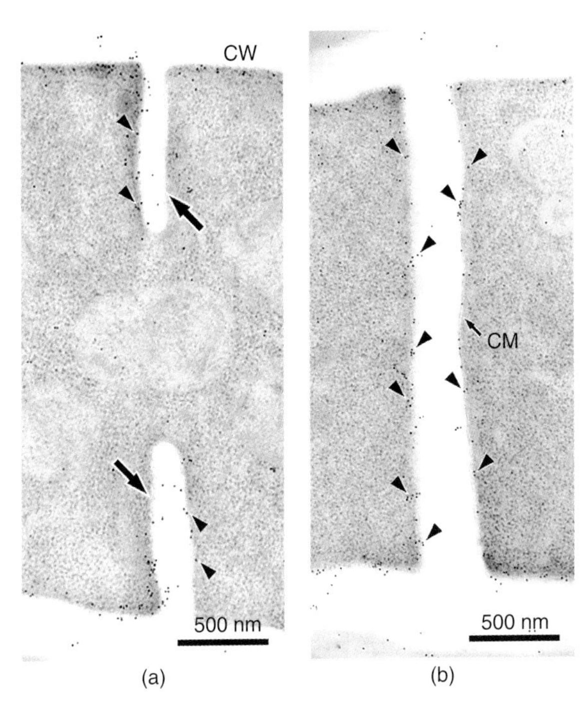

Figure 16.12 IEM images showing the *in situ* localization of α-1,3-glucan during septum formation. (a) Initiation of septum formation. (b) Completion of septum formation. Note that the α-1,3-glucans are labeled on the adjacent side of the cell membrane (arrowhead), which is the region of the secondary septum. From the initiation of the septum (arrow) to its completion, the α-1,3-glucan labeling remains in the same position (Osumi *et al.* 2006b).

the membrane contained many integral membrane particles distributed homogeneously, because cells were sampled during the exponential growth phase. The diameters of these particles are not uniform, but ranged from 10 to 15 nm. The resolution of integral particles is mostly similar to the image obtained by UHR LVSEM of uncoated yeast cells prepared by freeze-substitution and cryo-fracturing (Osumi *et al.* 1988a).

Nevertheless, these observations indicate that HPF provides a powerful technique for the preparation and observation of high-quality samples for SEM, and that it permits a more rapid analysis of the fine structure of *in situ* biological materials than the freeze-fracture method.

The cell wall of *S. pombe* is composed of β-1,3-glucan, β-1,6-glucan, α-1,3-glucan (which is a characteristic component of this yeast), and α-galactomannan (Figure 16.7) (Manners *et al.* 1973; Kopecka *et al.* 1995). A thin-sectioned image of the cell wall reveals that it is a three-layered structure that differs in electron density and is highly changeable depending on the prevailing genetic conditions (Osumi *et al.* 1998; Konomi *et al.* 2003; Ishijima *et al.* 1999). In earlier work using a protoplast regeneration system, we proposed that the actin cytoskeleton contributes to the formation of the cell wall and showed that this cytoskeleton was indeed associated with initial deposition of cell wall components (Osumi *et al.* 1989a,b; Takagi *et al.* 2003). Moreover, when the SEM (Osumi *et al.* 1989b, 1995) and TEM (Osumi 1998) analyses of the dynamics of the protoplast surface and the intracellular structure of the reverting *S. pombe* protoplast during cell wall formation were reviewed, the correlation between the cell wall formation and actin cytoskeleton was evident. A close spatial and temporal relationship between the actin cytoskeleton and cell wall formation was found by comparing wild-type *S. pombe* with the actin point-mutant *cps8* (Ishijima *et al.* 1999).

Cytokinesis and septum formation are the most critical steps in the cell cycle of yeast, in which continuous cell wall polymer remodeling is required for viability. Most studies have focused on mechanisms that regulate assembly of the septum (Simanis 2003; Guertin *et al.* 2002), but little morphological analysis has been conducted. The septum of *S. pombe* is a multilayered structure composed of a primary septum flanked by secondary septa on either side (Osumi and Sando 1969; Johnson *et al.* 1973). The primary septum mainly contains linear β-1,3-glucan and the secondary septa contain β-1,6 branched β-1,3-glucan and galactomannans; β-1,6-glucan is found only in the secondary septum (Sugawara *et al.* 2003; Humbel *et al.* 2001).

Correlation between the ULT LVSEM images of *S. pombe* septum formation and those obtained by SDS-FRL and IEM confirmed that the α-1,3-glucan synthase Mok1p is located on the cell membrane at the septum (new end) and on the swollen cell tip, which reflects the need for the synthesis of α-1,3-glucan in these areas. Specific localization of the synthase may involve lipid microdomains, or rafts (Rajaggopalan *et al.* 2003), which may serve to limit the localization of proteins that are required to function in different regions within the cell membrane. Thus, such rafts may restrict the machinery involved in cell division, including the enzymes that synthesize the components of the cell wall and septum to the relevant site(s) of action.

Our analysis also revealed that during septum formation, α-1,3-glucan is located on the adjacent cell membrane, when septum formation is initiated, and this localization prevails throughout septum formation. Linear β-1,3-glucan, the skeletal frame component of the cell wall (Osumi *et al.* 1989b, 1995), appears on indentations of the cell membrane opposite the septum, and exists in the center of the primary septum wall. However, at the completion of the septum, the densely stained rim of the α-galactomannan layer appears between the two daughter cells, and the linear β-1,3-glucan disappears from the primary septum. Moreover, after cleavage of the septum, it is not found in the outermost layer of the new end of the wall (Humbel *et al.* 2001). In contrast, the behavior of the β-1,6-glucan during septum formation differs from that of both β-1,3 and α-1,3-glucan, as it was not detected in the initial septum formation. However, it was detected after the invagination had progressed two-thirds of the cell width, and its expression in the septum continued until the secondary septum was formed (Sugawara *et al.* 2001). The α-1,3-glucan contributes not only to the formation of the secondary septum along with β-1,3-glucan but also to efficient cell separation, thereby exposing the secondary septa as the new ends of the daughter cells (Dekker *et al.* 2004).

Our previous studies on the regeneration system of the *S. pombe* protoplast revealed that during the formation of the cell wall framework, glucan fibrils form a network structure and develop into a bundle (Osumi *et al.* 1989b) and β-1,3-glucan is the main component of this structure, as shown by IEM (Figure 16.11) (Osumi 2012; Konomi *et al.* 2003; Osumi *et al.* 2006b; Takagi *et al.* 2003). When we analyzed the behavior of the linear β-1, 3-, β-1,6-, and α-1,3-glucans in formation of the glucan network, all three kinds were found in the network from the initial stages of its formation onward. Linear β-1,3-glucan was localized on the fine-fibril and bundle structures, while α-1,3-glucan and β-1,6-glucan were localized on the periphery of the regenerated protoplast. These observations support our hypothesis that the glucans act as the framework of the cell wall and are essential components in initiating cell wall formation.

Our analysis by ^{13}C-nuclear magnetic resonance (NMR) spectroscopy of the components of the cell wall of *S. pombe* showed that β-1,3-glucan may be linear or have a slightly branched structure and that β-1,6-glucan is highly branched (Sugawara *et al.* 2004). The β-1,6-glucan was defined as a highly β-1,3-branched β-1,6-glucan, and has many more branches in *S. pombe* than in *S. cerevisiae*. The α-1,3- and β-1,3-glucans were rigid, and thereby contribute to cell shape (Sugawara *et al.* 2004). Most of the α-1,3-glucan was found juxtaposed to the cell membrane (Sugawara *et al.* 2003). These observations and our present work strongly suggest that α-1,3-glucan, especially those in the new end wall of the daughter cell, contributes to maintain the cylindrical shape of *S. pombe* cells. This may explain why the budding yeast, which lacks this molecule in its cell wall, is ellipsoidal or round in shape. It is generally accepted that the septation apparatus also exists as an autonomous system in budding yeast (Cabib *et al.* 2001; Roh *et al.* 2002).

Moreover, it is now known that the mitotic exit network (MEN) proteins of *S. cerevisiae* and the septation initiation network (SIN) proteins of *S. pombe* contribute to the final events of mitosis in these organisms (Simanis 2003) and that there are differences and similarities between these protein systems (Guertin *et al.* 2002). The fission yeast *S. pombe* is a particularly suitable model for eukaryotic cytokinesis, and analyses of *S. pombe* mutants have just started to reveal how septum formation and cytokinesis are regulated spatially and temporally and, thus, are properly coordinated with other events in mitosis. There are more than 30 genes contributing to this event whose functions have been studied (Simanis 2003). However, further detailed morphological studies will be needed for our understanding of the mechanisms of septation and cytokinesis in *S. pombe*. Our studies here suggest that ULT LVSEM is a powerful technique in addressing these issues and HPF is useful for elucidation of the molecular anatomy of the cell.

16.5 DISCUSSION AND SUMMARY

1. The refined FE SEM, S-900LV, which gives better resolution, especially at LV below 5 keV, was developed to the UHR LVSEM. In resolution tests made at ×300 000 made using a gold-evaporated magnetic tape, the resolution was found to be about 1 nm at 2.5 keV. This performance at LV allowed the ultrastructure of the cell wall, especially reverting glucan network from the protoplast of *S. pombe*, to be revealed using this improved UHR LVSEM. The uncoated reverting protoplasts observed with this SEM revealed that the network was originally formed as secreted particles scattered on the protoplast surface secreted from the cell and these were subsequently stretched to microfibrils of about 2 nm thick. The microfibrils were twisted around each other and joined together forming a ribbon-shaped network of glucans about 16 nm thick, which covered the entire protoplast surface. The UHR LVSEM images of reverting protoplast treated with glucanase confirmed that the particles scattered on the protoplast surface in the initial stage of regeneration were glucan in nature.
2. When HPF was combined with ULT LVSEM using the cryo-system (the Gatan Alto 2500 Cryo Transfer system), fractured and coated yeast samples could be quickly prepared. These samples, following the generation of a fine fracture plane, revealed the ultrastructure of both external and internal cell components. We used this method to analyze the process of septum formation and cell separation in *S. pombe*. The images that we obtained provided a 3D view of these processes for the first time. We also showed that HPF in combination with IEM made it possible to preserve the antigenicity *in situ* localization and behavior of the cell wall component, α-1,3-glucan, and its synthase during septum formation.
3. Future directions: Three-dimensional (3D) analyses by FIB SEM and SBF SEM using LVSEM. To analyze the molecular architecture of cells, microsampling methods using an FIB, which are the main techniques for semiconductor device analysis, and SEM have

been applied to the 3D observation of resin-embedded yeast cells. This is the first case of applying FIB to the biological specimen (Kamino *et al.* 2004, 2006; Osumi *et al.* 2006a). Recently, FIB SEM tomography has become established as a novel and powerful approach for 3D imaging for biological research (Osumi *et al.* 2010; Yaguchi *et al.* 2010).

In the past, we constructed 3D images by successive serial sectioning of the biological specimen with subsequent imaging in the TEM (Osumi *et al.* 1998; Takagi *et al.* 2003). This approach was both technically challenging and time consuming. In contrast, the methods for reconstructing 3D images by SBF SEM (Denk and Horstmann 2004) and FIB SEM (Villinger *et al.* 2012; Wei *et al.* 2012) are powerful and quick techniques for this purpose. Recently, we have been able automatically to obtain 600 serial images of fission yeast, *S. pombe* cells, and cellulolytic fungus *Trichoderma reesei* overnight by FIB SEM and SBF SEM at 2.5 keV, and this is a powerful approach for successfully reconstructed 3D structure of *T. reesei* hypha (Shida *et al.* 2015) and the study on structural change in nuclear shape during mitosis (Nakano *et al.* 2013) as an application of LVSEM.

ACKNOWLEDGEMENTS

The author would like to express sincere thanks to Drs T. Nagatani and M. Sato of Hitachi High-Technologies Co. for their joint work on the achievement of LVSEM and is indebted to Dr T. Toda of Cancer Research UK for Mok1 and the Mok1p antibody, and to Dr N. Ohno of Tokyo Pharmaceutical University for the α-1,3-glucan-specific antibody. Thanks also to Dr Hax Werner of FEI Netherlands B.V., Mr S. Kikuchi of FEI Co. in Japan Ltd, and Mr T. Okada of Hitachi High-Technologies Co. for the development of the ULT-LVSEM method in combination with the Gatan Alto 2500 Cryo Transfer System. The author is grateful to Dr Mike Hayles of FEI Co. in The Netherlands and Dr Alan Robins of Oxford Ltd. for technical assistance in the experiments described in this report. Thanks also to Drs M. Konomi, T. Takagi, M. Sato, and M. Baba, and all members of Osumi's Lab., and Mr H. Okada and Ms Y. Osaki of NPO Integrated Imaging Research Support. This work was supported by a Grant-in-Aid for scientific research from the Ministry of Education, Culture, Sports, Science and Technology of Japan (No. 15570053) and the Open Research Center of Japan Women's University, established in private universities in Japan with the support of the Ministry of Education, Culture, Sports, and Science.

REFERENCES

Baba, M. and Osumi, M. (1987) Transmission and scanning electron microscopic examination of intracellular organelles in freeze-substituted *Kloeckera* and *Saccharomyces cerevisiae* yeast cells. *J. Electron Microsc. Tech.*, 5, 246–261.

Bahr, G.F. (1954) Osmium tetroxide and ruthenium tetroxide and their reactions with biologically important substances. *Exp. Cell Res.*, 7, 457–479.

Boyes, E.D. (1984) High-resolution at low voltage: The SEM philosopher's stone? in *Proc. of 42nd Ann. Meet. Electron Microsc. Soc. Am.*, pp. 446–450.

Cabib, E., Roh, D.H., Schmidt, M., Crotti, L.B., and Varma, A. (2001) The yeast cell wall and septum as paradigms of cell growth and morphogenesis. *J. Biol. Chem.*, 6, 19679–19682.

Carpenter, D.C. and Nebel, B.R. (1931) Ruthenium tetroxide as a fixative in cytology. *Science*, 74, 154–155.

Crew, A.V. (1985) Towards the ultimate scanning electron microscope. *Scanning Electron Miscrosc.*, II, 467–476.

Dekker, N., Speijer, D., Grun, C.H., van den Berg, M., de Haan, A., and Hochstenbach, F. (2004) Role of the α-glucanase Agn1p in fission-yeast cell separation. *Mol. Biol. Cell*, 15, 3903–3914.

Denk, W. and Horstmann, H. (2004) Serial block-face scanning electron microscopy to reconstruct three-dimensional tissue nanostructure. *PloS Biol.*, 2, e329. doi:10.137/journal.pbio.0020329.

Elder, H.Y., Gray, C.C., Jardine, A.G., Chapman, J.N., and Biddlecombe, W.H. (1982) Optimum conditions for the cryoquenching of small tissue blocks in liquid coolants. *J. Microsc.*, 126, 45–61.

Fujimoto, K. (1995) Freeze-fracture replica electron microscopy combined with SDS digestion for cytochemical labeling of integral membrane proteins. Application to the immunogold labeling of intercellular junctional complexes. *J. Cell Sci.*, 108, 3443–3449.

Guertin, D.A., Trautmann, S., McCollum, D. (2002) Cytokinesis in eukaryotes. *Microbiol. Mol. Biol. Rev.*, 66, 155–178.

Humbel, B.M., Konomi, M., Takagi, T., Kamasawa, N., Isijima, S.A., and Osumi, M. (2001) *In situ* localization of β-glucans in the cell wall of *Schizosaccharomyces pombe*. *Yeast*, 18, 433–444.

Ishijima, S.A., Konomi, M., Takagi, T., Sato, M., Ishiguro, J., and Osumi, M. (1999) Ultrastructure of cell wall of the *cps8* actin mutant cell in *Schizosaccharomyces pombe*. *FEMS Microbiol. Lett.*, 180, 31–37.

Johnson, B.F., Yoo, B.Y., and Calleja, G.B. (1973) Cell division in yeasts: movement of organelles associated with cell plate growth of *Schizosaccharomyces pombe*. *J. Bacteriol.*, 115, 358–366.

Joy, D.C. (1984) Resolution in low voltage SEM, in Proc. of 42nd Ann. Meet. Electron Microsc. Soc. Am., pp. 444–445.

Joy, D.C. (1989) Low voltage scanning electron microscopy. *Hitachi Instrument News*, 16, 3–11.

Kamino, T., Yaguchi, N., Sato, T., and Onishi, T. (2006) Application of FIB technique to 3D observation of resin embedded biological tissues, in *Proceedings of Microscopy Microanalysis* (Chicago), pp. 1232–1233.

Kamino, T., Yaguchi, T., Ohnishi, T., Ishitani, T., and Osumi, M. (2004) Application of a FIB-STEM system for 3D observation of a resin-embedded yeast cell. *J. Electron Microsc.*, 53, 563–566.

Klis, F.M. (1994) Review: Cell wall assembly in yeast. *Yeast*, 10, 851–869.

Kobori, H., Yamada, N., Taki, A., and Osumi, M. (1989) Actin is associated with the formation of the cell wall in reverting protoplasts of the fission yeast *Schizosaccharomyces pombe*. *J. Cell Sci.*, 94, 635–646.

Konomi, M., Fujimoto, K., Toda, T., and Osumi, M. (2003) Characterization and behaviour of α-glucan synthase in *Schizosaccharomyces pombe* as revealed by electron microscopy. *Yeast*, 20, 427–438.

Konomi, M., Kamasawa, N., Takagi, T., and Osumi, M. (2000) Immunoelectron microscopy of fission yeast using high pressure freezing. *Plant Morphol.*, 12, 20–31.

Kopecka, M., Fleet, G.H., and Phaff, H.J. (1995) Ultrastructure of the cell wall of *Schizosaccharomyces pombe* following treatment with various glucanases. *J. Struct Biol.*, 114, 140–152.

Manners, D.J., Masson, A.J., and Patterson, J.C. (1973) The structure of a β-(1$\rightarrow$3)-D-glucan from yeast cell walls. *Biochem. J.*, 135, 19–30.

Nagatani, T. and Saito, S. (1986) Instrumentation for ultra high resolution scanning electron microscopy. *Proc. of 11th Int. Congr. Electron Microsc. (Kyoto)*, 1, 2101–2104.

Nagatani, T., Nakaizumi, Y., Saito, S., and Yamada, M. (1983) An experiment on ultra-high resolution SEM. *Bio-Med. SEM*, 13, 8–9.

Nagatani, T., Saito, S., Sato, M., and Yamada, M. (1987a) Development of an ultrahigh resolution scanning electron microscope by means of a field emission source and in-lens system. in *Scanning Microscopy. Scanning* (eds A. Boyde and D.C. Joy), vol 1, International Inc., Chicago, IL, pp. 901–909.

Nagatani, T., Saito, S., Yamada, M., and Sato, M. (1987b) Development of an ultra high resolution scanning electron microscope by means of a field emission source and in-lens system. *Scanning Electron Microsc.*, 1 (3), 901–909.

Nagatani, T., Sato, M., and Osumi, M. (1990) Development of an ultra-high resolution low voltage (LV) SEM with an optimized "In-lens" design. *Proc. of 12th Int. Congr. Electron Microsc. (Seattle)*, 1, 388–389.

Naito, N., Yamada, N., Kobori, H., and Osumi, M. (1991) Contrast enhancement by ruthenium tetroxide for observation of the ultrastructure of yeast cells. *J. Electron Microsc.*, 40, 416–419.

Nakano, K., Okada, H., Morikawa, A., Tamichi, R., Murata, K., Miyazaki, N., Takagi, T., Sato, M., Tanizawa, H., Noma, K., and Osumi, M. (2013) FIB SEM is a powerful approach for studying structural change of nuclear shape during mitosis in fission yeast. *J. Electron Microsc.*, 62, B2-E20pm05.

Osumi, M. (1998) The ultrastructure of yeast: cell wall structure and formation. *Micron*, 29, 207–233.

Osumi, M. (2002a) On the ultrastructure of yeast cells (in Japanese). *Plant Morphol.*, 14, 54–67.

Osumi, M. (2002b) The significance of high pressure freezing technique in structural biology – with reference to a study of cell wall formation in fission yeast (in Japanese). *J. Japan Woman's Univ. Faculty of Science*, 10, 43–63.

Osumi, M. (2009) Research into yeasts focusing on its cell wall formation (in Japanese). *J. SJWS*, 10, 17–40.

Osumi, M. (2011) Coating method for ultra-high vacuum and ultralow temperature, in *SHIN-SOSA DENSHI KENBIKYO* (in Japanese) (ed. Kanto Branch of Japanese Society of Microscopy), Kyoritsu-Shuppan Co. Ltd, Tokyo, Japan, 182 pp.

Osumi, M. (2012) Visualization of yeast cells by electron microscopy. *J. Electron Microsc.*, 61, 343–365.

Osumi, M. and Nagatani, T. (1987) High-resolution low-voltage scanning electron microscopy of uncoated biological specimens fixed by the freeze substitution fixation method, in *Microbeam Analysis, Hawaii* (ed. R.H. Geiss), pp. 71–75.

Osumi, M. and Sando, N. (1969) Division of yeast mitochondria in synchronous culture. *J. Electron Microsc.*, 18, 47–56.

Osumi, M., Baba, M., and Yamaguchi, H. (1984) Electron microscopical study on the biosynthesis and assembly of yeast cell wall components, in *Microbial Cell Wall Synthesis and Autolysis* (ed. C. Nombela), Elsevier Science, Amsterdam, pp. 137–142.

Osumi, M., Baba, M., Naito, N., Taki, A., Yamada, N., and Nagatani T. (1988a) High resolution, low voltage scanning electron microscopy of uncoated yeast cells fixed by the freeze-substitution method. *J. Electron Microsc.*, 37, 17–30.

Osumi, M., Baba, M., Suzuki, T.,Watanabe, T., and Nagatani, T. (1985) Low accelerating voltage and high fidelity observation of yeast cells with high resolution scanning electron microscope. *Bio-Med. SEM*, 14, 47–51.

Osumi, M., Baba, M., Suzuki, T., Watanabe, T., and Nagatani, T. (1986) Low acceleration voltage and high-fidelity observation of microorganisms with high resolution scanning electron microscope. *Proc. 11th Int. Congr. Electron Microsc.* (Kyoto), 1, 3377–3378.

Osumi, M., Eguchi, T., Yaguchi, T., Sato, T., and Okada, H. (2010) Three dimensional observation of yeast cells and baculovirus pioneered by FIB-micro sampling. *Jpn J. Bacteriol.*, 65, 225.

Osumi, M., Kamino, T., Yaguchi, T., Sato, T., Ohnishi, T., and Ishitani, T. (2006a) A new application of the FIB-STEM system for the 3D observation of yeast cells, in *Proceedings of 16th International Microscopy Congress* (Sapporo), p. 435.

Osumi, M., Kobori, H., Yamada, N., Sato, M., Taki, A., and Naito, N. (1989a) Morphological aspect on cell wall formation and cytoskeleton in yeast, in *Current Problems of Opportunistic Fungal Infections*, Research Center for Pathogenic Fungi and Microbial Toxicoses, Chiba University, pp. 61–64.

Osumi, M., Konomi, M., Sugawara, T., Takagi, T., and Baba, M. (2006b) High-pressure freezing is a powerful tool for visualization of *Schizosaccharomyces pombe* cells: Ultra-low temperature and low-voltage scanning electron microscopy and immunoelectron microscopy. *J. ElectronMicrosc.*, 55, 75–88.

Osumi, M., Sato, M., Ishijima, S.A., Konomi, M., Takagi, T., and Yaguchi, H. (1998) Dynamics of cell wall formation in fission yeast, *Schizosaccharomyces pombe*. *Fungal Genet. Biol.*, 24, 178–206.

Osumi, M., Yamada, N., and Nagatani, T. (1988b) High-resolution, low-voltage SEM of cell wall regeneration of yeast *Schizosaccharomyces pombe* protoplasts, in Proc. of 46th Ann. Meet. Electron Microsc. Soc. Am., pp. 208–211.

Osumi, M., Yamada, N., Kobori, H., and Nagatani, T. (1988c) Coating for observation with high resolution and low voltage scanning electron microscope. *Bio-Med. SEM*, 17, 32–38.

Osumi, M., Yamada, N., Kobori, H., and Yaguchi, H. (1992) Observation of colloidal gold particles on the surface of yeast protoplasts with UHR-LVSEM. *J. Electron Microsc.*, 41, 392–396.

Osumi, M., Yamada, N., Kobori, H., Taki, A., Naito, N., Baba, M., and Nagatani, T. (1989b) Cell wall formation in regenerating protoplasts of *Schizosaccharomyces pombe*: Study by high resolution, low voltage scanning electron microscopy. *J. Electron Microsc.*, 38, 457–468.

Osumi, M., Yamada, N., Taki, A., Gotoh, M., and Nagatani, T. (1987) Observation of biological specimen by "in-lens" type field emission scanning electron microscope. *Bio-Med. SEM*, 16, 30–36.

Osumi, M., Yamada, N., Yaguchi, H., Kobori, H., Nagatani, T., and Sato, M. (1995) Ultrahigh-resolution low-voltage SEM reveals ultrastructure of the glucan network formation from fission yeast protoplast. *J. Electron Microsc.*, 44, 198–206.

Pawley, J. (1984) SEM at low beam voltage, in Proc. of 42nd Ann. Meet. Electron Microsc. Soc. Am., pp. 440–444.

Pawley, J.B. (1986) LVSEM: A new way of seeing biology, in Proc. of 45th Ann. Meet. Electron Microsc. Soc. Am., pp. 550–553.

Pawley, J.M. (1988) Low voltage scanning electron microscopy. *Electron Microscopy Society of America Bulletin*, 18 (1) (Spring), 61–64.

Rajaggopalan, S., Wschtler, V., and Blasubramanian, M. (2003) Cytokinesis in fission yeast: a story of rings, rafts and walls. *Trends Genet.*, 19, 403–408.

Ranvier, L. (1887) De l'emploi de l'acide perruthenique dans les recherchés histologiques, et de l'application de ce réactif à l'étude des vacuoles des cellules caliciformes. *C.R. Hebd. Seances Acad. Sci.*, 105, 145.

Roh, D.H., Bowers, B., Schmidt, M., and Cabib, E. (2002) The septation apparatus, an autonomous system in budding yeast. *Mol. Biol. Cell*, 13, 2747–2759.

Sato, M., Otsuka, S., Miyamoto, R., and Osumi, M. (1989) Development of low-voltage, high-resolution SEM I. Instrument. *Bio-Med. SEM*, 18, 1–3.

Shida, Y., Morikawa, A., Tamochi, R., Nango, N., Okada, H., Osumi, M., and Ogasawara, W. (2015) Ultrastructure of the cellulolytic fungus *Trichoderma reesei*. *Plant Morphol.*, 27, 15–20.

Simanis, V. (2003) Events at the end mitosis in the budding fission yeast. *J. Cell Sci.*, 116, 4263–4275.

Studer, D. and Muller, M.M. (1989) High pressure freezing comes of age. *Scanning Microsc. Suppl.*, 3, 253–269.

Sugawara, T., Sato, M., Takagi, T., Kamasaki, T., Ohno, N., and Osumi, M. (2003) *In situ* localization of cell wall α-1,3-glucan in the fission yeast *Schizosaccharomyces pombe*. *J. Electron Microsc.*, 52, 237–242.

Sugawara, T., Takagi, T., Sato, M., Ohno, N., and Osumi, M. (2001) Architecture and localization of the cell wall glucans of the fission yeast, *Schizosaccharomyces pombe*, in *Proceedings of MolecularMechanisms of Fungal CellWall Biogenesis*, p. 61. poster No.16 (Ascona, Switzerland).

Sugawara, T., Takahashi, S., Osumi, M., and Ohno, N. (2004) Refinement of the structures of cell-wall glucans of *Shizosaccharomyces pombe* by chemical modification and NMR spectroscopy. *Carbohydr. Res.*, 339, 2255–2265.

Takagi, T., Ishijima, S.A., Ochi, H., and Osumi, M. (2003) Ultrastructure and behavior of actin cytoskeleton during cell wall formation in the fission yeast *Schizosaccharomyces pombe*. *J. Electron Microsc.*, 52, 161–174.

Villinger, C., Gregorius, H., Kranz, C., Hohn, K., Munzberg, C., von Wichert, G., Mizaikoff, B., Wanner, G., and Walther, P. (2012) FIB/SEM tomography with TEM-like resolution for 3D imaging of high-pressure frozen cells. *Histochem. Cell Biol.*, 138, 549–556.

Wei, D., Jacobs, S., Modla, S., Zhang, S., Young, C.L., Cirino, R., Caplan, J., and Czymmek, K. (2012) High-resolution three-dimensional reconstruction of a whole yeast cell using focused-ion beam scanning electron microscopy. *Biotechniques*, 53, 41–48.

Yaguchi, N., Sato, T., Eguchi, T., Okada, H., and Osumi, M. (2010) Visualization of specific yeast cell wall by FIB-STEM. *Jpn J. Bacteriol.*, 65, 108.

Yamada, N., Nagatani, T., and Osumi, M. (1986) Low accelerating voltage and high fidelity observation of biological specimens by high resolution scanning electron microscope. *Bio-Med. SEM*, 15, 35–38.

Yamaguchi, H., Hiratani, T., Baba, M., and Osumi, M. (1985) Effect of aculeacin A, a wall-active antibiotic, on synthesis of the yeast cell wall. *Microbiol. Immunol.*, 29, 609–623.

Yamaguchi, H., Hiratani, T., Baba, M., and Osumi, M. (1987) Effect of aculeacin A on reverting protoplasts of *Candida albicans*. *Microbiol. Immunol.*, 31, 625–638.

17

Field Emission Scanning Electron Microscopy in Food Research

Johan Hazekamp and Marjolein van Ruijven
Unilever R&D Vlaardingen, Vlaardingen, The Netherlands

17.1 INTRODUCTION

With the growing interest of consumers in an active, healthy lifestyle, they are demanding high-quality products with fewer calories and less salt. To optimise food product formulations, functionality and interactions of the ingredients should be understood before changes in composition and processing can be made. When ingredients are replaced or reduced it may affect appearance, mouth feel and stability of the food products that should not be altered. It is the science area of food microstructure to correlate structure–function relations in (processed) foods.

These structure–function relations of food products are generally studied using a wide range of microscopic techniques providing structural data ranging from centimetres to nanometres, where wide-field microscopy, confocal microscopy and X-ray tomography image the bulk at ambient and dynamic conditions, and high-resolution information is derived from both transmission (TEM) and scanning electron microscopy (SEM). The hydrated nature of the often soft, condensed food material dictates a cryogenic approach in SEM and TEM.

17.2 A CLOSER LOOK AT FOOD MICROSTRUCTURE

With the variety of food products, there is also a huge variety in food structures, from relatively homogeneous liquids like tea, to multiphase semi-solids like margarine or ice cream. The building blocks of these multiphase systems, oil (or fat), water, protein, emulsifiers,

Biological Field Emission Scanning Electron Microscopy, First Edition.
Edited by Roland A. Fleck and Bruno M. Humbel.

thickeners, flavours, or even bulk materials like pieces of vegetables, can be very similar in various food products and yet have a significant different function in the overall microstructure and properties of the product. For instance, proteins added to mayonnaise are used to stabilise the oil-in-water emulsion and prevent coalescence of oil droplets whereas in fresh cheese products, the protein forms an extended protein network that defines firmness and water-holding capacity of the product. In addition, these building blocks will also vary in shape, size and distribution and will therefore influence the product properties such as its melt-down, texture and stability. These product characteristics lie in the microstructure of the materials used and are usually a result of a subtle balance of chemical and electrostatic interactions between proteins, carbohydrates, fatty acids and oily components, often in a low oil and hydrated continuous phase. Any change in microstructure may affect the product properties as described, and a firm understanding of the interaction of the components is therefore required to explain and steer the processing of new materials and design of new processed food products.

17.3 CRYO-PREPARATION AND OBSERVATION

Processed food, as described in the previous paragraph, is best characterised as hydrated condensed soft matter. The water phase typically is full of proteins, molecularly dispersed or aggregated by salt addition and/or heat treatments. Viscosity of the water phase can be enhanced by the addition of thickening agents. The water phase is usually continuous and solutes diffuse unrestricted as boundaries like the membranes in biomedical tissues are absent. A chemical treatment and subsequent dehydration of this continuous water phase would not only displace but most likely extract any small aggregates and soluble components.

As cryo-immobilisation routes for aqueous systems have come of age in the last decades, these procedures are preferred over chemical preservation. Although it is widely accepted that rapid cooling is the preferred cryo-immobilisation technique for structural and chemical preservation of hydrated specimens, it is also known that this approach may induce crystallisation and recrystallisation segregation artefacts. Segregation artefacts not only lead to a coarsening of the microstructure but may also have an effect on the distribution of the various soluble components. The structural and chemical resolution in the final sample is therefore a function of segregation artefacts and the optical resolution of the instruments used. Over the years, a broad cryo-toolbox has been developed, guaranteeing a vitrified state of the specimen whilst preparing freeze-fractured surfaces and a frost-free transfer (Cavalier, Spehner and Humbel, 2010; Echlin, 1992).

For food studies, a bulk freezing procedure by means of high-pressure freezing (HPF) results in a well-frozen volume of the sample that can be processed further, either by freeze-fracturing or cryo-planing. Where freeze-fracturing creates a topographic landscape following interfaces, the cryo-planing approach results in a completely cross-sectioned block-face, as described by Walther and Müller (1999) and Nijsse and van Aelst (1999). For freeze-fracturing, dedicated equipment is available from Leica Microsystems (Vienna, Austria) and Cressington Scientific Instruments (Whatford, UK) and the cryo-planing is typically done using a cryo-ultramicrotome sold by, for example, Leica Microsystems (Vienna) and RMC Boeckeler (JEOL JFD-V, Tuscon, USA). Figure 17.1 depicts an example of both techniques applied on one sample. Nowadays, the freeze-fracturing of cryo-specimens for SEM is done in the vacuum of a preparation chamber, attached to a cryo-SEM, whereas the

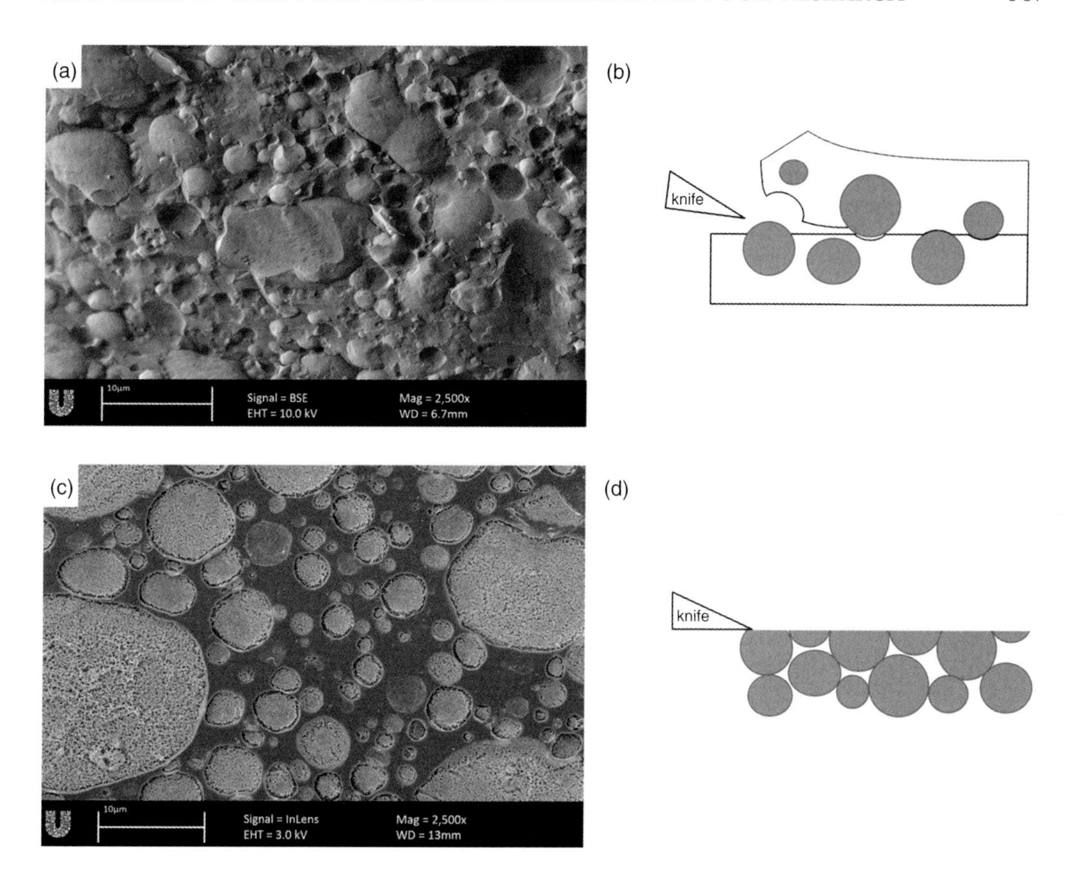

Figure 17.1 Freeze-fracturing techniques to create cross-fractured surface cryo-specimens. With freeze-fracturing (a, b) a true fracture is created over interfaces and membranes. The cryo-planing techniques (c, d) utilised by a diamond knife sections through all structural phases.

cryo-planing is typically done in a cryo-ultramicrotome requiring a transfer step between the microtome and the cryo-SEM.

Both freeze-fracturing and cryo-planning result in a large field of view in the cryo-field emission SEM (FEG-SEM) that is of benefit in food structural research, combining large-scale structural information with high-resolution detail of the location and the interactions of submicron-sized structures.

To preserve the original microstructure, high-pressure freezing is the preferred cooling method (Moor, 1987). It is, however, not suited for aerated systems, as gas cells will collapse under the high pressure applied in the HPF machine. Foams and aerated products are therefore frozen in melting ethane and the best preserved microstructure is limited to a fine superficial layer dependent on the heat transfer function of the specimen. Furthermore, the gas cells are good insulators, limiting the heat transfer that is required for vitrification. The microstructure observed in aerated materials, at best, is subject to a mild segregation and should be evaluated as such.

A critical step in the preparation of a cryo-SEM specimen is the removal of water by sublimation *in vacuo* from the frozen mass to create a surface topography. During this step the sample stage is heated to allow for sublimation of water from the frozen surfaces. The rate of sublimation is a function of temperature and vacuum conditions (Figure 17.2) and

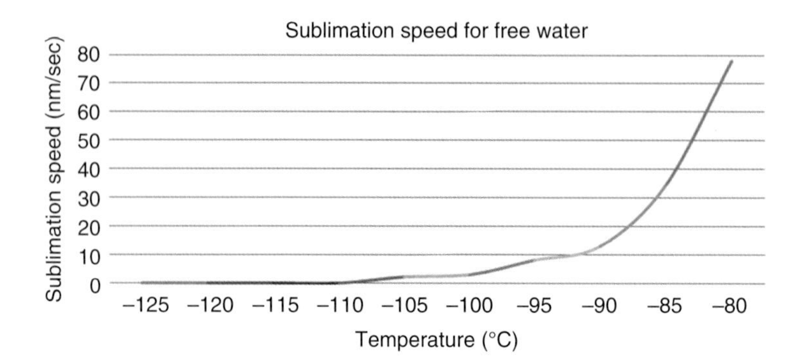

Figure 17.2 The sublimation speed of free water inside an electron microscope is a function of temperature and vacuum conditions (see Chapter 12 by Tacke, Lucas, Woodward, Gross and Wepf). In this specific example and as a rule of thumb, sublimation starts at −110 °C and rapidly increases at increasing temperatures. At varying vacuum conditions this graph needs to be recalibrated.

known to be very critical as the microstructure may collapse when specimens get over-dried by subliming too much water.

The cryo-stages in the preparation chamber of any cryo-transfer equipment are reproducible with respect to temperature control, but must be calibrated by the operator to get a feel for heating and cooling ramps and subsequent sublimation rates. Only when the sublimation rate of the configuration used is well characterised and controlled, can sublimation over fine, nm scale layers, be achieved.

As can be seen from the data displayed in Figure 17.3, when set to −85 °C, the specimen stage of our cryo-transfer system takes just over 5 minutes to actually reach the set temperature. The specimen, however, starts sublimation at temperatures above −110 °C and by the time the stage reaches the set temperature, the specimen has already been sublimed for over 4 minutes. These characteristics of the cooling/heating ramp will provide guidance in the controlled etching of a specimen and may need small adaptations for different specimens and specimen compositions.

Having complete control over the temperature settings of the cryo-preparation stage is crucial during preparation. During etching of free water, the temperature of the stage is

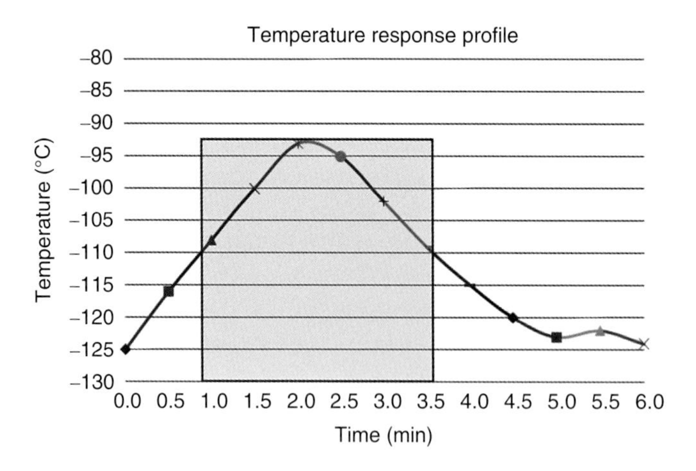

Figure 17.3 Temperature response profiles of a cryo-preparation stage in the sublimation range. The delay in response results in a slow warming during which uncontrolled sublimation is occurring. It is advised to record these profiles and characterise the temperature behaviour of the preparation stage.

set to a temperature at which water sublimes from its frozen state in the vacuum of the cryo-chamber. This temperature is above the recrystallization temperature of the specimen. For water, it is generally accepted that recrystallization effects during the short time span of etching do not introduce segregation artefacts traceable with the resolution of the instrument. No information is, however, available of the temperature effects on other components and effects that reversible or irreversible transitions may have on the microstructure. A deep etch usually results in, for example, collapse of fine structures or redistribution of fine granular or dispersed droplets simply moving downwards with a retracting waterfront. A well-calibrated temperature response profile potentially can fine-tune the etching capability, resulting in a reproducible and optimised specimen surface free of artefacts.

After sublimation the specimen is metal coated to reduce electron charging artefacts. For this, a sputtering head is provided in the cryo-preparation chamber capable of sputtering different metals to accommodate specific needs such as metal grain size. Finally, the specimen is transferred under the electron beam and imaged using low-dose conditions. During imaging the specimen is kept well below –125 °C and protected by an anti-contaminator (at least 10 degrees colder than the stage) to collect free water vapour.

17.4 APPLICATIONS OF FOOD MICROSCOPY

17.4.1 Ice Cream

The microstructure of ice cream is characterised as a smoothly textured semi-solid foam that is malleable and can be scooped. Ice cream is made from dairy products, such as milk and cream. Sugar is added and in some cases flavours, colourings and fruits are used. In the process of ice cream making, the mixture of ingredients is stirred slowly while cooling, in order to incorporate air and to prevent large ice crystals from forming.

The microstructure of a typical ice cream is shown in Figure 17.4. An ice cream sample was mounted in a pre-cooled OCT compound (optimal cutting temperature compound) mounting medium and loaded into the cryo-transfer system. The specimen was subsequently freeze-fractured, sublimed, metal coated and transferred into the cryo-SEM. The sublimation is mostly effective on the ice crystals (IC) as the continuous serum phase has formed a glass eutectic mixture as a result of the sugar and protein contents. This glass phase will not freeze-etch and show a freeze-fracture topography. The image shows the presence of air cells (A) and milk fat globules (F) clustered with casein micelles (arrows). The milk fat droplets and the casein micelles are easily recognised as a result of freeze-fracturing over the interfaces of these globular structures.

17.4.2 Foams

Whipped creams are oil-in-water emulsions stabilised by adsorbed milk protein; air is incorporated to create foam (Lomakina and Miková, 2006). For cryo-SEM studies, foam and aerated materials are the more challenging food samples since these samples are difficult to prepare as a result of the presence of air/solid interfaces. Because of the presence of large gas cells, these samples display a poor heat transfer upon freezing and high-pressure freezing cannot be applied as, under high-pressure conditions, the gas cells and hence the ultrastructure will collapse. Typically, these samples are quickly frozen on a stub/rivet in liquid ethane (Taatjes and Mossman, 2006) and freeze-fractured. The sample is slightly etched

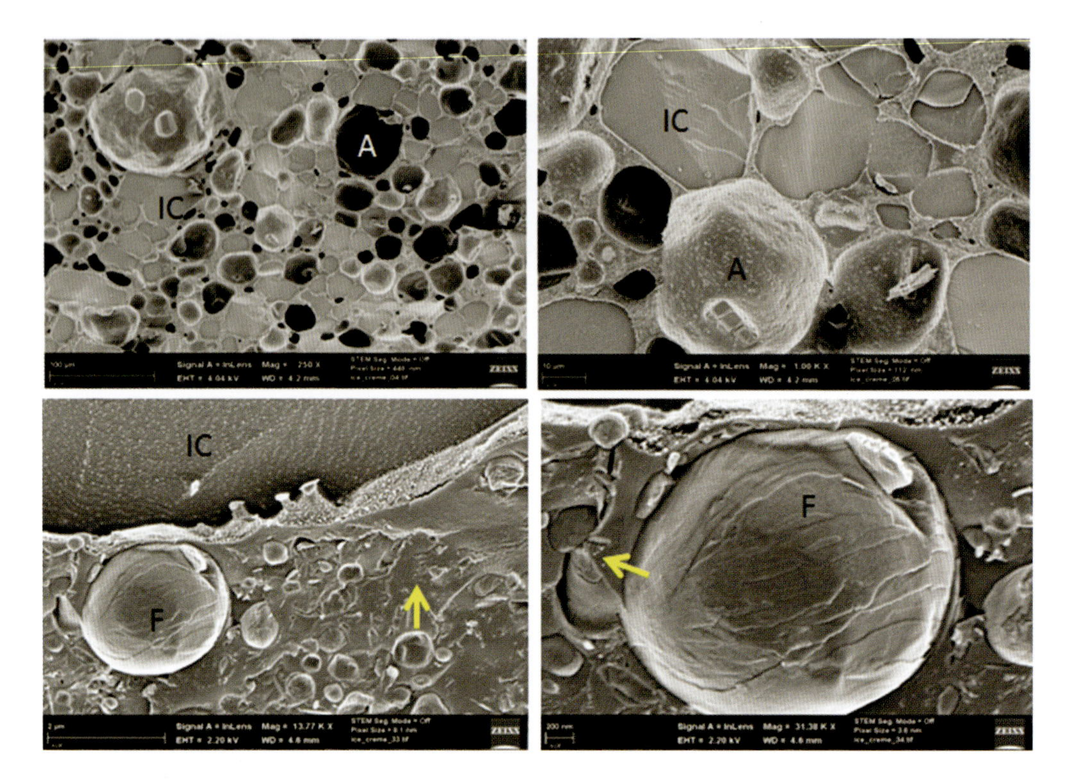

Figure 17.4 Cryo-SEM image of a freeze-fractured surface created in dairy ice cream. Ice cream is best characterised as a semi-solid foam consisting of air cells (A) and water ice crystals (IC). The serum phase (S) consists of sugar, soluble milk proteins, fat globules (F) and casein micelles (arrows).

(a sublimation of free water using the high-vacuum conditions at an elevated temperature of around –95 °C to create topography) and subsequently, to improve the details of the tiny fat platelets, double-layer coating techniques are regularly applied (Walther and Müller, 1997; Walther, 2003a, 2003b).

Figure 17.5B depicts the interface of an air bubble of a whipping cream, showing that the air interface is stabilised by the fat platelets of the oil droplets attached to the interface. In egg white foams (Figure 17.5C and D), the gas cells are stabilised by proteins rather than by oil and fat platelets. The sample shown in Figure 17.5C and D is not etched, since it will deteriorate the frozen microstructure and interface. The air bubble distribution is very homogeneous, as is apparent in Figure 17.5C. The appearance of the air bubble interface of the egg white foam is smooth since it is covered by egg albumin protein (Figure 17.5D).

17.4.3 Monoglyceride Networks

Mono- and diglycerides are commonly added to commercial food products in small quantities acting as emulsifiers, helping to mix ingredients such as oil and water that would not blend well. They are often found in bakery products, beverages, chewing gums, confections, margarine, shortening, ice cream and whipped toppings.

Monoglycerides, under specific processing, are capable of forming liquid crystalline phases with vast water-holding capacities. Figure 17.6 shows the microstructure of an

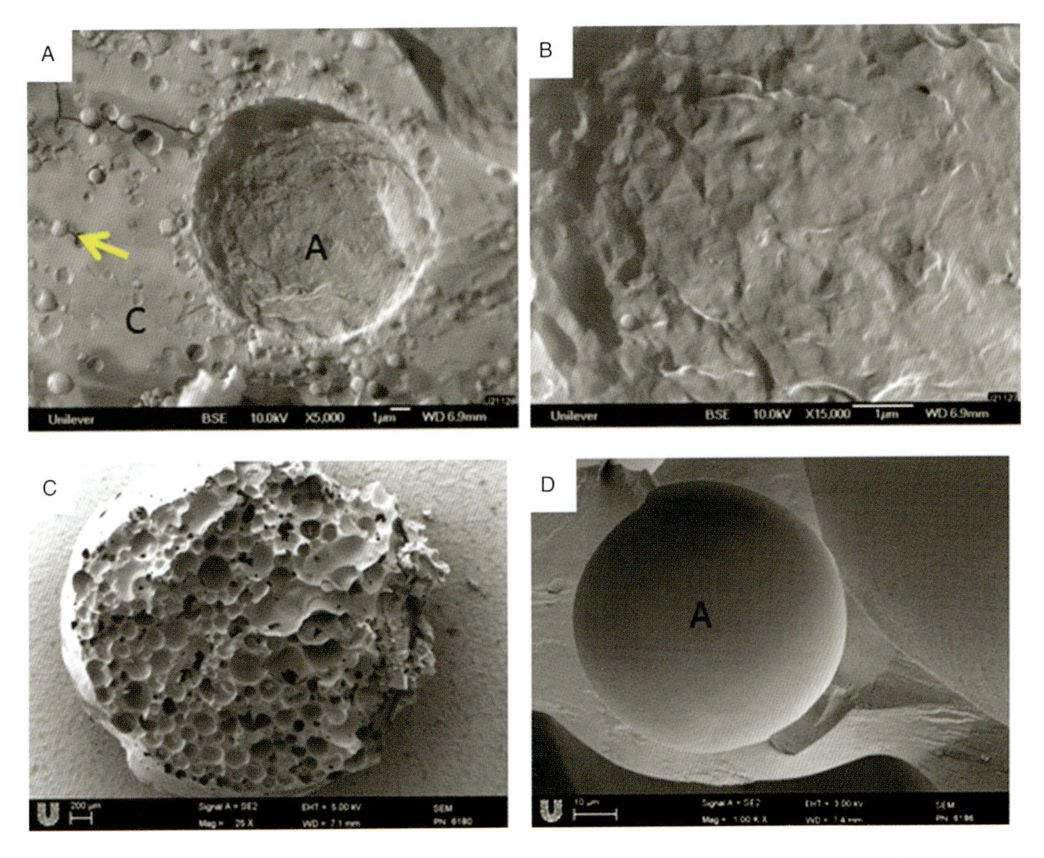

Figure 17.5 Cryo-SEM images of foams: whipping cream (A, B) and egg white foam (C, D). In the whipping cream, the continuous phase (C) is filled with oil droplets (arrow), which show prevalence to the air bubble interface. The interface is stabilised by the fat platelets of the oil droplets (B). The bubbles in an egg white foam are stabilised by albumen protein and results in a smooth interface.

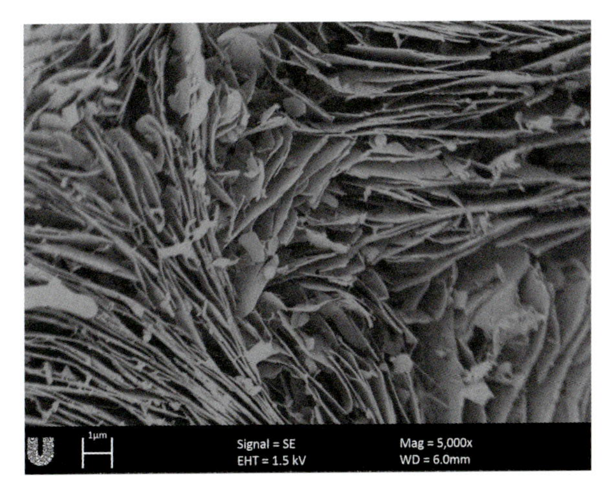

Figure 17.6 A crystal network consisting of 2% monoglyceride in water. The sample was deep-etched to image the 3D structure of the network.

ultra-light fat blend that may contain up to as much as 98% water. It is structured by a liquid crystalline phase of 2% monoglycerides forming a compact network of monoglyceride plates. Although high in water content, this product is stable and firm.

The SEM images were taken after freeze-fracturing a high-pressure frozen sample followed by extensive and deep-etching to reveal a three-dimensional organisation of the continuous monoglyceride phase. The deep-etching provides a deep view into the structure without the network collapsing. The monoglyceride crystal network appears to be insensitive for crystallisation artefacts of the free water and remains undisturbed.

17.4.4 Gelling Agents

Thickeners and gelling agents in processed food are added to structure the aqueous continuous phase either by water-binding activity (starch) or by network formation (proteins and carbohydrates). Commonly used gelling materials for networking action are gelatine, carageenan, agar, xanthan gum and locust bean gum. Nowadays, plant fibre material is used for a similar action and perceived as a more natural ingredient. It is this network formation, in usually hydrated systems, which pose a significant challenge in cryo-microscopic studies as water contents may be very high, even up to 98% water. Upon cryo-preservation of these networked aqueous systems, ice crystal growth induces segregation patterns with a network appearance that may be easily mistaken for the polymer network. The network material, carbohydrate or protein polymers, does not act as a barrier as in biological membranes and the migration of molecules is therefore unrestricted. This aspect dictates a cryogenic approach in sample preparation to preserve small aggregates and solutes as opposed to the artefacts known from chemical fixation and dehydration. Infusion of the specimen with cryo-protectants to reduce ice crystal growth changes the chemical and structural characteristic of the specimen and therefore cannot be applied. It is this water-rich nature of these gelled specimens that dictate a small sample volume at the moment of cooling to reach the high cooling rates required to vitrify the networked structure. However, in mixed systems, for example foams and emulsions, other components and structural dimensions may require a bulk specimen for observation of larger length scales that will not freeze properly. Segregation patterns as a result of ice crystallisation in those samples always interfere with network observation.

Figure 17.7 depicts a series of processed plant fibre material that was frozen using a metal mirror technique and freeze-fractured inside the SEM. The reason for this investigation was to indicate the fibre aspect ratios and the level of disintegration. This dictated the use of a large volume of sample although a small sample volume would have resulted in a better structural preservation.

Regarding aforementioned parameters, conclusions could be drawn about homogeneity between different levels of processing. At higher magnifications the fine fibrous structure is visible, showing an entangled network of fibres. It should be taken into account that this network may well be the result of freezing artefacts.

17.5 FACTS AND ARTEFACTS

The final image in cryo-SEM studies always reflects the quality of the preparation techniques used. During specimen preparation, sample material is extracted or relocated, fine detail may get obscured or even disappear in the vacuum, changing dimensions and distributions. Only

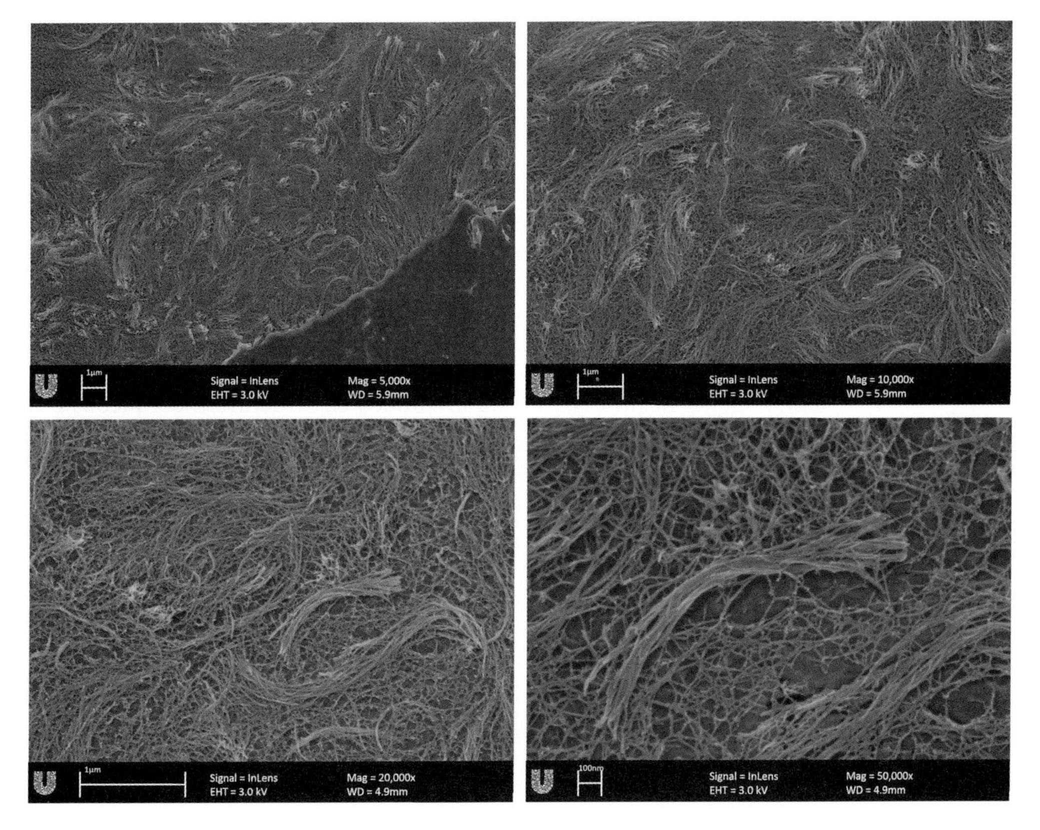

Figure 17.7 An aqueous slurry of processed plant fibre material (2.5%) after freeze-fracture cryo-SEM. The sample was frozen using a metal mirror protocol. Conclusions could be drawn regarding homogeneity and fibre aspect ratios. The fine network structure may be a segregation pattern as a result of insufficient cooling speed.

if the processes involved in the sample preparation are well known and taken into consideration, can these changes be kept to a minimum. A short list of the most common artefacts in cryo-SEM studies is listed in Figure 17.8. These artefacts are sample dependent and should always be optimised and standardised as minor adaptations in sublimation, fracturing and coating may lead to significant changes in the observed structure. It is good practice to compare the outcome of a study with complementary techniques that may partly overlap the resolution and image from a totally different perspective to support the cryo-SEM observations.

The dramatic effects of segregation as a result of low freezing rates is shown in Figure 17.9 depicting an image taken of a mixed system of two carbohydrate polymers, a concentrated particulate material dispersed in a continuous aqueous gel. This specimen was frozen in liquid ethane and cryo-planed using a glass knife. A gradient is visible in the structure that was induced by water crystallisation and indicated by the blue arrow. The best preserved material was found at the edge of the specimen and conclusions could be drawn about the dispersity, particle size distribution and perhaps even interconnectivity between the two phases. In the periphery of the specimen, the segregation patterns at low magnification may give the impression of a true network, but only after comparison with the well-preserved layers did it become obvious that the network was formed by ice crystals.

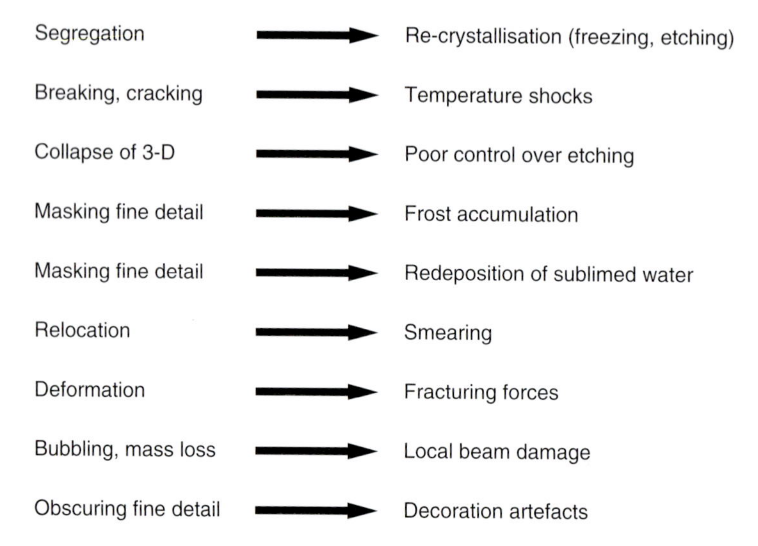

Figure 17.8 An overview of the most abundant artefacts observed in cryo-SEM studies as an effect of preparation and/or observation in the microscope.

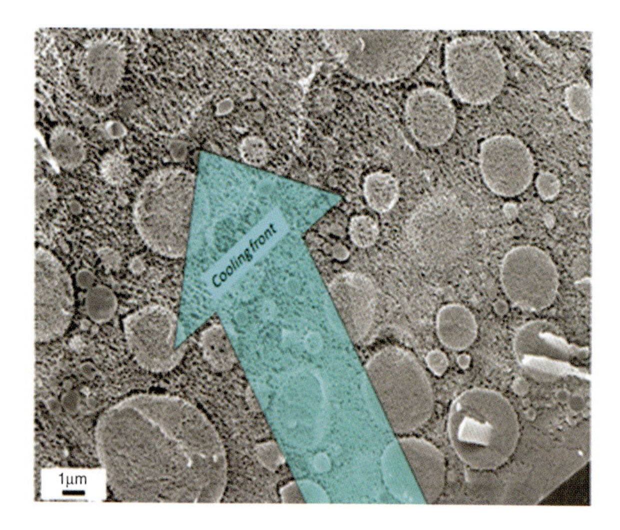

Figure 17.9 A two-phase gel system consisting of high concentrated particles dispersed in a continuous aqueous gel. The arrow indicated the cooling front as a result of immersion into liquid ethane. The periphery shows a network-like structure as a result of segregation and is far from the original structure near the specimen edge. Selection of the best area for imaging is done at low magnification.

As mentioned earlier, optimal cryo-preservation cannot always be used when large dimensions of structural parameters (e.g. gas cells in foams) dictate a bulk specimen preparation technique (Figure 17.5). In the foam specimen depicted in Figure 17.5, a low magnification observation to review the overall gas cell organisation is as important as the organisation of casein micelles and fat droplets at higher resolution. For the gas cell distribution, a larger sample volume was required although, as a result of low freezing rates, the centre of the specimen will show segregation artefacts at higher magnification. These observations were then made at the edges of the frozen specimen where the cooling rates are assumed to be maximal for this specimen.

17.6 CONCLUDING REMARKS/SUMMARY

Cryo-SEM takes us into the wonderful world of microstructure in its original state and becomes fascinating when knowledge of chemical composition, processing and sample preparation come together in the final result, the cryo-SEM image. The technique, in the hands of a trained cryo-electron microscopist, may seem 'doable' or even 'easy going', but perseverance and a deep understanding of the processes involved are the basis for cryo-SEM observation. In microscopy of processed food, this is even more the case as processing conditions and variations in composition play an additional role.

In this chapter we have discussed and outlined the technique of cryo-SEM and showed our good practice in the daily routine. We have shown that this is never a standard method nor a high-throughput technique and a science base and experience should be built up to image the right areas and to be able to make the right choices during preparation of cryo-specimens. Nevertheless, a suboptimal microstructure is occasionally the best we can get and conclusions should be adapted likewise.

ACKNOWLEDGEMENTS

The work displayed and discussed in this chapter represents the high level of expertise and broad knowledge of soft hydrated food material developed over the years since the introduction of cryo-SEM in Unilever R&D Vlaardingen. The examples presented here are the work of the Vlaardingen microscopy team and more specifically made by the SEM experts: Mr Peter Nootenboom, Mr Henrie van Aalst, Mrs Caroline Remijn and Dr Jaap Nijsse. They reflect years of development and creativity driven by the desire to image microstructure as close as possible to the original situation. The authors thank them for sharing their expertise in cryo-SEM and cryo-microscopy in general, as well as for in-depth and ongoing discussions.

REFERENCES

Cavalier, A., Spehner, D. and Humbel, B.M. (2010) *The Handbook of Cryo-preparation Methods for Electron Microscopy*, CRC Press.

Echlin, P. (1992) *Low-Temperature Microscopy and Analysis*, Plenum Press, New York and London.

Lomakina, K. and Miková, K. (2006) A study of the factors affecting the foaming properties of egg white – A review. *Czech J. Food Sci.*, 24, 110–118.

Moor, H. (1987) Theory and practice of high-pressure freezing, in *Cryo-techniques in Biological Electron Microscopy* (eds R.A. Steinbrecht and K. Zierold), Springer Verlag, pp. 175–191.

Nijsse, J. and van Aelst, A.C. (1999) Cryo-planing for cryo-scanning electron microscopy. *Scanning*, 21, 6.

Taatjes, D.J. and Mossman, B.T. (2006) *Cell Imaging Techniques: Methods and Protocols*, Humana Press.

Walther, P. (2003a) Cryo-fracturing and cryo-planing for in-lens cryo-SEM, using a newly designed diamond knife. *Microscopy and Microanalysis*, 9, 279–285.

Walther, P. (2003b) Recent progress in freeze-fracturing of high-pressure frozen samples. *Journal of Microscopy*, 212, 34–43.

Walther, P. and Müller, M. (1997) Double-layer coating for field-emission cryo-scanning electron microscopy – Present state and applications P. *Scanning*, 19, 343–348.

Walther, P. and Müller, M. (1999) Biological structure as revealed by high resolution cryo-SEM of block-faces after cryo-sectioning. *Journal of Microscopy*, 196, 279–287.

18

Cryo-FEGSEM in Biology

Paul Walther

Central Facility for Electron Microscopy, Ulm University, Ulm, Germany

18.1 INTRODUCTION

The scanned electron beam mode in an electron microscope (Knoll, 1935; von Ardenne, 1938) has advantages compared to the standard bright-field transmission electron microscopy mode. When using the scanning mode, signals can be recorded for image formation that cannot be easily focused with lenses, such as secondary electrons (SE), backscattered electrons (BSE) or even X-rays and cathodoluminescence signals and others. (In this chapter, only SE and BSE signals will be further discussed.) Since these signals are not focused with lenses, they are not subjected to chromatic or spheric aberration, as already considered by Manfred von Ardenne during the construction of the first SEM (reviewed in von Ardenne, 1996). In addition, some signals, such as secondary and backscattered electrons and X-rays can be measured above the sample. The sample, therefore, does not need to be electron transparent and also bulk samples can be investigated.

The major disadvantage of the scanning mode compared to the regular TEM mode has been the limited resolution. Resolution in SEM is a very complex issue and, according to Joy and Pawley (1992), 'is limited by at least three factors: the diameter of the electron probe, the size and shape of the beam/specimen interaction volume within the solid for the mode of imaging employed and the Poisson statistics of the detected signal'. The first factor, the diameter of the electron probe, is a technical parameter of the electron microscope used and has been constantly improved since the introduction of field emission cathodes (Crewe *et al.*, 1968). This allows for atomic resolution in the scanning transmission mode (Krivanek *et al.*, 2012) and (at least under special conditions) even in the secondary electron mode (Zhu *et al.*, 2009). To optimize the second and the third factor mentioned by Joy and Pawley is a major task of specimen preparation and will be further discussed in this chapter (Table 18.1).

For cryo-SEM the native, aqueous sample is cryo-fixed and afterwards prepared in the frozen state, where it behaves like a solid state sample. Cryo-preparation can

Biological Field Emission Scanning Electron Microscopy, First Edition.
Edited by Roland A. Fleck and Bruno M. Humbel.

overcome limitations of classical preparation methods based on chemical fixation, followed by dehydration with organic solvents and drying. The potential advantages of cryo-preparation methods for electron microscopy were already appreciated in the 1950s (e.g. Fernandez-Moran, 1960). A milestone of cryo-preparation for transmission electron microscopy (TEM) was the invention of the freeze-etch replica technique, pioneered by Steere (1957) and then developed into a widely applied routine method by Moor and Mühlethaler (1963). Thereby, a cryo-fractured sample was coated with heavy metal and carbon. After thawing the specimen, the metal-carbon replica was stripped off, cleaned and then viewed in the transmission electron microscope at ambient temperature. The first attempts to directly look at a frozen bulk sample mounted on a cold stage in the cryo-SEM were done by Thornley (1960). The method was further developed by Echlin and others (reviewed by Echlin, 1992). It was, however, limited by the limited resolution capacities of the SEMs available at the time. In the 1980s field emission gun SEMs (FEGSEMs) with smaller primary beam diameters became commercially available. It was realized that one could only make use of the improved instrumental performance by a better understanding of contrast mechanisms (Seiler, 1967; Reimer, 1978; Peters, 1982; Joy and Pawley 1992), beam specimen interactions (Pawley and Erlandsen, 1989) and by improving surface coatings (Peters, 1986; Hermann and Müller, 1991).

18.2 CRYO-PREPARATION AND CRYO-FEGSEM

Table 18.1 provides a diagram of the preparation steps used in this work.

18.2.1 Freezing

The first, most difficult and most limiting step in cryo-specimen preparation for electron microscopy is freezing. Water, which is the major component of biological systems, tends to form crystals during freezing. The frozen water in our daily life, such as snow, ice or even ice cream, exists in the form of hexagonal ice crystals. When water freezes, the water molecules tend to stick together in a hexagonal pattern, leading to crystal formation, thereby excluding other molecules such as proteins, lipids and sugars. Therefore, when the ice crystals grow, they push the non-water molecules to the periphery of the crystals and when finally the crystals meet, the non-water molecules accumulate at the border between the ice crystals. This causes a characteristic network-like pattern, destroying the natural structure of the biological system.

There are at least three parameters that can be optimized to keep ice crystal formation low (recently reviewed by Mielanczyk *et al.*, 2014).

The first is the addition of cryo-protectants. Small molecules such as sugars bind water molecules and reduce their mobility; during freezing this reduces their ability to integrate into a crystal. For the Tokuyasu cryo-sectioning method, for example, sucrose is added to the chemically pre-fixed samples to prevent ice crystal formation during freezing (Tokuyasu, 1973). Cryo-protectants, however, can affect or change the physiological state of a biological system, for example cause osmotic effects; therefore they are usually applied after chemical fixation. When a sample is chemically fixed before freezing, however, a major task of fast freezing is lost, namely that a biological system is immobilized within milliseconds from a defined physiological state. The second parameter influencing ice crystal formation is the cooling rate. The faster the cooling rate, the less time the ice crystals have to grow and the smaller they will stay. Sufficiently high cooling rates can be achieved at a sample surface by plunging it into ethane or propane cooled by liquid nitrogen. (Liquid nitrogen itself does not provide high cooling rates due to the 'Leidenfrost phenomenon'.) Cryo-plunging allows for

Table 18.1 Diagram of the preparation schedules used in this work. The colour codes refers to the figures in the chapter. Obviously other combinations of the preparation steps are also possible

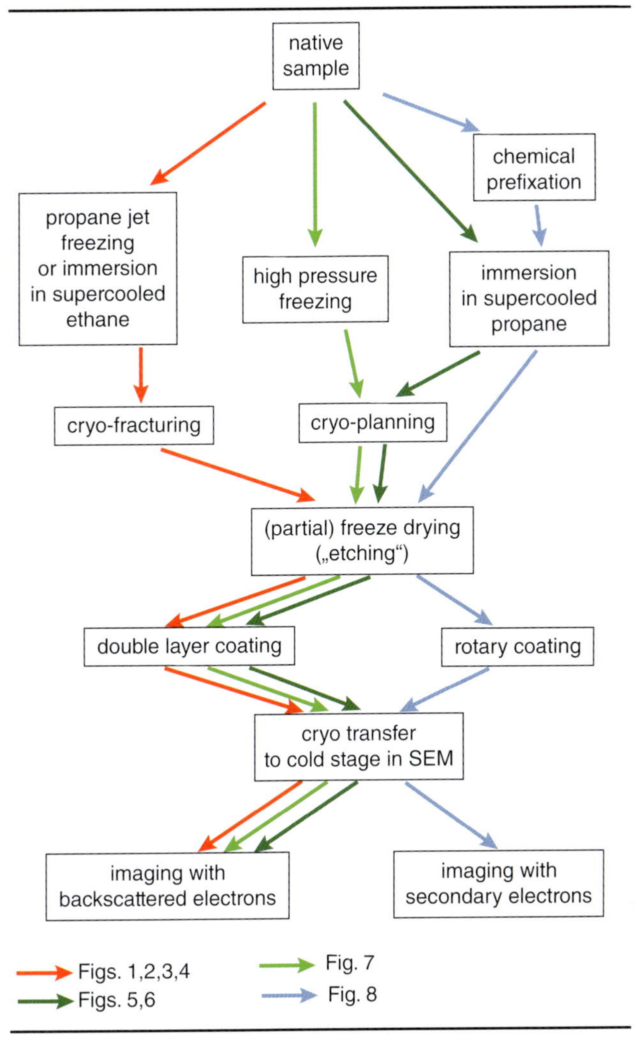

the vitrification (solidification in a glass-like state, without crystal formation) of thin films (50 nm to 1 μm) of water suspensions (Dubochet and McDowell, 1981) and it has been very successfully applied in cryo-TEM (reviewed by Dubochet, 2012). A variation of this approach is the propane-jet freezer, where one or two jets of cooled propane are directed against a low-mass biological sample (Müller, Meister and Moor, 1980). The limitation of these approaches is the poor thermal conductivity of H_2O causing a dramatic reduction of the cooling speed in the depth of the sample. In a depth of more than about 10 μm, the cooling speed is too slow for vitrification and ice crystal formation is inevitable, even if the cooling speed at the surface were to be infinitely fast (Studer *et al.*, 1995). The third factor influencing ice crystal formation is the pressure. When a pressure of about 2000 bar is applied to the specimen at the moment of freezing, ice crystal formation is reduced, even at relatively low cooling rates, because the water molecules are hindered to arrange in the hexagonal state that would request volume increase. This principle is used in high-pressure freezing machines (Moor and Riehle, 1968). It is generally agreed that with high-pressure freezing, samples up to a thickness of about 200 μm (depending on the water content) can be frozen with no major ice crystal segregation artefacts. High-pressure freezing is currently considered to be the gold standard for freezing of cells and tissue (Hohenberg *et al.*, 2003).

For cryo-FEGSEM, high-pressure freezing is the ideal cryo-fixation tool because it allows for cryo-fixation of relatively large samples. The geometry of sample and sample holder should be optimized for freezing, but one should also take into consideration the follow-up preparation steps, such as fracturing, sectioning and freeze-drying. These steps require stable mechanical mounting of the sample on a support.

18.2.2 Cryo-Stage and Cryo-Transfer to the SEM

After freezing, the sample needs to stay cold for all subsequent preparation steps and for SEM imaging. This requires a cold stage. In this study the samples were mounted on holders that fit on a Gatan cryo-holder 626. Similar results, however, can be achieved with other equipment. Good and careful sample preparation is the key for cryo-SEM. Especially important and critical is a good and stable mounting of the sample for the following reasons: appropriate heat transfer from the sample to the cold stage needs to be granted, so that the sample stays cold; the sample must not move during scanning in the SEM; and the sample and the cold stage must have an electrically conductive connection to prevent sample charge-up. Electrical conductivity is finally achieved by metal coating, as explained later, but it only works when the sample and cold stage are mechanically well connected, so that the metal layer is not interrupted between the sample and stage (Walther, 2003b).

18.2.3 Cryo-Fracturing

The standard method to look inside a frozen sample is cryo-fracturing. Thereby, the fractured sample is coated with a contrast-forming heavy metal layer (e.g. platinum) and an additional carbon layer for mechanical stability. Afterwards, all the organic components are removed by cleaning in sulfuric acid, so that a beam-transparent replica (a carbon film with a heavy metal layer) is left over that can be analyzed in TEM (Steere, 1957; Moor and Mühlethaler, 1963). This method has been very successful and widespread in the 1970s and 1980s and excellent cell biological research using the TEM replica labelling method as introduced by Fujimoto (1995) has also been published recently, such as Kamasawa *et al.* (2006). A limitation of the TEM replica method, however, is that biological samples frozen from the native state are very brittle and the replica tend to fall into small pieces during cleaning. This problem is circumvented in cryo-SEM, because the bulk, not beam-transparent sample can be imaged and the difficult replica cleaning process is unnecessary (reviewed by Echlin, 1992). Thus, freeze-fractured samples that would be difficult to replicate can be investigated over a wide magnification range.

18.2.4 Beam Sensitivity and Coating

Biological specimens are beam sensitive. Electrons from the electron beam are scattered in the biological sample. This is a prerequisite for contrast formation, on the one hand, but in the case of inelastic scattering caused by electron–electron interaction, energy from the beam is transferred to the sample. This can cause electrons to be kicked out of the electron shells and to become secondary electrons that can be used for image formation. In an electrically conductive environment, these electrons are immediately replaced. Biological samples, however, are insulators and removing electrons can cause breakage of covalent bonds producing free radicals that interact with other components of the sample, leading to mass loss (Talmon, 1987). This effect occurs whenever a secondary electron is produced, also at low accelerating voltages. Coating the sample with a thin conductive metal layer reduces (but does not fully prevent) beam damage. We found a double layer coating especially useful for cryo-fractured samples (Walther and Hentschel, 1989; Walther *et al.*, 1995). Similar as in the replica technique, the fracture face is first coated with a thin layer (1 to 3 nm) of heavy metal (e.g. platinum) that is in close apposition to the biological structures of interest. This layer is then stabilized with a 5 to 10 nm thick carbon layer that enhances electrical conductivity and mechanical stability. When imaged with secondary electrons, the overlying carbon layer is imaged and small biological structures remain hidden (snowed in) by the carbon coating. Therefore, backscattered electrons are used for image formation. They are mainly scattered by the heavy metal layer that mimics the structures of interest with high precision (Walther and Hentschel, 1989; Walther *et al.*, 1995; Figure 18.1) (see also Chapter 12 in Volume I by Tacke *et al.*).

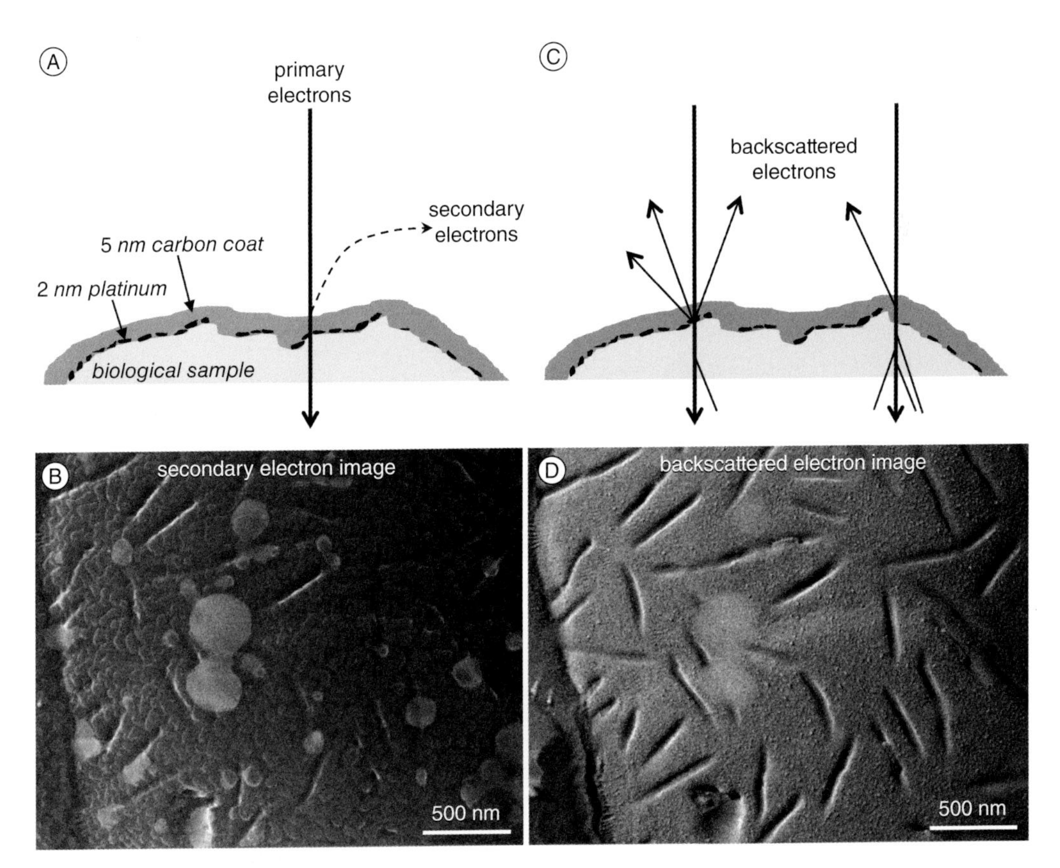

Figure 18.1 Double layer coating of a freeze-fractured yeast plasma membrane for observation in the fully hydrated state in the cryo-SEM. (A) Simplified schematic drawing of a fractured frozen hydrated sample coated with a thin (2 nm) platinum layer and an additional 5 nm carbon layer to increase mechanical stability and electrical conductivity. When imaged with secondary electrons (B) the surface of the carbon layer and of water vapour contaminations are imaged; small structural details are hidden. When imaged with backscattered electrons (C and D), however, electrons are mainly scattered at the platinum layer that is in close contact with the biological structure of interest and tiny surface structures such as intramembranous particles become visible.

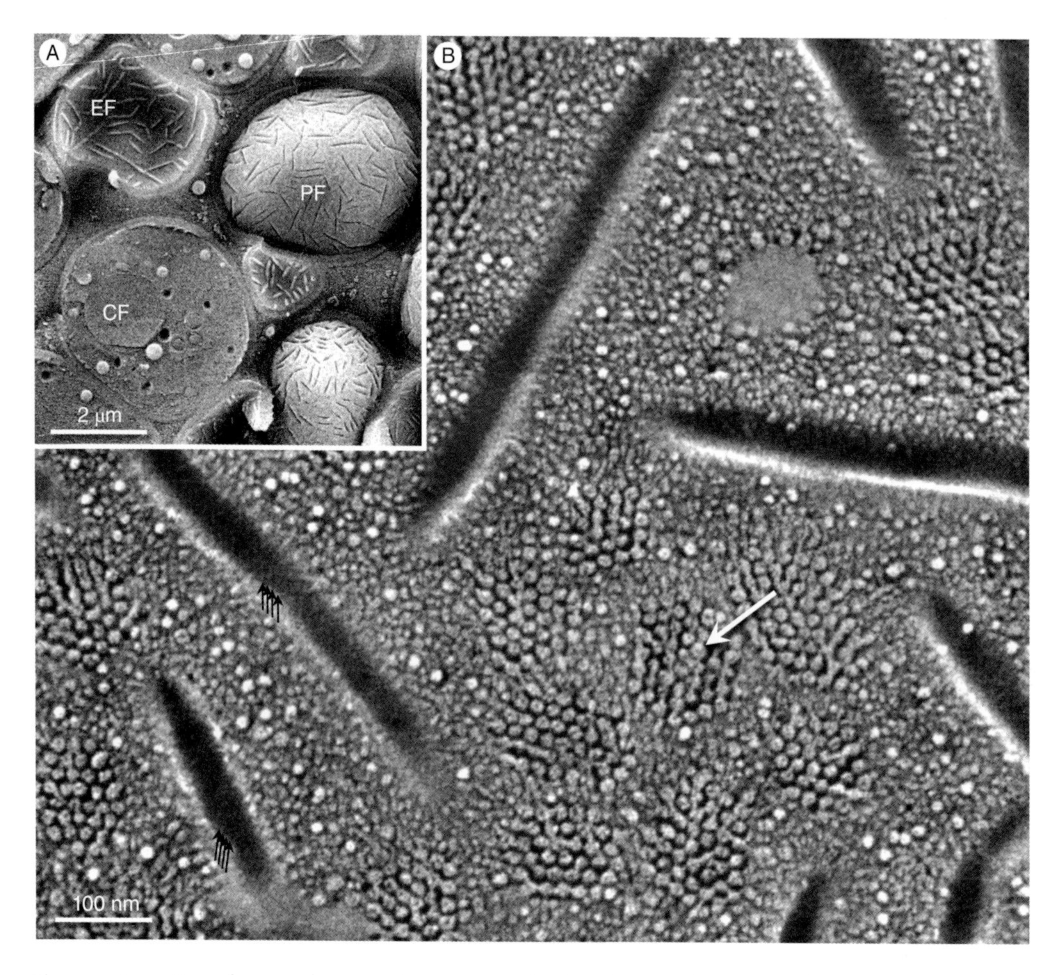

Figure 18.2 Cryo-fractured yeast cells imaged in the fully hydrated state with backscattered electrons in the cryo-FEGSEM. A) Overview of a cross-fractured cell (CF) and cells fractured along the plasma membrane. Protoplasmic fracture face (PF) and exoplasmic fracture face (EF). (B) High magnification of an area of a PF showing elongated invaginations with periodic substructures visible as parallel lines with a periodicity of about 4 nm (small black arrows). The white arrow depicts the H^+-ATPase proteins with a central depression arranged hexagonally with a periodicity of 15 nm.

A cryo-fractured yeast sample is shown in Figure 18.2. The overview (2A) shows three different fracturing aspects: CF is a cross-fractured cell and PF and EF depict two fractured cell membranes. According to Branton (1966), the fracture plane is not random; it preferentially (but not always) follows the inner, hydrophobic zone of the lipid bilayers. According to the nomenclature of Branton *et al.* (1975) the fracture face with the lipid monolayer facing the cytoplasm is called the protoplasmic fracture face (PF) and the other one is called the exoplasmic fracture face (EF). Figure 18.2B is a high magnification of the yeast cell membrane PF. The yeast cells have been starved before freezing, so they are in the stationary growth phase. In this stage they show two typical structures: oblongness invaginations of the membrane that reveal periodic structures, visible as parallel lines with a periodicity of about 4 nm (small black arrows), and hexagonally arranged particles (periodicity about 15 nm) with a central depression (large white arrow) as described by Gross, Bas and Moor (1978) on high resolution TEM replicas and by Walther *et al.* (1990, 1995) and by Hermann and Müller (1993) on sputter-coated high-resolution cryo-FEGSEM samples. These particles represent a storage form of the plasma membrane H^+-ATPase (Kühlbrandt, Zeelen and Dietrich, 2002). The image demonstrates the remarkable resolution that can be obtained from frozen hydrated samples with cryo-FEGSEM.

A cross-fractured yeast cell (*Saccharomyces cerevisiae*) is shown in Figure 18.3. The structure of the mitochondria (magnified in Figure 18.3B) is remarkable: the inner and outer mitochondrial membranes appear in very close apposition with a minimal intermembrane space. This is confirmed in Figure 18.3C where the fracture plane follows the mitochondrial membranes. The plane jumps from the inner (im PF) to the outer (om EF) mitochondrial membrane, so that a slab of the outer membrane is lying on top. This jumping of the fracture plane also indicates a close apposition of the two membranes (Lang and Bronk, 1978; Knoll and Brdiczka, 1983). The same close apposition of inner and outer mitochondrial membrane is also observed in cryo-fixed and freeze-substituted yeast mitochondria (Figure 18.3C; Walther and Ziegler, 2002) and in our more recent tomography data from macrophages (Höhn *et al.*, 2011).

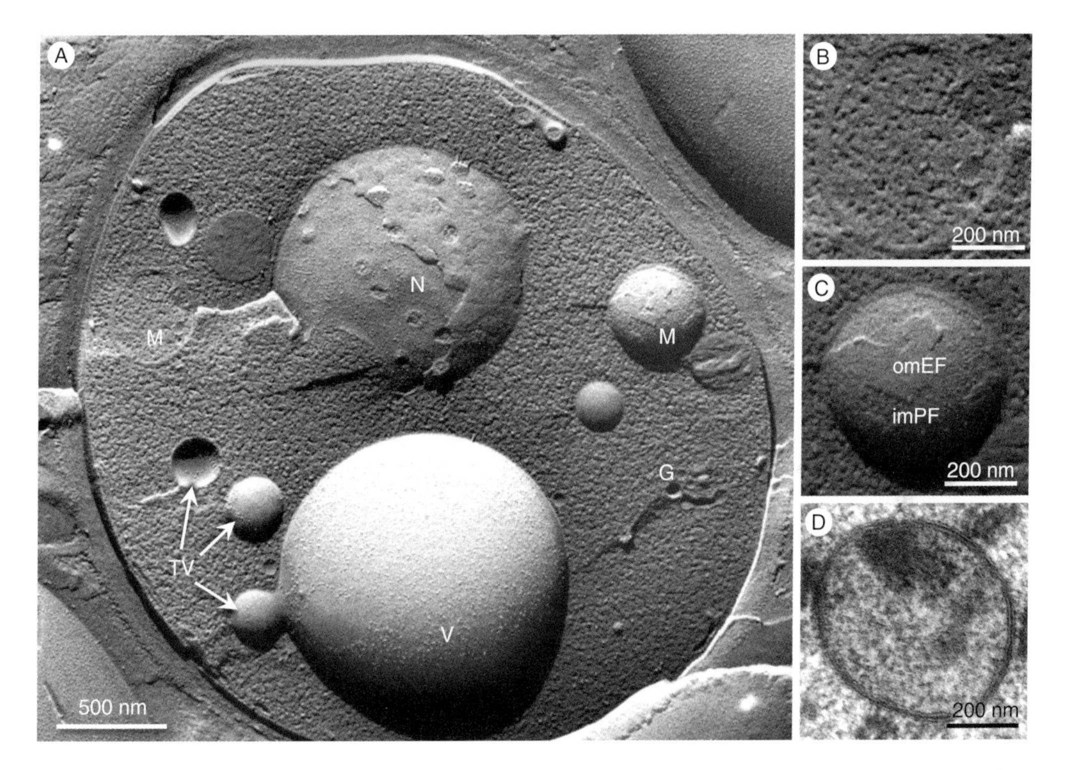

Figure 18.3 Cryo-fractured yeast cell (*Saccharomyces cerevisiae*) imaged with backscattered electrons, showing mitochondria (M), nucleus (N) with inner and outer nuclear membrane and nuclear pores, Golgi apparatus (G), vacuole (V) and transport vesicles (TV.) One transport vesicle is fusing with the vacuole. The two mitochondria are further enlarged in B and C. Inner and outer mitochondrial membrane are in close apposition, as is directly visible in B. In C the fracture plane jumps from the inner membrane's protoplasmic fracture face (imPF) to the outer membrane's extraplasmic fracture face (omEF) and back. This is only possible when the two membranes are in close apposition. These findings are in agreement with data obtained by freeze substitution (Figure 18.3D is from Walther and Ziegler, 2002).

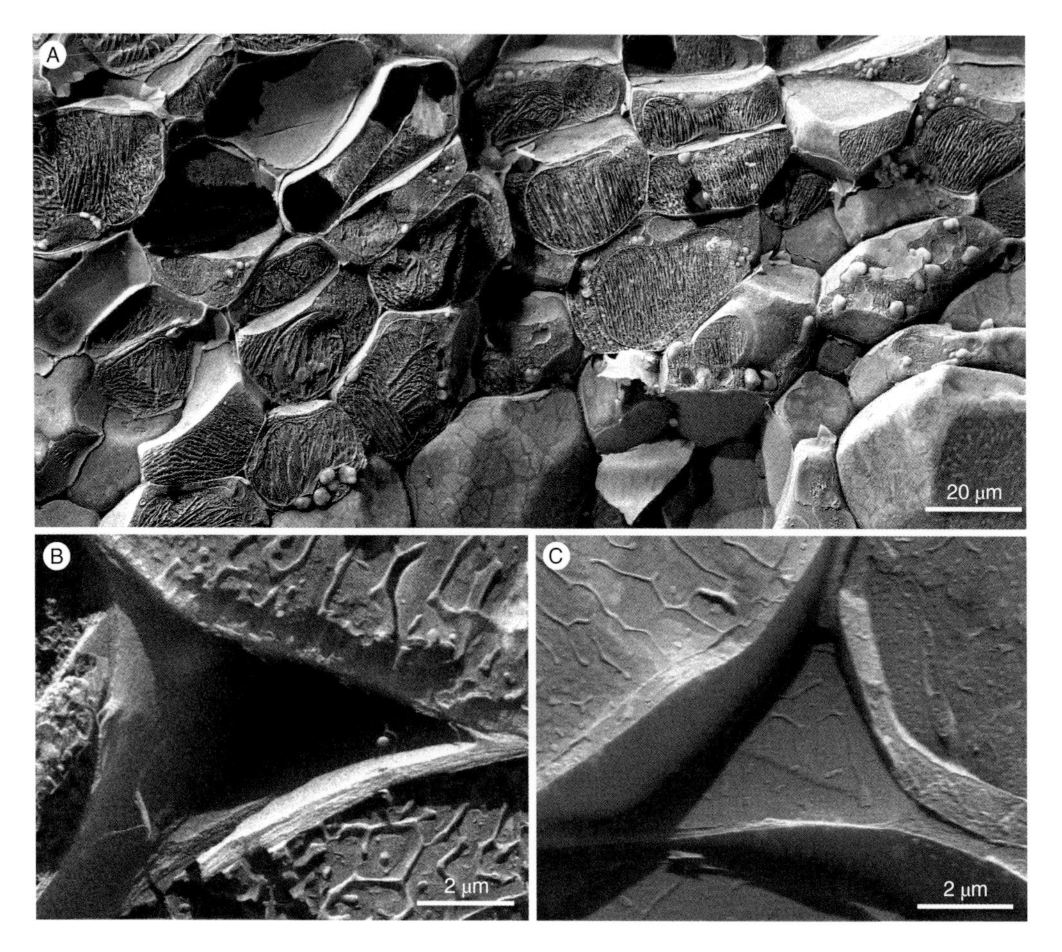

Figure 18.4 Cryo-fractured root nodules from lucerne (*Medicago sativa* L.). The topic of investigation was whether the intercellular spaces were filled with water or with air, a question that can be solved best by cryo-FEGSEM. A represents an overview of the cortex area of the nodule, while B and C show intercellular spaces of interest. B is empty but contained gas before cryo-preparation, whereas the intercellular space in C is full. Due to the fact that the same ice crystal segregation pattern is visible as in the surrounding cell, the material it is filled with must have a high water content. These images indicate that cryo-SEM allows a distinction to be made between water-filled and gas-filled intercellular spaces (from Weisbach *et al.*, 1999).

Lucerne root nodules are imaged in Figures 18.4 and 18.5 (from Weisbach *et al.*, 1999). In these plant organs, nitrogen fixation is performed by endosymbiotic bacteria from the family *Rhizobiaceae* using the enzyme nitrogenase. These bacteria are anaerobic. Therefore, oxygen transport must be strictly regulated and controlled in the nodules. It had been hypothesized that the plants control oxygen transport by regulating the water content of intercellular spaces. To test this, we cryo-fixed root nodules at different physiological stages and cryo-fractured or cryo-planed them and checked to what extend intercellular spaces were filled with water. Figure 18.4A is a cross-fractured area of a nodule. Numerous cells with cell organelles can be recognized, as well as the intercellular spaces in between the cells. Figure 18.4B and C represent two intercellular spaces at higher magnification. The space in Figure 18.4B is obviously empty as it was filled with gas before the sample was frozen. The space in Figure 18.4C is obviously filled with water. (Since the samples had to be frozen by plunge freezing and not high-pressure freezing, ice crystal segregation patterns are present in all three images.) These data prove that we can discern gas-filled and water-filled intercellular spaces. By quantitative evaluation of cryo-FEGSEM data of nodules at different physiological states we could reject the hypothesis that oxygen flow is controlled by the water content of the extracellular spaces (Weisbach *et al.*, 1999).

18.2.5 Cryo-planing

In order to obtain better orientation in the large nodules and in other large samples, we did cryo-planing of either the plunge frozen (Figure 18.5) or high-pressure frozen (Figures 18.6

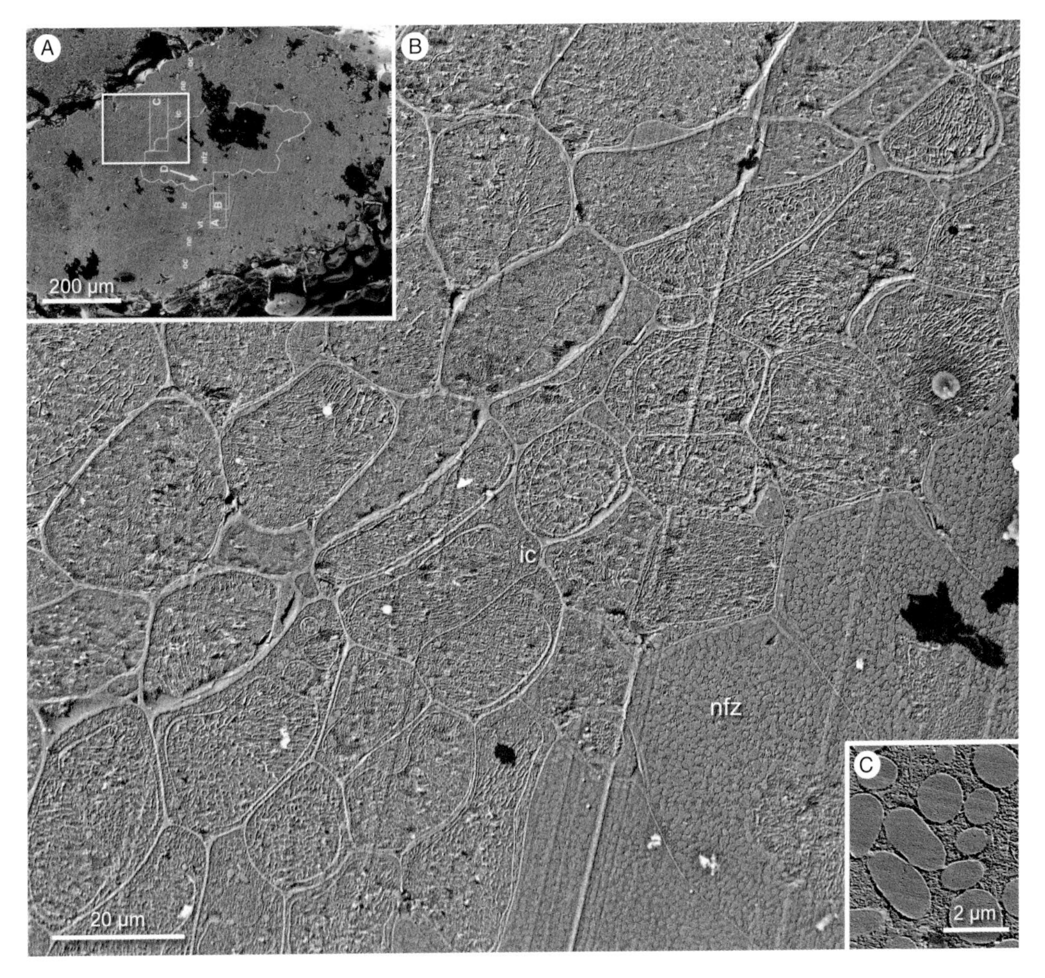

Figure 18.5 Block-face of a cryo-planed lucerne root nodule. (A) A complete root nodule with a surface of more than 1 mm^2 has been cryo-planed. The planed face is very smooth. At higher magnifications (B is the framed area of A) the different tissue layers become visible and the inner cortex cells (ic) can be discerned from the nitrogen fixing zone (nfz) where the cells contain the symbiotic bacteriae. (C) An enlargement of the nitrogen fixing zone with bacteria (from Weisbach *et al.*, 1999).

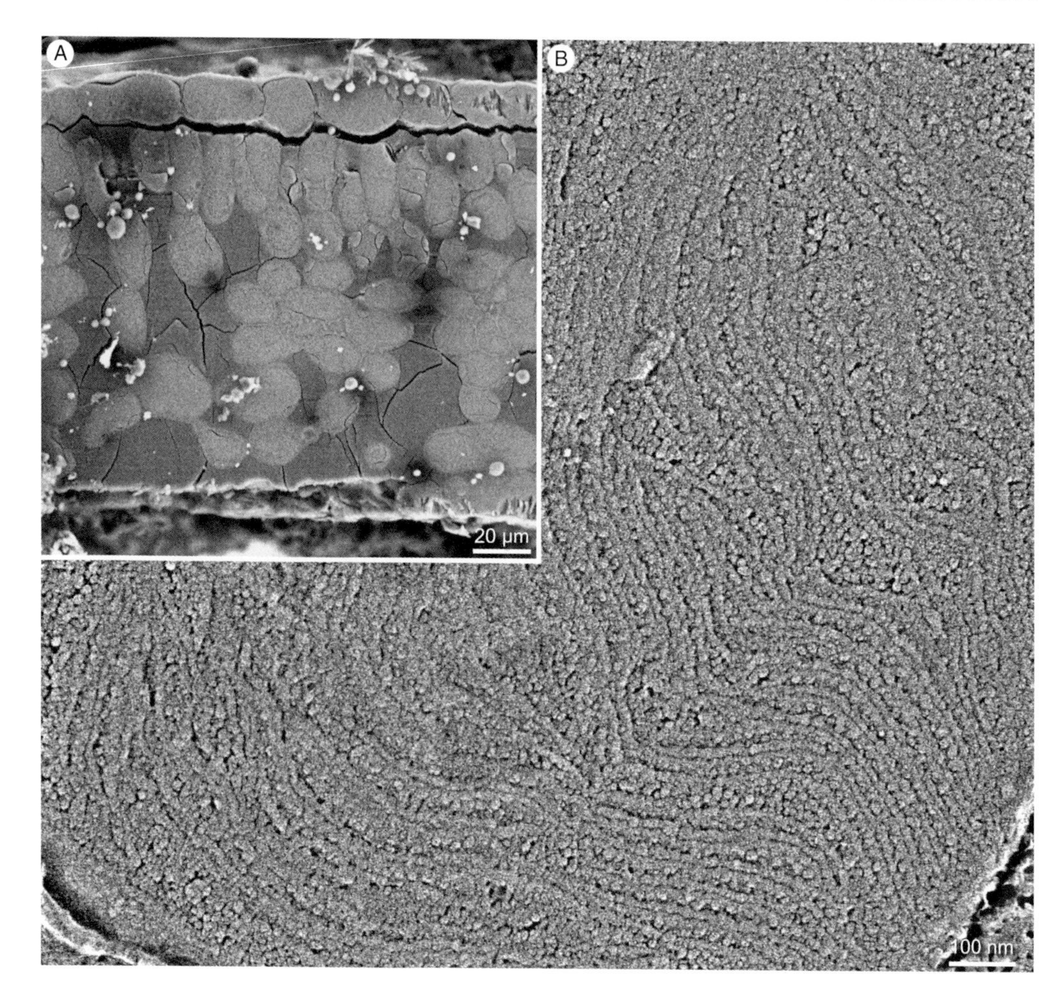

Figure 18.6 Cryo-planed apple leaf. (A) Cross-section through the whole leaf. The gaseous filled extracellular spaces had been filled with hexadecene to prevent collapse during high-pressure freezing. They appear slightly darker than the cells. (B) A portion of a chloroplast. The membrane stacks with the attached globular proteins become very visible (from Walther and Müller, 1999).

and 18.7) samples. Freeze-fracturing is limited because the path of the fracture plane cannot be predicted. The fracture plane preferentially (but not always) follows the inner, hydrophobic zone of the lipid bilayers (Branton, 1966). In a complex system, such as a cell, the fracture plane jumps in an unpredictable way from one bilayer to another. This sometimes impedes interpretation and especially morphometric analysis (e.g. Egelhaaf *et al.*, 1995). Therefore, an alternative method was investigated, namely to look at block faces of frozen and cryo-planed samples with the cryo-SEM (Walther and Müller, 1999). For this purpose the frozen samples were mounted in a cryo-ultramicrotome and were then trimmed with a cryo-diamond knife. It had been shown that cryo-sections of natively cryo-fixed biological material can be obtained (Zierold, 1982; Michel, Hillman and Müller, 1991; Al-Amoudi and Frangakis, 2013; and others). It is to be expected that the remaining block face contains similar biological information, since it contains the same biological material as the cryo-sections. As can be seen in Figures 18.5, 18.6 and 18.7, the block face is very smooth. When imaged at

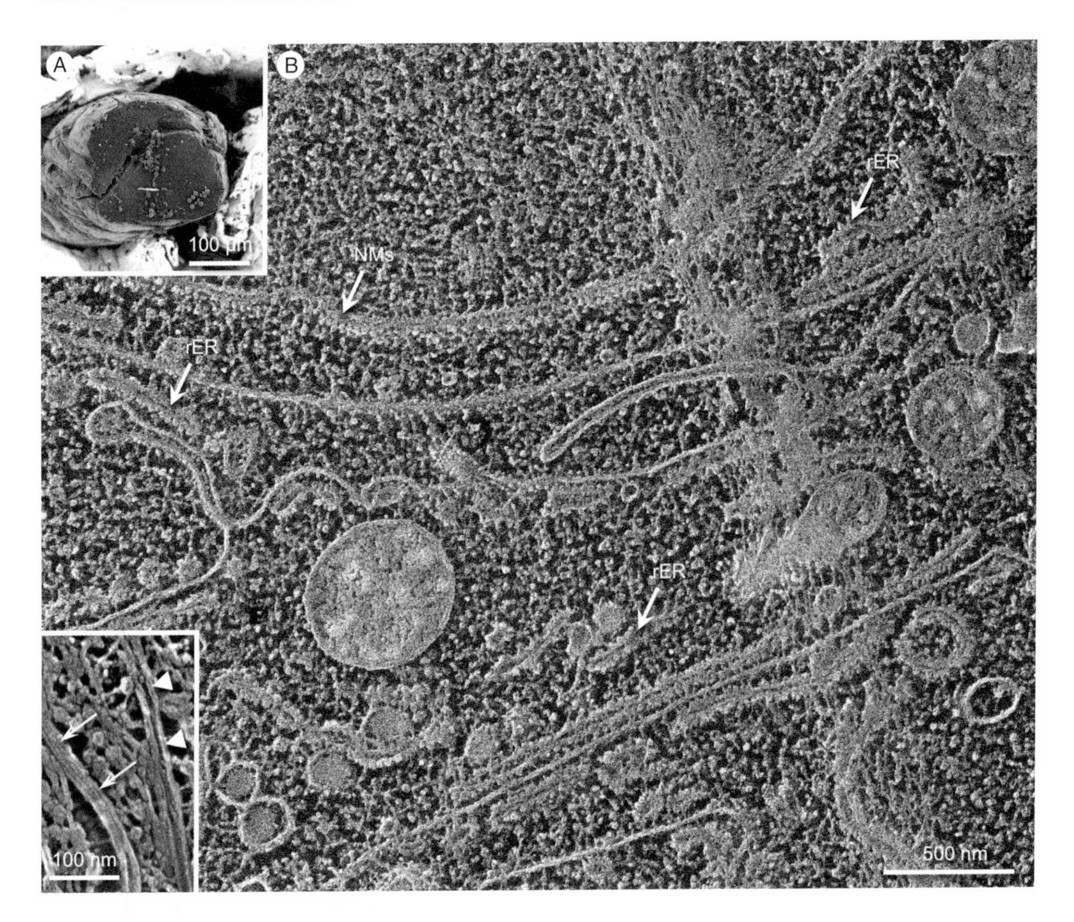

Figure 18.7 Larvae of the parasitoid wasp (*Hissopus pallidus*) after high-pressure freezing produces a flat block face of the whole larvae (A) by cryo-planing. The ultrastructure becomes visible at high magnification (B), showing nuclear membranes (NMs) and rough endoplasmic reticulum (rER). (C) An even higher magnification of the same sample showing two microtubules. The arrows point to an intact microtubule where the protofilaments are visible and the arrowheads point to a microtubule that had been cut in half during cryo-planing (from Walther and Müller, 1999).

higher magnifications, however, ultrastructural details become visible. The block face does not show the typical artefacts of cryo-sections such as crevasses and compression. It is also possible to obtain reasonably good block faces of samples that are not vitrified, but show ice crystal formation, as demonstrated in Figure 18.5, which is a plunge frozen sample. In addition, the size of the sample did not influence the quality of the block face, although it is recommended to use very small sized trimming pyramids for cryo-sections (Michel *et al.*, 1991). The lucerne root nodule imaged in Figure 18.5A, as an example, had a diameter of almost 1 mm!

A high-pressure frozen and cryo-planed apple leaf is shown in Figure 18.6. The cavities filled with air in the living leaf had been filled with hexadecene to allow for high-pressure freezing fixation. This frozen hexadecene is shown somewhat darker than the cells in Figure 18.6A. Figure 18.6B is a high magnification of a chloroplast in the same sample. Spherical proteins can be clearly identified on the membrane stacks.

Figure 18.7 is a larvae of the parasitoid wasp *Hissopus pallidus* after high-pressure freezing and cryo-planning. The ultrastructural details, especially the membranes, are very visible on the cryo-planed block face.

18.2.6 Partial Freeze Drying

In some cases the structures of interest are exposed to the surface before freezing. In order to prevent drying artefacts, however, they need to be covered with liquid that freezes during cryo-fixation. It then needs to be removed by sublimation at cold temperature in the vacuum. This process is usually referred to as partial freeze-drying or freeze-etching. In our hands partially dried samples were less affected by beam-induced damages and therefore the double layer coating approach was not necessary. We coated our freeze-dried samples with a layer of tungsten (averaged thickness about 1 nm) by electron beam evaporation (at a temperature of about –100 °C and a pressure of about 10^{-6} mbar) since, due to the high melting point, it produces coating layers with small grain sizes. Since the samples presented here were thin and adsorbed on silicon chips, charging was not an issue. Samples were then cryo-transferred to the SEM and imaged at a temperature of –100 °C and an accelerating voltage of the primary beam of 30 keV, detecting the secondary electron signal. By investigating the inner nuclear membrane of *Xenopus* oocytes we investigated the nuclear baskets as described by Ris (1991, 1997) and by Goldberg and Allen (1992). In these samples we also observed filamentous structures. When prepared by partial freeze-drying, it became visible that these filaments consisted of global subdomains that formed a right turned helix that perfectly fitted the structure of freeze-dried F-actin (Figure 18.8B). We therefore concluded that these structures must be nuclear F-actin (Figure 18.8A). This observation fits with findings by Bohnsack *et al.* (2006) that 'A selective block of nuclear actin export ... ' leads to actin accumulation in the nucleus that might polymerize and ' ... stabilizes the giant nuclei of *Xenopus* oocytes'.

18.3 DISCUSSION AND OUTLOOK

With cryo-FEGSEM, water containing bulk organic samples can be imaged at a resolution in the range of a few nm. In recent work we could show that even imprints of nanopatterned silicon surfaces *in pure water* can be visualized with cryo-SEM (Wiedemann *et al.*, 2013).

Recent developments in life science electron microscopy are correlative light and electron microscopy (e.g. Kukulski *et al.*, 2012) and 3D imaging approaches either by FIB-SEM tomography or by serial block face SEM tomography (Leighton, 1981; Denk and Horstmann, 2004). Possibly some insights from cryo-SEM could help to further develop these emerging fields. In our hands the fluorescence signal of dyes has been better retained in preliminary experiments when the samples were freeze-dried instead of classical critical point drying, which requires dehydration with organic solvents. Cryo-FIBSEM tomography has already been established (Schertel *et al.*, 2013). A fascinating question is whether cryo serial block face SEM could work. Two promising aspects of cryo-block face SEM are that the size of the samples can be relatively large (almost 1 mm in Figure 18.5) and that vitrification is not necessarily required. Major problems to solve are that the samples need to be electrically conductive. It would not be very feasible to do a coating after each sectioning cycle. Therefore, an alternative way of imaging the frozen block face must be found for cryo serial block face tomography.

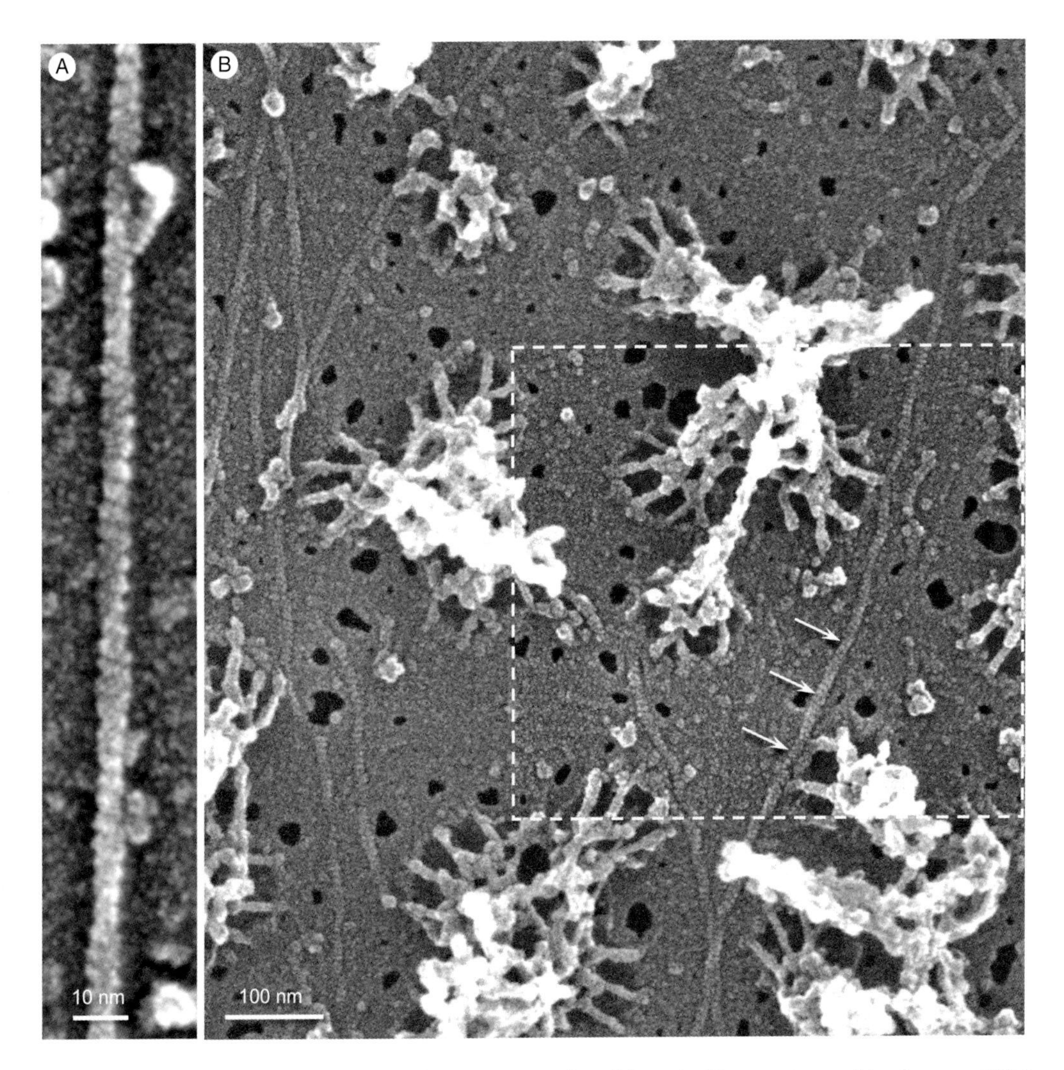

Figure 18.8 Actin in the cryo-SEM. (A) *In vitro* produced F-actin filament imaged in the cryo-SEM after partial freeze drying and coating with 1 nm tungsten. The helical arrangement of the subunits is clearly visible. (B) The inner nuclear membrane of a *Xenopus* oocyte after partial freeze-drying and coating with 1 nm tungsten. Besides the nuclear baskets of the nuclear pores, many filaments with a diameter of less than 10 nm are visible. The framed area was recorded with smaller pixel sizes and the globular subunits and their arrangement in a right-handed helix (arrows) becomes clearly visible. We concluded, therefore, that they represent nuclear F-actin (from Walther, 2008).

18.4 MATERIALS AND METHODS

In Figures 18.1, 18.2 and 18.3, Baker's yeast (*Saccharomyces cerevisiae*) were propane-jet frozen between two low-mass copper platelets (Müller *et al.*, 1980). These sandwiched samples were fractured in a freeze-etching device (BAF 300 from Bal-Tec, Principality of Liechtenstein, now Leica Microsystems, Vienna, Austria) at a vacuum of 1×10^{-7} mbar and a temperature of 123 K. Immediately after fracturing, the samples were coated by electron beam evaporation with about 2 nm of platinum–carbon (at an angle of 45°; unidirectional for Figures 18.1 and 18.3 and rotary coated for Figure 18.2) and with about 7 nm of carbon at an angle of 80°. During carbon evaporation the samples were rotated in order to obtain a more uniform coat. The cold samples were immediately put back into liquid nitrogen, where one of the copper platelets was mounted on a cryo-holder 626 (Gatan, Inc., Pleasanton, CA, USA) and cryo-transferred into an S-900 in-lens field emission SEM (Hitachi, Tokyo, Japan). During transfer the specimen was shielded by the shutter of the Gatan-holder. Specimens were investigated at 113 K. The primary accelerating voltage (V_0) was 30 keV for high magnifications and 10 keV for overviews. The beam current was 1–3 times 10^{-11} A, as measured with a Faraday cage. The backscattered electron images were recorded with a sensitive annular YAG-detector (Autrata, Hermann and Müller, 1992), except for Figure 18.1A, which is a secondary electron image.

Figure 18.4 shows the cryo-fracture of root nodules of lucerne (*Medicago sativa* L.). The plant roots were prepared under a water saturated N_2-stream at room temperature immediately after harvesting. The upper part of one undetached nodule (including the nitrogen-fixing zone and the meristem) was cut off, mounted with gum arabicum (0.2 g/ml H_2O) on an aluminium platelet (diameter 3 mm) and plunged into liquid ethane. The time from harvest to freezing of the root nodules was approximately 10 s. The frozen samples were afterwards fractured with the microtome of the BAF 300 and double layer coated as described above and imaged in the frozen state in the Hitachi S-900 FEGSEM with the backscattered electron signal as described for Figures 18.1 to 18.3 (Weisbach *et al.*, 1999)

Figures 18.5, 18.6 and 18.7 show cryo-planed samples. Root nodules (Figure 18.5) of lucerne (*Medicago sativa* L.) were frozen as described for Figure 18.4. Apple leaves (*Malus domestica*, variation Golden Delicious; Figure 18.6) and larvae of *Hissopus pallidus* (Figure 18.7) were high-pressure frozen in aluminium planchettes (diameter 3 mm with a central cavity 2 mm in diameter and 200 μm deep), as described by Studer, Michel and Müller (1989). The extracellular airways of the apple leaves were filled with hexadecene by applying a mild vacuum of about 20 mbar using a water jet pump as described by Michel, Gnägi and Müller (1991). The cavities between larvae and aluminium planchettes were also filled with hexadecene. The brittle frozen samples were then mounted in a special holder by clamping between two 150 μm thick indium foils. Cryo-sectioning was performed in a Reichert UC4 cryo-ultramicrotome (now distributed by Leica Microsystems, Vienna, Austria) using a cryo-diamond knife (Diatome AG, Biel, Switzerland) as described by Michel *et al.* (1992). The best results were obtained when the slowest possible sectioning speed was used (0.05 mm/s). When using faster sectioning speeds, for example 3 mm/s, smearing of the sectioned face in the cutting direction was observed. The bulk sample with its smooth block face was then cryo-transferred in liquid and gaseous nitrogen to a freeze-etching unit (Balzers BAF 301) warmed up to 163 K so that some ice sublimed from the surface, and then double layer coated as described for Figure 18.3. The block faces were then transferred and imaged as described for Figure 18.3 by using the Gatan cryo-stage and by recording the backscattered electron signal in a Hitachi S-900 FEGSEM.

For Figure 18.8, the F-actin (Figure 18.8A) and the inner nuclear membrane of *Xenopus laevis* oocyte stage VI samples (Figure 18.8B) were prepared as described in Walther (2008). The nuclei of the oocytes were attached on to silicon chips, the nuclear envelopes were rinsed only once very briefly and gently with isotonic buffer to remove the content of the nuclei and then transferred immediately into isotonic buffer containing 4% formaldehyde and 2% glutaraldehyde and afterwards frozen by plunge-freezing. The frozen samples were mounted on a holder that fits into the Gatan cryo-holder and cryo-transferred to a BAF 300 freeze-etching device (Bal-Tec, Principality of Liechtenstein). The samples were partially freeze-dried for 35 min at about 183 K and then rotary coated at the same temperature by electron beam evaporation with 1.5 nm of tungsten. Afterwards samples were mounted on a Gatan cryo-holder 626 (Gatan, Inc., Pleasanton, CA, USA) and transferred into an S-5200 in-lens FEGSEM (Hitachi, Tokyo, Japan). Specimens were investigated at a temperature of 173 K. The beam current was about $1\text{–}3 \times 10^{-11}$ A. The primary accelerating voltage (V_0) was 30 keV in order to use a beam diameter as small as possible (Hermann and Müller, 1991). Imaging was performed by collecting the secondary electron signal. Because the nuclear membranes represent a thin sample that was lying on the silicon support, electrical conductivity was good, so that double layer coating, as recommended for freeze-fracturing work (Walther, 2003a), was not necessary for these samples.

ACKNOWLEDGEMENTS

Most of the data presented in this work had been achieved during the author's employment at the Laboratory for Electron Microscopy 1, ETH Zürich and are part of his habilitation thesis. The help and support of his supervisor and mentor Martin Müller is gratefully acknowledged.

REFERENCES

Al-Amoudi, A. and Frangakis, A.S. (2013) Three-dimensional visualization of the molecular architecture of cell–cell junctions in situ by cryo-electron tomography of vitreous sections. *Methods Mol. Biol.*, 961, 97–117.

Autrata, R., Hermann, R. and Müller, M. (1992) An efficient single crystal BSE detector in SEM. *Scanning*, 14, 127–135.

Bohnsack, M.T., Stüven, T., Kuhn, C., Cordes, V.C. and Görlich D. (2006) A selective block of nuclear actin export stabilizes the giant nuclei of *Xenopus* oocytes. *Nat. Cell Biol.*, 8, 257–263.

Branton, D. (1966) Fracture faces of frozen membranes. *Proc. Natl. Acad. Sci. USA*, 55, 1048–1056.

Branton, D., Bullivant, S., Gilula, N.B., Karnovsky, M.J., Moor, H., Muhlethaler, K., Northcote, D.H., Packer, L., Satir, B., Satir, P., Speth, V., Staehlin, L.A., Steere, R.L. and Weinstein, R.S. (1975) Freeze-etching nomenclature. Science 190, 54–56.

Crewe, A.V., Eggenberger, D.N., Wall, J. and Welter, L.M. (1968) Electron gun using a field emission source. *Rev. Sci. Instrum.*, 39, 576–583.

Denk, W. and Horstmann, H. (2004) Serial block-face scanning electron microscopy to reconstruct three-dimensional tissue nanostructure. *PLoS Biol.*, November, 2 (11), e329.

Dubochet, J. (2012) Cryo-EM – The first thirty years. *J. Microsc.*, 245, 221–224.

Dubochet, J. and McDowell, A.W. (1981) Vitrification of pure water for electron microscopy. *J. Microsc.*, 124, 3–4.

Echlin, P. (1992) *Low-Temperature Microscopy and Analysis*, Plenum Press, New York and London.

Egelhaaf, S.U., Wehrli, E., Müller, M., Adrian, M. and Schurtenberger, P. (1995) Determination of the size distribution of lecithin liposomes: A comparative study using freeze fracture cryoelectron microscopy and dynamic light scattering. *J. Microsc.*, 184, 214–228.

Fernandez-Moran, H. (1960) Low temperature preparation techniques for electron microscopy of biological specimens based on rapid freezing with helium II. *Ann. NY Acad. Sci.*, 85, 689–713.

Fujimoto, K. (1995) Freeze-fracture replica electron microscopy combined with SDS digestion for cytochemical labeling of integral membrane proteins. Application to the immunogold labeling of intercellular junctional complexes. *J. Cell Sci.*, 108, 3443–3449.

Goldberg, M.W. and Allen, T.D. (1992) High resolution scanning electron microscopy of the nuclear envelope: Demonstration of a new, regular, fibrous lattice attached to the baskets of the nucleoplasmic face of the nuclear pores. *J. Cell Biol.*, 119, 1429–1440.

Gross, H., Bas, E. and Moor, H. (1978) Freeze fracturing in ultrahigh vacuum at –196 °C. *J. Cell Biol.*, 76, 712–728.

Hermann, R. and Müller, M. (1991) High resolution biological scanning electron microscopy: A comparative study of low temperature metal coating techniques. *J. Electron Microsc. Tech.*, 18, 440–449.

Hermann, R. and Müller, M. (1993) Progress in scanning electron microscopy of frozen-hydrated biological specimens. *Scanning Microsc.*, 7, 343–349.

Hohenberg, H.H., Müller-Reichert, T., Schwarz, H. and Zierold, K. (2003) Foreword, Special Issue on High Pressure Freezing. *J. Microsc.*, 212, 1–2.

Höhn, K., Sailer, M., Wang, L., Lorenz, L., Schneider, E.M. and Walther, P. (2011) Preparation of cryofixed cells for improved 3D ultrastructure with scanning transmission electron tomography. *Histochem. Cell Biol.*, 135, 1–9.

Joy, D.C. and Pawley, J.B. (1992) High-resolution scanning electron microscopy. *Ultramicrosc.*, 47, 80–100.

Kamasawa, N., Furman, C.S., Davidson, K.G., Sampson, J.A., Magnie, A.R., Gebhardt, B.R., Kamasawa, M., Yasumura, T., Zumbrunnen, J.R., Pickard, G.E., Nagy, J.I. and Rash, J.E. (2006) Abundance and ultrastructural diversity of neuronal gap junctions in the OFF and ON sublaminae of the inner plexiform layer of rat and mouse retina. Neuroscience, 142, 1093–1117.

Knoll, M. (1935) Aufladepotential und Sekundäremission elektronenbestrahlter Körper. *Z. Tech. Phys.*, 16, 467–475.

Knoll, G. and Brdiczka, D. (1983) Changes in freeze-fractured mitochondrial membranes correlated to their energetic state. Dynamic interactions of the boundary membranes. *Biochim. Biophys. Acta*, 733, 102–110.

Krivanek, O.L., Chisholm, M.F., Murfitt, M.F. and Dellby, N. (2012) Scanning transmission electron microscopy: Albert Crewe's vision and beyond. *Ultramicroscopy*, 123, 90–98.

Kühlbrandt, W., Zeelen, J. and Dietrich, J. (2002) Structure, mechanism, and regulation of the *Neurospora* plasma membrane H^+-ATPase. *Science*, 297, 1692–1696.

Kukulski, W., Schorb, M., Kaksonen, M. and Briggs, J.A. (2012) Plasma membrane reshaping during endocytosis is revealed by time-resolved electron tomography. Cell. 150, 508–520.

Lang, R.D.A. and Bronk, J.R. (1978) A study of rapid mitochondrial structural changes *in vitro* by spray-freeze-etching. *J. Cell. Biol.*, 77, 134–147.

Leighton, S.B. (1981) SEM images of block faces, cut by a miniature microtome within the SEM – A technical note. *Scan. Electron Microsc.*, 1981, 73–76.

Michel, M, Gnägi, H. and Müller, M. (1992) Diamonds are a cryosectioner's best friend. *J. Microsc.*, 166, 43–56.

Michel, M., Hillmann, T. and Müller, M. (1991). Cryosectioning of plant material frozen at high pressure. *J. Microsc.*, 163, 3–18.

Mielanczyk, L., Matysiak, N., Michalski, M., Buldak, R. and Wojnicz, R. (2014) Closer to the native state. Critical evaluation of cryo-techniques for transmission electron microscopy: Preparation of biological samples. *Folia. Histochem. Cytobiol.*, 52, 1–17.

Moor, H. and Mühlethaler, K. (1963) Fine structure in frozen-etched yeast cells. *J. Cell Biol.*, 17, 609–628.

Moor, H. and Riehle, U. (1968) Snap-freezing under high-pressure: A new fixation technique for freeze-etching. *Proc. 4th European Reg. Conf. on Electron Microscopy*, 2, 33–34.

Müller, M., Meister, N. and Moor, H. (1980) Freezing in a propane jet and its application in freeze-fracturing. *Mikroskopie*, 36, 129–140.

Pawley, J.B. and Erlandsen, S.L. (1989) The case for low voltage high resolution scanning electron microscopy of biological samples. *Scanning Microsc.*, 3, 163–178.

Peters, K.-R. (1982) Conditions required for high quality high magnification images in secondary electron-i scanning electron microscopy. *Scanning Electron Microscopy*, IV, 1359–1372.

Peters, K.-R. (1986) Rationale for the application of thin, continuous metal films in high magnification electron microscopy. *J. Microsc.*, 142, 25–34.
Reimer, L. (1978) Scanning electron microscopy – Present state and trends. *Scanning*, 1, 3–16.
Ris, H. (1991) The three-dimensional structure of the nuclear pore complex as seen by high voltage electron microscopy and high resolution low voltage scanning electron microscopy. *EMSA Bull.*, 21, 54–56.
Ris, H. (1997) High-resolution field-emission scanning electron microscopy of nuclear pore complex. *Scanning*, 19, 368–375.
Schertel, A., Snaidero, N., Han, H.M., Ruhwedel, T., Laue, M., Grabenbauer, M. and Möbius, W. (2013) Cryo FIB-SEM: Volume imaging of cellular ultrastructure in native frozen specimens. *J. Struct. Biol.*, 184, 355–360.
Seiler, H. (1967) Einige aktuelle Probleme der Sekundärelektronenemisison. *Z. Angew Phys.*, 22, 249–263.
Steere, R.L. (1957) Electron microscopy of structural detail in frozen biological specimens. *J. Biophys. Biochem. Cytol.*, 3, 45–60.
Studer, D., Michel, M. and Müller, M. (1989) High-pressure freezing comes of age. *Scanning Microsc.*, 3 (Suppl. 3), 253–268.
Studer, D., Michel, M., Wohlwend, M., Hunziker, E.B. and Buschmann, M.D. (1995) Vitrification of articular cartilage by high-pressure freezing. *J. Microsc.*, 179, 321–332.
Talmon, Y. (1987) Electron beam radiation damage to organic and biological cryo-specimens, Chapter 3 in *Cryotechniques in Biological Electron Microscopy* (eds R.A. Steinbrecht and K. Zierold), Springer Verlag, Berlin, pp. 64–84.
Thornley, R.F.M. (1960) Recent developments in scanning electron microscopy, in *Proceedings of EUREM*, pp. 173–176.
Tokuyasu, K.T. (1973) A technique for ultracryotomy of cell suspensions and tissues. *J. Cell Biol.*, 57, 551–565.
von Ardenne, M. (1938) Das Elektronen-Rastermikroskop. Praktische Ausführung. *Z. Tech. Phys.*, 19, 407–416.
von Ardenne, M. (1996) Reminiscences on the origins of the scanning electron microscope and the electron microprobe. *Advances in Imaging and Electron Physics*, 96, 635–652.
Walther, P. (2003a) Cryo-fracturing and cryo-planning for in-lens cryo-SEM, using a newly designed diamond knife. *Microscopy and Microanalysis*, 9, 279–285.
Walther, P. (2003b) Recent progress in freeze fracturing of high-pressure frozen samples. *J. Microsc.*, 212, 34–43.
Walther, P. (2008) High-resolution cryo-SEM allows direct identification of F-actin at the inner nuclear membrane of *Xenopus* oocytes by virtue of its structural features. *J. Microsc.*, 232, 379–385.
Walther, P. and Hentschel, J. (1989) Improved representation of cell surface structures by freeze substitution and backscattered electron imaging. *Scanning Microsc.*, 3 (Suppl. 3), 201–211.
Walther, P. and Müller, M. (1999) Biological ultrastructure as revealed by high resolution cryo-SEM of blockfaces after cryo-sectioning. *J. Microsc.*, 196 (3), 279–287.
Walther, P. and Ziegler, A. (2002) Freeze substitution of high-pressure frozen samples: The visibility of biological membranes is improved when the substitution medium contains water. *J. Microsc.*, 208, 3–10.
Walther, P., Hentschel, J., Herter, P., Mueller, T. and Zierold, K. (1990) Imaging of intramembranous particles in frozen hydrated cells (*Saccharomyces cerevisiae*) by high resolution cryo SEM. *Scanning*, 12, 300–307.
Walther, P., Wehrli, E., Hermann, R. and Müller, M. (1995) Double layer coating for high resolution low temperature SEM. *J. Microsc.*, 179, 229–237.
Weisbach, C., Walther, P., Hartwig, U.A. and Nösberger, J. (1999) Electron microscopical investigation of water occlusions in intercellular spaces in the inner cortex of lucerne nodules. *J. Struct. Biol.*, 126, 59–71.

Wiedemann, S., Plettl, A., Walther, P. and Ziemann, P. (2013) Freeze fracture approach to directly visualize wetting transitions on nanopatterned superhydrophobic silicon surfaces: More than a proof of principle. *Langmuir*, 29, 913–919.

Zhu, Y., Inada, H., Nakamura, K. and Wall, J. (2009) Imaging single atoms using secondary electrons with an aberration-corrected electron microscope. *Nature Materials*, 8, 808–812.

Zierold, K. (1982) Preparation of biological cryosections for analytical electron microscopy. *Ultramicrosc.*, 10, 45–53.

19

Preparation of Vitrified Cells for TEM by Cryo-FIB Microscopy

Yoshiyuki Fukuda[1,2], **Andrew Leis**[3] **and Alexander Rigort**[1,4]

[1]*Department of Structural Biology, Max Planck Institute of Biochemistry, Am Klopferspitz 18, 82152 Martinsried, Germany*
[2]*Department of Cell Biology and Anatomy, Graduate School of Medicine, The University of Tokyo, 7-3-1 Hongo, Bunkyo-ku, Tokyo 113-0033, Japan*
[3]*Bio21 Molecular Science and Biotechnology Institute, The University of Melbourne, 30 Flemington Road, Parkville, Victoria 3010, Australia*
[4]*Thermo Fisher Scientific, FEI Deutschland GmbH, Fraunhoferstrasse 11B, 82152 Martinsried, Germany*

19.1 INTRODUCTION

Studying the molecular machinery of cells from atomic detail to the cellular context and beyond is a great challenge for cell biology. It requires an understanding beyond the operation of individual proteins towards whole, dynamic and stochastically variable systems. Remarkably little is known about the supramolecular organization of molecular machines within the cell, yet this organization is of critical importance for all biochemical pathways to function efficiently. Dealing with this complexity requires innovative imaging tools and manipulation devices and methods coming not only from biophysics and systems biology but also from other fields of science and engineering.

Many seminal discoveries in cell biology have been underpinned by breakthroughs in the physical sciences and, in particular, electron microscopy (EM), which played a key role in gaining fundamental insights into cellular organization and ultrastructure. Developments such as the vitrification of samples (Dubochet et al. 1988) allowed faithful preservation of molecular and macromolecular structures in their native hydrated environment. Recent years have witnessed rapid progress in methodological and technical developments for transmission electron microscopy (TEM) of vitreous specimens (for reviews, see Lucic et al. 2013; Yahav et al. 2011). Biochemically isolated and purified molecular complexes or conveniently

Biological Field Emission Scanning Electron Microscopy, First Edition.
Edited by Roland A. Fleck and Bruno M. Humbel.

small samples such as the majority of 'cell-free' viruses can be vitrified directly as a thin film; here, the cross-section is sufficiently small so as to offer useful resolution. A major limitation for studying intact cells by electron tomography is that only a limited sample thickness range (< 0.5–1 μm) is accessible with today's intermediate voltage electron microscopes. Sample thinning is a way to overcome this restriction, and here the development of the microtome (Porter and Blum 1953) and later its application to suitable frozen-hydrated samples, termed vitreous sectioning (McDowall et al. 1983), was arguably the most important development towards enabling the preparation of sufficiently thin samples. Nevertheless, mechanical deformations caused by cryo-sectioning – non-uniform compression and crevassing – represent a substantial limitation of this approach (Al-Amoudi et al. 2005) and attempts to eliminate or at least minimize these artefacts have been only moderately successful. An alternative to mechanical sectioning with a diamond knife became available with the focused ion beam (FIB) microscope. In contrast to ultramicrotomy, the FIB system ablates the sample using a beam of heavy ions rather than mechanical shear, thereby leaving the specimen unaffected by compression forces. In combination with cryogenic instrumentation, FIB technology can be used to precisely modify the thickness of frozen-hydrated samples for electron tomography studies. Importantly, it fulfils the key prerequisite of maintaining the amorphous form of the vitreous water (Marko et al. 2006). Providing controlled access to the cell interior opens up new opportunities for performing structural studies *in situ* within otherwise unperturbed cellular environments (Figure 19.1). Only a few systems lend themselves to such analysis without thinning. The best examples are suspensions of viruses and other colloidal systems (Danino 2012; Hryc et al. 2011). Monodisperse systems also lend themselves to the preparation of quite uniform ice layers that are consistent in

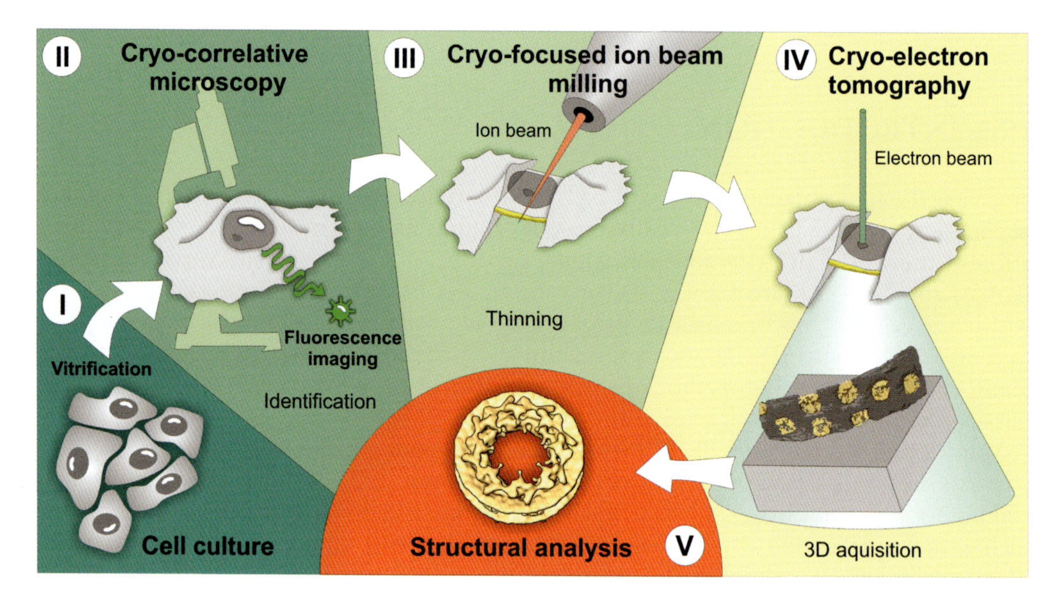

Figure 19.1 Schematic representation of the workflow for performing structural biology *in situ*. (I) Cell culture. Various cell types can be grown on or deposited on EM grids. Cryo-fixation by vitrification ensures that the cells are preserved in their native, hydrated environment. (II) Cryo-correlative microscopy. Assists in identifying and navigating to the area of interest. (III) Cryo-FIB milling. A focused ion beam removes frozen cellular material above and below a target region and leaves behind an electron-transparent lamella. (IV) Cryo-electron tomography. A three-dimensional dataset is acquired from the electron-transparent lamella. (V) Structural analysis. Computational methods are used for the structural characterization of macromolecular complexes (example shown is a nuclear pore complex). Reproduced from Rigort (2016).

cross-section. More complex and variable examples include the relatively thin regions of neurons (Asano et al. 2015), thinner regions of a select few cell lines (Brandt et al. 2010; Koning et al. 2008), some prokaryotes (Brandt et al. 2009) and exceptionally small, unicellular eukaryotes (Henderson et al. 2007). Clearly, a universal approach to fabricating uniform sample geometries is required. The FIB system has the potential to mature into a key instrument within the workflow for cellular structural biology. Its ability to precisely target and prepare electron-transparent volumes that harbour molecular machines within their functional context is becoming increasingly important for ultrastructural imaging in cell biology, analogous to the impact that micro- and nanomachining techniques have had on the progress of materials science (Overwijk et al. 1993).

This chapter provides a basic introduction to the operating principle of the FIB instrument and its use in the biological sciences, followed by a brief overview of the technical requirements for working with vitrified biological specimens. We introduce a procedure for the preparation of electron-transparent lamellae from vitrified cells and show how this method facilitates tomographic studies on any chosen part of the cell. We conclude with a discussion on the role of the cryo-FIB microscope for performing structural studies *in situ* and its fit within an integrated workflow for structural biology (Figure 19.1).

19.2 OPERATING PRINCIPLE OF THE FIB INSTRUMENT

19.2.1 Interactions of the Ion Beam and Sample – Sputtering and Milling

Site-specific removal of material with an FIB system is achieved by generating and focusing a stream of high-energy ionized atoms of a relatively 'heavy' (high mass) element on to the specimen surface (for a detailed introduction to the FIB technique see Giannuzzi and Stevie 2005; Yao 2007). The impacting ions can expel surface layer atoms from their positions by collisions in a process known as sputtering. Typically, Ga^{+} is preferred because it also fulfils the requirement for the practical manufacture of a liquid–metal ion source (LMIS, see Section 19.2.2). Progressively scanning the ion beam over the specimen surface allows successive, layer-by-layer removal of material (Figure 19.2a). This bulk removal process is called 'milling'. Both the large size of primary ions and the high kinetic energy (up to 30 keV) give rise to a high momentum, which is transferred to the atoms on the specimen surface during ion–atom collisions. Sputtering occurs when the energy, which is transferred from the ions to the target atoms, leads to a 'collision cascade', causing atoms on or close to the surface to be ejected from the specimen (Figure 19.2a). Ion–atom interactions can be either elastic or inelastic. The large number of scattering events occurring during collision cascades leads to a distribution between elastic and inelastic processes. Elastic or nuclear events result from interactions of energetic ions with nuclei of the target atoms. In the case of inelastic or electronic interactions, the energy of the charged particle is transferred to electrons.

19.2.2 Beam Generation and Shaping in the Ion Column

During FIB milling, a focused beam of ions removes atoms from the specimen surface in a precise and controlled manner. The LMIS is located at the top of the ion beam column (Figure 19.2b). Ions are extracted from this source via the process of field evaporation and are focused into a beam by electrostatic lenses. A typical LMIS consists of a capillary tube, which houses the reservoir of heavy-metal atoms, and a tungsten needle with a sharp tip that runs through it. The reservoir is heated to near evaporation, causing the liquid metal to flow along the needle and wet the tip. An extraction electrode, called the extractor, is

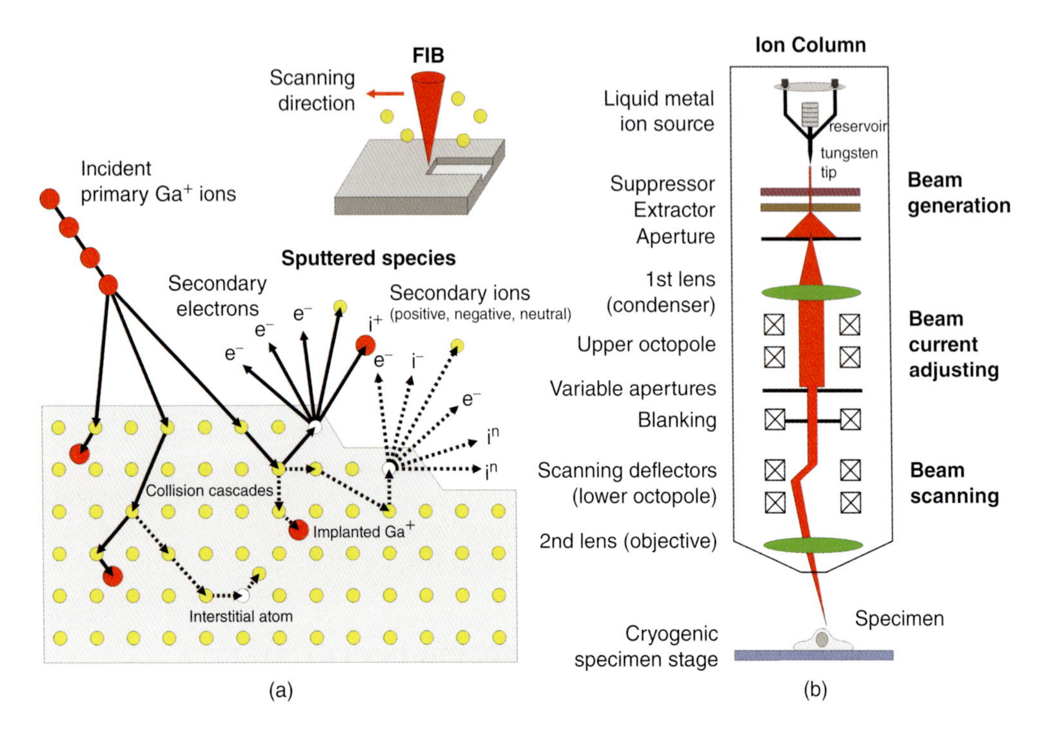

Figure 19.2 Milling principle and ion beam generation in the FIB column. (a) Scheme depicting the effects of focusing a high-energy ion particle beam onto a target surface. Multiple ion–atom collisions occur. When a surface atom gains a kinetic energy momentum that is sufficient to overcome the surface binding energy of the target material, it is ejected as a sputtered species. Scanning the beam multiple times over the target surface leads to progressive removal of material (inset in (a)). (b) Operating principle of an FIB column. The ion beam arises from a liquid metal ion source at the top of the column. It is condensed and focused onto the specimen by a set of electrostatic lenses and deflection coils. Adapted from Rigort and Plitzko, 2015 (Rigort and Plitzko 2015), Elsevier Rights.

positioned close to the tip and produces a strong electrostatic force, which causes the liquid metal meniscus to form a sharply peaked cone, known as a Taylor cone. Positively charged ions are pulled out of the cone apex by the electric field (up to +9 kV) and are accelerated towards the extractor. The ion source is operated at low extraction currents of 1–3 μA to maintain a stable beam and to reduce the energy spread of the beam. In order not to alter the source tip surface condition and for the overall stability of the FIB system, it is important to keep the extraction current stable. This is achieved by biasing the voltage of the suppressor electrode, usually operating with an applied electric field of up to +2 kV.

Before the charged particle beam is scanned over the sample surface it must be focused within the ion column to the desired spot size by a series of apertures and electrostatic lenses, as shown in Figure 19.2b. A spray aperture ('Aperture', Figure 19.2b) limits the 'brightness' of the generated beam and blocks the majority of ions for the case where their respective velocity vectors are not pointed towards the direction of the beam. The ions are then focused and collimated into a parallel beam by the upper (condenser) lens. Inside the ion column, electrostatic lenses are used because magnetic lenses would need to be impractically large to provide enough power for focusing the massive gallium ions.

Beam astigmatism is adjusted by the upper octopole lens, by eliminating the ions that are not directed vertically. The beam current is adjusted by beam-limiting apertures (using a variable aperture mechanism). For cryo-FIB applications typical beam energies are between

10 and 30 keV and beam currents vary from 1 pA up to 10 nA. Smaller aperture sizes are used in conjunction with lower beam currents; these are suitable for milling sensitive samples (e.g. cryo specimens). In contrast, bigger apertures allow selection of higher beam currents, which can be used for faster milling and removal of bulk material. To limit the beam from reaching the specimen, it can be deflected from the optical axis. This is accomplished by the blanking deflector and aperture. Scanning the beam over the sample in a user-defined pattern is achieved by the beam scanning deflectors of the lower octopole lens. Finally, the lower (objective) lens is used for focusing and reducing the spot size of the beam, enabling imaging resolutions in the sub 5 nm range.

Various metal (e.g. Al, Zn, In, Cs, Au) or alloy sources (e.g. Si-Be-Au (ion produced: Si) or B-Ni-Pt (ion produced: B), Ge-Au (ion produced: Ge)) can be used as the sputtering source in an FIB system (for further reading, see Orloff et al. 2003). However, the manufacture of such sources can be difficult due to the physical properties of the materials (e.g. very high melting points, reactivity at high temperatures and high vapour pressure at melting points). Alloy sources can facilitate the use of such materials, owing to their lower melting points and/or lower vapour pressure, but they require an additional mass separator in the ion column to allow only ions with a fixed mass-to-charge ratio to pass. Gallium (Ga^+) is the preferred and most widespread used ion species due to its low melting point, volatility and vapour pressure. These properties allow easier handling and operation, as lower temperatures are required for heating the source and keeping it in the liquid phase during operation. The low volatility at the melting point ensures that the supply of liquid metal lasts longer, increasing source lifetime.

19.2.3 Ion Beam-Induced Damage and Artefacts

Sputtering the sample by colliding high energy Ga^+ ions onto its surface atoms causes various damage effects to occur. Concerning frozen-hydrated biological specimens, two damage mechanisms are of particular importance and must be controlled: local heating via the transfer of thermal energy to the interior of the specimen, and the implantation of Ga^+ ions into the near-surface region of the specimen. Both effects strongly depend on the interplay between the accelerating voltage, ion current, angle of incidence used for milling and the chemical composition of the target. If local heating were to occur, it would cause the formation of damaging ice crystals via a phase transition as the specimen temperature was warmed above the devitrification temperature of water, $-137\,°C$ (Dubochet et al. 1988); however, in pioneering experiments (Marko et al. 2006), it was shown that FIB milling of vitreous ice does not induce heating sufficient to cause such a phase change. The amorphous state of the vitreous ice and its subsequent transformation to a crystalline state via deliberate heating could be verified by selected area electron diffraction.

The penetration depth of ions into the sample is typically restricted to a layer that extends a few tens of nanometres into the milled surface. The size of this Ga^+ implantation zone can be estimated by Monte Carlo (MC) simulations of ion trajectories in matter using the SRIM software package (Ziegler et al. 2010). The ranges for penetration of 30 keV ions into vitreous ice (assuming a target density of 0.93 g/cm^3; Karuppasamy et al. 2011) are represented schematically in Figure 19.3. The most significant implantation depth is achieved when directing the ion beam perpendicular to the specimen surface. Using 30 keV ions, the theoretical depth of the ion–sample interaction volume can reach 70 nm (Figure 19.3a). This can be minimized by further reducing the ion impact angle. Selecting grazing angles of incidence (i.e. 80° or below from the grid surface normal) drastically decreases the volume

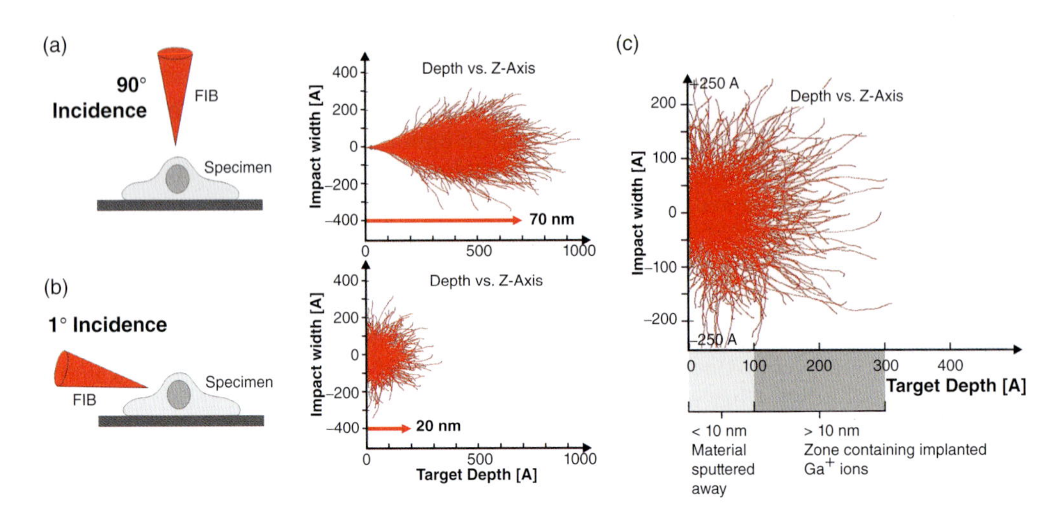

Figure 19.3 Monte Carlo simulation of impacting gallium ions. Monte Carlo simulations (Ziegler et al. 2010) (SRIM software) of 30 keV Ga^+ ion trajectories in the target 'vitreous water' (assumed density: 0.93 g/cm^3): (a) incidence angle perpendicular to the surface (90°) and (b) at shallow (1°) incidence angle. The number of trajectories drawn for each plot is 500. (c) Detailed view of the scatterplot from (b). The first 10 nm layer is sputtered away immediately by the progressively scanned ion beam. The remaining 10–15 nm layer represents a zone that contains implanted Ga^+ ions.

affected by direct ion–sample interactions to approximately 20 nm (Figure 19.3b). The MC simulations do not take into account the effect of sputtering. As a matter of fact, energy from the impacting ions that is transferred to near-surface atoms causes these atoms to overcome the surface binding energy of the target material and to escape from the specimen. This leads to a 10 nm surface layer that is immediately sputtered away, while the ion beam is progressively scanned over the specimen surface (Figure 19.3c). A small fraction of Ga^+ ions is retained in the newly created surface and this narrow implantation zone is limited to within approximately 10 nm of the milled surfaces (Marko et al. 2007).

Other ion beam radiation-induced effects concerning mainly materials science specimens (e.g. crystalline materials) are surface amorphization, ion channelling, the production of interstitials, lattice defects and atomic mixing. In addition, a common effect observed in practice is the redeposition of the sputtered material on the milled surfaces. For frozen-hydrated specimens, this can be lessened by positioning an anticontaminator device (a surface that is kept at colder temperatures than the specimen itself) close to the vitrified sample.

19.3 CRYO-FIB APPLICATIONS IN BIOLOGY

Focused ion beam technology has been used in the materials science and semiconductor industries for more than a decade, mainly for failure analysis, circuit edit and mask repair (Giannuzzi and Stevie 1999; Volkert and Minor 2007). Due to its unsurpassed site-specific preparation abilities, FIB milling has become a routine method for sample preparation in this discipline (Kirk et al. 1989; Reyntjens and Puers 2001).

Combining a FIB instrument with a scanning electron microscope (SEM) allows simultaneous monitoring of the milling results with the electron beam and, more importantly, relatively gentle (non-destructive) imaging of the specimen surface, as compared to ion beam imaging. Nowadays, a number of two-beam (FIB-SEM) systems from various manufacturers (e.g. JEOL, Hitachi, Japan; TESCAN, Czech Republic; Zeiss, Germany; FEI, USA) are available. Some of these systems are branded as *cross-beam* (Zeiss), *dual-beam* (FEI) or *multibeam* (JEOL) microscopes. The combination of an imaging tool with a precision machining tool expanded the possibilities considerably and opened up new opportunities not only for materials science but also for biological applications (Ballerini et al. 2001). Thus, the combination with the SEM allowed the acquisition of three-dimensional (3D) information from resin-embedded biological samples by sequentially imaging the milled blockfaces via detection of either secondary electron or backscattered electron signals (see also Chapter 24 by Giannuzzi and Chapter 26 by Kizilyaprak *et al.*). This is distinct from the TEM tomography/lamella approach described in this chapter. In these so-called 'FIB-SEM tomography' (Holzer and Cantoni 2012) or 'slice and view' (Heymann et al. 2006; Knott et al. 2008) applications, the layer-by-layer removal of material is accomplished by FIB milling. Although the resolution provided by this technique is currently modest compared to TEM, substantially larger areas can be explored. Recently, FIB-SEM tomography was applied to high-pressure frozen mouse optic nerve tissue and allowed the visualization of cellular structures with the in-lens secondary electron (SE) detector (Schertel et al. 2013). An interesting alternative to the FIB-SEM approach is the so-called Denk microscope (Denk and Horstmann 2004), where an ultramicrotome housed within the microscope chamber is used to remove material sequentially, analogous to the function of the ion beam. Although cryo-ultramicrotomes are used to furnish vitreous sections, to our knowledge the Denk technique has not been used for frozen-hydrated specimens. The knife would require cooling so as not to devitrify the specimen during cutting.

Applying the FIB technique to frozen-hydrated biological specimens requires hardware and protocols adapted to cryogenic needs (see the next section). The first successful cryo-FIB experiment on biological specimens was performed in 2003 using a cryo-stage specifically adapted to the geometric requirements of an FIB-SEM system (Mulders 2003). It demonstrated that milling into the frozen substrate to create site-specific cross-sections can be carried out in a straightforward manner and that the obtained cross-sections show site-specific information. However, no effort was made to determine the state of the ice, either before or after FIB milling. This was done three years later in a study investigating whether cryo-FIB thinning of frozen-hydrated specimens causes devitrification (Marko et al. 2006). The results presented in this study proved that the ice was vitreous prior to commencement of FIB milling and that milling of vitreous ice does not induce heating sufficient to cause devitrification. Consequently, these findings led the way to the first TEM cryo-tomographic experiments of vitrified biological material (Marko et al. 2007; Rigort et al. 2010) and to further developments, such as the *in situ* cryo-FIB-lamella preparation technique for plunge frozen cells on EM grids (Rigort et al. 2012a), described in this chapter (and applied also by Wang et al. 2012 and Engel et al. 2015). Other studies demonstrated the preparation of thin lamellae by FIB milling of high-pressure frozen samples (Hayles et al. 2010) and sought to distinguish vitreous from crystalline ice in high-pressure frozen samples by recording transmission electron backscatter diffraction (EBSD) patterns (de Winter et al. 2013).

19.4 INSTRUMENTATION FOR CRYO-FIB MILLING

Cryo-preparation systems enable the transfer of vitrified specimens onto a stable cryo-stage within the high vacuum chamber of the FIB-SEM instrument for further observation or manipulation (Figure 19.4). A number of challenges are faced when applying the cryo-FIB technique to frozen-hydrated material. The specimen must remain below the devitrification temperature of approximately −137 °C at all times, while significant heating during milling

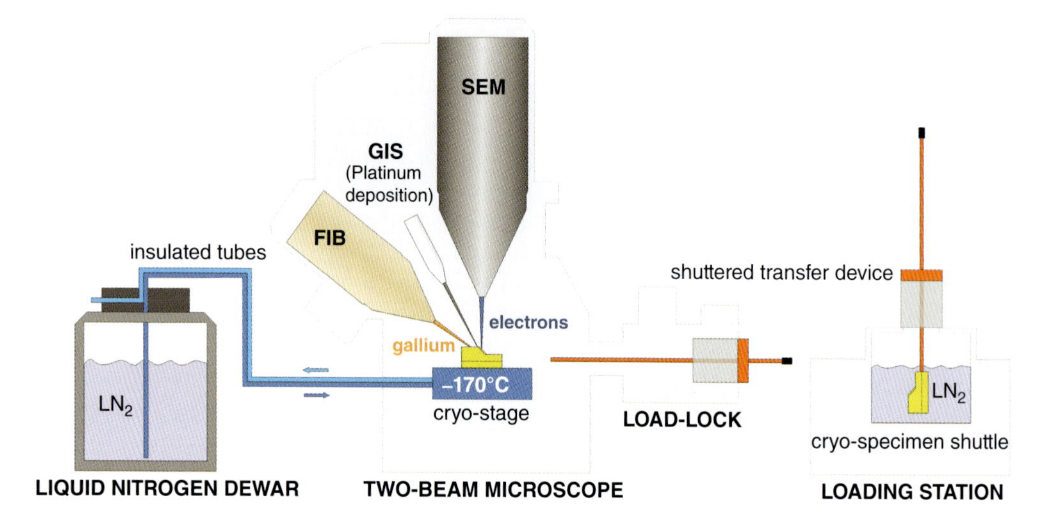

Figure 19.4 Instrumentation for cryo-FIB milling. Schematic illustration of the cryogenic setup for an FIB microscope. In an external loading station, vitrified specimens are mounted under liquid nitrogen into a cryo-transfer shuttle. The shuttle is transferred to the high-vacuum chamber of the two-beam microscope by a shuttered transfer device, which is docked to a load-lock system. A nitrogen-cooled cryo-stage takes up the cryo-shuttle. The focused ion beam is used for milling the frozen-hydrated specimens. Adapted from Rigort *et al.*, 2012c, Elsevier Rights.

(e.g. by using high milling currents) must be avoided. In addition, the sample must remain free from frost or other contaminants during the multiple transfers of EM grids. This is especially an issue for projects that require cryo-correlative microscopy, which adds further manipulation steps to the cryogenic workflow (see Figure 19.1) and thus further risk of contamination or inadvertent warming.

Vitrified samples are then mounted in a cryo-transfer holder ('cryo-shuttle') in an external loading station and transferred to the high vacuum chamber of the FIB-SEM instrument via a load-lock and transfer system. Inside the FIB-SEM microscope, a thermally isolated, cooled cryo-stage is attached to the FIB-SEM stage and an anticontaminator device (the cold trap) is positioned in close vicinity to this stage to prevent contamination of the sample with hydrocarbons. Depending on the system used, cooling is achieved directly by purging cold nitrogen gas through the cryo-stage (e.g. Quorum Polarprep system) or indirectly by using copper braided tapes (e.g. Leica VCT system). The latest generation of these commercial cryo-systems comes with options for automatic refilling of nitrogen, which will allow extended runs and avoid vibrations resulting from manual refilling of the Dewar vessels. During milling experiments, the cryo-stage is operated at temperatures of −150 °C to −170 °C, while the anticontaminator device is maintained at a temperature of approximately −180 °C. For high-precision milling, a reliable and stable cryo-stage is mandatory, as mechanical vibrations and thermal drift effects can adversely affect the milling result. Moreover, proper accessibility of the specimen with respect to the incident ion beam is of critical importance. The use of a cryo-holder with a pre-tilted specimen uptake device facilitates milling at grazing angles of incidence, in order to minimize potential ion beam-induced damage effects (Rigort et al. 2010). Once the milling experiment is completed, the cryo-transfer holder is retrieved with the transfer rod and returned to the loading station for downstream imaging (e.g. electron cryo-tomography). A variety of cryo-transfer holders is commercially available. These comprise holders for EM grids or carriers for high-pressure frozen samples.

19.5 THE MANUFACTURE OF *IN SITU* CRYO-TEM LAMELLAE FROM VITRIFIED CELLS ON EM GRIDS

The FIB-SEM instrument is used to prepare frozen-hydrated cells for cryo-TEM investigations. Figure 19.5 displays an SEM micrograph of frozen-hydrated *Dictyostelium discoideum* cells attached to the carbon support film of an EM grid. The cells are embedded in vitreous ice, which becomes apparent during TEM imaging. Operating the two-beam instrument in the SEM mode allows imaging and screening for a target cell or specific region thereof, and to select it for thinning by FIB milling (Figure 19.6a). Imaging with the SEM is performed at accelerating voltages of 1–10 keV. Milling takes place by focusing 30 keV gallium ions at beam currents of 30–50 pA directed at shallow tilt angles on to the specimen. For the preparation of an *in situ* lamella, a rectangular sector below and above the target region is drawn computationally (within the graphical user interface of the FIB control software) to define the milling region, and the sectors are then exposed to the rastered gallium beam (Figure 19.6b and c). Sputtering of frozen-hydrated material within the milling patterns leaves behind a thin lamella supported on both sides by the remaining bulk ice. The thin lamella, cut out from the cellular volume, is electron-transparent and its geometry is suitable for imaging by electron cryo-tomography (Figure 19.6d and Figure 19.7).

19.5.1 FIB Protocol for Cryo-TEM Lamella Preparation

The *in situ* cryo-FIB-thinning method is applicable to all kinds of eukaryotic cells that can be grown on biocompatible EM specimen supports (carbon films on gold bars). To achieve vitrification, the cells are subjected to 'plunge' freezing by rapid immersion into a secondary cryogen (e.g. liquid ethane) using a vitrification device (Dubochet et al. 1988; Dubochet and McDowall 1981; Frederik et al. 2009). Once frozen, EM grids with vitrified cells are

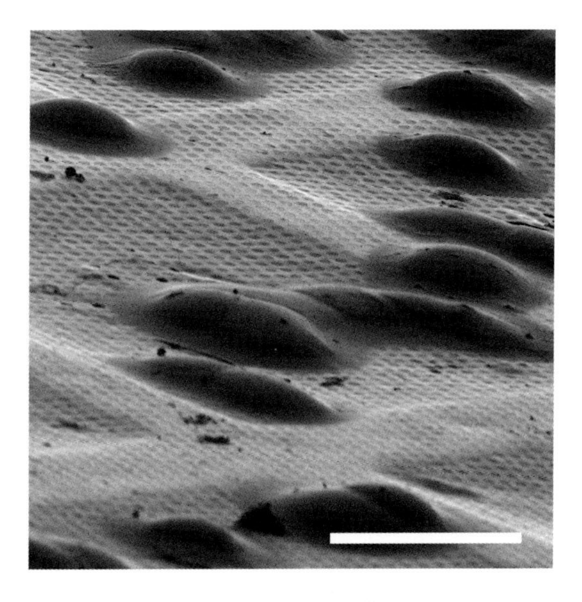

Figure 19.5 Vitrified eukaryotic cells. SEM micrograph of cells embedded in vitreous ice and attached to the carbon support film of an EM grid. The cells were prepared by plunge freezing. Scale bar: 30 μm. Adapted from Rigort *et al.*, 2012c, Elsevier Rights.

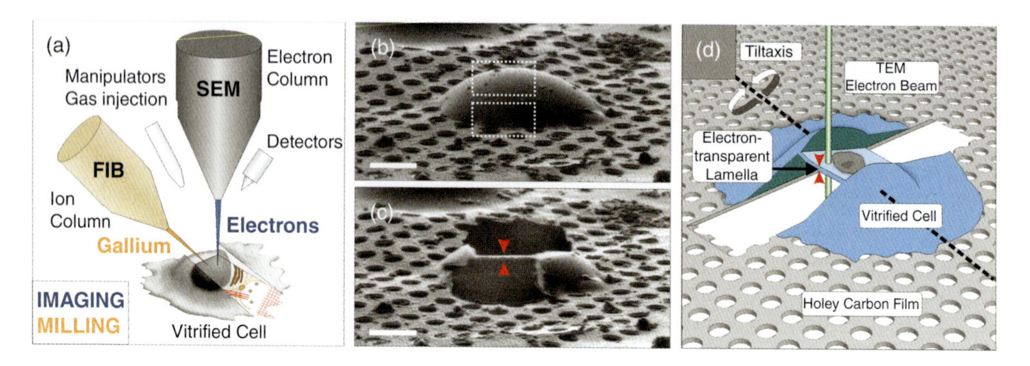

Figure 19.6 Manufacture of thin lamellae by cryo-FIB milling. (a) Thinning of the frozen-hydrated cellular sample is achieved in a process termed 'milling', i.e. sputtering the unwanted material with a focused beam of gallium ions. The electron beam is used to guide the process as well as for non-destructive imaging. (b) Cryo-SEM micrograph of a vitrified cell adhered to a holey carbon film on an EM grid. The rectangular patterns indicate the milling regions. Vitreous lamellae are cut at a shallow angle to the supporting grid. (c) Same region after ion milling. Frontal view of the resulting lamella (red arrowheads) in the milling direction. (d) 3D scheme depicting the principle of the *in situ* lamella milling approach. Lamellae extending over tens of micrometres of cellular space can be prepared by FIB milling. Scale bars: (b and c) 6 μm. Adapted from Lučić, Rigort and Baumeister, 2013, Rockefeller University Press Rights.

mounted inside a loading station into a dedicated cryo-holder and are transferred onto the cooled (−150 °C to −170 °C) cryo-stage within the FIB-SEM microscope. Target cells are identified from the secondary electron image generated by the electron beam. The selection of target regions can be further facilitated by cryo-correlative light microscopy, by localizing features of interest via fluorescently labelled molecules (Rigort et al. 2010; Sartori et al. 2007; Schwartz et al. 2007; van Driel et al. 2009).

The basic idea of the *in situ* milling approach is to prepare cellular samples for TEM investigations directly on the EM specimen grid, avoiding the necessity for a further manipulation step involving, for example, lift-out of the delicate and thin lamellae. Such lift-out approaches are commonly used in materials science (Giannuzzi and Stevie 2005; Yao 2007) and involve the preparation of thin, electron-transparent lamellae by FIB milling, followed by their lift-out with a micromanipulator device and placement on an EM grid, such that the TEM beam is normal to the cross-sectional face of the lamella. Lift-out approaches are difficult to realize with samples that must be kept below the devitrification temperature ($<$ −137 °C) because the manipulator also requires suitable cooling. Nevertheless, approaches for realizing lift-out at cryogenic temperatures have been reported (Rubino et al. 2012; Mahamid et al. 2015). Thinning the specimen directly on the EM grid has the additional advantage that multiple regions of interest (usually up to 10 for one grid) can be prepared before transfer to the TEM.

Before the milling process is started, the target cell must be properly oriented in order to fulfil the geometrical requirements for shallow or grazing angles of incidence (compare Figure 19.3). Typical angles of incidence for the ion beam are 5° to 15° from the grid surface. Next, an overview image of the designated target cell in the direction of the incident ion beam is acquired by imaging secondary electrons generated by the Ga^+ beam (Figure 19.7d). On this image, two rectangular milling patterns, an upper and a lower pattern, are defined flanking the intended area for lamella preparation. The upper milling pattern (Figure 19.7b) will cause the ion beam to remove material from the (nominal) top

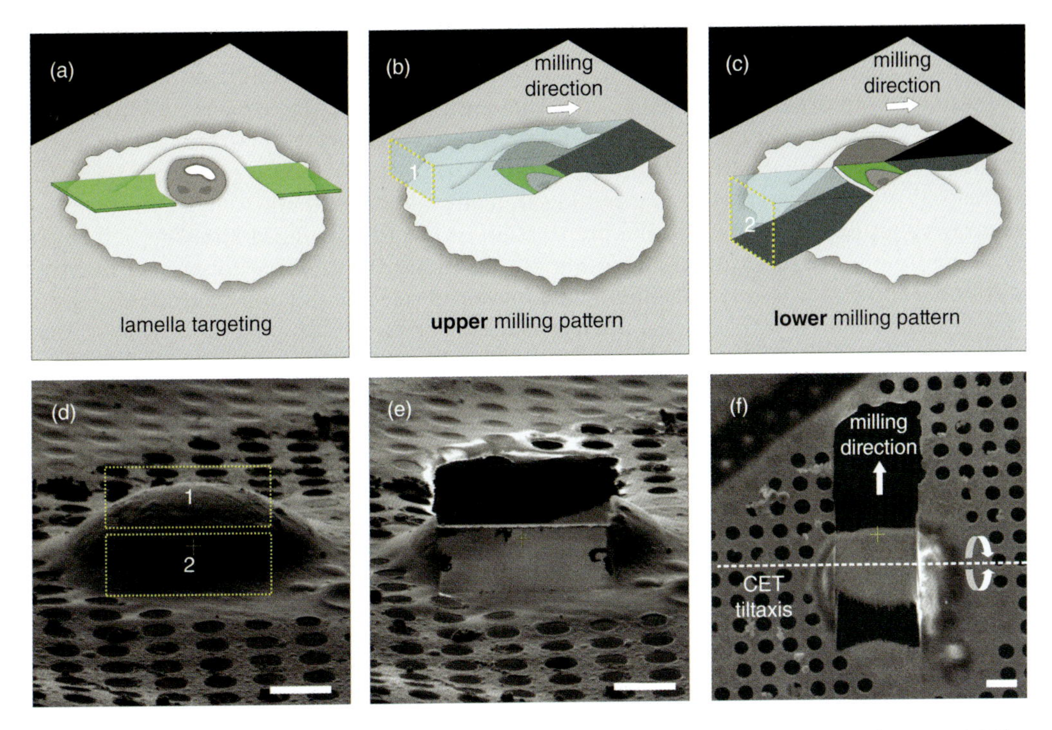

Figure 19.7 Illustration and example of the *in situ* cryo-FIB-lamella preparation technique. (a) Targeting of a region of interest to make it accessible for cryo-TEM. (b) Material removal by ion milling above the target region. The yellow dashed rectangle indicates the milling pattern (1) above the target region. (c) Corresponding milling pattern (2) below the target region. (d) Cryo-SEM micrograph of a vitrified cell on an EM grid. Frontal view in the direction of the incident ion beam. The yellow rectangles correspond to the upper (1) and lower (2) milling patterns (also indicated in b and c). Pattern milling is carried out in parallel mode (by alternately scanning both patterns). (e) Milling result. Thinned target region exposing several square micrometres of cellular space for subsequent investigations by cryo-EM. Frontal view of the ~300 μm thick lamella in the ion beam direction. (f) Top view of the lamella from (e). The white, dashed line indicates the tilt axis for tomographic acquisition in the TEM. Scale bars: (d to f) 4 μm. Reproduced from Rigort and Plitzko, 2015, Elsevier Rights.

side of the cell, as well as part of the carbon support film. The distance between the lower pattern (Figure 19.7c) and the upper pattern governs the final thickness of the lamella. Milling the lower pattern removes part of the carbon film and the cellular material beneath the lamella. The two patterns can be milled in a consecutive or parallel manner, the latter by rapidly rastering the ion beam iteratively across both areas. The final outcome of a thinning experiment can be observed by imaging the top view of the resulting lamella with the electron beam (Figure 19.7f).

The width of the milling patterns is governed by the respective cell dimensions. In order to provide enough lateral support for the lamella, a pattern should not cover more than two-thirds of the cell's visible width. The height of each pattern must cover the entire visible cellular area, to make sure that all cellular material above and below the lamella is removed. The *in situ* cryo-FIB-lamella preparation method can be carried out in two steps to achieve greatest efficiency. In a first step, for coarse milling, an ion beam current of 300 pA is used and a relatively large distance of approximately 1–2 μm between both patterns is selected. The higher currents used during coarse milling allow for a faster removal of bulk frozen-hydrated material. In a second step, fine milling is carried out by placing and redefining the rectangular patterns closer to each other, leaving 200–500 nm between them – the intended final thickness of the lamella. Lowering the milling currents to 30–50 pA minimizes ion beam-induced damage effects on the milled surfaces.

Milling times largely depend on the pattern sizes being used, which in turn are determined by the sizes of the target cells. Coarse milling typically takes 10–20 min, whereas fine milling requires about twice as long. For the preparation of *in situ* cryo-TEM lamellae, cells must be located within the central region of a mesh (compare Figure 19.5) to prevent obstruction of the ion beam by the grid bars and to ensure maximal accessibility for subsequent tomographic imaging in the TEM.

19.6 ELECTRON CYRO-MICROSCOPY AND TOMOGRAPHY OF LARGE FIB-MILLED WINDOWS

19.6.1 Providing Windows into the Cell's Cytoplasm

Cryo-FIB prepared lamellae are transferred into the TEM with a shuttered holder system, which ensures thermal stability and protects the vitrified specimen from frost contamination caused by ambient humidity (Rigort et al. 2012a). A supporting framing structure provides stability to the delicate gold EM grids and prevents damage of the thin ice layers by mechanical stress during the transfer steps (Rigort et al. 2012c). The windows made accessible for cryo-TEM imaging can cover several square micrometres of cellular space. Figure 19.8 shows a 300 nm thin cryo-lamella prepared from a vitrified *Dictyostelium discoideum* cell covering an area of more than 30 μm^2. The thin lamella is supported on both sides by the remaining bulk ice (Figure 19.8a and b). For tomographic imaging, the sample is subjected to tilting in the electron microscope and a 3D density is reconstructed from the projections recorded at successively tilted views (for reviews, see Lucic et al. 2013; Yahav et al. 2011). To maintain accessibility to the regions of interest, even at high tilt angles, the sample is oriented in such a way that the axis of the electron microscope stage is oriented perpendicular to the milling direction. This arrangement also facilitates the setup of automated acquisition, in particular the placement of preparatory states like autofocusing and tracking, which must not overlap with the acquisition/exposure position if signal-to-noise characteristics are to be maximized (Figure 19.8b). Figure 19.8c shows cryo-TEM images recorded under low-dose conditions that were stitched together to obtain an overview image of the lamella. The overview reveals a cytoplasmic region containing numerous organelles. Within the cytoplasm, mitochondria, rough endoplasmic reticulum and vesicular and tubular components of the vacuolar and endosomal systems can be discerned, interspersed with a myriad of macromolecular complexes. The nucleoplasm reveals only little variations in density. It is delineated by the nuclear envelope containing nuclear pore complexes. A tomographic dataset acquired from the border region between nucleoplasm and cytoplasm and a corresponding surface rendering is shown in Figure 19.8d and e. It captures a 0.6 μm^2 portion of the nuclear envelope containing 10 nuclear pore complexes and part of the adjacent cytoplasm.

The *in situ* lamella milling approach allows dissection of regions located deep inside the cytoplasm of cells. In this way, the 3D architecture of cytoskeletal elements (e.g. microtubules) can be investigated by cryo-tomography. In Figure 19.9a, a tomographic

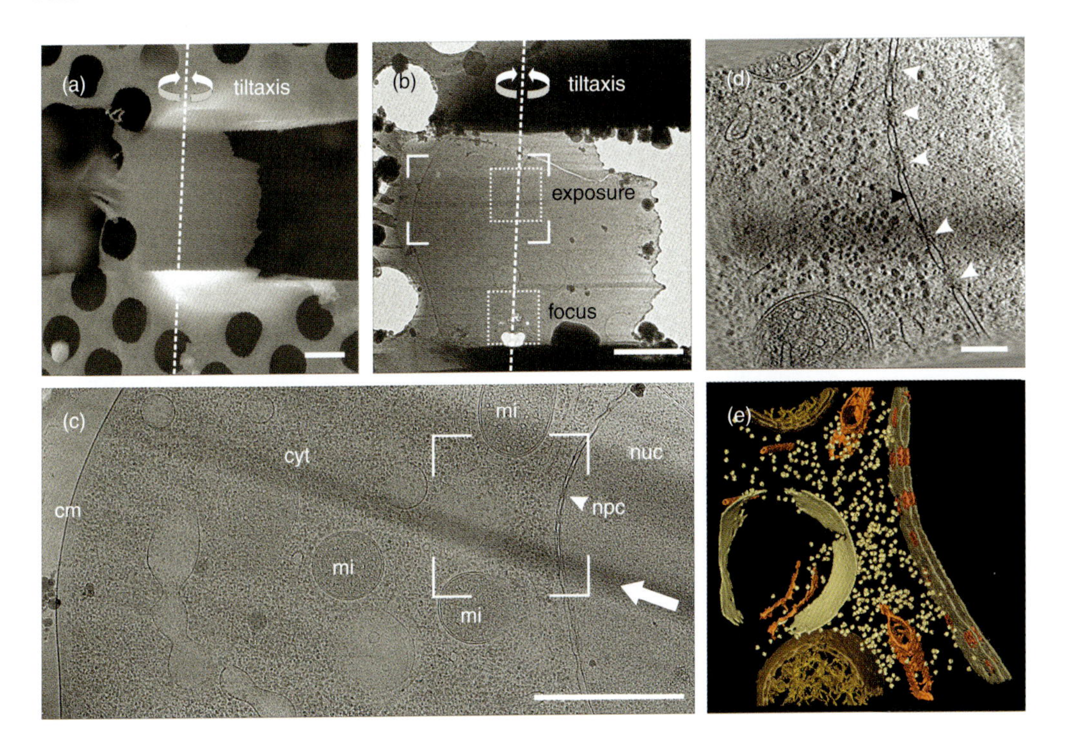

Figure 19.8 Cryo-microscopy and tomography of large FIB-milled windows. (a) SEM top view of a milled lamella across a vitrified *D. discoideum* cell. The lamella is supported on the sides by the remaining bulk ice material. (b) Corresponding TEM micrograph after cryotransfer of the lamella into the TEM; imaged at lower magnification after recording of a tomographic tilt series. The dashed square regions denote the areas for tomographic exposure and focus, distributed along the tilt axis. (c) TEM projection micrographs (corresponding to the framed region in (b)) stitched together to obtain an overview image of the lamella. On the left side, the border region between the ice and cell membrane (cm) can be recognized. Various cross-sectioned mitochondria (mi) with tubular cristae are visible. Within the lamella, the nuclear envelope with nuclear pore complexes (npc) can be clearly discerned, separating the nucleoplasm (nuc) from the cytoplasm (cyt). The white arrow indicates the milling direction and highlights a prominent curtaining streak across the lamella. (d) Slice through the x, y plane of a tomographic reconstruction (area corresponds to the framed region in (c) and exposure region in (b)), showing the nuclear envelope (black arrowhead) with nuclear pore complexes (white arrowheads). (e) Surface rendered visualization of the tomographic volume from (d), displaying the nuclear envelope, endoplasmic reticulum, mitochondria, microtubules, vacuolar compartment and putative ribosomes. Scale bars: (a and b) 2 μm, (c) 1 μm, (d) 200 nm. Reproduced from Rigort and Plitzko, 2015, Elsevier Rights.

slice through an approximately 200 nm thick cytoplasmic volume, traversed by several microtubules, is shown. The microtubules are spatially oriented in transverse and longitudinal directions as well as all possible variants. Accordingly, they appear to be cut length- and cross-wise in the tomographic slice. The corresponding surface-rendered visualization shows their 3D arrangement (Figure 19.9d). Between the cytoskeletal structures, a dense network of interconnected tubular ducts and vesicles can be recognized. The vesicular structures are spherical, as one would expect them to appear in the absence of compression forces typical of microtome sections.

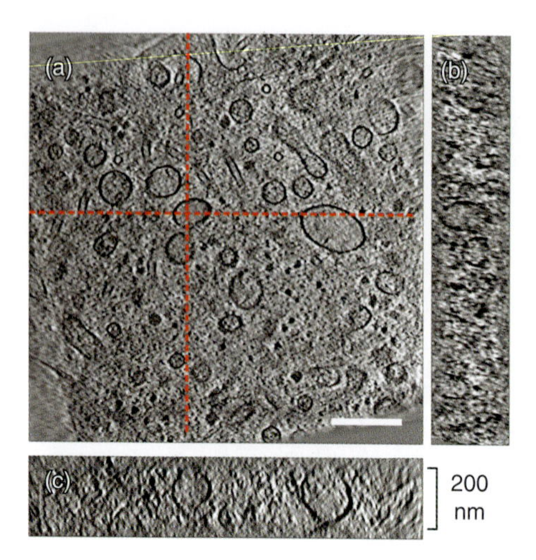

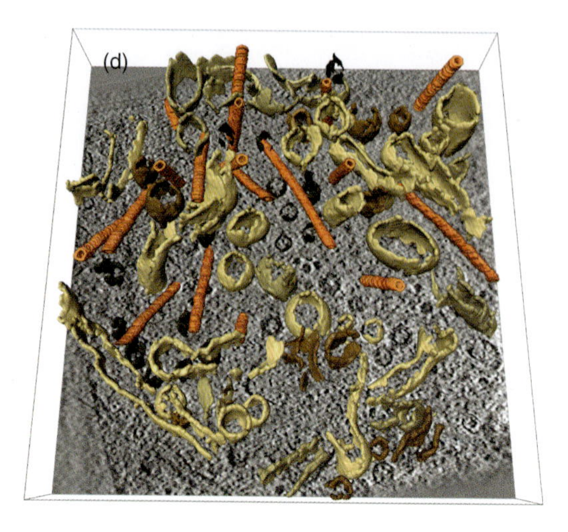

Figure 19.9 Exposing the cytoplasm for tomographic studies. (a) Tomographic slice along the x, y plane showing a cytoplasmic volume traversed by several microtubules. (b and c) Corresponding x, z and y, z planes (sectional planes are indicated by the red dashed lines in (a)). The overall volume has a thickness of approximately 200 nm. (d) Surface-rendered visualization of the tomographic volume, showing microtubules (orange) traversing the volume in various directions. A dense network of interconnecting tubular structures and vesicles can be seen. Scale bars: (a) 200 nm. Adapted from Rigort *et al.*, 2012a, PNAS Rights.

19.6.2 Study of FIB-Prepared Primary Neuronal Cells

In order to investigate synaptic function, it is important to study its structural architecture. Morphological analysis using TEM has contributed fundamentally to our understanding of synaptic structure and function (Harris and Weinberg 2012). However, structural studies of synapses require that the specimen is investigated in a close-to-physiological state under hydrated conditions. Studying synaptic structures derived from primary neuronal cell cultures by cryo-EM is a challenging task, not only because the specimen must be thinned before structural analysis but also due to the delicate nature of the cells and prolonged culturing periods. Neuronal cells investigated here are derived from embryonic or neonatal rat hippocampi. For primary neuronal cell culture, co-cultivation with astroglial cell monolayer as feeder cells supports neuronal cell growth by releasing trophic factors (Kaech and Banker 2006). During cultivation *in vitro*, dissociated neuronal cells extend axons and dendrites, and form synaptic connections with other neuronal cells. A cultivation period of 17–18 days is required for the formation and maturation of the complex synaptic network architecture. For analysis by cryo-EM, primary neuronal cells can be cultured on poly-L-lysine-coated carbon-filmed, standard gold EM grids. For vitrification, the grids are subjected to plunge-freezing as shown previously for *Dictyostelium*.

Figure 19.10 shows cryo-SEM images of vitrified primary neuronal cells cultured on EM grids. The cells exhibit a complex network-like architecture with numerous interconnected neuronal processes. From the central mesh regions shown in Figure 19.10a, a cryo-lamella was prepared (Figure 19.10b). The thinned region proceeds across a neuronal process and allows the investigation of its synaptic structures by cryo-TEM. So far, these structures have been analysed in synaptosomal fractions obtained from crude preparations of neuronal tissue (Fernandez-Busnadiego et al. 2010). Such so-called 'synaptosomes' contain synaptic vesicles, as well as pre- and post-synaptic terminals, and possess the intact molecular

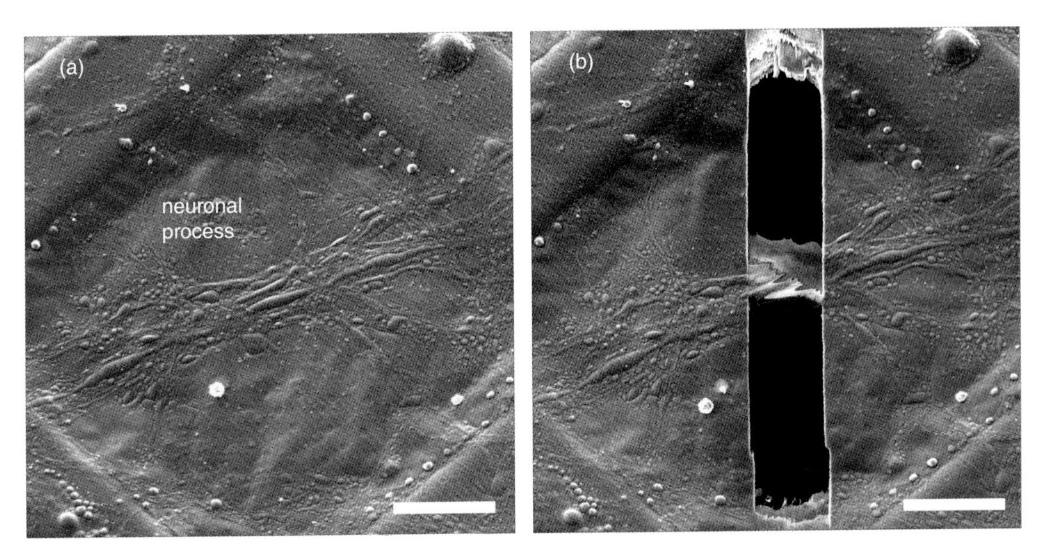

Figure 19.10 Cryo-SEM images of neuronal processes derived from plunge-frozen primary cultured mammalian neuronal cells. (a and b) Cryo-SEM micrographs of rat hippocampal neuronal cells cultured on an EM grid exhibiting neuronal processes before (a) and after (b) lamella milling by cryo-FIB. Scale bars: (a and b) 20 μm.

machinery for synaptic function, including neurotransmitters and functional ion channels. The ability to prepare lamellae with a defined thickness from functional synapses allows higher resolution structures of its molecular constituents to be attained, by identifying macromolecules with the help of computational pattern recognition methods (see the next section).

Synaptic structures from FIB milled neuronal cells, imaged by electron cryo-tomography are displayed in Figure 19.11. In the upper panel tomographic slices from reconstructed tomograms are shown (Figure 19.11a to c). The corresponding 3D segmentation of these tomograms is visualized in the lower panel. Figure 19.11a and b represent neuronal processes containing synaptic vesicles, microtubules and mitochondria. A synapse formed between two neuronal processes is shown in Figure 19.11c. Besides microtubules, mitochondria, synaptic vesicles, smooth endoplasmic reticulum and dense core vesicles, the synaptic cleft and post-synaptic density can be clearly discerned.

19.6.3 *In Situ* Mapping of Macromolecular Complexes

The preparation of self-supporting lamellae from frozen-hydrated cells grown on EM grids facilitates the identification and localization of macromolecules and their subsequent structural analysis by averaging techniques, such as subtomogram averaging and classification (Forster et al. 2005). In subtomogram averaging, a higher resolution structure of a particle (e.g. a protein or a protein complex) is obtained by aligning and averaging 3D subvolumes extracted from tomograms. In a first (identification) step, a cross-correlation function between the tomogram and a template is computed. Here, a generic-shaped template (e.g. a sphere) helps to reduce *a priori* assumptions about the structures of interest. From the corresponding correlation peaks, 3D subvolumes can be extracted and classified using missing-wedge corrected multireference classification procedures. The class averages can

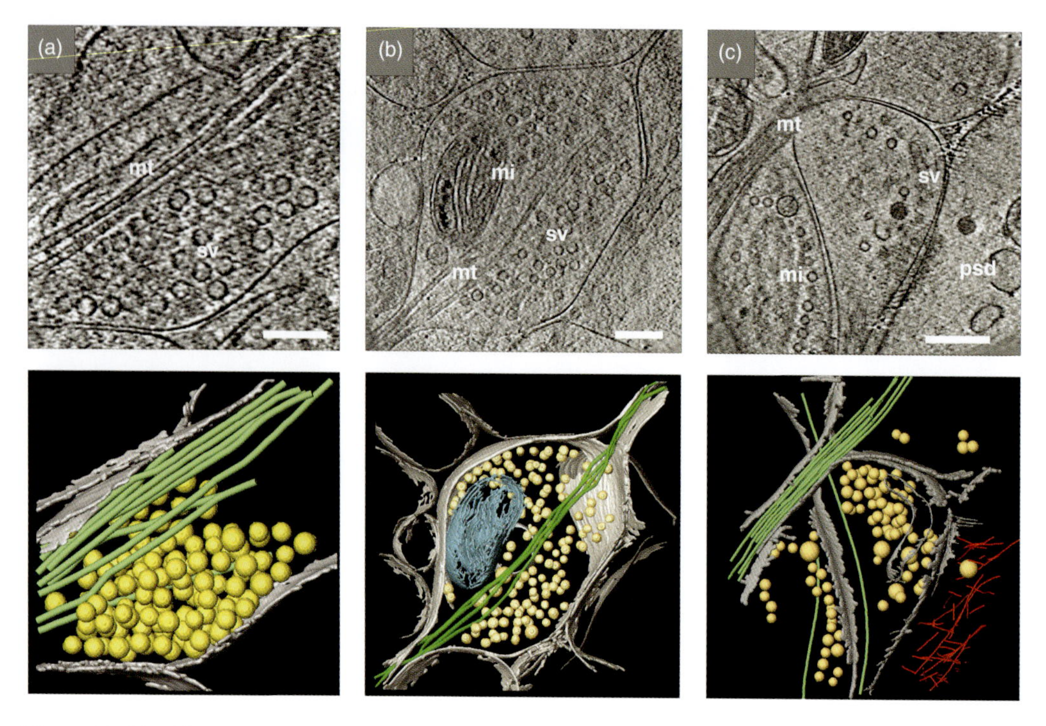

Figure 19.11 Electron cryo-tomography of FIB milled primary cultured neuronal cells. Top: tomographic slice. Bottom: isosurface representation. (a and b) Neuronal process with microtubules and synaptic vesicles. (c) Synapse formed between two neuronal processes. The post-synaptic density (psd) can be clearly discerned. 3D visualization: plasma membrane (grey), microtubules (mt, green), actin (red), mitochondrion (mi, cyan) and synaptic vesicles (sv, yellow). Scale bars: (a, b and c) 200 nm. Adapted from Lučić, Rigort and Baumeister, 2013, Rockefeller University Press Rights.

be further refined and reconstructed to higher resolution structures in an iterative process using the initial template and the data from the class averages. The procedure is conceptually similar to single-particle EM analysis (see Lau and Rubinstein 2013).

So far, approaches aiming to identify individual molecules or protein complexes in tomograms have been restricted to viruses, small prokaryotic cells or very thin regions of eukaryotic cells (Asano et al. 2015; Brandt et al. 2009, 2010; Forster et al. 2005) or had to be performed on specimens that had been thinned by vitreous sectioning (Al-Amoudi et al. 2007; Pierson et al. 2010). Thin sections of vitrified material have allowed investigations of structures located in the thickest regions of intact cells, such as chromatin (Eltsov et al. 2008). One of the most promising results demonstrated so far in applying template matching to vitreous sections revealed the molecular architecture of cadherins in desmosomal contacts (Al-Amoudi et al. 2007). The results showed that vitreous sectioning, although being a challenging method to perform, can obtain images that contain high levels of detail. However, a major drawback of vitreous sectioning is the compression occurring along the cutting direction, an artefact that is difficult to quantify and correct (McDowall et al. 1983; Al-Amoudi et al. 2005; Hsieh et al. 2006). Compression hinders the statistical exploitation of data, especially in cases where merging of multiple datasets is required to increase the amount of particles amenable to statistical analysis. The unpredictable degree and

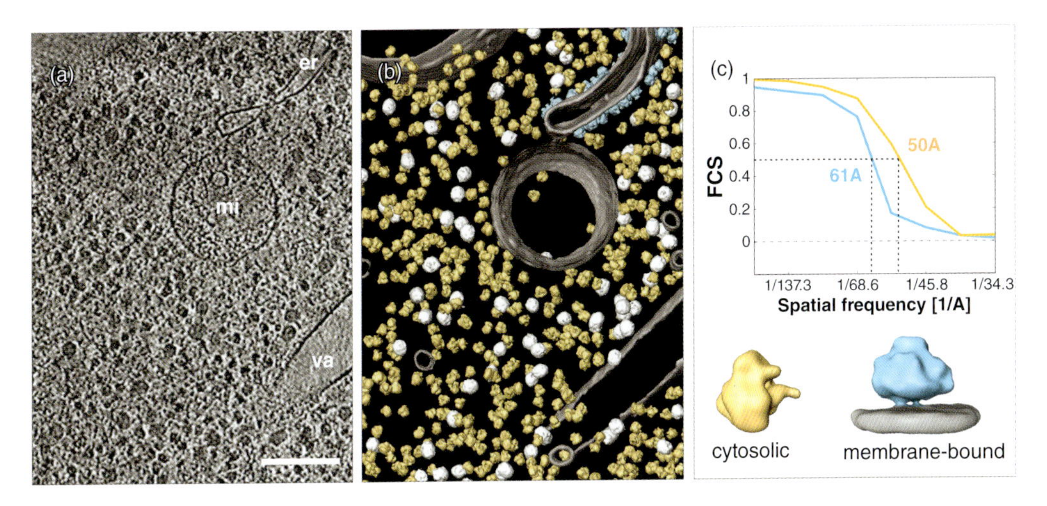

Figure 19.12 Visualization of the crowded cytoplasm. (a) Tomographic slice showing the dense packing of macromolecular complexes in the cytoplasm (endoplasmic reticulum (er); vacuolar compartment (va); mitochondrion (mi)). (b) Corresponding surface rendering, colour coded for membrane-bound ribosomes (blue), cytoplasmic ribosomes (yellow) and putative storage vesicles (white), obtained by subtomogram averaging and classification. The displayed 200 nm thick tomographic volume covers 0.31 μm^3 of cellular space; 3.2% of this volume is occupied by ribosomes and 1.7% by larger macromolecular complexes of similar size (e.g. putative storage vesicles). Membrane-bound ribosomes account for only 0.41% of the mapped volume and 1.4% space is taken up by the rough endoplasmic reticulum (cross-section). (c) Class averages and Fourier shell correlation for cytosolic and membrane-bound ribosomes. The resolution as determined by the widely used 0.5 Fourier shell correlation (FSC) criterion is 5 nm for cytosolic ribosomes and 6 nm for membrane-bound ribosomes. Scale bar: (a and b) 200 nm. Reproduced from Rigort and Plitzko, 2015, Elsevier Rights.

anisotropic nature of sample compression makes it difficult, if not impossible, to correct such data by computational methods and it restricts the usefulness of this preparation technique to rigid complexes such as ribosomes (Pierson et al. 2011). Importantly, it is not clear if differential compression takes place on this scale.

Controlling the sample thickness by cryo-FIB micromachining offers new opportunities for performing structural studies *in situ*. The lack of compression forces allows merging of multiple datasets, which increases the particle pool that is available for statistical analysis. Hence, resolution limits can be improved, similar to what is commonly done in single-particle cryo-EM, and (anisotropic) compression is not an issue. Figure 19.12a shows a slice from a tomogram of a cytoplasmic region exhibiting a dense packing of macromolecules. The approximately 200 nm thick lamella harbours electron-dense 'globular' particles, evenly distributed throughout the cytoplasm. These particles were analyzed by subtomogram averaging and classification. The corresponding surface rendering (Figure 19.12b) visualizes the results and was used to quantify the spatial distribution of membrane-bound ribosomes, cytoplasmic ribosomes and putative storage vesicles. The resolution of the averaged membrane-bound and cytoplasmic ribosomes was assessed according to the Fourier shell correlation (FSC), yielding approximately 6 nm for membrane-bound ribosomes and approximately 5 nm for cytoplasmic ribosomes at the arbitrary 0.5 threshold criterion (Figure 19.12c).

19.7 OUTLOOK: ENABLING STRUCTURAL BIOLOGY *IN SITU* BY CRYO-FIB PREPARATION

The cryo-FIB microscope is a precision manipulation tool that allows targeting of frozen hydrated cells and preparing subcellular regions from them without violating the structural integrity of the region to be visualized. Such TEM-transparent regions make it possible to study the molecular machinery of cells within the functional cellular context. Macromolecules or cellular assemblies can be localized, identified and analyzed *in situ* by computational methods such as subtomogram averaging and classification (for reviews, see Lucic et al. 2013; Briggs 2013). A major advantage of ion beam milling over vitreous sectioning is the absence of compression forces. This allows merging of multiple tomographic datasets, which in turn increases the pool of particles (i.e. macromolecules under scrutiny) amenable to statistical analysis. Having control over the thickness of the vitrified sample allows pushing the resolution to the limits, as the attainable resolution in TEM increases with decreasing sample thickness up to the point permitted by the path length and beam cross-section. However, it is the dimension of the biological structure of interest that decides the chosen thickness of a lamella. A large assembly such as the nuclear pore complex requires a sufficient depth of view to be captured within a lamella. This means that a compromise between best resolution and structural completeness must be decided upon for some structures. Still, the maximum thickness should not exceed the mean free path for inelastically scattered electrons, which is approximately 200 to 400 nm in vitreous ice, depending on the electron microscope's accelerating voltage.

Cryo-FIB technology is still at an early stage and improved devices and methodologies need to be developed to optimize its application. Locating objects and events of interest within the cell and capturing them in a lamella is by no means a trivial task. In principle, every region within a cell can be targeted by properly positioning the milling patterns and adjusting the angle of incidence accordingly (Figure 19.13). For the preparation of an electron-transparent lamella comprising an object of interest, its 3D coordinates within the cell must be known. Ideally, this information comes from fluorescence microscopy (Fukuda et al. 2014). The better spatially resolved this information is the more precise the targeting accuracy. In this context, super-resolution light microscopy techniques based on structured illumination (Gustafsson et al. 2008) or on single-molecule localization methods (e.g. PALM and STORM; Betzig et al. 2006; Rust et al. 2006) are promising. Although still at an early stage, the first super-resolution cryo-fluorescence light microscopy studies have been reported and the field is currently gaining momentum (for a review, see Kaufmann et al. 2014a). A correlated cryo-PALM and electron cryo-tomography study by Chang *et al.* showed that the photo switching capabilities remain intact at cryogenic temperatures (Chang et al. 2014). Kaufmann *et al.* demonstrated super-resolution imaging (cryo-STORM) of vitrified biological samples with a structural resolution of $\sim$125 nm and an average single-molecule localization accuracy of $\sim$40 nm (Kaufmann et al. 2014b). A main advantage of cryo-fluorescence approaches is that they maintain cellular ultrastructure throughout the entire workflow. They do not have to compromise between ultrastructural or fluorophore preservation, as is the case for traditional (resin-based) workflows (Kopek et al. 2012). Cryo-conditions fully preserve the fluorophores and offer enhanced photostability during imaging. Future correlative approaches will focus on closing the gap between dynamic observation by light microscopy and ultrastructural imaging by EM. It is also possible to utilize lamellae for analytical techniques such as X-ray fluorescence spectroscopy or electron energy loss spectroscopy (EELS), noting that the dose required for electron-based techniques precludes the acquisition of correlative image data (i.e. from the same region)

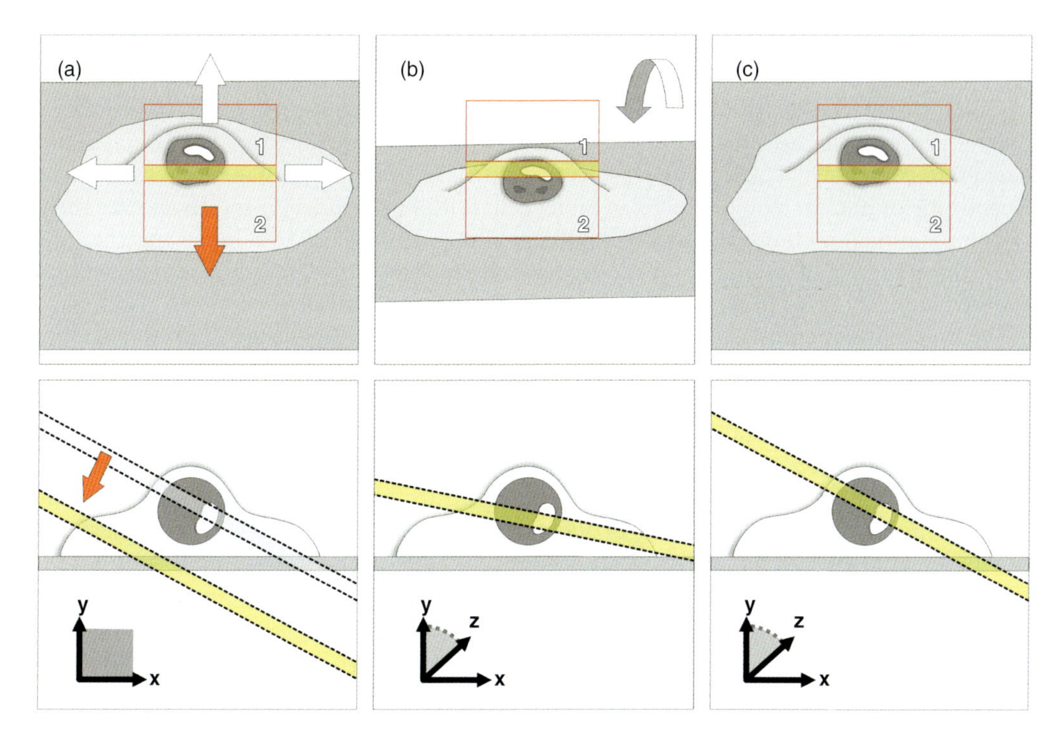

Figure 19.13 Versatility of the FIB lamella milling approach. Schematic drawing visualizing the options for targeting distinct cellular regions. (a) Horizontal and vertical positioning of the milling patterns allows different cellular areas to be targeted (below: example of how changing the vertical positioning influences the outcome of the milling experiment and the corresponding lamella). (b and c) Lowering the angle of incidence furnishes larger cellular areas within a lamella (below: examples visualizing the effect of changing the angle of incidence for milling). Lower panels: lateral cross-sectional view of the cell, with the lamella area indicated in yellow.

and that comparatively thin lamellae (on the order of 50–100 nm) are required for many spectroscopic techniques. Narrowing the time interval between observation and freezing will provide temporal resolution to electron cryo-tomography, e.g. allowing us to observe synaptic release on sub-second timescales or detecting cellular features that are present only transiently or under certain cellular conditions. Nevertheless, before such time-resolved freezing approaches become reality, suitable methods and tools will have to be developed.

Sample preparation by cryo-FIB micromachining is a key step in the workflow for structural biology (compare Figure 19.1). It provides controlled access to structural features buried deep inside cellular systems and makes them accessible to study by electron cryo-tomography. The resolution in electron cryo-tomography is primarily affected by two conflicting requirements. On the one hand, the overall electron dose for a tomogram must be kept below the threshold of visible radiation damage, which begins with the subtle erasing of fine detail, and, on the other hand, the electron dose must be maximized so as to record data with maximal signal-to-noise characteristics. Nevertheless, recent years have seen major improvements in the efficiency of detectors (McMullan et al. 2009) and developments in instrumentation such as phase plates (Danev et al. 2014; Danev and Nagayama 2010) promise to greatly expand the possibilities for cryo-EM in signal contrast technologies. In contrast to the charge-coupled device-based detectors that are still in widespread use, the new direct electron detectors have a higher sensitivity and make more efficient use of the allowed electron exposure, increasing the signal-to-noise ratio and the attainable resolution. Reliable phase plates enhance the contrast and allow imaging closer to focus, thus capturing the high-resolution information. The concomitant increase in contrast makes it more feasible to detect macromolecular machines, such as the proteasome or nuclear pore complex in intact cells with higher fidelity, and to analyze them by computational methods (Asano et al. 2015; Mahamid et al. 2016). Taken together, the improvements suggest that it may become possible to study structures that were previously 'too small' (<200 kDa) to be imaged by standard defocus-based phase contrast methods.

The successful integration of different methods and technologies into a robust and reliable workflow is one of the key challenges for performing structural biology investigations *in situ*. Such a workflow must be able to integrate dynamic and structural information coming from different ranges of spatial resolution (e.g. from fluorescence microscopy to X-ray imaging to EM). Software solutions enabling the transfer of 3D coordinate information from light microscopy to the FIB and electron microscope are required, as well as user-friendly routines allowing a reliable transfer of the vitrified specimen across the different imaging modalities. Studying the molecular machinery of cells at molecular resolution (Figure 19.14) will rely on 'hybrid approaches', combining different experimental techniques such as light microscopy, electron microscopy, X-ray diffraction and NMR spectroscopy (Steven and Baumeister 2008). In such an approach, the cryo-FIB instrument has the potential to mature to a stage where it becomes a common laboratory research tool for the preparation of vitrified cells for cryo-EM.

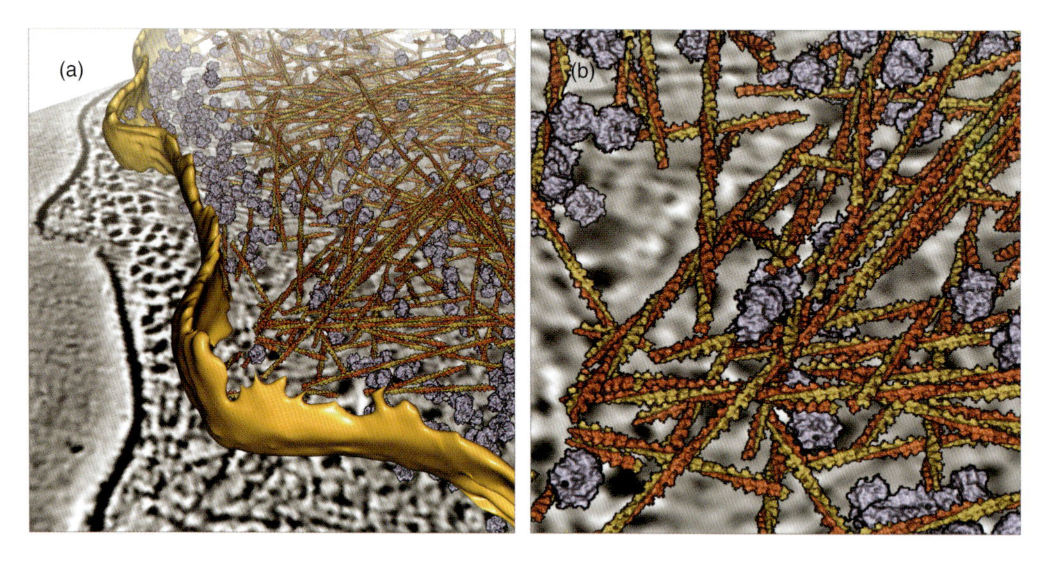

Figure 19.14 Artistic representation of the principle of performing structural biology *in situ*. Cellular cryo-tomogram with overlaid molecular structures of actin filaments and ribosomes ((a) overview with cell membrane; (b) zoomed-in view). Actin filaments were segmented automatically using a method described in Rigort et al. (2012b). For visualization purposes the molecular structure of actin was superimposed over the computationally extracted filament network. Ribosomes were identified using template matching procedures. 3D visualization: cell membrane (yellow), ribosomes (blue), actin (red-orange). Adapted from Lučić, Rigort and Baumeister, 2013, Rockefeller University Press Rights.

ACKNOWLEDGEMENTS

The authors thank Norbert Lindow and David Guenther for help in preparing Figure 19.14 and Wolfgang Baumeister for support.

REFERENCES

Al-Amoudi, A., Diez, D.C., Betts, M.J. and Frangakis, A.S. The molecular architecture of cadherins in native epidermal desmosomes. *Nature*, 2007, 450 (7171), 832–837.

Al-Amoudi, A., Studer, D. and Dubochet, J. Cutting artefacts and cutting process in vitreous sections for cryo-electron microscopy. *Journal of Structural Biology*, 2005, 150 (1), 109–121.

Asano, S., Fukuda, Y., Beck, F., Aufderheide, A., Forster, F., Danev, R. *et al.* Proteasomes. A molecular census of 26S proteasomes in intact neurons. *Science*, 2015, 347 (6220), 439–442.

Ballerini, M., Milani, M., Batani, D. and Squadrini, F. (eds). Focused ion beam techniques for the analysis of biological samples: A revolution in ultramicroscopy, *Proc. SPI*, 2001, 4261, 92–104.

Betzig, E., Patterson, G.H., Sougrat, R., Lindwasser, O.W., Olenych, S., Bonifacino, J.S. *et al.* Imaging intracellular fluorescent proteins at nanometer resolution. *Science*, 2006, 313 (5793), 1642–1645.

Brandt, F., Carlson, L.A., Hartl, F.U., Baumeister, W. and Grunewald, K. The three-dimensional organization of polyribosomes in intact human cells. *Molecular Cell*, 2010, 39 (4), 560–569.

Brandt, F., Etchells, S.A., Ortiz, J.O., Elcock, A.H., Hartl, F.U. and Baumeister, W. The native 3D organization of bacterial polysomes. *Cell*, 2009, 136 (2), 261–271.

Briggs, J.A. Structural biology *in situ* – the potential of subtomogram averaging. *Current Opinion in Structural Biology*, 2013, 23 (2), 261–267.

Chang, Y.W., Chen, S., Tocheva, E.I., Treuner-Lange, A., Lobach, S., Sogaard-Andersen, L. *et al.* Correlated cryogenic photoactivated localization microscopy and cryo-electron tomography. *Nature Methods*, 2014, 11 (7), 737–739.

Danev, R. and Nagayama, K. Phase plates for transmission electron microscopy. *Methods in Enzymology*, 2010, 481, 343–369.

Danev, R., Buijsse, B., Khoshouei, M., Plitzko, J.M. and Baumeister, W. Volta potential phase plate for in-focus phase contrast transmission electron microscopy. *Proceedings of the National Academy of Sciences of the United States of America*, 2014, 111 (44), 15635–15640.

Danino, D. Cryo-TEM of soft molecular assemblies. *Current Opinion in Colloid and Interface Science*, 2012, 17 (6), 316–329.

Denk, W. and Horstmann, H. Serial block-face scanning electron microscopy to reconstruct three-dimensional tissue nanostructure. *PLoS Biology*, 2004, 2 (11), e329.

van Driel, L.F., Valentijn, J.A., Valentijn, K.M., Koning, R.I. and Koster, A.J. Tools for correlative cryo-fluorescence microscopy and cryo-electron tomography applied to whole mitochondria in human endothelial cells. *European Journal of Cell Biology*, 2009, 88 (11), 669–684.

Dubochet, J. and McDowall, A.W. Vitrification of pure water for electron-microscopy. *Journal of Microscopy*, 1981, 124 (Dec.), Rp3–Rp4.

Dubochet, J., Adrian, M., Chang, J.J., Homo, J.C., Lepault, J., McDowall, A.W., *et al.* Cryo-electron microscopy of vitrified specimens. *Quarterly Reviews of Biophysics*, 1988, 21 (2), 129–228.

Eltsov, M., Maclellan, K.M., Maeshima, K., Frangakis, A.S. and Dubochet, J. Analysis of cryo-electron microscopy images does not support the existence of 30-nm chromatin fibers in mitotic chromosomes *in situ*. *Proceedings of the National Academy of Sciences of the United States of America*, 2008, 105 (50), 19732–19737.

Engel, B.D., Schaffer, M., Kuhn Cuellar, L., Villa, E., Plitzko, J.M. and Baumeister, W. Native architecture of the *Chlamydomonas* chloroplast revealed by *in situ* cryo-electron tomography. *eLife*, 2015, 4.

Fernandez-Busnadiego, R., Zuber, B., Maurer, U.E., Cyrklaff, M., Baumeister, W. and Lucic, V. Quantitative analysis of the native presynaptic cytomatrix by cryoelectron tomography. *The Journal of Cell Biology*, 2010, 188 (1), 145–156.

Forster, F., Medalia, O., Zauberman, N., Baumeister, W. and Fass, D. Retrovirus envelope protein complex structure *in situ* studied by cryo-electron tomography. *Proceedings of the National Academy of Sciences of the United States of America*, 2005, 102 (13), 4729–4734.

Frederik, P.M., de Haas, F. and Storms, M.M.H. Controlled vitrification, in *Handbook of Cryo-Preparation Methods for ElectronMicroscopy* (ed. A. Cavalier), CRC Press, Taylor&Francis Group, Boca Raton, FL,: 2009, pp. 71–102.

Fukuda, Y., Schrod, N., Schaffer, M., Feng, L.R., Baumeister, W. and Lucic, V. Coordinate transformation based cryo-correlative methods for electron tomography and focused ion beam milling. *Ultramicroscopy*, 2014, 143, 15–23.

Giannuzzi, L.A. and Stevie, F.A. *Introduction to Focused Ion Beams: Instrumentation, Theory, Techniques and Practice*, Springer, 2005, 357 pp.

Giannuzzi. L.A. and Stevie, F.A. A review of focused ion beam milling techniques for TEM specimen preparation. *Micron*, 1999, 30 (3), 197–204.

Gustafsson, M.G., Shao, L., Carlton, P.M., Wang, C.J., Golubovskaya, I.N., Cande, W.Z. *et al.* Three-dimensional resolution doubling in wide-field fluorescence microscopy by structured illumination. *Biophysical Journal*, 2008, 94 (12), 4957–4970.

Harris, K.M. and Weinberg, R.J. Ultrastructure of synapses in the mammalian brain. *Cold Spring Harbor Perspectives in Biology*, 2012, 4 (5).

Hayles, M.F., Matthijs de Winter, D.A., Schneijdenberg, C.T., Meeldijk, J.D., Luecken, U. and Persoon, H. et al. The making of frozen-hydrated, vitreous lamellas from cells for cryo-electron microscopy. *Journal of Structural Biology*, 2010, 172 (2), 180–190.

Henderson, G.P., Gan, L. and Jensen, G.J. 3-D ultrastructure of O. *tauri*: Electron cryotomography of an entire eukaryotic cell. *PLoS One*, 2007, 2 (8), e749.

Heymann, J.A., Hayles, M., Gestmann, I., Giannuzzi, L.A., Lich, B. and Subramaniam, S. Site-specific 3D imaging of cells and tissues with a dual beam microscope. *Journal of Structural Biology*, 2006, 155 (1), 63–73.

Holzer, L. and Cantoni, M. Review of FIB-tomography, Chapter 11 in *Nanofabrication Using Focused Ion and Electron Beams: Principles and Applications* (eds I. Utke, S.A. Moshkalev and P. Russell), Oxford University Press, New York, USA. 2012, pp. 410–435.

Hryc, C.F., Chen, D.H. and Chiu, W. Near-atomic-resolution cryo-EM for molecular virology. *Current Opinion in Virology*, 2011, 1 (2), 110–117.

Hsieh, C.E., Leith, A., Mannella, C.A., Frank, J. and Marko, M. Towards high-resolution three-dimensional imaging of native mammalian tissue: Electron tomography of frozen-hydrated rat liver sections. *Journal of Structural Biology,* 2006, 153 (1), 1–13.

Kaech, S. and Banker, G. Culturing hippocampal neurons. *Nature Protocols*, 2006, 1 (5), 2406–2015.

Karuppasamy, M., Karimi Nejadasl, F., Vulovic, M., Koster, A.J. and Ravelli, R.B. Radiation damage in single-particle cryo-electron microscopy: Effects of dose and dose rate. *Journal of Synchrotron Radiation*, 2011, 18 (Pt 3), 398–412.

Kaufmann, R., Hagen, C. and Grunewald, K. Fluorescence cryo-microscopy: Current challenges and prospects. *Current Opinion in Chemical Biology*, 2014a, 20, 86–91.

Kaufmann, R., Schellenberger, P., Seiradake, E., Dobbie, I.M., Jones, E.Y., Davis, I. *et al.* Super-resolution microscopy using standard fluorescent proteins in intact cells under cryo-conditions. *Nano Letters*, 2014b, 14 (7), 4171–4175.

Kirk, E.C.G., Williams, D.A., and Ahmed, H. Cross-sectional transmission electron-microscopy of precisely selected regions from semiconductor-devices. *Institute of Physics Conference Series*, 1989 (100), 501–506.

Knott, G., Marchman, H., Wall, D. and Lich, B. Serial section scanning electron microscopy of adult brain tissue using focused ion beam milling. *The Journal of Neuroscience*, 2008, 28 (12), 2959–2964.

Koning, R.I., Zovko, S., Barcena, M., Oostergetel, G.T., Koerten, H.K., Galjart, N. *et al.* Cryo electron tomography of vitrified fibroblasts: Microtubule plus ends *in situ. Journal of Structural Biology*, 2008, 161 (3), 459–468.

Kopek, B.G., Shtengel, G., Xu, C.S., Clayton, D.A. and Hess, H.F. Correlative 3D superresolution fluorescence and electron microscopy reveal the relationship of mitochondrial nucleoids to membranes. *Proceedings of the National Academy of Sciences of the United States of America*, 2012, 109 (16), 6136–6141.

Lau, W.C. and Rubinstein, J.L. Single particle electron microscopy. *Methods in Molecular Biology*, 2013, 955, 401–426.

Lucic, V., Rigort, A. and Baumeister, W. Cryo-electron tomography: The challenge of doing structural biology *in situ*. *The Journal of Cell Biology*, 2013, 202 (3), 407–419.

Mahamid, J., Pfeffer, S., Schaffer, M., Villa, E., Danev, R., Cuellar, L.K. *et al.* Visualizing the molecular sociology at the HeLa cell nuclear periphery. *Science*, 2016, 351 (6276), 969–972.

Mahamid. J., Schampers, R., Persoon, H., Hyman, A.A., Baumeister, W. and Plitzko, J.M. A focused ion beam milling and lift-out approach for site-specific preparation of frozen-hydrated lamellas from multicellular organisms. *Journal of Structural Biology*, 2015, 192 (2), 262–269.

Marko, M., Hsieh, C., Moberlychan, W., Mannella, C.A. and Frank, J. Focused ion beam milling of vitreous water: Prospects for an alternative to cryo-ultramicrotomy of frozen-hydrated biological samples. *Journal of Microscopy*, 2006, 222 (Pt 1), 42–47.

Marko, M., Hsieh, C., Schalek, R., Frank, J. and Mannella, C. Focused-ion-beam thinning of frozen-hydrated biological specimens for cryo-electron microscopy. *Nature Methods*, 2007, 4 (3), 215–217.

McDowall, A.W., Chang, J.J., Freeman, R., Lepault, J., Walter, C.A. and Dubochet, J. Electron microscopy of frozen hydrated sections of vitreous ice and vitrified biological samples. *Journal of Microscopy*, 1983, 131 (Pt 1), 1–9.

McMullan, G., Clark, A.T., Turchetta, R. and Faruqi, A.R. Enhanced imaging in low dose electron microscopy using electron counting. *Ultramicroscopy*, 2009, 109 (12), 1411–1416.

Mulders, H. The use of a SEM/FIB dual beam applied to biological samples. *GIT Imaging and Microscopy*, 2003 (2), 8–10.

Orloff, J., Swanson, L., Utlaut, M.W. High resolution focused ion beams : *FIB and its applications: the physics of liquid metal ion sources and ion optics and their application to focused ion beam technology*. New York: Kluwer Academic/Plenum Publishers; 2003. x, 303 pp.

Overwijk, M.H.F., Vandenheuvel, F.C. and Bullelieuwma, C.W.T. Novel scheme for the preparation of transmission electron-microscopy specimens with a focused ion-beam. *Journal of Vacuum Science & Technology*, 1993, 11 (6), 2021–2024.

Pierson, J., Fernandez, J.J., Bos, E., Amini, S., Gnaegi, H., Vos, M., *et al.* Improving the technique of vitreous cryo-sectioning for cryo-electron tomography: Electrostatic charging for section attachment and implementation of an anti-contamination glove box. *Journal of Structural Biology*, 2010, 169 (2), 219–225.

Pierson, J., Ziese, U., Sani, M. and Peters, P.J. Exploring vitreous cryo-section-induced compression at the macromolecular level using electron cryo-tomography; 80S yeast ribosomes appear unaffected. *Journal of Structural Biology*, 2011, 173 (2), 345–349.

Porter, K.R. and Blum, J. A study in microtomy for electron microscopy. *The Anatomical Record*, 1953, 117 (4), 685–710.

Reyntjens, S. and Puers, R. A review of focused ion beam applications in microsystem technology. *Journal of Micromechanics and Microengineering*, 2001, 11 (4), 287.

Rigort, A. and Plitzko, J.M. Cryo-focused-ion-beam applications in structural biology. *Archives of Biochemistry and Biophysics*, 2015, Sept. 1, 581, 122–130.

Rigort, A. Recent developments in FEI's in situ cryo-electron tomography workflow. *Nature Methods* 2016, 13 (11).

Rigort, A., Bauerlein, F.J., Villa, E., Eibauer, M., Laugks, T., Baumeister, W. *et al.* Focused ion beam micromachining of eukaryotic cells for cryoelectron tomography. *Proceedings of the National Academy of Sciences of the United States of America*, 2012a, 109 (12), 4449–4454.

Rigort, A., Bauerlein, F.J.B., Leis, A., Gruska, M., Hoffmann, C., Laugks, T. *et al.* Micromachining tools and correlative approaches for cellular cryo-electron tomography. *Journal of Structural Biology*, 2010, 172 (2), 169–179.

Rigort, A., Gunther, D., Hegerl, R., Baum, D., Weber, B., Prohaska, S. *et al.* Automated segmentation of electron tomograms for a quantitative description of actin filament networks. *Journal of Structural Biology*, 2012b, 177 (1), 135–144.

Rigort, A., Villa, E., Bauerlein, F.J., Engel, B.D. and Plitzko, J.M. Integrative approaches for cellular cryo-electron tomography: correlative imaging and focused ion beam micromachining. *Methods in Cell Biology*, 2012c, 111, 259–281.

Rubino, S., Akhtar, S., Melin, P., Searle, A., Spellward, P. and Leifer, K. A site-specific focused-ion-beam lift-out method for cryo transmission electron microscopy. *Journal of Structural Biology*, 2012, 180 (3), 572–576.

Rust, M.J., Bates, M. and Zhuang, X. Sub-diffraction-limit imaging by stochastic optical reconstruction microscopy (STORM). *Nature Methods*, 2006, 3 (10), 793–795.

Sartori, A., Gatz, R., Beck, F., Rigort, A., Baumeister, W. and Plitzko, J.M. Correlative microscopy: Bridging the gap between fluorescence light microscopy and cryo-electron tomography. *Journal of Structural Biology*, 2007, 160 (2), 135–145.

Schertel, A., Snaidero, N., Han, H.M., Ruhwedel, T., Laue, M., Grabenbauer, M. *et al.* Cryo FIB-SEM: Volume imaging of cellular ultrastructure in native frozen specimens. *Journal of Structural Biology*, 2013, 184 (2), 355–360.

Schwartz, C.L., Sarbash, V.I., Ataullakhanov, F.I., McIntosh, J.R. and Nicastro, D. Cryo-fluorescence microscopy facilitates correlations between light and cryo-electron microscopy and reduces the rate of photobleaching. *Journal of Microscopy*, 2007, 227 (Pt 2), 98–109.

Steven, A.C. and Baumeister, W. The future is hybrid. *Journal of Structural Biology*, 2008, 163 (3), 186–195.

Volkert, C.A. and Minor, A.M. Focused ion beam microscopy and micromachining. *MRS Bulletin*, 2007,32 (05), 389–399.

Wang, K., Strunk, K., Zhao, G., Gray, J.L. and Zhang, P. 3D structure determination of native mammalian cells using cryo-FIB and cryo-electron tomography. *Journal of Structural Biology*, 2012, 180 (2), 318–326.

de Winter, D.A., Mesman, R.J., Hayles, M.F., Schneijdenberg, C.T., Mathisen, C. and Post, J.A. *In-situ* integrity control of frozen-hydrated, vitreous lamellas prepared by the cryo-focused ion beam-scanning electron microscope. *Journal of Structural Biology*, 2013, 183 (1), 11–18.

Yahav, T., Maimon, T., Grossman, E., Dahan, I. and Medalia, O. Cryo-electron tomography: Gaining insight into cellular processes by structural approaches. *Current Opinion in Structural Biology*, 2011, 21 (5), 670–677.

Yao, N. *Focused Ion Beam Systems: Basics and Applications*, Cambridge University Press, 2007, 408 pp. 1818-1823.

Ziegler, J.F., Ziegler, M.D. and Biersack, J.P. SRIM - The stopping and range of ions in matter. *Nucl. Instrum. Meth. B*, 2010, 268 (11–12), 1818-1823.

20

Environmental Scanning Electron Microscopy

Rudolph Reimer, Dennis Eggert and Heinrich Hohenberg
Heinrich Pette Institute, Leibniz Institute for Experimental Virology, Hamburg, Germany

20.1 INTRODUCTION

Water is an important structural component of biological specimens. Tissues contain more than 80% of intracellular water, suspended material even more. The total removal of this water for electron microscopical investigations results in dramatical alterations of the cellular structure and produces a wide spectrum of specific structural artefacts inducing shrinkage, redistribution and loss of cellular structure. Moreover, some samples like hydrogels, fats, hydrated protein macromolecules or intact membrane and protein networks cannot be imaged at all.

Many attempts have been made to circumvent these artefacts by preserving the environmental factor 'cellular water' in order to guarantee a 'native-like' imaging of biological material, even at higher resolution in the electron microscope. However, specimens containing volatile components like water can normally not be investigated without physically isolating them from the microscope's vacuum or applying elaborate cryotechniques to keep the water in the frozen state due to the high vacuum conditions inside the microscope's sample chamber of scanning electron microscopes (SEM).

Recently several solutions were developed where the separation of the liquid-containing sample from the microscope high vacuum is done by using electron-transparent ultrathin film windows. Such film windows have to be sufficiently stable to withstand pressure differences of several orders of magnitude. This can be achieved either by the use of stable enough material like silicon nitride or by stabilization with metal grids, as in the case of polymer (e.g. polyimide) films. The thickness of the electron-transparent window has to be small enough to allow the pass-through of the electrons from the column/sample

Biological Field Emission Scanning Electron Microscopy, First Edition.
Edited by Roland A. Fleck and Bruno M. Humbel.

chamber to the vacuum-isolated sample. This principle is being applied in commercially available microscopy systems, such as ClairScope (JEOL) or airSEM (B-nano) and also in the WETSEM capsules developed by Quantomix (Figure 20.1). The WETSEM capsules isolate the water-containing sample from the microscope's high vacuum using an additional vacuum-tight small specimen chamber with electron-transparent ultrathin film window. For proper imaging the hydrated sample must be thin, fixed, contrasted (uranium acetate and/or osmium) and placed directly under the membrane.

The physical separation of the microscope high vacuum from the atmospheric pressure sample by a window has its pros and cons. The major advantage is the unique possibility to apply SEM at true atmospheric pressure environment. The main problem of SEM imaging through a window is the loss of electrons due to the window thickness and/or air in the atmospheric environment and therefore a low signal-to-noise ratio. Moreover, the possibility of imaging 3D structures is limited because of the necessity to place the sample as close as possible to the window. In practice the depth of BSE imaging, at least of hydrated encapsulated samples, is also very small and proximal to the film (Figure 20.1).

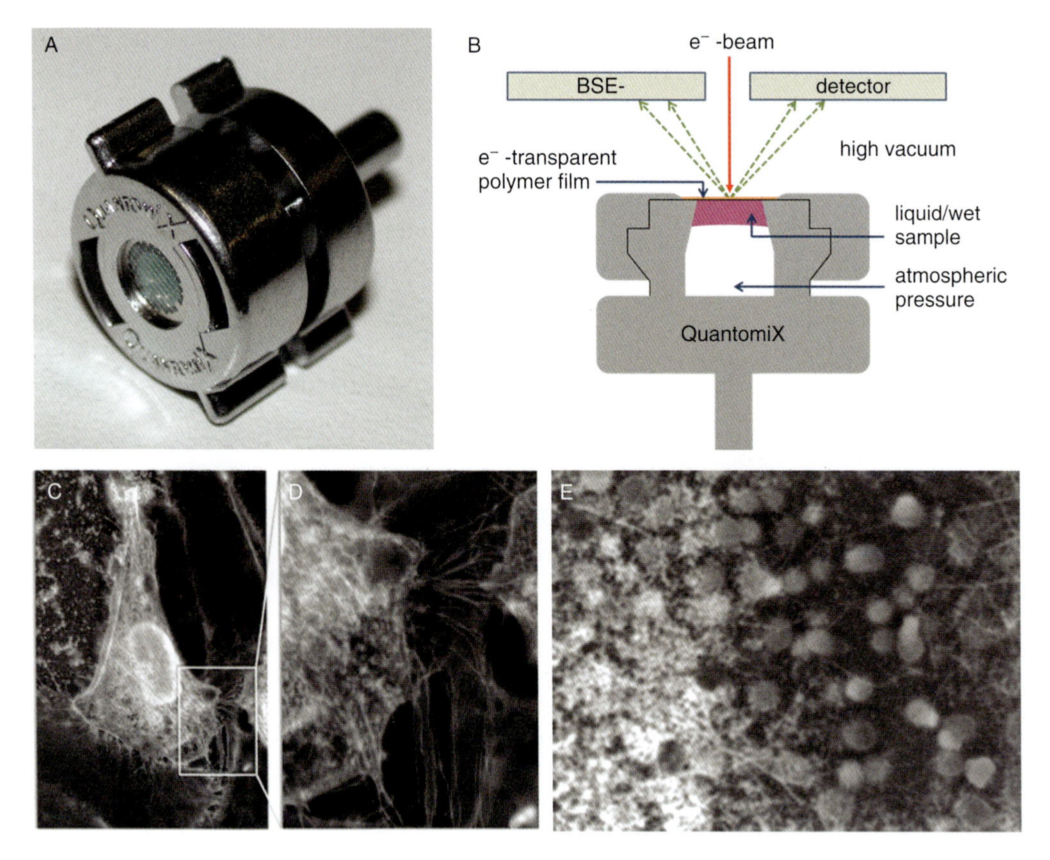

Figure 20.1 Quantomix WetSEM capsule (A) and its operation principle (B). The sample is isolated from the high vacuum in the microscope by a polyimide window stabilized with a metal grid. This allows the imaging of liquids or cells grown on the film 'through the window'. (C and D) C6 glioblastoma cells grown as a monolayer on the polymer film, chemically fixed and stained with uranyl acetate. (E) Clotted human blood stained with uranyl acetate. Fine structures like filopodia (D) or fibrin fibers (E) are clearly visible.

The environmental scanning electron microscope (ESEM), in contrast, enables the investigation of hydrated or vacuum-unstable samples at elevated residual gas pressure (i.e. in a low vacuum) directly inside the microscope sample chamber (Danilatos and Robinson, 1979). The focus of our contribution lies not in giving an overview of the development of the ESEM, its modification and instrumentation, but instead concentrate extensively and – where necessary – in detail on the development of tailored methodical steps for the application of environmental scanning electron microscopy, with water as the imaging gas, to investigate fully hydrated, native-like biomedical tissues. Therefore the reader is referred to the relevant literature describing historical and technical developments of the ESEM in general or in detail (e.g. Danilatos and Robinson, 1979; Stokes, 2008).

20.2 SIGNAL GENERATION IN THE ESEM

For the correct use of the ESEM as an imaging instrument in biology and medicine, the understanding of signal generation is of high importance, in particular if a biomedical specimen with different amounts of water content, various topography and structural density have to be imaged and analysed.

When an electron beam irradiates the specimen, it causes different types of interactions with its surface (Figure 20.2). The most commonly utilized mode for environmental SEM is the secondary electron imaging mode. Hereby the signal is formed by the low-energy

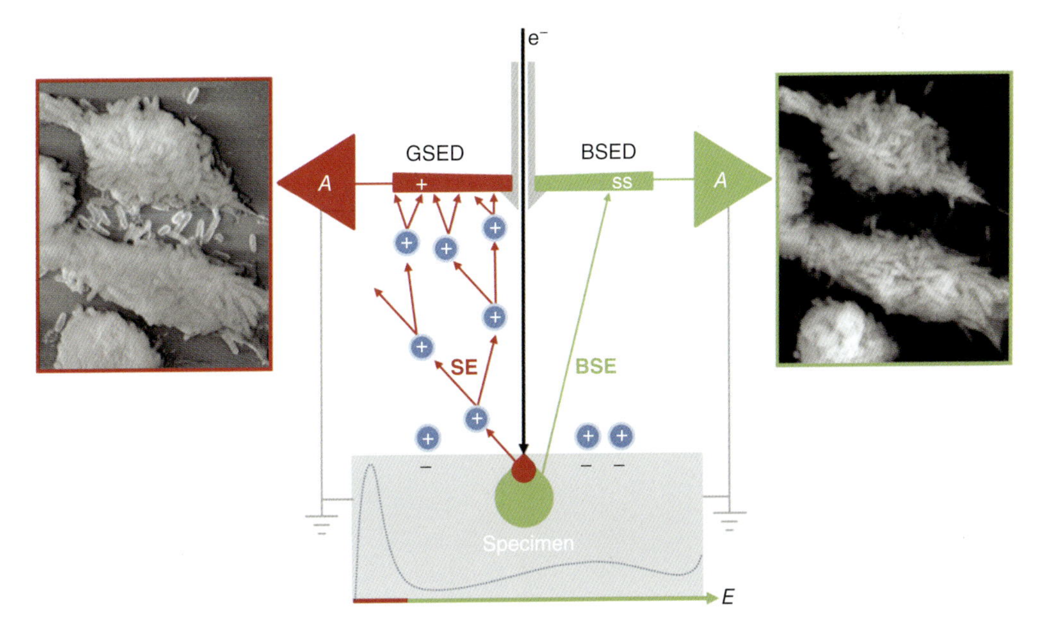

Figure 20.2 Signal generation in the ESEM (simplified). Left part, in red: secondary electron (SE) imaging with the gaseous secondary electron detector (GSED). The positively biased detector attracts the secondary electrons that ionize residual gas molecules by collisions. This electron cascade leads to SE-signal amplification. The charging of the sample is compensated by the positive ions. Right part, in green: backscattered electrons (BSE) are detected by a solid state backscattered electron detector (BSED). The distribution of the SE (<50 eV) and the BSE (>50 eV) is shown in the energy diagram. Example images show *Yersinia enterocolitica*-infected macrophages in SE and BSE imaging.

secondary electrons (< 50 eV) that leave the uppermost layer of the specimen and therefore contain information about the surface topography. The residual gas inside the sample chamber (e.g. air, nitrogen, water vapour) contributes to the amplification of the signal. A positive bias of a few hundreds of volts is applied to the detector. This causes an acceleration of the secondary electrons towards the detector. On this path the electrons hit the gas molecules of the residual gas, leading to their ionization and the formation of new signal electrons. The ESEM signal is mainly based on these novel electrons generated in the amplification cascade. The corresponding image represents the topography of the specimen. The emerging positively charged gas ions are accelerated towards the specimen, neutralizing charges that could arise at non-conductive surfaces (Figure 20.2).

The so-called gaseous secondary electron (GSE) detector is insensitive to light and temperature changes. This enables the investigation of physical processes like freezing or melting in material sciences and the application of photo stimulation in the investigation of living specimens in ESEM.

More detailed discussions of ESEM signal generation and the influence of the related parameters, for example the type and pressure of the imaging gas, chamber temperature, beam current, etc., can be found in the dedicated literature (Danilatos and Robinson, 1979; Stokes, 2008)

20.2.1 Imaging of Non-conductive Samples

In the ESEM the positively charged gas ions neutralize charges that in principle arise at the surface of non-conductive specimens, inducing an unspecific secondary electron signal. In the conventional SEM this surface charging is usually prevented by applying thin metal or carbon films. The ESEM-mode allows non-conductive samples like insects, plants or plant elements to be investigated (see Figure 20.3) directly, without any artificial surface coating. The topography of relatively dry and stable surfaces like exoskeletons of insects/arachnids or plant leaf surfaces can best be visualized with the GSE detector (Figure 20.3). Depending on the sample, this can be done in a wide pressure range from 100 to 2000 Pa with or

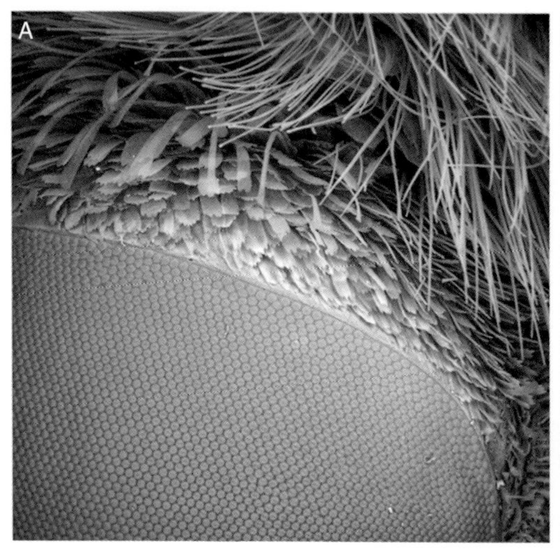

Figure 20.3 GSE imaging of non-conductive biological samples in the ESEM. (A) Moth-eye and (B) olive leaf trichomes. Imaged with the GSE detector at 200 and 400 Pa respectively. Image courtesy of Robert Getzieh, HPI.

Figure 20.4 BSE Imaging of non-conductive biological samples in the ESEM. Preselection of specific tissue areas for ultrathin sectioning. (B) Detailed view of the selected area (SA). Sample: diamond-trimmed epoxy resin block with embedded OsO_4-stained mouse liver tissue. Image courtesy of Martin Warmer, HPI.

without cooling. The signal-to-noise ratio is better at smaller pressures but the charge compensation by ionized gas molecules can drop down and charging of the sample can occur. Water-containing samples (e.g. plant leaves) will also dehydrate much faster at a lower pressure.

In a further application, an overview or the pinpoint preselection of tissue material (e.g. endothelia) embedded in epoxy resin before and after trimming is possible. After the ESEM imaging of the trimmed pyramid a section of the preselected area (Figure 20.4, SA) can be imaged at high resolution in the TEM and correlated with the ESEM. Moreover, the imaged lateral faces (Figure 20.4, LF) of the pyramid allow the material density to be estimated for the Z-direction in the follow-up sectioning process. Such estimations are very helpful if serial sectioning and 3D reconstruction is intended. Furthermore, specific stainings (see Section 20.4.6) and their distribution can directly be determined, allowing the reorientation and correlation of ESEM and TEM images.

20.2.2 Imaging of Hydrated Samples

A common problem that occurs during the investigation of totally native hydrated and more or less soft biological material is the presence of thin water films at the surface of sectioned or suspended specimens. Water is opaque in the ESEM (Figure 20.5), preventing the visualization of fine topographic details of the sample. Therefore, the imaging of wet specimens with the GSE detector is often not feasible as their surfaces are covered with this opaque water layer. Even a controlled evaporation of the water by raising the temperature or lowering the pressure inside the sample chamber results in a dissatisfying visualization of surface details of cells or tissues due to artefacts caused by the effects of high water surface tension. Figure 20.2 shows the typical flattening of cell topography due to water evaporation (cells in the red frame). Stabilization of the minute surface structures is possible, for example, with silica (Kaehr *et al.*, 2012) but is laborious, time consuming and not applicable for animal tissues.

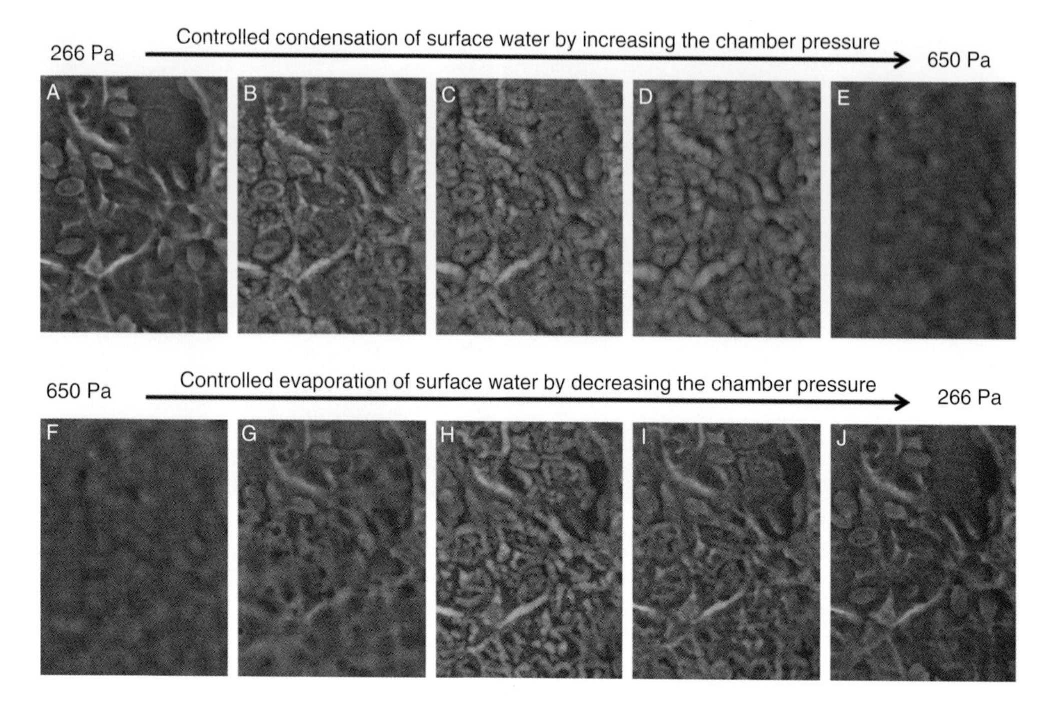

Figure 20.5 Controlled condensation and evaporation of surface water in the ESEM. The image shows different hydration states of the same area after changing of the chamber pressure from 266 Pa to 650 Pa and back to 266 Pa. The opaque water layer completely covers the surface structures. The image shows the surface of a duck Glisson's capsule. The elongated bird erythrocytes remain in their position during the procedure.

The utilization of backscattered electrons for ESEM helps to circumvent these problems by focusing on the interior structures of cells and tissues. Backscattered electrons arise from deeper layers of the sample (Figure 20.2, green). They represent the so-called material contrast. In native or fixed but non-contrasted biological material, only very few backscattered electrons are generated (due to the lack of heavy elements). On their way to the detector, these electrons are scattered by the residual gas molecules inside the sample chamber. Therefore, only a weak signal can be detected at the BSE-detector, causing noisy images. The combination of a heavy metal staining with the controlled evaporation of the bulk surface water enables the generation of BSE images with a good signal-to-noise ratio in an ESEM (Figure 20.2, green frame).

20.3 RADIATION DAMAGE

The experience with a wide spectrum of biomedical specimens has shown that less or non-hydrated material or material without a hydration supporting matrix (see Figure 20.7) is more susceptible for beam damage. As mentioned above and shown in Figure 20.5, it is possible to perform and image directly the dynamics of controlled hydration and dehydration processes in the ESEM. Therefore the controlled removal of water films from many specimens was possible and here especially material with less or no structural bound water showed distinct beam damaging patterns at higher magnification. The example in

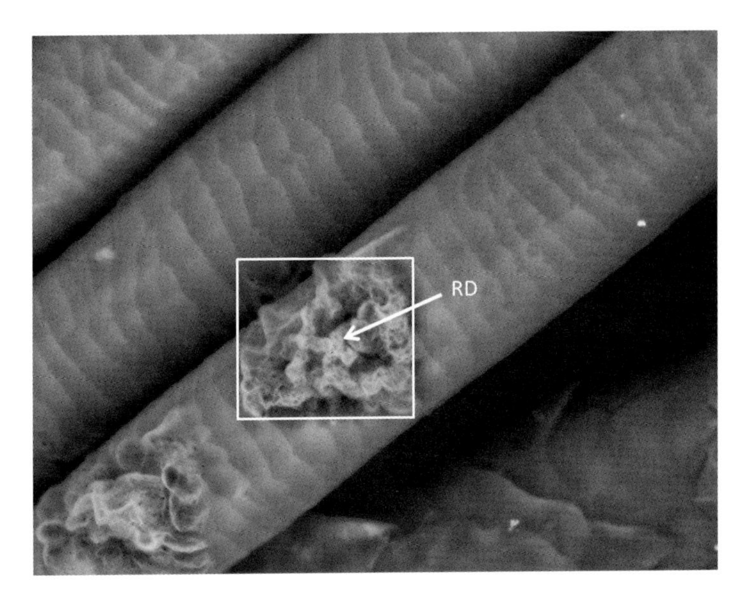

Figure 20.6 Human hair in the ESEM mode. The nearly dry mouse hairs look intact. Inside the white box, where one fibre was investigated at higher magnification, the radiation damage (RD) is obvious.

Figure 20.6 shows massive structural damage (arrow) on the surface of human hair in the ESEM inside the scanning frame (RD).

20.4 APPLICATION OF ESEM FOR INVESTIGATION OF HYDRATED BIOMEDICAL SAMPLES

The very first application of the ESEM by its inventor Gerasimos Daniel Danilatos in the late 1970s was the investigation of wool fibres in their native state (Danilatos and Robinson, 1979). Nowadays environmental scanning electron microscopy is a common technique in material sciences as non-conductive materials can be investigated without conductive coatings. This facilitates the electron microscopic investigation and the quantification of dynamic processes like crack formation in plastics or the drying and hardening of concrete, respectively cement (Zankel *et al.*, 2007, Sakalli and Trettin, 2015).

In biological research environmental scanning electron microscopy is mainly utilized in botany and zoology. It allows the study of alive and/or native-like plant tissues and arthropods.

Recently the ESEM is also being applied in biomedical research. The ESEM offers the unique possibility to image soft or water- and lipid-containing samples without former microstructure-changing preparation steps, thus opening completely new perspectives in preclinical research.

20.4.1 Water in Biomedical Material: Hydrated Samples in the ESEM

The water present in biological material has a major influence on the structure and shape of biological specimens. The total or partial removal of this important molecule – determining structural integrity of cellular material – causes drastic alterations of the sample's morphology, fine structure and also its electric conductivity. Since it was shown that water has different states of organisation on the cellular and particularly on the macromolecular level, it is necessary to discuss why and how these different states may influence the investigation of hydrated material in the ESEM. For more and detailed information about the state of water in the cell and its influence on cellular structure and function we recommend *Life's Matrix* from Phillip Ball (1999) and the overview given by Wiggins (1990) and Negendank *et al.* (1986). Summing up the existing models of the organization and specific function of water in the cell it can be hypothesized that water is a biomolecule itself involved in every structural and functional process at different levels of molecular organization. Moreover, the organization of water in the cell is highly complex. (a) There is 'bulk water', which is not or less influenced by the macromolecules, it has a high mobility, freezes at 0 °C and represents up to 90% of the cellular water. (b) The 'structural bound water' forms a 'hydration shell' of pure water around molecules. It is influenced by the surface charge of the specific molecules or macromolecular complexes, freezes at −30 °C and represents 2–80% of the cellular water. Here the influence of the molecule's surface charge on the water molecules depends on their respective distances, but the resulting mobility of the influenced water is clearly reduced. (c) The 'structure-integrated water' present in the monomolecular water layers around the hydrophilic groups of biopolymers has nearly no mobility, freezes around −70 °C and represents 0.2–10% of the cellular water. In a nutshell: water needs more comprehensive consideration if we want to apply it as a quality factor for native-like imaging.

If hydrated material is investigated in the ESEM it should be taken into account that the sort of water present in the investigated specimen or mainly on its surface often determines the quality of the resulting image or limits the imaging process itself drastically. To give an example: very dense biofilms (see Figure 20.8) contain a very high amount of water-binding structures resulting in mainly strong structural bound water even at the surface, which cannot be removed from the specimen by any ESEM-specific parameter (e.g. higher vacuum or specimen heating). The removal of this water, for example by applying a high vacuum, results in the total drying and the structural collapse of the biofilm and its components. Whereas the controlled condensation or evaporation of 'bulk' surface water in the ESEM is possible without detectable structural damages (see Figure 20.6). Different surface hydration states of the same specimen surface area could be imaged after changing the chamber pressure. In one state (E and F) the opaque water layer completely covers the surface structures. The imaging of surface details is impossible even though the tissue structures are stained with osmium. After the total removal of the 'bulk' water, the image shows the surface details of the duck Glisson's capsule. The elongated bird erythrocytes remain in their position during the procedure. This removal of the highly mobile 'bulk water' can be accomplished many times without causing detectable structural damage because the surface fine structure of the tissue is stabilized by its inherent structural bound water.

Investigating clinical material we often have to image the surfaces of very thin tissue sections (liver, kidney, muscle, etc.). These thin layers tend to lose their water (even if structurally bound) very fast in the course of the investigation, because there was no more water around on the specimen support and the added water drop (bulk water) disappeared very fast. Applying the excellent water storage capacity of specific materials (agar matrices, matrix-integrated molecular sieves or superadsorbers), we installed a hydration supporting matrix under the hydrated specimen and placed the sample on top of this matrix. The water in this artificial water reservoir is structurally bound and present there for a longer time period, preventing the fast dehydration of the thin tissue section on top of the supporting matrix. We could measure that there was a constant thin layer of water present for up to 30 minutes. In Figure 20.7 the factors that influence the image quality of biological material in the ESEM are listed.

On the basis of the skin sample and its preparation sequence described in Figure 20.8 it is easy to understand that the native-like imaging of, for instance, clinical material is based on the strict consideration of the factors listed in Figure 20.7.

In Figure 20.8 the typical preparation sequence of clinical material (human skin) is represented. To preserve the structural integrity of the clinical material a sterile punch biopsy is taken (A) and the biopsy cylinder with its native surface (B) is directly transferred into a sterile culture medium agar-matrix (C), followed by investigation in the ESEM, after transferring the specimen holder with the agar layer into the specimen chamber. In the ESEM image it is directly noticeable that there is a surface layer on top of the skin biopsy (D), which is not present on the surface of the dehydrated skin (E). This surface layer represents the lipid biofilm protecting the skin. After conventional preparation this biofilm is totally removed by the applied dehydration alcohol. After ESEM investigation the specimen can be further prepared, for instance with the help of cryo-techniques using high pressure freezing, freeze substitution and embedding. After ultrathin sectioning the native-like skin can be investigated at high resolution in the TEM.

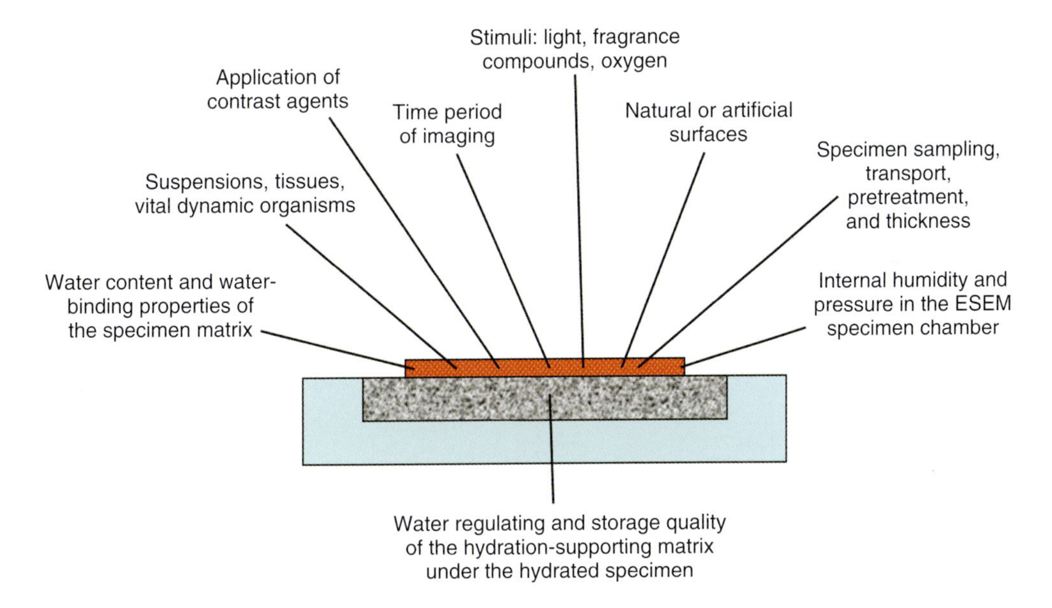

Figure 20.7 Factors influencing the imaging quality of biomedical material in the ESEM.

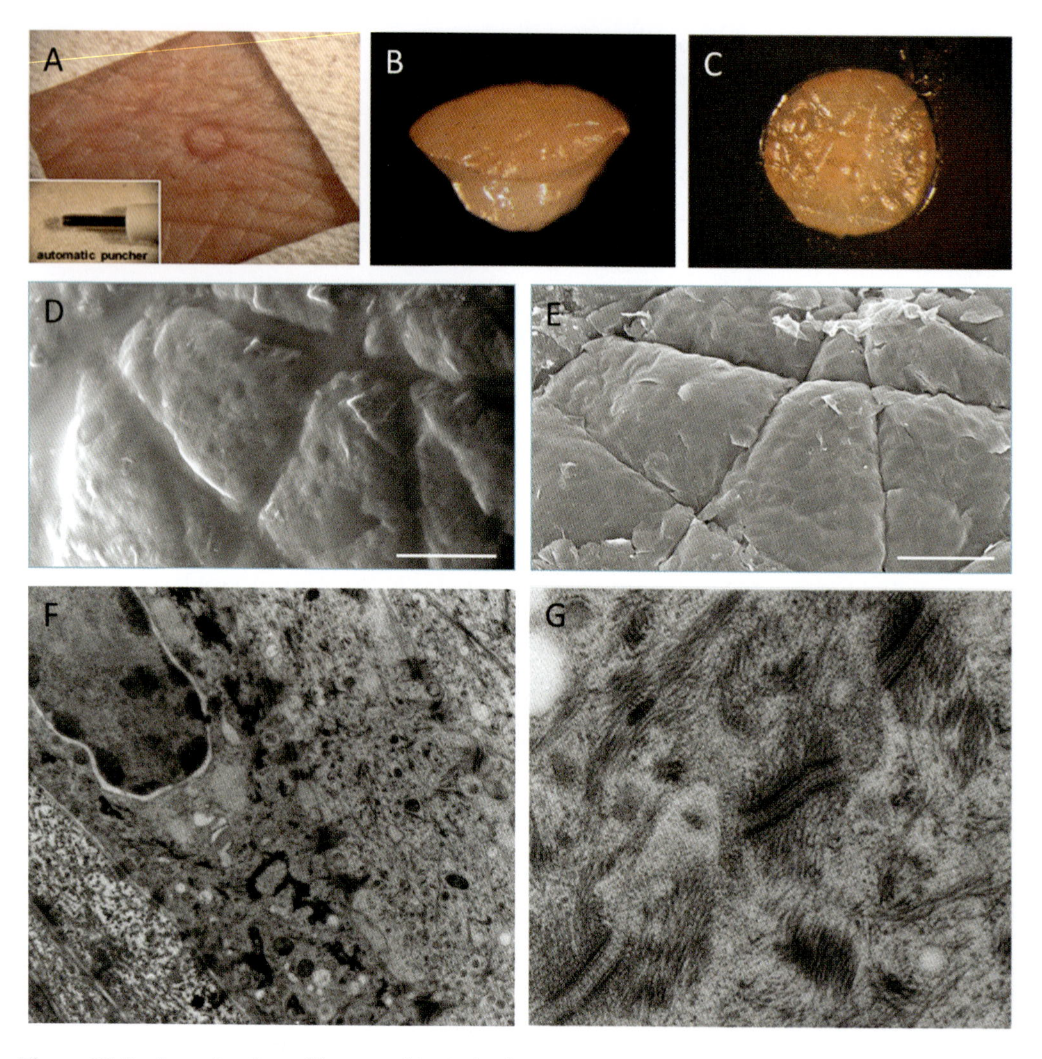

Figure 20.8 Investigation of human skin in both ESEM and TEM in a native-like state. (A and B) A punch biopsy is taken from the patient. (C) The biopsy is put into a culture medium agar supply. (D) ESEM GSED image of the native skin surface containing water and lipids in comparison to a critical-point dried skin (E). (F and G) High resolution TEM images of the very same sample (D) after high pressure freezing and freeze substitution. The ultrastructure of the desmosomes is very well preserved (G).

20.4.2 Biofilms

Samples containing high amounts of water, like algal, fungal or bacterial biofilms, can be imaged without dehydration in their native state. The extracellular polymeric substance matrix remains in its shape allowing the *in situ* investigation of the populating microorganisms (Figure 20.9). Very short preparation times enable fast investigation cycles, providing the possibility of the use of an ESEM as a clinical diagnosis instrument (see Figure 20.10).

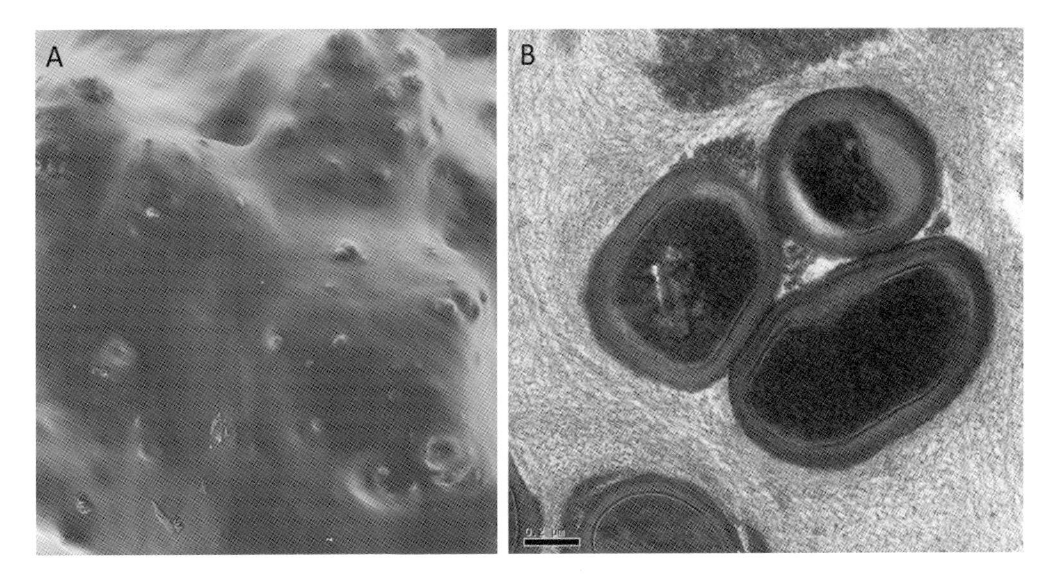

Figure 20.9 *Pseudomonas aeruginosa* biofilm in the ESEM (A). The polysaccharide matrix is totally hydrated and thereby non-transparent for the electrons. (B) The same sample, high-pressure frozen and freeze-substituted, shows the dense polysaccharide matrix surrounding the bacteria.

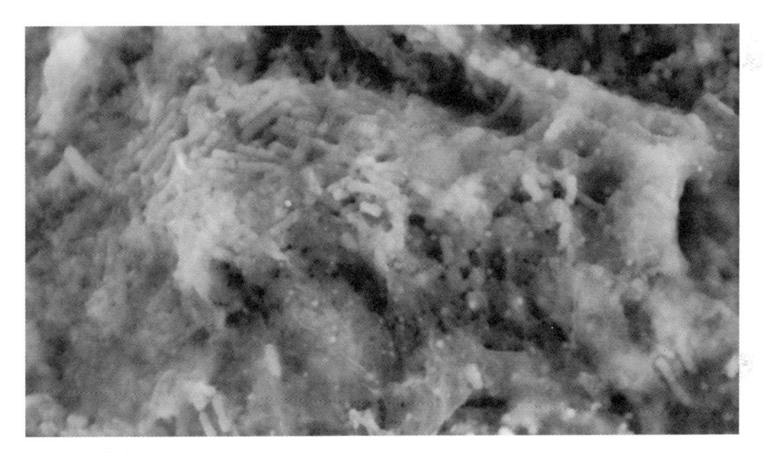

Figure 20.10 ESEM image of a bacterial biofilm on a human vocal fold. Rod-shaped bacteria embedded in an extracellular polymeric substance matrix are clearly visible.

20.4.3 Lipids

Lipids or lipid-rich tissues are also hard to image in conventional scanning electron microscopy. The dehydration steps that are necessary for conventional electron microscopy include the treatment with organic solvents like ethanol or acetone that dissolve, remove or alter lipid structures. As the preparation steps for ESEM do not include dehydration or treatment with organic solvents, cells or tissues containing a high amount of lipids like brown or white adipose tissue can be imaged in their native-like state (Figure 20.11).

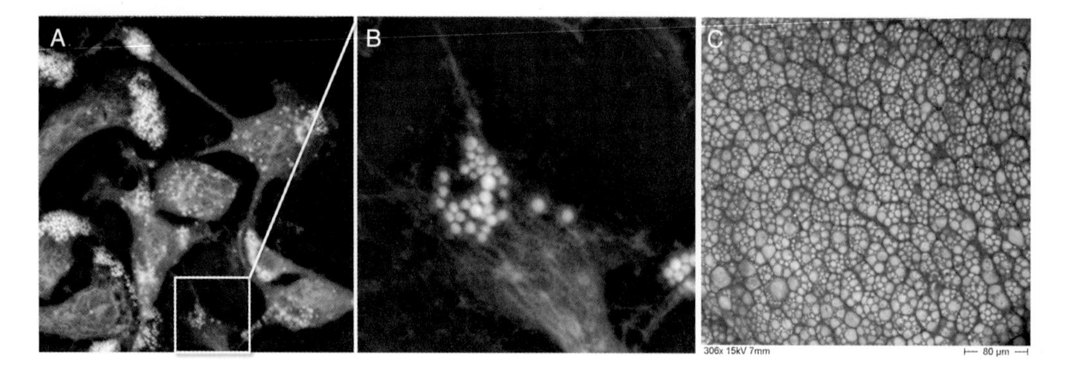

Figure 20.11 Lipid droplets in cells and tissue in the ESEM. (A and B) HT29 colon adenocarcinoma cells grown as a monolayer in a Petri dish; BSE, 20 kV. (C) Mouse brown adipose tissue; BSE, 12 kV. Lipid droplets inside the cells are clearly visible as in monolayer cells, as well as in solid tissue. Image (C) shows more topography due to the reduced acceleration voltage.

20.4.4 Tissue Surfaces

Biological materials contain high amounts of water. This water has a major influence on the structure and shape of biological samples. The dehydration that is necessary for conventional electron microscopy causes drastic structural alterations of the sample. These shrinking artefacts make proper visualization of larger areas of the tissue surface oftentimes impossible.

In the ESEM samples stay hydrated during the whole imaging process. Therefore shrinking is minimized, making the artefact-free imaging as well as measurements of large areas of the sample possible (Figure 20.12).

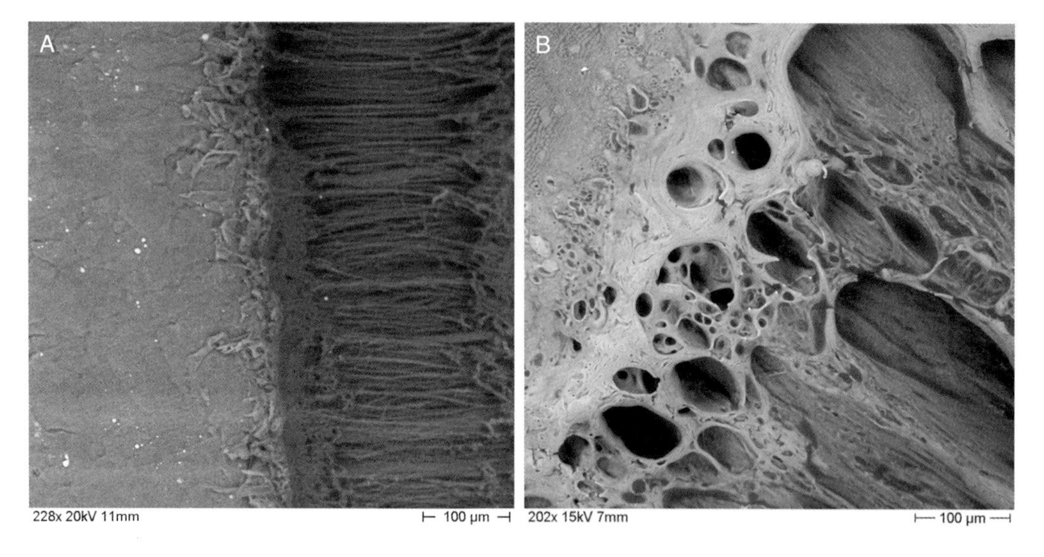

Figure 20.12 Comparison of the tissue damage caused by different types of chirurgical lasers. Left image: tissue (skin) sectioned by pulsed infrared laser (PIRL). The PIRL heats up the tissue so fast that it evaporates directly (so called ablation). The surrounding tissue stays unaffected, causing minimal scar formation. Right image: tissue (skin) sectioned by a CO_2-laser. The thermal load of the tissue caused huge damages (burns), especially visible at the edges of the laser section.

20.4.5 Tissue Microanatomy

Another application of the ESEM in biomedical research is the visualization of internal structures in tissues or organs. To visualize internal structures they must be prepared appropriately using a scalpel or a vibratome. Depending on the used cutting instrument, it is possible to get different types of surface topography and therefore different information. Cutting with very sharp blades or specialized cutting devices leads to very flat surfaces, comparable to microtome sections. The final image of such samples in the ESEM is directly comparable with classical histological sections (Figure 20.13, A and B). Cutting with a scalpel leads to a tearing rather than sectioning and allows certain structures like nerve or muscle fibres to be accentuated (Figure 20.13, C and D). The direct and immediate three-dimensionality of the ESEM images gives a completely new plastic view on many tissues that was only possible previously with the help of cartoon reconstructions.

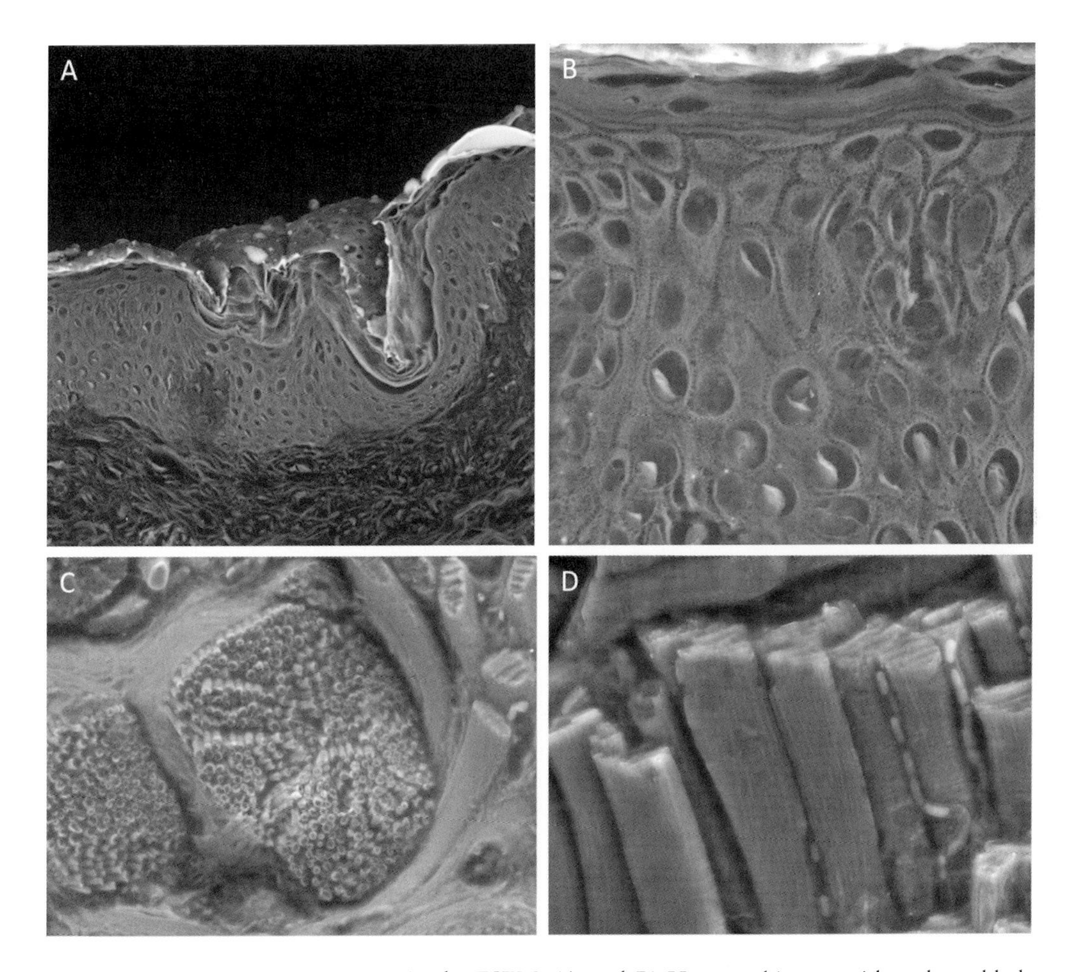

Figure 20.13 Tissue cross-sections in the ESEM. (A and B) Human skin cut with a sharp blade. (C and D) Mouse tongue cut with a scalpel. Single axons surrounded by myelin sheaths (C) and single myofibrils (D) are clearly visible. (C and D) Images courtesy of Carl Zeiss Microscopy GmbH.

20.4.6 Specific Staining

The resolution of the ESEM is not high enough to visualize low density labellings with gold-conjugated antibodies. Although an amplification (e.g. silver enhancement) is thinkable, the final yield of backscatter electrons will still be quite low, depending of course on the detector/amplifier quality but also on imaging conditions like working distance, chamber pressure or beam voltage/current.

A promising alternative to immunogold labelling for ESEM is the use of 3,3′-diaminobenzidine (DAB) oxidation with subsequent osmium staining for generation of a specific contrast. The utilization of targeted oxidation of the 3,3′-diaminobenzidine (DAB) producing an osmiophilic polymer allows specific staining of biological structures for environmental SEM microscopy. The DAB oxidation step can be carried out either by photo oxidation (Grabenbauer, 2005) or enzymatic oxidation in the presence of hydrogen

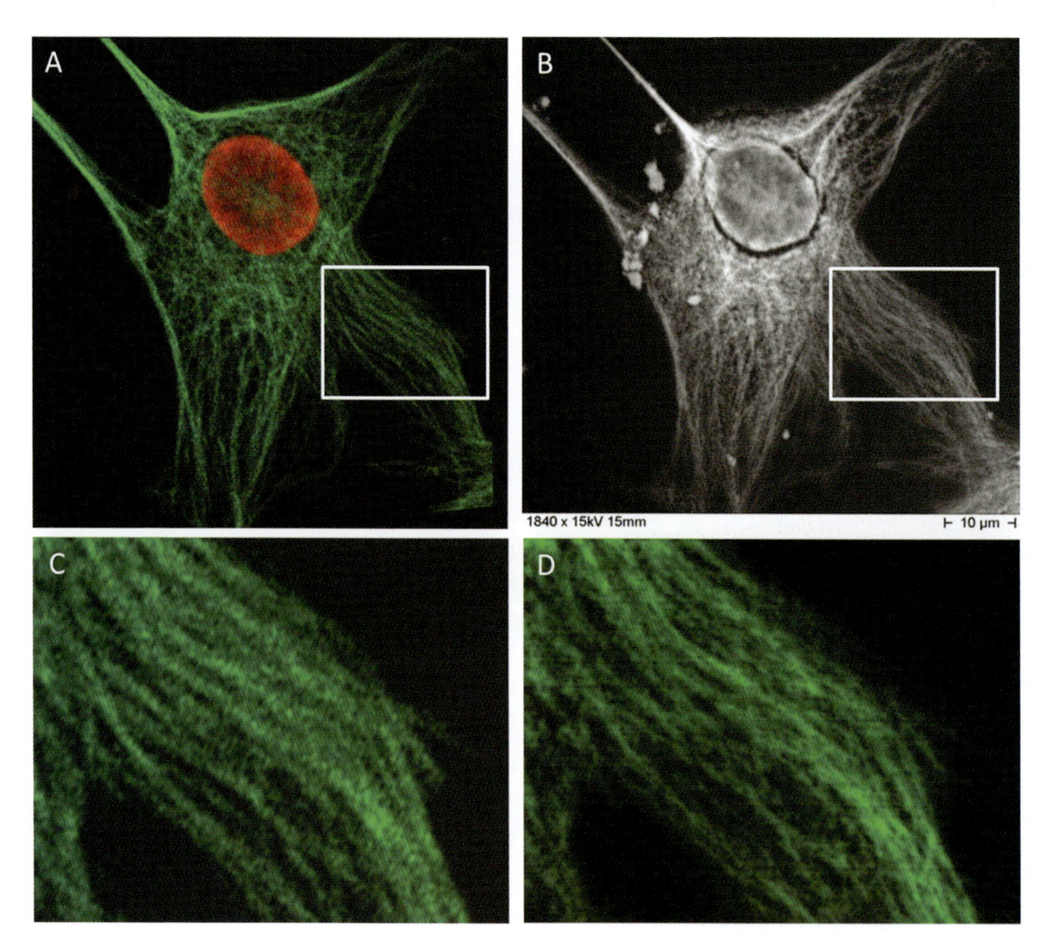

Figure 20.14 Specific contrast generation in the ESEM. Tubulin labelling in a permeabilized human foreskin fibroblast cell, visualized by fluorescence microscopy (A) and by ESEM (B). In the first case an Alexa 488-conjugated secondary antibody was used (green, DNA staining with DRAQ5 in red). For the ESEM imaging the sample was additionally labelled with an HRP-conjugated secondary antibody, incubated with DAB (30 min, until a visible browning occurred) and stained with reduced OsO_4 (5 min, 1% OsO_4 reduced with 1.5% $K_4Fe(CN)_6$). BSE acquisition at 15 kV (B). (C and D) Magnified areas of (A) and (B). The ESEM image (D) was coloured green for a better comparison. The same structures are labelled in both cases, although the ESEM image has a higher resolution.

peroxide. This allows the application of fluorescent protein markers or horseradish peroxidase (HRP)-conjugated antibodies for specific staining (Figure 20.14). The following example illustrates the specificity of this type of staining.

20.4.7 Correlative Fluorescence Light and Environmental Scanning Electron Microscopy

Correlative light and electron microscopy (CLEM) is the method of choice to add structural information from an SEM to the specific fluorescence labelling of the structures of interest. As the shrinkage of thoroughly prepared ESEM samples is negligible (depends on magnification), re-localization of regions of interest is possible after the investigation by fluorescent light microscopy. The major advantage of CLEM with the ESEM is the ability to investigate thick or optically opaque samples because the imaging can be done from the same side/direction (in contrast to systems where the light and SEM paths are separated to opposite sides). This also allows a more precise overlay in the imaging software for such specimens.

For CLEM the sample has to be prepared for fluorescence microscopy first. Then the sample is placed on a Petri dish with a bottom suited for high resolution light microscopy (thickness 170 μm, either microscopy-suited plastic or glass) and imaged using an inverted fluorescence microscope. To image the complete sample at high resolution the use of image stitching/montages can be useful.

After the fluorescence microscopy the sample is contrasted with 1% OsO_4 in PBS, washed in dd H_2O and scanned in the ESEM, keeping the same orientation of the sample.

The fluorescence image facilitates localization of interesting areas in the ESEM (e.g. groups of infected cells). The regions can be relocalized either by prominent features that are visible in both, the fluorescence microscope and the ESEM (Figure 20.15) or by the use of fiducial markers (e.g. fluorescent latex beads or composite nanoparticles).

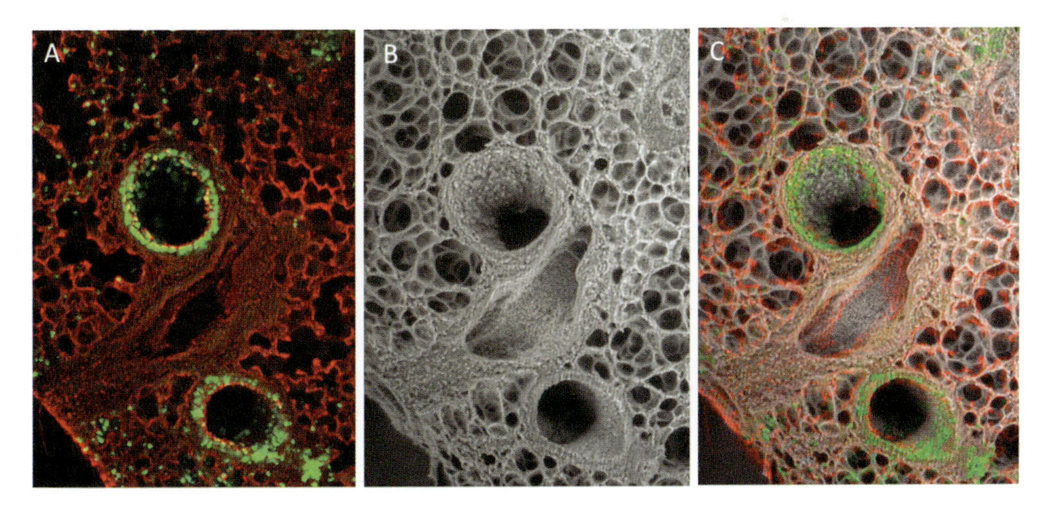

Figure 20.15 Correlative fluorescence light and environmental scanning electron microscopy. (A) Confocal laser scanning microscope (CLSM) image of a vibratome section of a mouse lung 24 h after infection with influenza A viruses. The infected cells expressing GFP are displayed in green. (B) Same region of the lung in the ESEM after contrasting with OsO_4. (C) Overlay of the CLSM and ESEM images.

20.4.8 Living Specimens

Not only fixed and specially prepared but also native unfixed living biological specimens can be investigated in the ESEM. This enables the study of small insects or arachnids and their movements in a scanning electron microscope. However, living specimens (Figures 20.16 and 20.17) need some stimuli (see Figure 20.7). If transferred on top of the agar matrix the small animals remained motionless. After placing a piece of Mortadella sausage on to the matrix the maggots started to move and at least to eat. This observation could be reproduced with cheese mites, adding a piece of cheese and with nematodes. Here we served bacterial suspensions.

Figure 20.16 Living *Lucilia sericata* maggot (A) and cheese mite *Tyroglyphus casei* in the ESEM (B). The images are taken from a time series acquired with approximately 3 frames/sec. Higher frame rates impair the signal-to-noise ratio.

20.5 PREPARATION STEPS FOR ESEM

20.5.1 Investigation of Living Specimens

Living samples do not need a special preparation procedure. Plant tissues can be imaged directly. Arthropods can be imaged alive and in movement. They survive the conditions in the ESEM and therefore can even be taken out from the microscope still alive after imaging. To persuade the animal to move, special stimuli like light or scents are needed. Temperature

Figure 20.17 Wuerchwitz mite cheese. The mites inhabit the bran surrounding the cheese.

control of the sample holder with a Peltier device is also very important. To evaluate the exact parameters of the microscope, like pressure and temperature for the imaging of living samples, a reference sample like the cheese mite (*Tyroglyphus casei,* available from the 'Wuerchwitz mite cheese', www.milbenkäse.de) can be useful.

An approximately 1 cm piece of cheese is placed on the sample holder. At a pressure range of 600–700 Pa the sample chamber is filled with water vapour in 'manual' operating mode. The imaging is performed at a sample chamber pressure between 650 and 2000 Pa. The cheese mites *Tyroglyphus casei* are very robust and can be imaged under these conditions in the ESEM while moving for prolonged times.

20.5.2 Investigation of Hydrated Tissues

For the investigation of tissues for which the BSE modus is best suited see Section 20.2.2, Imaging of Hydrated Samples.

1. Organs, organ parts or tissues are fixed with 4% formaldehyde or with 2.5% glutaraldehyde.
2. Specimens are washed in PBS for 3×5 min.
3. Under a stereomicroscope the organs, organ parts or tissues are sectioned, preferably to small slices, using a suitable scalpel or blade.
4. The tissue slices are contrasted by incubation in 1% OsO_4 in PBS for 15–60 min (depending on the type of tissue).
5. After osmification the samples must be washed thoroughly in dd H_2O to remove excess OsO_4 that might degas from the samples. This step is very important as the degassing of OsO_4 would contaminate the microscope column.
6. The specimen holder (a small Petri dish) is filled with 2% agar in dd H_2O. After stiffening the agar. small holes have to be punched in. This will prevent the bursting of the agar due to water outgassing. The agar serves as a hydrating matrix keeping the sample hydrated and preventing the sample from drying out (Figure 20.7).
7. The sample is placed directly on the agar and can be imaged in the ESEM is this state. A light microscopic image of the sample on the agar can help to orientate during ESEM imaging.
8. The vacuum system of the sample chamber of the ESEM is operated in manual mode. At 600–700 Pa the sample chamber is filled with water vapour. Then the pressure is slowly lowered to 266–400 Pa. This prevents spontaneous freezing of the wet sample caused by evaporation cooling of the water. When the chamber pressure of 266 Pa is reached the working distance is set to approximately 5 mm and the sample is focused.

Every sample behaves differently concerning the hydration conditions in the ESEM; therefore the parameters still need to be optimized for every sample. A Peltier device for cooling is not necessary because the evaporation of the water is permanently cooling the sample. The sample itself is a cold trap when its temperature is below the surrounding temperature. This enables a fast hydration due to the increase of the water vapour pressure inside the sample chamber (see Figure 20.5).

20.6 SUMMARY AND OUTLOOK

ESEM allows the fast and efficient imaging of biomedical specimens after very few and uncomplicated preparation steps. The ESEM is not a high resolution imaging method compared to other established high-vacuum SEM systems. However, it allows fast and easy 3D SEM studies of the surface of hydrated material and the investigation of dynamic processes and living organisms beyond LM resolution. Additionally, ESEM offers the analysis of processes combined with adsorption and desorption of different kinds of water. In principle no total chemical fixation is necessary for ESEM, as only the surface layer should be fixed and contrasted. The dramatic structural changes due to total dehydration for SEM can be totally avoided. Moreover, no artificial surface layer (e.g. heavy metal films) has to be added in order to guarantee conductivity of the specimen and to prevent extensive surface charging.

It was shown that the usual wet-mode pump cycles can partially or totally dehydrate biologic material in the ESEM. Less strongly bound bulk water on the specimen surface tends to evaporate. Uncontrolled ad- or desorption of water may lead to temperature variations on surfaces, including possible specimen freezing.

Our studies point out that tissues, matrices and organisms with more or less strongly bound structural and associated water survive longer investigation periods only in combination with water-binding matrices introduced in our lab. Special water reservoirs (laminated agar matrices, matrix integrated molecular sieves or superadsorbers) prevent or at least delay dehydration processes of biomaterial. This allows reinvestigation and recultivation of identical material that is controlled constantly and reliably hydrated. Besides controlled humidity and fine-regulated ESEM internal pressure and temperature, other stimulation and environment factors (light, natural environment matrices) are also of importance for the native-like imaging of living specimen.

Using native material (e.g. clinical specimens), the sampling methods (avoidance of cutting artefacts, etc.), time periods from sampling through to final ESEM investigation (fast transfer) and the stability of the humid environment are of great importance in respect to preserving the fine structural quality of the specimen. In special cases structural bound water is directly present at the outer surface of the specimen, for example in bacteria biofilms. Here the removal of water from the surface is not possible because a unique strong water-binding matrix is present all over the biofilm.

Moreover, the ESEM in its wet mode allows fast *in situ* microanatomy investigations and analysis of underlying – not directly accessible – specimen structures supported by micromanipulator driven cutting instruments, because no charging effects occur while removing the overlying layers (see Figure 20.18).

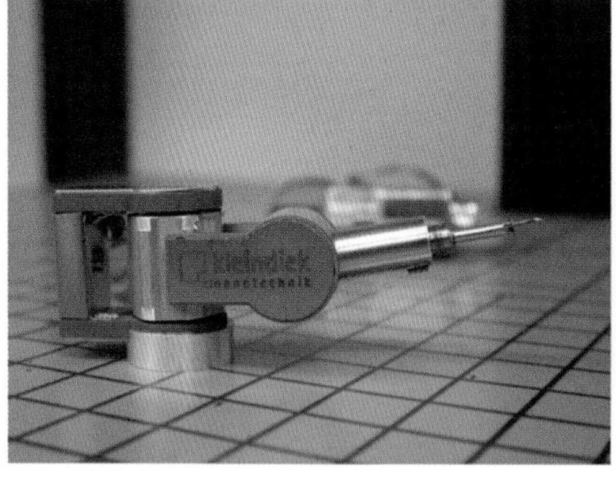

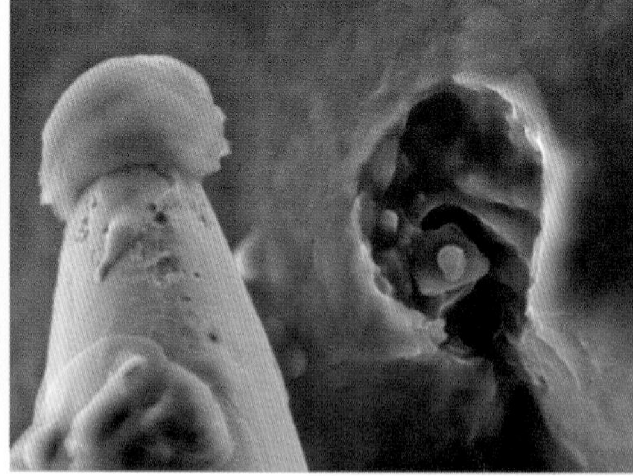

Figure 20.18 Nanomanipulator for SEM with piezoelectric actuators. Tip of the manipulator with the dissected sample of a mouse tumour in the ESEM. Image courtesy of Robert Getzieh, HPI.

Backscattered electron contrast, applied for imaging wet samples under water vapour conditions, allow high contrast sample details to be imaged of different tissues (skin, brain, liver, lung biopsies). Special preparation cycles integrate correlative imaging approaches: (a) LM applications for the overview and orientation in the case of larger specimens or to preselect specimen details of interest and (b) the combination of ESEM investigation with follow-up cryo-methods for the controlled imaging of surfaces and internal specimen structures of identical material in the TEM at high resolution after low temperature dehydration and embedding. Now and in the future, special markers such as composite nanoparticles, which can be identified in light and electron microscopy, are absolutely necessary and important tools to correlate identical areas in the same specimen.

Natural surfaces of every organ (e.g. skin) can directly be investigated after biopsy, inclusive adhering biofilms. If the material has to be cross-sectioned for the investigation the fixation and/or osmification process can directly take place in the course of the cutting process. This guarantees a direct and very fast fixation of the freshly cut surfaces.

Analytical methods applied during ESEM imaging could allow 3D histology with a wide spectrum of classical histological dyes as they are often based on different metal salts (see Figure 20.19).

We were able to clearly show that the ESEM is a fantastic instrument to investigate clinical material in 3D without the need for complicated and time-consuming preparation procedures. For the future it would be highly recommended for microscope manufacturers to produce less complicated and semiautomatic ESEM variants as tabletop devices for high-quality and very fast clinical research and diagnosis, for example during surgery or for tumour staging with the help of rapid sectioning or biopsy methods. 3D microdissection and pinpoint removal of native material in the ESEM would open completely new perspectives in microanatomy or molecular biology.

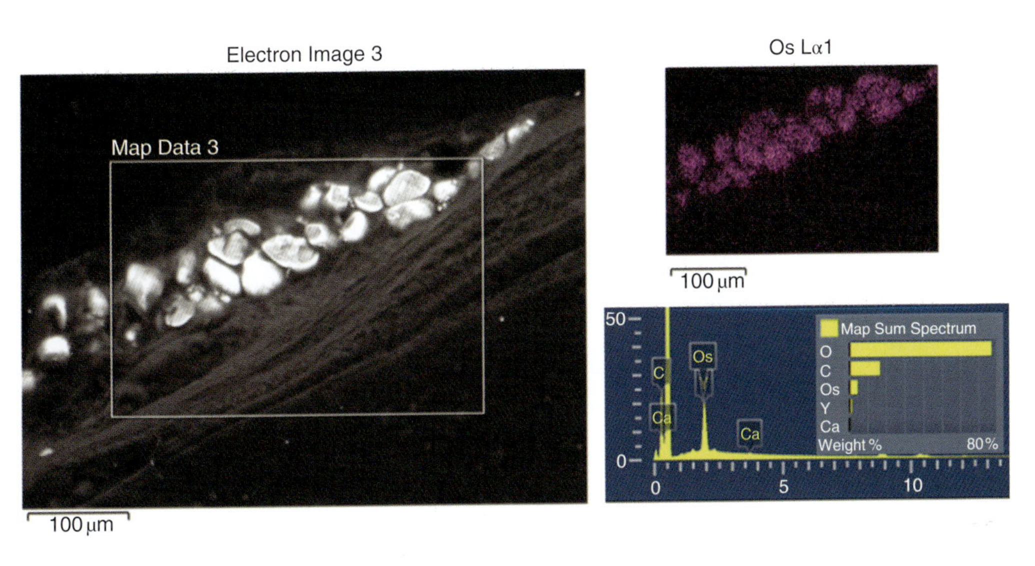

Figure 20.19 Energy dispersive analytics in the ESEM. (A) Mouse adipocytes, stained with OsO_4 and imaged at 15 kV and 400 Pa with a BSE detector. (B) Energy dispersive spectroscopy (EDS) image of the same area showing the osmium staining. The signal spectrum is shown below. Image courtesy: Carl Zeiss Microscopy GmbH.

ACKNOWLEDGEMENTS

Although the images were generated by the authors they were described within the projects outlined by Bartelt *et al.* (2011), Böttcher *et al.* (2013, 2015), Hess *et al.* (2013), Jowett *et al.* (2014), Linke *et al.* (2014, 2015) and Resa-Infante *et al.* (2014).

REFERENCES

Ball, Ph. (1999) *Life's Matrix*, University of California Press.

Bartelt, A., Bruns, O.T., Reimer, R., Hohenberg, H., Ittrich, H., Peldschus, K., Kaul, M.G., Tromsdorf, U.I., Weller, H., Waurisch, C., Eychmüller, A., Gordts, P.L., Rinninger, F., Bruegelmann, K., Freund, B., Nielsen, P., Merkel, M. and Heeren, J. (2011) Brown adipose tissue activity controls triglyceride clearance. *Nat. Med.*, 17 (2), 200–205.

Böttcher, A., Jowett, N., Kucher, S., Reimer, R., Schumacher, U., Knecht, R., Wöllmer, W., Münscher, A. and Dalchow, C.V. (2013) Use of a microsecond Er:YAG laser in laryngeal surgery reduces collateral thermal injury in comparison to superpulsed CO_2 laser. *Eur. Arch. Otorhinolaryngol.*, 271 (5), 1121–1128.

Böttcher, A., Kucher, S., Jowett, N., Krötz, P., Reimer, R., Schumacher, U., Anders, S., Knecht, R., Münscher, A., Dalchow, C.V. and Miller, D. (2015) Reduction of thermocoagulative injury via use of a picosecond infrared laser (PIRL) in laryngeal tissues. *Eur. Arch. Otorhinolaryngol.*, 272 (4), 941–948.

Danilatos, G.D. and Robinson, V.N.E. (1979) Principles of scanning electron microscopy at high specimen chamber pressures. *Scanning*, 2, 72–82.

Grabenbauer, M., Geerts, W.J., Fernadez-Rodriguez, J., Hoenger, A., Koster, A.J. and Nilsson, T. (2005) Correlative microscopy and electron tomography of GFP through photooxidation. *Nat. Methods, November*, 2 (11), 857–862. PubMed PMID: 16278657.

Hess, M., Hildebrandt, M.D., Müller, F., Kruber, S., Kroetz, P., Schumacher, U., Reimer, R., Kammal, M., Püschel, K., Wöllmer, W. and Miller, D. (2013) Picosecond infrared laser (PIRL): An ideal phonomicrosurgical laser? *Eur. Arch. Otorhinolaryngol.*, 270 (11), 2927–2937.

Jowett, N., Wöllmer, W., Reimer, R., Zustin, J., Schumacher, U., Wiseman, P.W., Mlynarek, A.M., Böttcher, A., Dalchow, C.V., Lörincz, B.B., Knecht, R. and Miller, R.J. (2014) Bone ablation without thermal or acoustic mechanical Injury via a novel picosecond infrared laser (PIRL). *Otolaryngol. Head Neck Surg.*, 150 (3), 385–393.

Kaehr, B., Townson, J.L., Kalinich, R.M., Awad, Y.H., Swartzentruber, B.S., Dunphy, D.R. and Brinker, C.J. (2012) Cellular complexity captured in durable silica biocomposites. *Proc. Natl Acad. Sci. USA*, Oct. 23, 109 (43), 17336–17341; doi: 10.1073/pnas.1205816109. Epub 2012 October 8. PubMed PMID: 23045634; PubMed Central PMCID: PMC3491527.

Linke, S.J., Ren, L., Frings, A., Steinberg, J., Wöllmer, W., Katz, T., Reimer, R., Hansen, N.O., Jowett, N., Richard, G. and Dwayne Miller, R.J. (2014) Perspectives of laser-assisted keratoplasty: Current overview and first preliminary results with the picosecond infrared laser ($\lambda = 3$ μm) (in German). *Ophthalmologe*, 111 (6), 523–530.

Linke, S., Frings, A., Ren, L., Gomolka, A., Schumacher, U., Reimer, R., Hansen, N.O., Jowett, N., Richard, G. and Miller, R.J. (2015) A new technology for a plantation free corneal trephination: The picosecond infrared laser (PIRL). *PLoS One*, March 17, 10 (3), e0120944. doi: 10.1371/journal.pone.0120944.

Negendank, W. (1986) The state of water in the cell, in *Science of Biological Specimen Preparation* (eds M. Müller, R. Becker, A. Boyd and J. Wolosewick), SEM Inc., AMF O'Hare, Chicago, IL, USA.

Resa-Infante, P., Thieme, R., Ernst, T., Arck, P.C., Ittrich, H., Reimer, R. and Gabriel, G. (2014) Importin-α7 is required for enhanced influenza A virus replication in the alveolar epithelium and severe lung damage in mice. *J. Virol.*, 88 (14), 8166–8179.

Sakalli, Y. and Trettin, R. (2015) Investigation of C(3) S hydration by environmental scanning electron microscope. *J. Microsc.*, April 16; doi: 10.1111/jmi.12247. [Epub ahead of print] PubMed PMID: 25882158.

Stokes, D. (2008) *Principles and Practice of Variable Pressure: Environmental Scanning Electron Microscopy (VP-ESEM)*, John Wiley & Sons Ltd.

Wiggins, P.M. (1990) Role of water in some biological processes. *Microbiol. Rev.*, December, 54 (4), 432–449. Review. PubMed PMID: 2087221; PubMed Central PMCID: PMC372788.

Zankel, A., Poelt, P., Gahleitner, M., Ingolic, E. and Grein, C. (2007) Tensile tests of polymers at low temperatures in the environmental scanning electron microscope: An improved cooling platform. *Scanning*, Nov.–Dec., 29 (6), 261–299. PubMed PMID: 18076055.

21

Correlative Array Tomography

Thomas Templier and Richard H.R. Hahnloser
Institute of Neuroinformatics, University of Zurich and ETH Zurich, Neuroscience Center Zurich, Zurich, Switzerland

21.1 INTRODUCTION

21.1.1 Array Tomography and Its Tradeoffs

The understanding of structure–function relationships in biological tissues necessitates the visualization of both proteins and their three-dimensional physical context. Several microscopy imaging techniques can visualize either the former or the latter. However, the quest for a single technique that can readily capture both remains open. Array tomography (AT), introduced in 2007 (Micheva and Smith 2007) is an approach to resolve conflicting requirements for the simultaneous volumetric ultrastructural observation of biological samples with the resolution of electron microscopy together with the analysis of antigens by means of fluorescent light microscopy (LM). In the following, we review diverse AT protocols and discuss their strengths and weaknesses.

It is well known that the preservation of both antigenicity and ultrastructure are two largely incompatible aims of current tissue preparation protocols (Shu et al. 2011; Rostaing et al. 2004; Grabenbauer et al. 2005; Fiserova and Goldberg 2010; Salio et al. 2011; Morgan and Lichtman 2013; Newman et al. 1983). This incompatibility prevents the simultaneous observation of both molecular and physical architectures. To address this issue, three main array tomography sample preparation methods have been introduced. The original approach aimed to visualize the molecular composition of brain tissues (Micheva and Smith 2007; Micheva et al. 2010a; Busse and Smith 2013; Kopeikina et al. 2011; Allen et al. 2012; Micheva and Bruchez 2012) and employed a sample preparation protocol tuned for antigenicity preservation (e.g., no glutaraldehyde fixation, no heavy metal staining, bench embedding, and resin infiltration) at the expense of the loss of ultrastructure quality. A first variation of this original protocol has been recently introduced by the same laboratory (Collman et al. 2015) to provide a better ultrastructure quality while

Biological Field Emission Scanning Electron Microscopy, First Edition.
Edited by Roland A. Fleck and Bruno M. Humbel.

maintaining a comparable LM quality. That improved protocol relies mainly on freeze substitution and the absence of osmium tetroxide staining. The second variation to the original AT sample preparation, correlative array tomography (CAT), which we detail in this chapter, aims at tissue ultrastructure preservation necessary for assessment of neural connectivity, at the expense of prohibiting access to the endogenous molecular architecture (Oberti et al. 2010, 2011). CAT makes use of the fixative agent glutaraldehyde and requires heavy metal staining for ultrastructural contrast. CAT offers several advantages over many electron microscopy (EM)-only techniques, namely convenient volumetric data acquisition, easy simultaneous handling, staining and storage of hundreds of sections, and, most importantly, suitability for imaging by correlative light and electron microscopy procedures.

21.1.2 Volumetric Electron Microscopic Imaging: To Handle, Stain, and Store Hundreds of Ultrathin Sections

A key component of AT relies on the production of arrays of ultrathin sections from resin-embedded biological samples. Ultrathin serial sectioning substantially increases the resolution along the depth axis from an photonic resolution of at best about 700 nm (Micheva and Smith 2007) to the physical sectioning resolution in the 30–200 nm range. Subsequent sample collection provides the ability to create libraries of sectioned samples that can be processed at any time (Hayworth et al. 2006) (note, however, that LM imaging should be performed shortly after staining (see the supplementary information in Micheva et al. 2010b). For high-resolution EM imaging, a microscope operator has the choice of either acquiring complete (imaging of all sections) or partial datasets from a portion of the sectioned tissue and return to specific areas for more detailed analysis at a later date. Image acquisition can be performed with advanced sample navigation tools (Terasaki et al. 2013; Hayworth et al. 2014).

Collection of a large number of thin sections by ultramicrotomy and their mounting on a single rigid physical substrate provides a convenient means for thin section handling, loading in light or scanning electron microscopes (SEMs), and sample storage, compared, for example, with dexterous manual handling of fragile grids required in transmission electron microscopy (TEM). As we describe in detail in Section 9.2.2, AT on a rigid substrate allows for on-section immunohistochemistry, which is achieved by depositing the labeling solution on the flat substrate. In this way, all sections can be simultaneously stained, avoiding the need both for EM grid staining machines used in TEM (Van Damme et al. 2006) and for time-consuming and error-prone manual handling and loading of TEM grids (Anderson et al. 2009, 2011; Cardona et al. 2010). Rigid substrates usually fit through the airlock opening of SEMs (large substrates of up to 10 cm × 10 cm can also be readily loaded via the chamber door), allowing simple loading of dozens up to potentially thousands of serial sections at a time.

By contrast, non-AT approaches to volumetric imaging such as focused ion beam (FIB) (Knott et al. 2008) and serial block face (SBF) (Briggman et al. 2011) scanning electron microscopy (SEM) both irreversibly and systematically destroy the tissue after imaging (destructive techniques), forcing the experimenter to either take the risk of missing regions of interest or to image the entire exposed area, the latter of which slows down the acquisition process and introduces challenges around postprocessing and evaluation of very large datasets.

21.1.3 Correlative Light and Electron Microscopy

The need for correlative light and electron microscopy lies in the intrinsic properties of biological tissues, namely the intricate relationship between the molecular and physical architectures. In neuroscience, the gold standard imaging technique for analysis of morphology and connectivity of neural tissue at the level of single synapses and organelles is undoubtedly electron microscopy (Morgan and Lichtman 2013; Helmstaedter 2013). Nevertheless, the last decade has seen a significant increase in the number of correlative microscopy studies in biology (Brown et al. 2009; Giepmans et al. 2005; McDonald 2009; Watanabe et al. 2011). Mainly two combinations of light and electron microscopy imaging modalities have been explored: (1) confocal light microscopy and subsequent focused ion beam SEM (Sonomura et al. 2013; Canty et al. 2013); (2) two-photon microscopy followed by either serial block face SEM (Briggman et al. 2011), TEM (Bock et al. 2011), or FIB-SEM (Maco et al. 2013). Compared to AT, all these methods suffer from the inability to combine the two modalities on the exact same sample and at the same stage of the processing pipeline. This inability entails that additional efforts are required to achieve the desired correlation of LM-EM modalities. These alternative methods not only necessitate extremely careful sample handling and preparation in order to conserve regions of interest for subsequent EM imaging, but they also lead to difficult computer vision problems arising from the much smaller spatial resolution of LM versus EM imagery.

We report here only on AT approaches based on wide field fluorescence microscopy. In principle, the arrays of sections collected on a rigid support could also be imaged with new generation subdiffraction light microscopes such as STED (Punge et al. 2008) or STORM (Nanguneri et al. 2012), the latter of which achieves an impressive volumetric resolution of 28 nm × 28 nm × 40 nm (Nanguneri et al. 2012).

21.1.4 Workflow

The CAT workflow presented in this chapter is sketched in Figure 21.1; all steps are described and discussed throughout the core of this chapter.

21.2 CAT SAMPLE PREPARATION PROTOCOLS

The sample preparation protocol should be carefully chosen depending on the goal of the experiment. We describe here the protocol optimized for circuit tracing developed in our laboratory and briefly review the one optimized for proteometric analysis, as originally introduced with AT in 2007. The differences in these protocols reflect the well-known compromise between ultrastructure and antigenicity preservation (Fiserova, J. and Goldberg 2010; Salio et al. 2011; Muehlfeld and Richter 2006; Ghrebi et al. 2007; Stierhof and Kasmi 2010).

21.2.1 Fixation and Embedding

Fixation, dehydration, and resin embedding are necessary steps in order to visualize biological tissue in electron microscopes. In the following, we summarize the key differences between CAT protocols for circuit tracing (Oberti et al. 2010, 2011) and for proteometric analysis (Micheva and Smith 2007; Micheva et al. 2010a).

21.2.1.1 CAT Sample Preparation for Circuit Tracing

The sample preparation protocol optimized for correlative circuit tracing developed in our laboratory contains two different heavy metals (1% osmium tetroxide and 1% uranyl acetate) to ensure strong staining of membranes (see Oberti et al. 2010 for a detailed protocol).

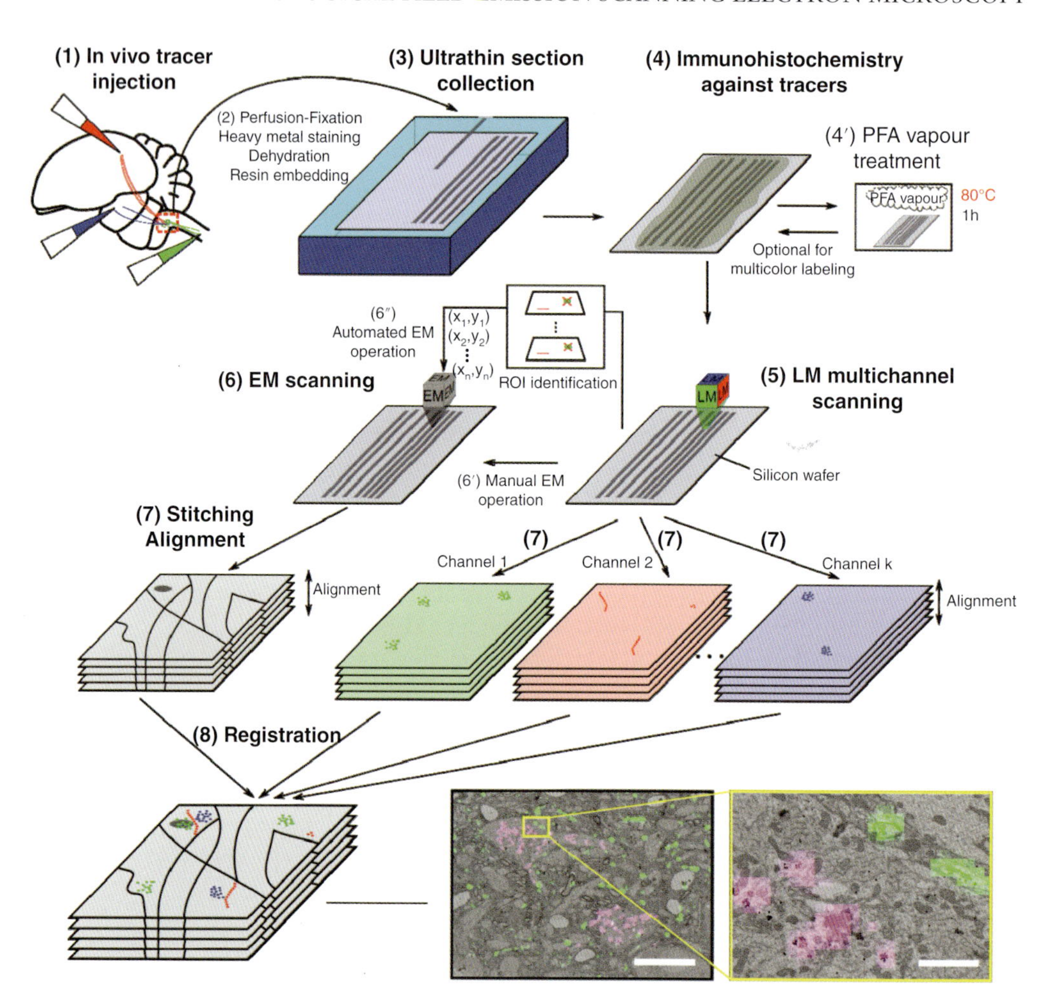

Figure 21.1 Workflow of correlative array tomography (CAT) for identification of neuron types in their ultrastructural context. (1) *In vivo* injection of neuroanatomical tracers to label structures of interest. (2) The animal is perfused with fixative for optimal fixation. The brain is dissected out, cut manually or with a vibratome to extract a region of interest. This region is subsequently stained with heavy metals, dehydrated, infiltrated with epoxy resin, and cured at 52 °C. (3) Sectioning of ultrathin sections of the resin-hardened sample and collection on a flat conductive silicon wafer. (4) Immunohistochemistry is performed on the silicon wafer by placing drops of staining solution on the substrate. (4′) Optional treatment with formaldehyde vapor to destroy free remaining binding sites of secondary antibodies; this treatment allows the staining of different antigens with two different antibodies stemming from the same species. (5) The silicon wafer is cover-slipped and scanned in a wide-field fluorescent microscope. (6) The wafer is subsequently scanned in the EM. The locations to scan in the EM are either defined manually by the EM operator (6′) or calculated from the location of objects of interest, identified in the LM (6″). (7) Images acquired in the LM and the EM are then automatically stitched and aligned using custom software, resulting in several aligned volumetric image stacks. (8) The EM and LM datasets are automatically aligned using custom software. White scale bars bottom left and bottom right: 15 μm and 2 μm.

Table 21.1 Neuroanatomical tracers tested for antigenicity preservation in our embedding protocol. – = no survival, + = some survival

Carrier	Hapten	Antigenicity	Fluorescence	Product number (Life Technologies)	Antibody species	Product number
Dextran	BDA	+	–	D-1956	Mouse Streptavidin	Jackson Immu. : 200-002-211 Life Tech.: S-11223
Dextran	488	+	–	D-22910	Rabbit Rat	Life Tech.: A-11094 Biotem: custom
Dextran	Texas Red	+	+	D-3328	Rabbit Goat	Life Tech.: A-6399 Vector Labs: SP-0602
Dextran	Fluorescein	+	+	D-1820	Rabbit	Life Tech.: A-889
Dextran	Lucifer Yellow	+	–	D-1825	Rabbit	Life Tech.: A-5750
Dextran	647	–	–	D-22914	Mouse Guinea Pig	Abcam: ab52060 Biotem: custom
Dextran	Tetramethyl-rhodamine	–	–	D-1817	Rabbit	Life Tech.: A-6397

Similarly to published protocols that yield good ultrastructure quality and that to some extent enable postembedding immunohistochemistry (IHC) (Heck et al. 2002; Kraehenbuhl and Jamieson 1973), fixation in our protocol is performed with 4% formaldehyde (FA) and 0.075% glutaraldehyde (GA) diluted in 0.1 M phosphate buffer at pH 7.4. This protocol yields good ultrastructure but destroys antigenicity of endogenous proteins. Namely, none of the following endogenous molecules could be visualized using immunohistochemistry in our laboratory: acetylcholine, parvalbumin, synapsin, and PSD-95 (data not shown); similar findings are reported in Collman et al. (2015). The Looger laboratory has recently developed endogenous tags that survive to some extent mild EM embedding protocols (Paez-Segala et al. 2015; Viswanathan et al. 2015). It is not known though whether these tags would also survive harsher protocols optimized for connectomics (Mikula and Denk 2015; Hua et al. 2015; Tapia et al. 2012).

Fortunately, we found that some exogenous molecules such as biotin and some fluorophores carried by neuroanatomical tracers conserve their antigenicity and even their fluorescence in some cases after embedding for electron microscopy (see Table 21.1).

21.2.1.2 AT Sample Preparation for Proteometric Analysis

AT was introduced in 2007 for high-dimensional proteometric analysis using fluorescence microscopy. The original AT sample preparation has been tuned to maximize antigenicity at the expense of good ultrastructure. In this paragraph, we detail the steps of AT sample preparation together with explanations. The choice of the fixative mixture has a crucial impact on the ultrastructure quality and antigenicity preservation of endogenous molecules.

As described in complete detail in Micheva and Smith (2007), the brain is first fixed in 4% formaldehyde only, without any glutaraldehyde. It is thought that glutaraldehyde fixation is harsh and leads to alteration of the 3D conformation of relatively large molecules, whereas it can retain small molecules such as metabolites (Anderson et al. 2011). Some antigens

(GABA, glutamate, PIP2) were visualized (Micheva and Smith 2007) only when glutaraldehyde was added to the fixative solution. The effects of fixation on the ability of an antibody to bind to its target are complex, and we refer the reader to the following publications (Hayat 1986; Glauert and Lewis 1999; Skepper and Powell 2008) for further reading.

Osmium (with proteolytic activity) is omitted in AT, because it greatly alters the three-dimensional conformation of many endogenous proteins, thus making them inaccessible to IHC (Collman et al. 2015). Osmium is, however, an excellent staining agent for electron microscopy (Mikula and Denk 2015; Hua et al. 2015; Tapia et al. 2012). Uranyl acetate is also omitted even though it is considered less harsh than osmium tetroxide in terms of alteration of three-dimensional conformation (Collman et al. 2015; Phend et al. 1995; Terzakis 1968; Berryman and Rodewald 1990).

Also, in the original AT study, the tissue dehydration prior to resin infiltration was pursued only up to 95%. The resin LR White was preferred over other resins because it preserves antigenicity (Newman et al. 1983; Luby-Phelps et al. 2003); LR White polymerizes at a temperature of 50 °C. Epon would probably have led to optimal sectioning quality; however, Epon is a hydrophobic resin that tends to react more with biological molecules (Hayat 1995). Synapse counts reported with this AT protocol are consistent with synapse densities obtained with stereological methods using electron microscopy. Many antibodies against endogenous proteins have been successfully used, including well-known synaptic proteins such as Synapsin, Synaptophysin, VGluT1, VGluT2, PSD-95, NMDAR, GAD and Gephyrin (Micheva and Smith 2007; Micheva et al. 2010a).

21.2.1.3 *Other Variations of CAT Sample Preparation Protocols*

We review in this section several recent studies that have introduced variations to the original AT sample preparation protocol. The variations reflect the tight compromise in sample preparation to achieve both preservation of antigenicity and ultrastructure.

Stemming from the laboratory that originally developed AT, one study (Collman et al. 2015) reports on "conjugate array tomography", an AT variation that better preserves ultrastructure while maintaining good antigenicity of endogenous molecules. This feat is mainly due to the following modifications: perfusion fixation is done with 2% FA, 2% glutaraldehyde at pH 6.8, uranyl acetate staining is done at –90 °C with a high concentration of 2–4%, infiltration with the Lowicryl resin HM20 is done at –45 °C and polymerization is performed with UV radiation at room temperature.

In Rah et al. (2013), thalamocortical input on to layer 5 pyramidal neurons in a mouse was investigated. Authors used genetic lines and viral vectors to express endogenous fluorescent proteins in pre- and postsynaptic neurons of interest. In order to validate the location of putative synapses identified with AT, the authors correlated LM imagery with EM imagery. To this end, they developed a protocol that provided enough ultrastructural contrast for synapse identification, while antigenicity of endogenous proteins was retained. Their sample preparation protocol includes 0.2% of glutaraldehyde in the perfusion fixation solution, 0.001% osmium tetroxide staining and a low-temperature (–20 °C) infiltration and polymerization of the hydrophilic resin LR White.

In a large collaborative effort, new fluorescent probes have been designed (Viswanathan et al. 2015) that can be targeted with IHC after perfusion fixation with 4% formaldehyde and 0.2% glutaraldehyde, staining with 1% osmium tetroxide and freeze-substitution embedding with the HM20 resin. It remains to be tested whether these new probes can be processed using protocols aimed at ultrastructure preservation (Mikula and Denk 2015; Hua et al. 2015; Tapia et al. 2012).

21.2.2 Section Cutting and Collection

We briefly review several AT-compatible techniques for the collection of ultrathin sections of resin-embedded tissue. These can be classified based on the type of substrate on which the sections are collected: flat conductive substrate or conductive flexible tape.

21.2.2.1 Flat Conductive Substrate

There exist several flat conductive substrates for correlative microscopy, including indium tin oxide coated (ITO) glass (e.g., coverslips or LM slides) and silicon wafers (Ted Pella, #16015 Type P <100> or #21610-6). We found that the latter substrate presents several advantages (Oberti et al. 2010, 2011; Horstmann et al. 2012). (1) Silicon wafers are naturally conductive and they do not require any chemical pretreatment (an acid sulfuric and perhydrol pretreatment can be performed to permanently hydrophilize the substrate (Horstmann et al. 2012), but we prefer to hydrophilize temporarily with a simple glow discharge treatment because we later make use of the hydrophobicity for staining). We have observed that samples collected on silicon wafers can be imaged with a high current/probe in SEMs (3 nA) without a charging problem, whereas such high currents are unusable with samples collected on ITO slides because of charge buildup at the surface of the substrate. (2) Collected sections are visible to the naked eye and can be imaged using bright-field reflection microscopes. (3) Silicon wafers are easily cleavable with a diamond scriber into rectangles of any shape (for silicon wafers with a <100> crystal orientation). (4) The hydrophobicity of silicon wafers allows straightforward immunohistochemical (IHC) staining, because drops of solution stay in place, which substantially reduces the amount of solution (and cost) needed for staining. In the example of Figure 21.2, we used only 150 μl of solution for IHC during each staining step to label approximately 600 sections, representing a ratio of 0.25 μl/section. (5) Silicon wafers reflect light; therefore the fluorescence signal emitted in the direction of the substrate is reflected towards the objective, yielding a stronger signal than ITO glass, for example. Finally, (6) silicon wafers are nearly perfectly flat and thus are suitable for new-generation multibeam FEGSEM because all beams can simultaneously be in focus, whereas samples collected on tape might exhibit stronger height variations.

In the next paragraphs we focus on techniques for section collection.

(a) Ribbon pick-up by wafer retraction. To acquire volumetric tissue information it is important to reliably cut and collect large numbers of consecutive ultrathin sections from the same sample, for example using so-called "histo-Jumbo" diamond knifes provided by Diatome (Blumer et al. 2002). These diamond knives are operated over a large water boat in which entire microscope slides can be immersed. The approach proposed initially in Blumer et al. (2002) and enhanced in Micheva et al. (2010c) is to obliquely insert a flat substrate into the water boat of a diamond knife so that the front part is immersed and the back part remains dry above the water. Thereafter, ribbons of consecutive sections are produced (see Micheva et al. 2010c for details), detached from the knife edge, and moved to the substrate with an eyelash. When the beginning of the ribbon reaches the non-immersed part of the substrate, it adheres to it and anchors the whole chain of sections. The substrate is then slowly retracted out of the water, which can be done manually using forceps or with a custom substrate holder such as the one introduced in Horstmann et al. (2012) or the more elaborate one in Spomer et al. (2015).

(b) Ribbon pick-up by water removal. Instead of partly immersing the silicon wafer (substrate) at an oblique angle to the surface, we prefer to immerse it entirely prior to cutting. During and after the cutting process, the series of ribbons can be moved on the water surface using eyelashes. The water is then slowly removed with a custom-made flexible syringe to deposit the sections on the silicon wafer. As soon as the surface is dry enough to prevent the sections from flowing off the wafer, the substrate is carefully removed and placed on a heating plate at 45 °C for 10 to 30 minutes to uniformly dry the sections without creating folds. To avoid damaging the sections, care should be taken that the knife edge remains dry during the few minutes needed to remove the water (the sections could get stuck on the edge).

We believe that collecting sections on a silicon wafer is ideal for staining and imaging of large numbers of consecutive sections, as shown in Figure 21.2, in which a wafer is shown that carries 589 consecutive sections (two sections are missing). The density of sections on the wafer can be high because ribbons can be moved very close to each other.

21.2.2.2 Flexible Tape

A method for automated section collection is the ATUM (automated tape collection ultramicrotome) developed by K. Hayworth and colleagues et al. (2006). The ATUM consists of a conveyer belt that loads a flexible tape into the boat of a conventional diamond knife.

Figure 21.2 A total of 589 consecutive ultrathin sections (two sections are missing) have been collected on a single silicon wafer. (a) Bright-field mosaic of the wafer surface. The sections are clearly visible as dark regions. (b) All sections have been segmented with custom software. The algorithm detects the section borders first and then identifies section corners. (c) Magnified region of (b) showing three ribbons of ultrathin sections. The yellow frames have been automatically inserted on all subsequent sections after manual definition of a region of interest within the first section, demonstrating automated access to corresponding subregions. (d) Liquid deposited on a silicon wafer. The hydrophobicity of the substrate allows the liquid to stay in place.

Sections are produced with a conventional ultramicrotome and are then collected on the tape and subsequently stored in a reel. An operator can then unroll the tape, cut parts of appropriate lengths (about 1 section per 1 mm), and glue the ribbon parts on to a carbon adhesive tape, which itself is glued on to a silicon wafer. Each silicon wafer holds approximately 200 sections that can be imaged in a normal SEM with a backscattered electron detector. Most tapes come with the drawback of being strongly autofluorescent, thus preventing their use in light microscopy. A possible solution for correlative microscopy is to collect sections on a ribbon made of thin glass (Richard Schalek, Lichtman's laboratory, personal communication).

21.2.3 Postembedding On-Section Immunohistochemistry

21.2.3.1 Neuroanatomical Tracers Retaining Fluorescence and/or Antigenicity

The CAT approach in Oberti et al. (2010, 2011) aims at preserving antigenicity not of endogenous proteins (e.g., synaptic proteins) but of exogenous compounds (e.g., fluorophores of neuroanatomical tracers). We have identified a set of exogenous neuroanatomical tracers (listed in Table 21.1) whose antigenicity and sometimes whose fluorescence survive the harsh embedding protocol. So far we successfully used four tracers in anterograde labeling experiments (biotinylated dextran amine (BDA), Texas Red, Fluorescein, and Dextran 488) and four tracers in retrograde labeling experiments (Texas Red, Fluorescein, Dextran 488, and Lucifer Yellow). We performed tracer localization in ultrathin embedded sections using conventional on-section immunohistochemical staining (Schwarz and Humbel 2007; Fabig et al. 2012) that is composed of the following steps: short etching, blocking, primary antibody labeling, washing, secondary antibody labeling, and final washing (see Oberti et al. 2010 for details).

For example, we have found that the antigenicity of Dextran-647 does not survive the embedding at all. We tried two different commercial antibodies and a custom-made antibody that all successfully stained Alexa647 in fixed wet sections but did not give any positive signal in postembedding IHC. In contrast, the fluorescence of the two fluorophores Texas Red and Fluorescein is preserved after embedding (as shown in Figure 21.3). However, the signal is very weak, and comparing native fluorescence with the signal after IHC shows that most of the fluorophores lost their fluorescence but retained their antigenicity.

We have found that the anterograde tracer biotinylated dextran amine (BDA) exhibits excellent postembedding antigenicity. We recommend using BDA with IHC rather than with avidin–biotin complexes (ABCs) for ultrastructure studies, for the following reasons. (1) In IHC there is no limitation of sample size beyond the limitation set by the penetration depth of reagents during the embedding protocol. In contrast, the ABC technique requires sections thinner than 60–70 μm to allow the reagents to penetrate the wet section. (2) BDA antigens can be labeled either with fluorophores, electron dense gold particles, or both. Note that the labeling of BDA with black diaminobenzidin (DAB) deposit used in the ABC method often obstructs visualization of fine ultrastructural details, whereas immunogold staining is easily adjustable (gold particle size, antibody concentration, incubation duration, and silver enhancement duration) to also visualize surrounding structures of labeled cells. (3) IHC localizes BDA antigens accurately and does not co-label adjacent structures, as can be the case with the ABC method. (4) Using IHC, the tissue can be immediately processed for embedding after microtomy, whereas the ABC procedure necessitates approximately half a day, which may compromise ultrastructure quality in EM.

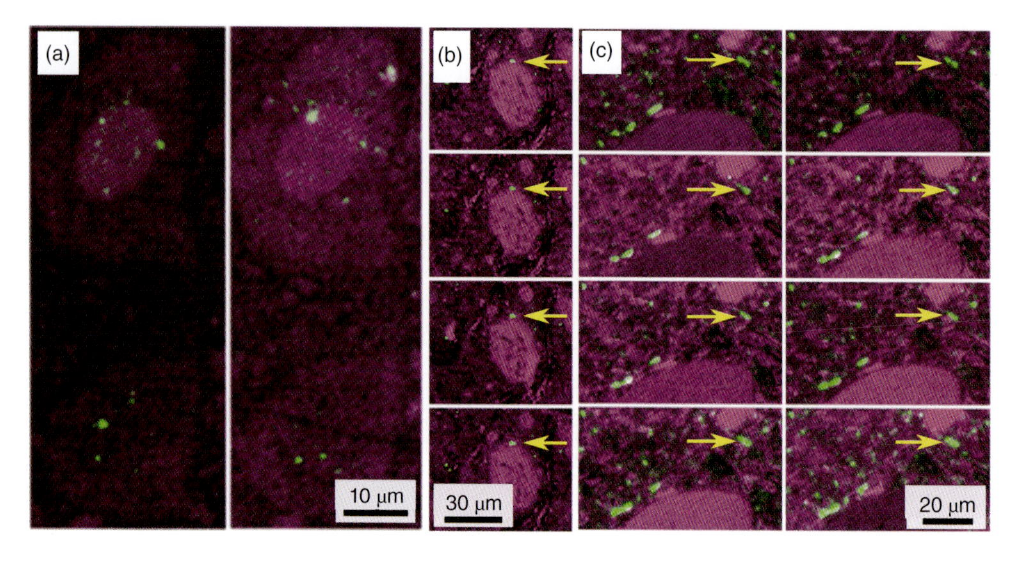

Figure 21.3 Postembedding antigenicity and/or fluorescence preservation of several neuroanatomical tracers. (a) Two consecutive sections showing the survival of fluorescence of Fluorescein (green) in two adjacent retrogradely labeled neurons. (b) Four consecutive sections showing the survival of Fluorescein antigenicity in anterogradely labeled axons. The yellow arrows point to an axon that can be clearly followed over the consecutive sections. (c) Eight consecutive sections showing antigenicity survival of the fluorophore Alexa 488 in anterogradely labeled axons. The arrows indicate an axon that can be clearly followed over consecutive sections.

21.2.3.2 Postembedding Multicolor Imaging

Here we discuss various strategies for simultaneous visualization of different neuroanatomical tracers. The main obstacle towards achieving multicolor imaging is the fact that most commercially available antibodies against fluorophores are raised in the same species and are based on the same isotype (rabbit IgG, see Table 21.1), limiting IHC to only one primary antibody.

(a) Custom antibodies. The first option that we have investigated is the elaboration of custom antibodies raised in species other than the rabbit. The first question arising when producing antibodies is the choice between monoclonal and polyclonal antibodies. Polyclonal antibodies might in theory be less sensitive to 3D conformation alteration of their antigen as they recognize many antigenic epitopes, increasing the probability of recognizing an unmodified epitope. The development of polyclonal antibodies is usually faster, cheaper, and more likely to succeed, compared with the development of monoclonal antibodies. A monoclonal antibody tends to be more sensitive to epitope alterations undergone during EM embedding. In our case, we have successfully used both monoclonal and polyclonal antibodies to target fluorophores and biotin. It should be noted that our targets (fluorophores and biotin) are so small that they probably exhibit a single epitope, similarly to digitonin or polysaccharides, for example, chitin. This implies that monoclonal antibodies might work for our targets against which a polyclonal antibody has proven successful.

    Custom antibodies were produced by the company Biotem (Apprieu, France) within roughly 4 months. We produced polyclonal antibodies raised in three rats against the fluorophore AlexaFluor 488. Given the small size (600 Da) and the known lack of immunogenicity of AlexaFluor 488, we decided to attach it to a keyhole limpet hemocyanin (KLH) protein to trigger an immune reaction, with the result that the produced antibodies recognized specifically the target Alexa Fluor 488 but not the other fluorophores listed in Table 21.1. This custom antibody allowed us to increase the number of simultaneously usable fluorescent channels from 3 to 4 (Rabbit/Lucifer Yellow, Mouse/BDA, Goat/Texas Red, and additionally Rat/488) in same-section multilabeling experiments.

(b) Direct immunohistochemistry. Another strategy to achieve multicolor imaging is to use direct immunohistochemistry, that is, to use labeled primary antibodies. Such an approach allows simultaneous use of primary antibodies raised in the same species. However, direct labeling is achieved at the expense of labeling sensitivity, because available antigens at the surface of the ultrathin section are sparse. Moreover, the conjugation of primary antibodies with fluorophores might be laborious, prone to errors, and might alter the antibody binding properties. These drawbacks led us to not investigate this option.

We note, however, that we satisfactorily used fluorophore-labeled Streptavidin to label biotin. The relatively high sensitivity we achieved is probably due to the affinity of (strept)avidin to biotin, which is very high compared with affinities of classic antibodies. Streptavidin is also convenient because it is readily commercially available with almost any fluorescent or electron-dense tag (fluorophore, gold, fluoronanogold), though gold–streptavidin has the downside of being a poor detection system (B. Humbel, personal communication).

21.2.3.3 Markers for Postembedding On-Section Immunohistochemistry

Many markers have been used in the past decades for visualization of antigens on ultrathin sections. Some are visible in the electron microscope, some in the fluorescence microscope, and some in both. We review in this section markers of interest for AT.

The most broadly used markers for on-section labeling were originally colloidal gold particles (Robenek 1990) coupled to either Immunoglobulin G (IgG) or protein A. Gold particles of different sizes can be used to distinguish various antigens (Wang and Larsson 1985) and are often silver enhanced to augment their visibility in the electron microscope.

Fluorescent markers used for on-section post-embedding labeling allow the visualization of antigens in conventional bright field (Micheva and Smith 2007; Oberti et al. 2010; Watanabe et al. 2011; Schwarz and Humbel 2007; Schwarz 1994) and new generation superresolution fluorescence microscopes. They provide a high versatility by allowing multichannel fluorescence imaging. Bright field microscopes provide a maximum resolution of 200 nm and superresolution microscopes a resolution down to 20 nm (Huang et al. 2009). The scanning speed we achieved with a modern motorized bright-field fluorescence microscope (Zeiss Axioobserver Z2) is about 30 s/30 × 30 μm^2 with four fluorescence channels (1 s exposure time each) and a bright-field illumination channel.

Attempts to visualize antigens in both LM and EM lead to the development of dual markers such as fluoronanogold (Takizawa et al. 1998) and fluorophore-coupled colloidal gold. These markers consist of an immunoglobulin, decorated with both fluorophores and gold particles. We have successfully used colloidal gold Alexa 488 IgG (Life technologies, A-31561) as shown later in Figures 9.6 and 9.7. A simple technique for antigen labeling in both LM and EM is sequential labeling with a fluorophore followed by gold particles (see Fabig et al. 2012 for interesting diverse variants). It should be noted that the close proximity of gold particles and fluorochromes can lead to a decrease in fluorescence (Humbel et al. 1998). In this technique, the incubation time of the secondary antibody is split into two phases: the first one contains gold-labeled IgGs and the second one contains fluorophore-labeled IgGs. The binding affinity of gold-decorated IgGs tends to be smaller than that of fluorophore-decorated IgGs; therefore the gold-labeling step takes place first and is usually longer (we obtained good results with 1 h gold followed by 30 min fluorophore labeling).

Quantum dots are relatively recent markers that have been introduced for their use in biology in Chan and Nie (1998) and popularized in Giepmans et al. (2005) and Michalet et al. (2005). We have used quantum dots coupled to secondary antibodies for on-section immunolabeling and found that the labeling is satisfactory, as also reported in Nisman et al. (2004). That is, the relative brightness (as assessed by measuring signal intensity with constant imaging settings including exposure times) of quantum dots compared with classic fluorophores in our application is approximately the same as in Nisman et al. (2004). The main advantage of quantum dots for multilabeling experiments is their narrow emission spectrum along with a constant broad excitation spectrum. Appropriate emission filters (but not excitation filters) are required to visualize them.

It has been shown (Dahan et al. 2003) that quantum dots are visible as electron dense aggregates in electron micrographs. However, we did not succeed in obtaining a satisfactory signal, even after silver enhancement of variable duration. This limitation might come from the relatively strong background signal of our samples, which is not present in Dahan et al. (2003).

Cathodoluminescent materials exhibit the property of sending photons when hit by electrons. They are excellent candidate markers for CAT because the multidimensional fluorescence signals can be collected simultaneously with the EM signal inside the imaging chamber of a single microscope. Efforts are being undertaken to produce small, spectrally well-separated cathodoluminescent probes that can be used as tags in conventional immunohistochemistry (Glenn et al. 2012; Zhang et al. 2014).

It is worth mentioning the existence of new dual markers called plasmonic fluorophores. They consist of gold nanoparticles and fluorophores being brought into one single construct. The electron-dense compound has a rod shape, providing the interesting property of being distinguishable from gold particles in electron micrographs. However, we have not obtained any positive signal with anti-rabbit antibodies decorated with these markers.

Finally, we note that singlet oxygen generators (Shu et al. 2011) represent a powerful alternative method for visualizing endogenous proteins in the EM by generating singlet oxygen that catalyzes a polymerization reaction of diaminobenzidine into an electron dense product. However, currently we are not aware of successful extensions of the singlet oxygen method, which yield high membrane contrast in the EM (but see Atasoy et al. 2014 for the identification of molecularly defined synapse types).

21.2.4 Data Acquisition

In this section, we detail imaging procedures for the two modalities. LM imaging consists of acquiring first a low-resolution overview and then of scanning all regions of interest at high resolution. EM imaging is performed on regions of interest identified in LM imagery.

21.2.4.1 LM

The first step in acquiring imaging data for correlative array tomography is to scan the samples in a light microscope (e.g., Axio Observer, Zeiss). This order is preferred because the electron beam readily quenches fluorescence (Fabig et al. 2012). Moreover, prior LM scanning can greatly ease the EM operation afterwards, as shown later in this chapter.

First, we acquire a low-resolution mosaic overview of the complete piece of wafer (typically with a 5× air objective) in the bright-field channel. This step takes only a few minutes for a wafer such as that in Figure 21.2. We subsequently stitch the image tiles with the plugin "Grid/CollectionStitching" (Preibisch et al. 2009) available in Fiji.

We wrote custom scripts for automating the recognition of ultrathin sections and for extraction of their locations, using only minimal user input for proper ribbon numbering. The result is shown in Figure 21.2b and c, where the section numbers are overlaid with the sections. This section numbering provides the list of coordinates of all sections on the wafer. At this point, we can manually define a region of interest (ROI) in one of the sections and automatically compute the corresponding ROIs on all other sections (yellow frames in Figure 21.2b and c). These ROIs can be read by the light microscope software and are subsequently scanned automatically (in the example of the wafer in Figure 21.2, it would take a long time to manually define 589 regions to be scanned).

For high-resolution light microscopic imaging, Fluoromount DAPI mounting medium (Life Techologies, S36939) is applied to the wafer, which is then covered with a 0.17 mm cover glass. Each ROI is scanned in the prescribed list of fluorescence channels, giving rise to

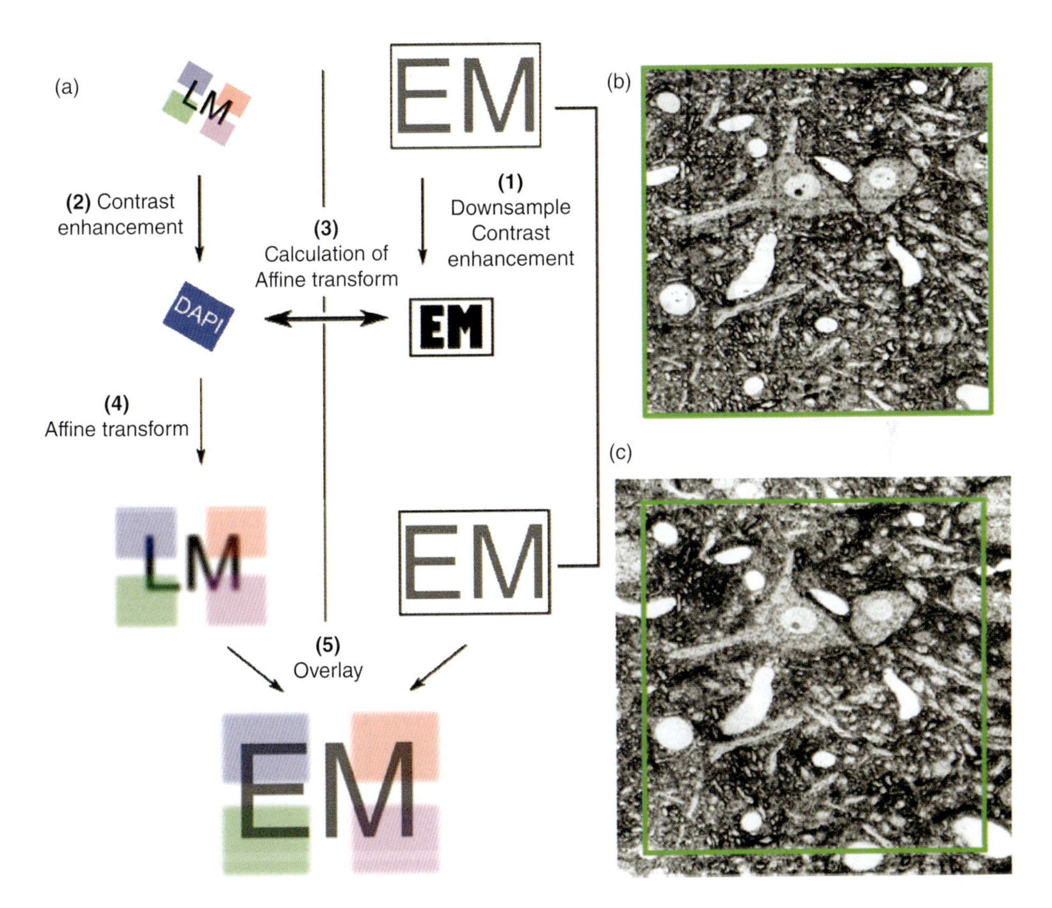

Figure 21.4 Registration of low-resolution LM images to high-resolution EM micrographs. (a) workflow of the registration procedure. The different steps are described in detail in the text. (b) Downsampled and contrast-enhanced electron micrograph. (c) Up-sampled and contrast-enhanced DAPI channel. The green frames in b and c highlight the same sample region as identified after the affine transformation.

a multidimensional mosaic that we stitch together using the tools freely available from the Smith laboratory (http://smithlabsoftware.googlecode.com) and custom software scripts. We then align imagery from consecutive sections using the contrast-enhanced DAPI channel that provides ultrastructural details at low resolution (see Figure 21.4). This alignment is achieved with the TrakEM2 SIFT alignment algorithm "elastic mosaic alignment" (Saalfeld et al. 2012), resulting in an aligned multidimensional image stack in which each stack section corresponds to one physical section.

21.2.4.2 EM

After LM imaging, the coverslip is removed, the wafer is washed during 2×10 minutes in double distilled water, and the sections undergo a silver enhancement treatment for 15 minutes (Nanoprobes, HQ Silver). Then the wafer is washed again, treated with 1% uranyl acetate, followed by Reynold's lead citrate, and mounted on a 100 mm pin mount (Ted Pella, #16111) with carbon sticks (Ted Pella, #16084-3).

We used the secondary electron detector (Merlin, Zeiss) for fast navigation and for manual positioning of the electron beam on the large wafer. We acquired images at high magnification using the backscattered electron detector with a dwell time of 10 μs and at a resolution of 5 nm/pixel. Although we have not investigated thoroughly every possible parameter combination, we found the following parameters to be satisfactory: 2 keV acceleration voltage, 2 nA probe current, 1 keV energy threshold for the backscattered electron filter, and about 3.5 mm working distance.

(a) Targeted EM imaging. The growing need for volumetric EM imaging necessitates image acquisition from large numbers of consecutive sections. Neuronal processes can be contained in thousands or tens of thousands of ultrathin sections because they exhibit cross-section areas that vary between a few dozens of nanometers up to a few dozens of micrometers (Helmstaedter 2013), and because their length can grow up to the millimeter range. Such numbers call for automated electron microscope operation, as demonstrated in Potter et al. (1999) and Suloway et al. (2005) for TEM and in Terasaki et al. (2013) for SEM imaging.

A volume of 200 × 200 × 200 μm^3 necessitates scanning of 4000 consecutive 50-nm thick sections. To manually assign 4000 locations in the electron microscope would be extremely time consuming; therefore we wrote custom scripts that implement the following semiautomatic imaging workflow: (1) Define reference points in the LM coordinate system, which are also visible in the EM (corners of sections, glass scribe markings on wafer). (2) A human operator either (a) defines a reference region within one single ultrathin section (Figure 9.2c, yellow frame) or (b) scrolls through the aligned multidimensional LM stack and selects a region of interest in each consecutive section. (3) The operator locates in the EM the reference points previously defined in the LM. (4) The custom script generates the locations and imaging parameters that are read by the EM scanning software (Atlas 4, Fibics Inc.).

We have worked so far with a simple manual identification in step 2b, but we seek to provide automated methods to extract objects of interest from the LM volumetric data. Thus, we foresee that the entire LM and EM image acquisition processes could be automated after manual initialization.

21.2.4.3 Registration of LM and EM Imagery

We combine the high versatility of LM with high-resolution EM by overlay of LM and EM imagery. We have written custom scripts to allow automated registration in TrakEM2 of low-resolution LM pictures and high-resolution EM micrographs. Figure 21.4a gives an overview of the process, which is based on a few steps. (1) The stitched EM high-resolution image is down-sampled and contrast-enhanced (Figure 21.4a) using the local contrast enhancement implemented in Fiji (Zuiderveld 1994). (2) The DAPI channel of the low-resolution LM image is up-sampled and contrast-enhanced. The contrasted LM and EM pictures are astonishingly similar. (3) The landmark-based registration algorithm implemented in TrakEM2 is applied (Saalfeld et al. 2012) (found under "elastic registration mosaic"). This algorithm computes an affine transformation that maps the DAPI channel to the down-sampled EM picture. (4) All fluorescence channels are then affine-transformed according to the calculated transformation. The resulting registered DAPI and EM images are shown in Figure 21.4b and c. (5) Finally, the low-resolution fluorescence channel pictures are up-sampled and overlaid with the original high-resolution EM picture (Figure 21.4a).

To assess the accuracy of the registration, we labeled discrete structures (BDA-filled axons) on-section using immunohistochemical staining with fluorocolloidal gold. The dual marker is visible both in the LM and in the EM and serves as a ground truth for the registration. In one ultrathin section, we manually selected 64 labels distributed over the entire section (in a regular 6×6 grid spanning a 350×350 μm^2 area, two labels per grid unit if available), and manually marked their positions in the LM and EM images. An example is provided in Figure 21.5. We then computed the registration error as the distance between the centers of mass of the two manually labeled regions. The root-mean-square (RMS) error was 0.53 μm and the largest error was only 1.06 μm. This RMS error corresponds to the size of approximately 1.5 LM pixels (LM imaging has been performed with a 20× air objective, providing 0.32 μm/pixel). Our method is thus very accurate. Such high accuracy of the LM–EM registration allowed us to easily navigate the sections and identify corresponding structures across the two modalities. However, some neuronal processes exhibit much smaller dimensions than the maximum RMS error reported. Therefore, if the object is not dually labeled in the two modalities and that neighboring structures exhibit similar shapes,

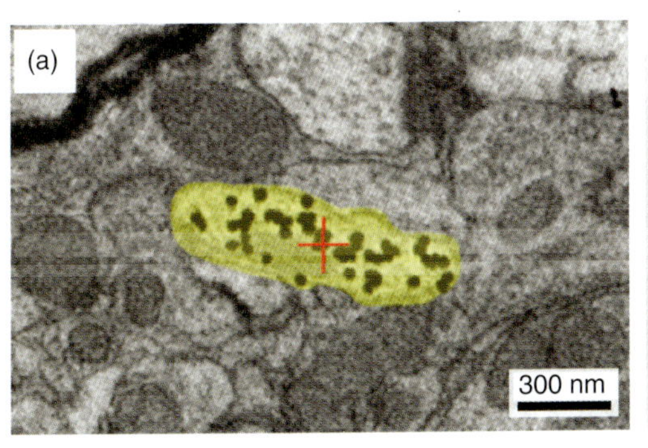

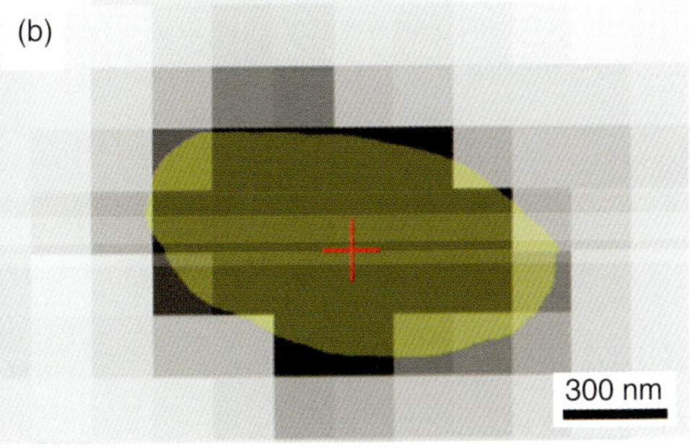

Figure 21.5 Electron (a) and fluorescent light (b) micrographs showing a BDA-labeled axon marked by dual silver enhanced colloidal gold-Alexa 488. The yellow regions have been manually drawn on the electron dense and fluorescent labels. The red crosses mark their centers of mass that have been used to estimate registration error. Scale bars: 300 nm.

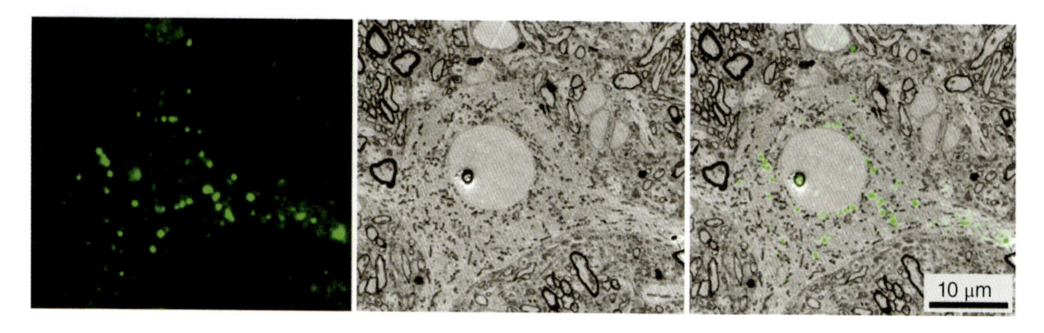

Figure 21.6 Simultaneous LM and EM imaging with an integrated LM/EM (samples imaged at Delmic BV, Delft, Netherlands with an integrated SECOM platform (LM 100× oil objective) and Quanta 250 FEG, FEI Company, Eindhoven). Left: LM; middle: EM; right: LM/EM merge. Shown are motoneurons in the hypoglossal nucleus labeled during postembedding IHC with rabbit anti-Alexa 488 and Alexa 546 anti-rabbit. The tracer Alexa 488 was retrogradely transported from the syrinx of a zebra finch into hypoglossal motoneurons.

then the assignment of the LM label to its corresponding EM micrograph position cannot be unambiguously done within a single section. Reliable assignment would require the analysis of a sufficient number of consecutive sections in order to distinguish between the labeled process and neighboring similar ones. However, we never encountered an ambiguous case; we were always able to identify a dually labeled antigen of interest. We therefore conclude that our registration procedure is highly satisfactory.

21.2.4.4 Integrated LM/EM

Note that another promising approach for correlative microscopy is to incorporate a high-numerical aperture light microscope into the SEM (Figure 21.6) (Liv et al. 2013; Zonnevylle et al. 2013) or to acquire cathodoluminescent signals within a SEM chamber stemming from new generation dual markers such as nanodiamonds excited by the electron beam (Glenn et al. 2012). The registration step could be omitted thanks to direct acquisition of LM data inside the EM. However, as long as the LM acquisition time is much smaller than the EM acquisition time and as long as the tissue exhibits sufficient contrast in LM and EM imagery for automated accurate registration, we feel there is almost no inconvenience to first imaging the specimen in the LM and subsequently in the EM.

21.3 APPLICATION: IDENTIFICATION OF PROJECTION NEURON TYPE IN ULTRASTRUCTURAL CONTEXT

In the last part of this chapter we present an application that demonstrates the power of CAT applied to the analysis of brain circuits. Our animal model is the zebra finch, a songbird whose brain contains a set of discrete and interconnected brain nuclei dedicated to song production and learning. We injected BDA in a motor cortical region, the nucleus Robustus of the Arcopallium (RA), in order to anterograde label descending axon terminals innervating motoneurons of the syringeal muscles, the muscles of the vocal organ. We also injected fluorescent tracers *in vivo* into different syringeal muscles, leading to retrograde labeling of motoneurons.

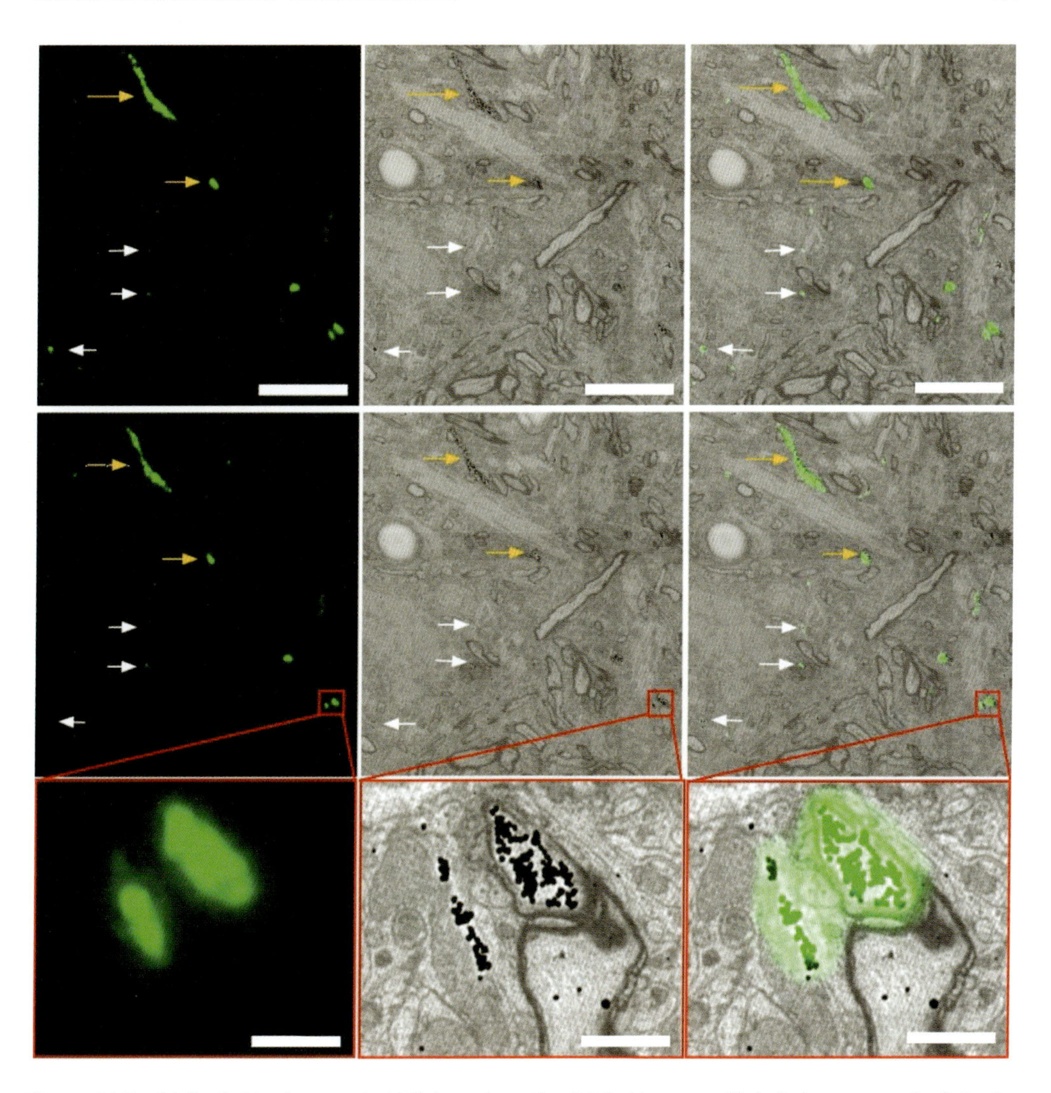

Figure 21.7 Light (left), electron (middle), and overlay (right) images of labeled axon terminals in the hypoglossal nucleus originating from the cortical-like motor area RA. Immunogold markers are indicated by arrows in the electron micrographs. (a, b) Low-resolution overview of two consecutive 50 nm thick ultrathin sections. Consecutive sections serve as control for assessing the reliability of the fluorescence signal. The left parts in a and b are almost identical (arrows). Many labeled axons (arrows) are present and even very small ones are clearly visible (white arrows). (c) Higher magnification view of the region delineated by red in b, showing a labeled unmyelinated and a labeled myelinated axon. The two close axons are clearly distinguishable in the LM channel. The density of electron dense dots inside labeled axons is 8.2 d/μm^2, whereas it is 0.2 d/μm^2 in non-labeled areas. Single dots have an average diameter of 34 nm. Scale bars in a and b: 10 μm; in c: 1 μm.

Figure 21.7 shows light and electron micrographs of two consecutive ultrathin sections taken from the termination site of labeled cortical axons. The secondary antibody used for labeling BDA on ultrathin sections was decorated with the dual marker Colloidal Gold-Alexa 488 (Life Technologies, A31561) and colloidal gold enhanced with silver (Nanoprobes, HQ Silver).

As can be seen in the light micrographs of Figure 21.7a and d, the green signal (Alexa 488) present in one section is also present in the consecutive section. Note that the conventional processing of BDA-labeled axons with the ABC procedure produces electron-dense deposits within single neuronal processes. These deposits are visible consistently in every consecutive section (da Costa and Martin 2009; Lang et al. 2011). However, it is not known whether this consistent staining is due to the continuous presence of tracer molecules in every section or due to the ABC reaction spreading and filling gaps in tracer-free portions of the neuronal processes. To resolve this question, the data in Figure 21.7 (see also Figure 21.8) suggest that tracer molecules are present in significant amounts on every consecutive section. The density of gold particles in labeled structures is 8.2 d/μm^2, compared to 0.2 d/μm^2 in non-labeled structures, demonstrating that postembedding visualization of BDA by means of on-section immunohistochemistry is as sensitive as the conventional preembedding ABC procedure.

Figure 21.8 shows light and electron micrographs of the hypoglossal (motor) nucleus containing motoneurons and descending cortical axons. Which muscle does a given cell

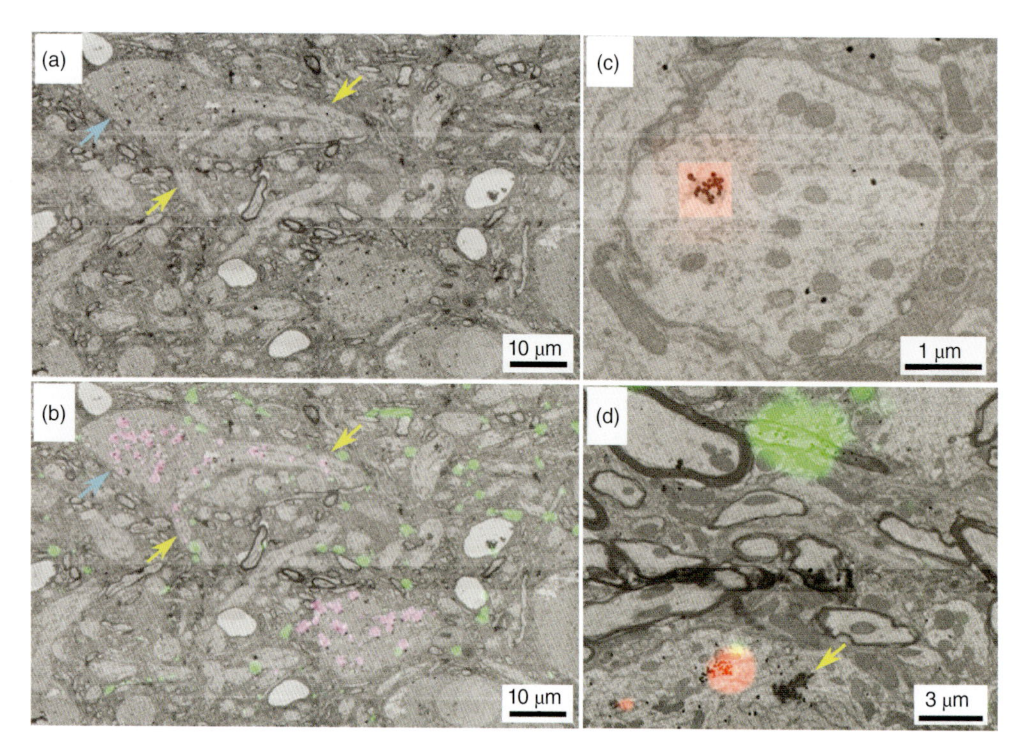

Figure 21.8 Multicolor array tomography. (a) Large EM field of view of part of the avian vocal motor nucleus (hypoglossal nucleus). Part of a cell body is visible on the top left, with two dendritic processes extruding from it (marked with yellow arrows). (b) Overlay of EM and LM imagery. The magenta label indicates the presence of Texas Red molecules and the green label indicates the presence of biotin molecules. (c) A motoneuron dendrite containing a lysosome filled with Texas red molecules, as revealed by IHC. Sequential gold labeling followed by fluorophore labeling reveals a label simultaneously visible in the EM (black, electron-dense dots) and in the LM (red, fluorescent signal). The LM channel was acquired with an objective with lower resolution than usual, yielding pixelated signal and accentuating the discrepancy between LM and EM resolution). (d) ROI showing the close apposition of labeled structures. A cortical axon is labeled in green (BDA) and a motoneuron soma is labeled in red (Texas Red). Both axons and motoneurons are also labeled with gold (black dots). The yellow arrow shows an artifact of silver enhancement.

body innervate? Thanks to the injection of different tracers into different muscles, we can retrieve the identity of the innervated cell body by overlaying the LM and EM pictures of this same region (Figure 21.8b). In this figure, the magenta label identifies Texas Red tracer molecules that had been injected *in vivo* into the Ventralis Syringealis (VS) vocal muscle. The retrograde tracer is present in cell bodies (cyan arrows in Figure 21.8a), proximal dendrites (yellow arrows in Figure 21.8a), and in some distal dendrites, as shown in Figure 21.8c. The green label localizes descending cortical axons as depicted in Figure 21.7. The overlay of the two modalities not only provides the identity of the structures but it also offers a convenient way to browse the dataset. Namely, in TrakEM2 (Cardona et al. 2012), a software optimized for the visualization of large image datasets, a user can quickly navigate and zoom into ROIs identified in LM imagery, thus greatly speeding up the analysis of the EM imagery.

21.4 CONCLUSION

In this chapter we have introduced correlative array tomography (CAT), a correlative light and electron microscopy technique. We hope to have convincingly demonstrated that CAT enhances the observation of ultrastructural details in EM imagery by harnessing the power of multidimensional light microscopy.

CAT is driven by our desire of attaining biological understanding by relating structure to function. We have detailed the sample preparation and visualization procedures required for CAT. Currently, CAT works well with chemical fluorophores but not yet with fluorescent proteins or endogenous proteins. To be able to clearly visualize endogenous molecules in their ultrastructural context, including high membrane contrast, would constitute an immense breakthrough for biological structure–function studies.

ACKNOWLEDGEMENTS

The authors thank Bruno Humbel and Roland Fleck for helpful comments on this chapter. This work was supported by the Swiss National Science Foundation (Grant 31003A_127024) and by the European Research Council under the European Community's Seventh Framework Programme (FP7/2007-2013/ERC Grant AdG 268911). The authors acknowledge support from the Scientific Center for Optical and Electron Microscopy ScopeM of the Swiss Federal Institute of Technology ETHZ.

REFERENCES

Allen, N.J., Bennett, M.L., Foo, L.C., Wang, G.X., Chakraborty, C., Smith, S.J., *et al.* Astrocyte glypicans 4 and 6 promote formation of excitatory synapses via GluA1 AMPA receptors. *Nature [Internet], Nature Publishing Group, a division of Macmillan Publishers Limited*, All Rights Reserved, 2012, June, 486 (7403), 410–414. Available from: 10.1038/nature11059.

Anderson, J.R., Jones, B.W., Watt, C.B., Shaw, M.V., Yang, J.-H., Demill, D., *et al.* Exploring the retinal connectome. *Mol. Vis.* [Internet], *Emory University*, 2011, 17 (41), 355–379. Available from: http://www.pubmedcentral.nih.gov/articlerender.fcgi?artid=3036568&tool=pmcentrez&rendertype=abstract.

Anderson, J.R., Jones, B.W., Yang, J.-H., Shaw, M.V., Watt, C.B., Koshevoy, P., *et al.* A computational framework for ultrastructural mapping of neural circuitry. *PLoS Biol.* [Internet], 2009, March, 7 (3), e1000074. Available from: 10.1371/journal.pbio.1000074.

Atasoy, D., Betley, J.N., Li, W.-P., Su, H.H., Sertel, S.M., Scheffer, L.K., *et al.* A genetically specified connectomics approach applied to long-range feeding regulatory circuits. *Nat. Neurosci.* [Internet], 2014, 17, 1830–1839. Available from: http://www.ncbi.nlm.nih.gov/pubmed/25362474.

Berryman, M.A. and Rodewald, R.D. An enhanced method for post-embedding immunocytochemical staining which preserves cell membranes. *J. Histochem. Cytochem.*, 1990, 38, 159–170.

Blumer, M.J.F., Gahleitner, P., Narzt, T., Handl, C., and Ruthensteiner, B. Ribbons of semithin sections: An advanced method with a new type of diamond knife. *J. Neurosci. Methods*, 2002, October, 120 (1), 11–16.

Bock, D.D., Lee, W.-C.A., Kerlin, A.M., Andermann, M.L., Hood, G., Wetzel, A.W., *et al.* Network anatomy and *in vivo* physiology of visual cortical neurons. *Nature* [Internet], 2011, March, 471 (7337), 177–182. Available from: 10.1038/nature09802.

Briggman, K.L., Helmstaedter, M., and Denk, W. Wiring specificity in the direction-selectivity circuit of the retina. *Nature* [Internet], Nature Publishing Group, a division of Macmillan Publishers Limited, All Rights Reserved, 2011, March, 471 (7337), 183–188. Available from: 10.1038/nature09818.

Brown, E., Mantell, J., Carter, D., Tilly, G., and Verkade, P. Studying intracellular transport using high-pressure freezing and Correlative Light Electron Microscopy. *Semin. Cell Dev. Biol.* [Internet], 2009, 20 (8), 910–919. Available from: http://www.sciencedirect.com/science/article/pii/S1084952109001530.

Busse, B. and Smith, S. Automated analysis of a diverse synapse population. *PLoS Comput. Biol.* [Internet], Public Library of Science, 2013, 9 (3), e1002976. Available from: 10.1371/journal.pcbi.1002976.

Canty, A.J., Huang, L., Jackson, J.S., Little, G.E., Knott, G., Maco, B., *et al.* In-vivo single neuron axotomy triggers axon regeneration to restore synaptic density in specific cortical circuits. *Nat. Commun.* [Internet], Nature Publishing Group, a division of Macmillan Publishers Limited, All Rights Reserved, 2013, June, 4 . Available from: 10.1038/ncomms3038.

Cardona, A., Saalfeld, S., Preibisch, S., Schmid, B., Cheng, A., Pulokas, J., *et al.* An integrated micro- and macroarchitectural analysis of the Drosophila brain by computer-assisted serial section electron microscopy. *PLoS Biol.* [Internet], 2010, 8 (10). Available from: 10.1371/journal.pbio.1000502.

Cardona, A., Saalfeld, S., Schindelin, J., Arganda-Carreras, I., Preibisch, S., Longair, M., *et al.* TrakEM2 software for neural circuit reconstruction. *PLoS One* [Internet], Public Library of Science, 2012, 7 (6), e38011. Available from: 10.1371/journal.pone.0038011.

Chan, W.C.W. and Nie, S. Quantum dot bioconjugates for ultrasensitive nonisotopic detection. *Science [Internet]*, 1998, 281 (5385), 2016–2018. Available from: http://www.sciencemag.org/content/281/5385/2016.abstract.

Collman, X.F., Buchanan, J., Phend, K.D., Micheva, K.D., Weinberg, R.J., and Smith, S.J. Mapping synapses by conjugate light-electron array tomography, *J. Neurosci.*, 2015, 35(14), 5792–807.

da Costa, N.M. and Martin, K.A.C. Selective targeting of the dendrites of corticothalamic cells by thalamic afferents in area 17 of the cat. *J. Neurosci.* [Internet], 2009, November, 29 (44), 13919–13928. Available from: 10.1523/JNEUROSCI.2785-09.2009.

Dahan, M., Lévi, S., Luccardini, C., Rostaing, P., Riveau, B., and Triller, A. Diffusion dynamics of glycine receptors revealed by single-quantum dot tracking. *Science* [Internet], 2003, October, 302 (5644), 442–445. Available from: 10.1126/science.1088525.

Fabig, G., Kretschmar, S.,Weiche, S., Eberle, D., Ader, M., Kurth, T. Chapter 5 - Labeling of ultrathin resin sections for correlative light and electron microscopy, in *Correlative Light and Electron MIcroscopy* (eds T. Müller-Reichert and P. Verkade) [Internet], Academic Press, 2012, pp. 75–93. Available from: http://www.sciencedirect.com/science/article/pii/B9780124160262000054.

Fiserova, J. and Goldberg, M. Immunoelectron microscopy of cryofixed freeze-substituted *Saccharomyces cerevisiae*, in *Immunoelectron Microscopy* (eds S.D. Schwartzbach and T. Osafune) [Internet], Humana Press, 2010, pp. 191–204. Available from: 10.1007/978-1-60761-783-9_15.

Ghrebi, S.S., Owen, G.R., and Brunette, D.M. Triton X-100 pretreatment of LR-white thin sections improves immunofluorescence specificity and intensity. *Microsc. Res. Tech.* [Internet], Wiley Subscription Services, Inc., A Wiley Company, 2007, 70 (7), 555–562. Available from: 10.1002/jemt.20422.

Giepmans, B.N.G., Deerinck, T.J., Smarr, B.L., Jones, Y.Z., and Ellisman, M.H. Correlated light and electron microscopic imaging of multiple endogenous proteins using Quantum dots. *Nat. Methods* [Internet], 2005, October, 2 (10), 743–749. Available from: 10.1038/nmeth791.

Glauert, A.M. and Lewis, P. Biological Specimen Preparation for Transmission Electron Microscopy *[Internet]*, Princeton University Press, 1999, 316 pp. Available from: http://press.princeton.edu/titles/6666.html.

Glenn, D.R., Zhang, H., Kasthuri, N., Schalek, R., Lo, P.K., Trifonov, A.S., *et al.* Correlative light and electron microscopy using cathodoluminescence from nanoparticles with distinguishable colours. *Sci. Rep.* [Internet], 2012, 2, 865. Available from: 10.1038/srep00865.

Grabenbauer, M., Geerts, W.J.C., Fernadez-Rodriguez, J., Hoenger, A., Koster, A.J., and Nilsson, T. Correlative microscopy and electron tomography of GFP through photooxidation. *Nat. Meth.* [Internet], 2005, November, 2 (11), 857–862. Available from: http://dx.doi.org/10.1038/nmeth806.

Hayat, M.A. Glutaraldehyde: Role in electron microscopy. *Micron Microsc. Acta*, 1986, 17, 115–135.

Hayat, M.A. *Immunogold–Silver Staining: Principles, Methods, and Applications*, CRC Press, 1995.

Hayworth, K.J., Kasthuri, N., Schalek, R., and Lichtman, J.W. Automating the collection of ultrathin serial sections for large volume TEM reconstructions. *Microsc. Microanal.* [Internet], 2006, 12 (Suppl. S02), 86–87. *Available from*: 10.1017/S1431927606066268.

Hayworth, K.J., Morgan, J.L., Schalek, R., Berger, D.R., Hildebrand, D.G.C., and Lichtman, J.W. Imaging ATUM ultrathin section libraries with WaferMapper: A multi-scale approach to EM reconstruction of neural circuits. *Front Neural Circuits* [Internet], 2014, January [cited 2014, Aug. 31], 8 (June), 68. Available from: http://www.pubmedcentral.nih.gov/articlerender.fcgi?artid=4073626&tool=pmcentrez&rendertype=abstract.

Heck, W.L., Slusarczyk, A., Basaraba, A.M., and Schweitzer, L. Subcellular localization of GABA receptors in the central nervous system using post-embedding immunohistochemistry. *Brain Res. Brain Res. Protoc.* [Internet], 2002, June, 9 (3), 173–180. Available from: http://www.ncbi.nlm.nih.gov/pubmed/12113777.

Helmstaedter, M. Cellular-resolution connectomics: Challenges of dense neural circuit reconstruction. *Nat. Methods* [Internet], Nature Publishing Group, a division of Macmillan Publishers Limited, All Rights Reserved, 2013 June, 10 (6), 501–507. Available from: 10.1038/nmeth.2476

Horstmann, H., Körber, C., Sätzler, K., Aydin, D., and Kuner, T. Serial section scanning electron microscopy (S(3)EM) on silicon wafers for ultra-structural volume imaging of cells and tissues. *PLoS One* [Internet], 2012, 7 (4), e35172. Available from: 10.1371/journal.pone.0035172.

Hua, Y., Laserstein, P., and Helmstaedter, M. Large-volume en-bloc staining for electron microscopy-based connectomics. *Nat. Commun.* [Internet], Nature Publishing Group, 2015, 6, 7923. Available from: http://www.nature.com/doifinder/10.1038/ncomms8923.

Huang, B., Bates, M., and Zhuang, X. Super-resolution fluorescence microscopy. *Annu. Rev. Biochem.*, 2009, 78, 993–1016.

Humbel, B.M., De Jong, M.D.M., Müller, W.H., and Verkleij, A.J. Pre-embedding immunolabeling for electron microscopy: An evaluation of permeabilization methods and markers. *Microsc. Res. Tech.*, 1998, 42, 43–58.

Knott, G., Marchman, H., Wall, D., and Lich, B. Serial section scanning electron microscopy of adult brain tissue using focused ion beam milling. *J. Neurosci.* [Internet], 2008, 28 (12), 2959–2964. Available from: http://www.jneurosci.org/content/28/12/2959.short.

Kopeikina, K.J., Carlson, G.A., Pitstick, R., Ludvigson, A.E., Peters, A., Luebke, J.I., *et al.* Tau accumulation causes mitochondrial distribution deficits in neurons in a mouse model of tauopathy and in human Alzheimer's disease brain. *Am. J. Pathol.* [Internet], 2011, October, 179 (4), 2071–2082. Available from: http://www.sciencedirect.com/science/article/pii/S0002944011006559.

Kraehenbuhl, J.P. and Jamieson, J.D. Localization of intracellular antigens using immunoelectron microscopy. *Electron Microsc. Cytochem.*, 1973, 181–192.

Lang, S., Drouvelis, P., Tafaj, E., Bastian, P., and Sakmann, B. Fast extraction of neuron morphologies from large-scale SBFSEM image stacks. *J. Comput. Neurosci.* [Internet], 2011, November, 31 (3), 533–545. Available from: 10.1007/s10827-011-0316-1.

Liv, N., Zonnevylle, A.C., Narvaez, A.C., Effting, A.P.J., Voorneveld, P.W., Lucas, M.S., *et al.* Simultaneous correlative scanning electron and high-NA fluorescence microscopy. *PLoS One* [Internet], 2013, 8 (2), e55707. Available from: 10.1371/journal.pone.0055707.

Luby-Phelps, K., Ning, G., Fogerty, J., and Besharse, J.C. Visualization of identified GFP-expressing cells by light and electron microscopy. *J. Histochem. Cytochem.* [Internet], 2003, 51 (3), 271–274. Available from: http://jhc.sagepub.com/content/51/3/271.abstract.

Maco, B., Holtmaat, A., Cantoni, M., Kreshuk, A., Straehle, C.N., Hamprecht, F.A., *et al.* Correlative *in vivo* 2 photon and focused ion beam scanning electron microscopy of cortical neurons. *PLoS One* [Internet], Public Library of Science, 2013, 8 (2), e57405. Available from: 10.1371/journal.pone.0057405.

McDonald, K.L. A review of high-pressure freezing preparation techniques for correlative light and electron microscopy of the same cells and tissues. *J. Microsc.* [Internet]. Blackwell Publishing Ltd, 2009, 235 (3), 273–281. Available from: 10.1111/j.1365-2818.2009.03218.x.

Michalet, X., Pinaud, F.F., Bentolila, L.A., Tsay, J.M., Doose, S., Li, J.J., *et al.* Quantum dots for live cells, *in vivo* imaging, and diagnostics. *Science [Internet]*, 2005, 307 (5709), 538–544. Available from: http://www.sciencemag.org/content/307/5709/538.abstract.

Micheva, K.D. and Bruchez, M.P. The gain in brain: novel imaging techniques and multiplexed proteomic imaging of brain tissue ultrastructure. *Curr. Opin. Neurobiol.* [Internet], 2012, February, 22 (1), 94–100. Available from: 10.1016/j.conb.2011.08.004.

Micheva, K.D. and Smith, S.J. Array tomography: a new tool for imaging the molecular architecture and ultrastructure of neural circuits. *Neuron* [Internet], 2007, July, 55 (1), 25–36. Available from: http://dx.doi.org/10.1016/j.neuron.2007.06.014.

Micheva, K.D., Busse, B., Weiler, N.C., O'Rourke, N., and Smith, S.J. Single-synapse analysis of a diverse synapse population: proteomic imaging methods and markers. *Neuron* [Internet], 2010a, November, 68 (4), 639–653. Available from: 10.1016/j.neuron.2010.09.024.

Micheva, K.D., Busse, B., Weiler, N.C., O'Rourke, N., and Smith, S.J. Single-synapse analysis of a diverse synapse population: proteomic imaging methods and markers. *Neuron* [Internet], Elsevier Inc., 2010b, Nov. 18 [cited 2013, Nov. 11], 68 (4), 639–653. Available from: http://www.pubmedcentral.nih.gov/articlerender.fcgi?artid=2995697&tool=pmcentrez&rendertype=abstract.

Micheva, K.D., O'Rourke, N., Busse, B., and Smith, S.J. Array tomography: Production of arrays. *Cold Spring Harb. Protoc.*, 2010c, November, 2010 (11), pdb.prot5524.

Mikula, S. and Denk, W. High-resolution whole-brain staining for electron microscopic circuit reconstruction. *Nat. Methods* [Internet], 2015 (April). Available from: http://www.nature.com/doifinder/10.1038/nmeth.3361.

Morgan, J.L. and Lichtman, J.W. Why not connectomics? *Natural Methods* [Internet], Nature Publishing Group, a division of Macmillan Publishers Limited, All Rights Reserved, 2013, June, 10 (6), 494–500. Available from: 10.1038/nmeth.2480.

Muehlfeld, C. and Richter, J. High-pressure freezing and freeze substitution of rat myocardium for immunogold labeling of connexin 43. *Anat. Rec. Part A: Discov. Mol. Cell Evol. Biol.* [Internet]. Wiley Subscription Services, Inc., A Wiley Company, 2006, 288A (10), 1059–1067. Available from: 10.1002/ar.a.20380.

Nanguneri, S., Flottmann, B., Horstmann, H., Heilemann, M., and Kuner, T. Three-dimensional, tomographic super-resolution fluorescence imaging of serially sectioned thick samples. *PLoS One* [Internet], 2012, 7 (5), e38098. Available from: 10.1371/journal.pone.0038098.

Newman, G.R., Jasani, B., and Williams, E.D. A simple post-embedding system for the rapid demonstration of tissue antigens under the electron microscope. *Histochem. J.* [Internet], Kluwer Academic Publishers, 1983, 15 (6), 543–555. Available from: 10.1007/BF01954145.

Nisman, R., Dellaire, G., Ren, Y., Li, R., and Bazett-Jones, D.P. Application of quantum dots as probes for correlative fluorescence, conventional, and energy-filtered transmission electron microscopy. *J. Histochem. Cytochem.*, 2004, January, 52 (1), 13–18.

Oberti, D., Kirschmann, M.A., and Hahnloser, R. Correlative microscopy of densely labeled projection neurons using neural tracers. *Front. Neuroanat.*, 2010, 4 (0).

Oberti, D., Kirschmann, M.A., and Hahnloser, R.H.R. Projection neuron circuits resolved using correlative array tomography. *Front. Neurosci.* [Internet], 2011, 5 (0). Available from: http://www.frontiersin.org/Journal/Abstract.aspx?f=55&name=neuroscience&ART_DOI=10.3389/fnins.2011.00050.

Paez-Segala, M.G., Sun, M.G., Shtengel, G., Viswanathan, S., Baird, M.A., Macklin, J.J., *et al.* Fixation-resistant photoactivatable fluorescent proteins for CLEM. *Nat. Methods [Internet]*, 2015, Jan. 12 (November 2014). Available from: http://www.nature.com/doifinder/10.1038/nmeth.3225.

Phend, K.D., Rustioni, A., and Weinberg, R.J. An osmium-free method of epon embedment that preserves both ultrastructure and antigenicity for post-embedding immunocytochemistry. *J. Histochem. Cytochem.*, 1995, 43, 283–292.

Potter, C.S., Chu, H., Frey, B., Green, C., Kisseberth, N., Madden, T.J., *et al.* Leginon: a system for fully automated acquisition of 1000 electron micrographs a day. *Ultramicroscopy* [Internet], 1999, 77 (3–4), 153–161. Available from: http://www.sciencedirect.com/science/article/pii/S0304399199000431.

Preibisch, S., Saalfeld, S., and Tomancak, P. Globally optimal stitching of tiled 3D microscopic image acquisitions. *Bioinformatics* [Internet], 2009, 25 (11),1463–1465. Available from: http://bioinformatics.oxfordjournals.org/content/25/11/1463.abstract.

Punge, A., Rizzoli, S.O., Jahn, R.,Wildanger, J.D., Meyer, L., Schönle, A., *et al.* 3D reconstruction of high-resolution STED microscope images. *Microsc. Res. Tech.* [Internet], Wiley Subscription Services, Inc., A Wiley Company, 2008, 71 (9), 644–650. Available from: 10.1002/jemt.20602.

Rah, J.C., Erhan, B., Colonell, J., Mischchenko, Y., Karsh, B., Fetter, R.D., Myers, E.W., Chklovskii, D.B., Svoboda, K., Harris, T.D., Isaac, J.T.R. Thalamocortical input onto layer 5 pyramidal neurons measured using quantitative large-scale array tomography. *Frontiers in Neural Circuits*, 2013, 7 (November), 177, 1–16.

Robenek, H. *Colloidal Gold: Principles, Methods, and Applications*, vols I and II (vol. III in preparation) (ed. M.A. Hayat), Academic Press, Inc., New York, 1989 ISBN 0–12–333927–8, vol. I, 536 pages, ISBN 0–12–333928–6, vol. II, 484 pages. *Scanning* [Internet],Wiley Periodicals, Inc.; 1990; 12 (4), 244. Available from: 10.1002/sca.4950120410.

Rostaing, P., Weimer, R.M., Jorgensen, E.M., Triller, A., Bessereau, J.-L. Preservation of immunoreactivity and fine structure of adult *C. elegans* tissues using high-pressure freezing. *J. Histochem. Cytochem.* [Internet], 2004, 52 (1), 1–12. Available from: http://jhc.sagepub.com/content/52/1/1.abstract.

Saalfeld, S., Fetter, R., Cardona, A., and Tomancak, P. Elastic volume reconstruction from series of ultra-thin microscopy sections. *Nat. Meth.* [Internet], Nature Publishing Group, a division of Macmillan Publishers Limited, All Rights Reserved, 2012, July, 9 (7), 717–720. Available from: 10.1038/nmeth.2072.

Salio, C., Lossi, L., and Merighi, A. Combined light and electron microscopic visualization of neuropeptides and their receptors in central neurons, in *Neuropeptides* (ed. A. Merighi) [Internet],Humana Press, 2011, pp. 57–71. *Available from*: 10.1007/978-1-61779-310-3_3.

Schwarz, H. and Humbel, B. Correlative light and electron microscopy using immunolabeled resin sections, in *Electron Microscopy* (ed. J. Kuo) [Internet], Humana Press, 2007, pp. 229–256. *Available from*: 10.1007/978-1-59745-294-6_12.

Schwarz, H. Immunolabeling of ultrathin sections for fluorescence and electron microscopy. *Electron Microsc.*, 1994, 3, 255–256.

Shu, X., Lev-Ram, V., Deerinck, T.J., Qi, Y., Ramko, E.B., Davidson, M.W., *et al.* A genetically encoded tag for correlated light and electron microscopy of intact cells, tissues, and organisms. *PLoS Biol* [Internet], 2011, April, 9 (4), e1001041. Available from: http://dx.doi.org/10.1371/journal.pbio.1001041.

Skepper, J.N. and Powell, J.M. Ultrastructural immunochemistry. *Cold Spring Harb. Protoc.*, 2008, 3, 1–6.

Sonomura, T., Furuta, T., Nakatani, I., Yamamoto, Y., Unzai, T., Matsuda, W., *et al.* Correlative analysis of immunoreactivity in confocal laser-scanning microscopy and scanning electron microscopy with focused ion beam milling. *Front Neural Circuits* [Internet], 2013, 7 (26). Available from: http://www.frontiersin.org/neural_circuits/10.3389/fncir.2013.00026/abstract.

Spomer, W., Hofmann, A., Wacker, I., Ness, L., Brey, P., Schroder, R.R., *et al.* Advanced substrate holder and multi-axis manipulation tool for ultramicrotomy. *Microsc. Microanal.* [Internet], 2015, 21 (Suppl. S3), 1277–1278. Available from: http://journals.cambridge.org/article_S1431927615007175.

Stierhof, Y.-D. and Kasmi, F. El. Strategies to improve the antigenicity, ultrastructure preservation and visibility of trafficking compartments in *Arabidopsis tissue. Eur. J. Cell Biol.* [Internet], 2010, 89 (2–3), 285–297. Available from: http://www.sciencedirect.com/science/article/pii/S0171933509003677.

Suloway, C., Pulokas, J., Fellmann, D., Cheng, A., Guerra, F., Quispe, J., *et al.* Automated molecular microscopy: the new Leginon system. *J. Struct. Biol.* [Internet], 2005, July, 151 (1), 41–60. Available from: 10.1016/j.jsb.2005.03.010.

Takizawa, T., Suzuki, K., and Robinson, J.M. Correlative microscopy using fluoronanogold on ultrathin cryosections: Proof of rinciple. *J. Histochem. Cytochem.* [Internet], 1998, 46 (10), 1097–1102. Available from: http://jhc.sagepub.com/content/46/10/1097.abstract.

Tapia, J.C., Kasthuri, N., Hayworth, K.J., Schalek, R., Lichtman, J.W., Smith, S.J., *et al.* High-contrast en bloc staining of neuronal tissue for field emission scanning electron microscopy. *Nat. Protoc.* [Internet], Nature Publishing Group, a division of Macmillan Publishers Limited, All Rights Reserved, 2012, February, 7 (2): 193–206. Available from: 10.1038/nprot.2011.439.

Terasaki, M., Shemesh, T., Kasthuri, N., Klemm, R.W., Schalek, R., Hayworth, K.J., *et al.* Stacked endoplasmic reticulum sheets are connected by helicoidal membrane motifs. *Cell [Internet]*, 2013, 154 (2), 285–296. Available from: http://www.sciencedirect.com/science/article/pii/S0092867413007708.

Terzakis, J.A. Uranyl acetate, a stain and a fixative. *J. Ultrastruct. Res.*, 1968, 22, 168–184.

Van Damme, D., Coutuer, S., De Rycke, R., Bouget, F.-Y., Inzé, D., and Geelen, D. Somatic cytokinesis and pollen maturation in *Arabidopsis* depend on TPLATE, which has domains similar to coat proteins. *Plant Cell Online [Internet]*, 2006, 18 (12), 3502–3518. Available from: http://www.plantcell.org/content/18/12/3502.abstract.

Viswanathan, S., Williams, M.E., Bloss, E.B., Stasevich, T.J., Speer, C.M., Nern, A., *et al.* High-performance probes for light and electron microscopy. *Nat Methods* [Internet]. 2015 (April). Available from: http://www.nature.com/doifinder/10.1038/nmeth.3365.

Wang, B.L. and Larsson, L.I. Simultaneous demonstration of multiple antigens by indirect immunofluorescence or immunogold staining. Novel light and electron microscopical double and triple staining method employing primary antibodies from the same species. *Histochemistry*, 1985, 83 (1), 47–56.

Watanabe, S., Punge, A., Hollopeter, G., Willig, K.I., Hobson, R.J., Davis, M.W., *et al.* Protein localization in electron micrographs using fluorescence nanoscopy. *Nat. Methods* [Internet], 2011, January, 8 (1), 80–84. Available from: 10.1038/nmeth.1537.

Zhang, H., Aharonovich, I., Glenn, D.R., Schalek, R., Magyar, A.P., Lichtman, J.W., *et al.* Silicon-vacancy color centers in nanodiamonds: Cathodoluminescence imaging markers in the near infrared. *Small*, 2014, 10, 1908–1913.

Zonnevylle, A.C., van Tol, R.F.C., Liv, N., Narvaez, A.C., Effting, A.P.J., Kruit, P., *et al.* Integration of a high-NA light microscope in a scanning electron microscope. *J. Microsc.* [Internet], 2013. Available from: 10.1111/jmi.12071.

Zuiderveld, K. *Graphics Gems*, vol. IV (ed. P.S. Heckbert), Academic Press Professional, Inc., San Diego, USA, 1994, pp. 474–485. Available from: http://dl.acm.org/citation.cfm?id=180895.180940.

22

The Automatic Tape Collection UltraMicrotome (ATUM)

Anwen Bullen
UCL Ear Institute, London, UK

22.1 INTRODUCTION

The importance of detailed three-dimensional structural data for understanding biological systems has led to the development of a variety of systems for accessing this information at the electron microscopic level. This is particularly true in the field of neural connectomics, where high-resolution information is also required to encompass vast volumes, far beyond those that have been previously achieved. The inability of electrons to pass through large volumes of tissue makes serial sectioning of samples the most practical way to achieve this. The volumes required are not practically achievable by traditional serial section transmission electron microscopy and therefore new methods of sample preparation and imaging have been developed.

Serial section electron microscopy has traditionally required a high level of skill and often a great deal of time on the part of the operator in order to collect sections without significant losses. However, it remains one of the few ways to collect three-dimensional information in the electron microscope covering regions larger than those that can be accessed by electron tomography. In the SEM, the use of backscatter or secondary electron imaging has allowed the imaging of many more sections than could be fitted on a conventional TEM grid, using the array tomography (AT) technique (Micheva and Smith, 2007). Imaging can be achieved on both glass slides and silicon chips (Horstmann *et al.*, 2012). However, array tomography still requires the cutting of sections by hand and collection on to sample supports, and therefore still suffers problems of section loss (although less so than the collection on to TEM grids) and a requirement for a great deal of time and skill on the part of the operator. An alternative to manual sectioning has been the development of automated sectioning

Biological Field Emission Scanning Electron Microscopy, First Edition.
Edited by Roland A. Fleck and Bruno M. Humbel.

and imaging inside the SEM, either by a diamond knife mounted on a microtome inside the SEM (serial block face SEM) or by focused ion beam milling (FIB-SEM) (See Chapter 23 by Genoud and Chapter 26 by Kizilyaprak *et al.*). Both techniques completely automate the sectioning and imaging process, but they both also destroy the sample as it is imaged. In contrast AT allows repeated imaging of the same set of sections at different resolutions as required, subject to the limitations of sample erosion (beam induced damage) from beam-specimen interactions.

The automatic tape collection ultramicrotome (ATUM) combines these desirable features by automating serial section collection from a standard ultramicrotome and producing an array of sections on a silicon wafer that may be imaged in the SEM. The ATUM was developed by a team led by Professor Jeffrey Lichtman at Harvard University, as part of a project with the long-term goal of producing a connectome for an entire animal brain, a task that will eventually require a fully automated preparation, imaging and reconstruction platform (Schalek *et al.*, 2011). The group have recently published results showing the reconstruction of a 0.13 mm^3 section of mouse neocortex from 2250 29 nm ATUM sections (Kasthuri *et al.*, 2015). The ATUM is being commercially developed under licence from Harvard University by RMC Boeckeler Instruments, in combination with their PowerTome ultramicrotome system, and supplied together as the ATUMtome.

The ATUM is fitted to a standard ultramicrotome as seen in Figure 22.1a. The device consists of a pair of tape spools and a tape track that passes over a nose that is inserted into

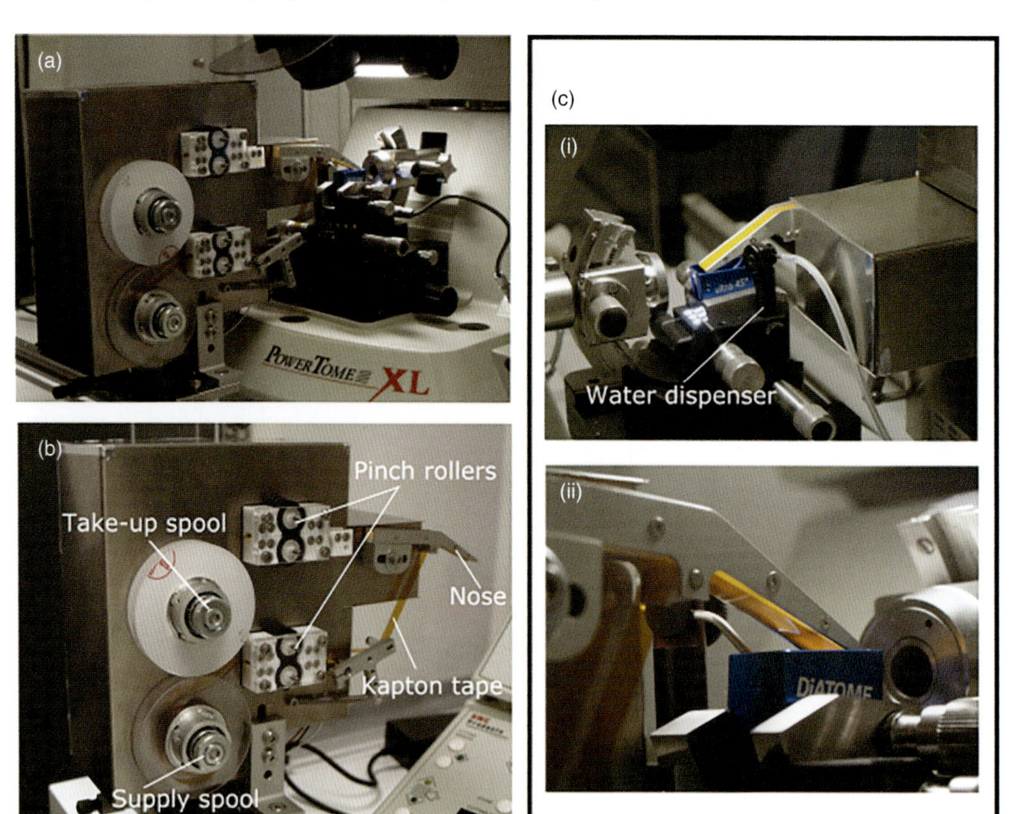

Figure 22.1 The ATUM system. (a) The ATUM mounted on an RMC Powertome Ultramicrotome. For clarity, the sample block and chuck have been removed. (b) The ATUM tape path. (c) (i) Automatic water dispenser set up to maintain the water level in the knife boat during cutting. (ii) Placement of the dispenser in the knife boat.

the boat of a diamond knife (Figure 22.1a). A motor drives tape from the supply spool to the take-up spool. As the knife cuts the sections, they float on the surface of the water in the boat and are carried on to the tape that is passing over the nose of the ATUM. Tension is maintained on the tape by a pair of pinch rollers either side of the nose (Figure 22.1b). The water level in the boat is monitored by the computer and maintained by an automatic water dispenser (Figure 22.1c). The ATUM can run unattended for extended periods. Rates of section loss for the ATUM are reported to be less than 0.1% and collections of 10 000 sections have been made with ATUM running for 24 hours (Schalek *et al.*, 2012; Hayworth *et al.*, 2014).

22.2 SAMPLE PREPARATION FOR ATUM

22.2.1 Staining

As the ATUM is fitted to a standard ultramicrotome, sample preparation is similar to that for conventional transmission electron microscopy. However, as the sections are imaged by secondary electron or backscatter techniques, protocols that include enhanced heavy metal contrast are often preferred. Tapia *et al.* (2012) identified staining protocols using a reduced osmium–thiocarbohydrazide–osmium (ROTO) protocol followed by *en bloc* staining with uranyl acetate and lead aspartate or copper sulphate and lead citrate depending on the structures to be examined. *En bloc* osmium and uranyl acetate staining can also provide sufficient contrast for backscatter imaging and 1% osmium tetroxide can also be sufficient for good contrast. Staining of tissue for the ATUM may require staining over much larger depths than are usually required. Although tissue may be sectioned (e.g. by vibratome) prior to staining, this may not always be desirable, particularly for neural circuit reconstruction where information may be lost at the section planes. To combat this problem Mikula and Denk (2015) have developed a protocol for staining the whole brain without sectioning. In this protocol extracellular space is preserved by preventing osmotic imbalance during aldehyde fixation, improving the penetration of reduced osmium fixatives by the addition of formamide, and substitution of thiocarbohydrazide with pyrogallol to prevent the liberation of nitrogen bubbles in the tissue during staining. Post-sectioning stains such as uranyl acetate and lead citrate may also be applied to ATUM sections, although this is not always necessary for visualisation, depending on the staining protocol used and the structures and level of contrast required.

22.2.2 Sectioning

Once samples are stained and embedded in resin, they may be prepared for conventional ultramicrotomy. The size of the block face can vary significantly for ATUM processing, although larger block faces tend to produce more reliable spacing in sections on the tape, and block sizes of 2 mm wide by 3 mm tall are in use for ATUM sectioning. Small block sizes are also used, but small sections are more likely to form ribbons as they are cut and are also more likely to rotate as they float from the knife to the tape. Changes in section spacing and rotation can both cause problems for automated image collection.

The ATUM can be used with a variety of tapes depending on the properties required. For electron microscopy, the most common tape in use for ATUM sectioning is polyimide film (Kapton®). Kapton tape combines the required flexibility with sufficient tensile strength to make it possible to handle the tape for mounting. The Kapton tape is often washed with alcohol before use to remove oils from the surface, and the hydrophilicity of the tape increased by glow discharging or plasma treating the tape before use (Kasthuri *et al.*, 2015). It is also suggested that the conductivity of the tape may be improved by a ~5 nm carbon coating applied to the surface, particularly for secondary electron detection (Hayworth *et al.*, 2014). However, the glow discharge and carbon coating of tape in a standard apparatus is a difficult and often impractical process, as the surface of the tape must be treated. Pre-treated and coated tape can now be bought commercially, and specialised coaters and coater attachments to allow automatic treatment of tape are in development.

In order to ensure consistency in the sections collected by the ATUM, the environment of the ATUM and ultramicrotome is an important consideration. Fluctuations in heat and humidity, air currents and vibrations may all affect cutting and collecting of sections. The ATUM is best positioned in a room free of draughts (or air currents from air conditioning systems) and with a constant temperature and relative humidity of around 45% rh. The ATUM is placed on an air table to minimise vibration and is best left undisturbed without human interaction once cutting; therefore it may be advisable to place the ATUM in a room with low foot traffic. RMC products produce a fact sheet of pre-installation considerations for the ATUM (http://www.rmcboeckeler.com/download/manuals/ATUMtome_pre-installation_considerations-1.pdf).

To cut sections using the ATUM, tape is first loaded on to the ATUM by passing the tape from the lower spool through the tape path to the upper spool. Some tape may be run through the instrument prior to use to ensure the tape path is running smoothly and under tension. The ultramicrotome is aligned as it would be for standard ultramicrotomy. When the knife is correctly positioned, the ATUM nose is introduced to the knife boat. The nose is placed a short distance back from the knife-edge to allow sections to float away from the knife and on to the tape. If the nose of the ATUM is too close to the knife it may disturb cutting but if it is too far away sections may rotate before they are picked up by the ATUM tape. When the nose has been correctly positioned the water level must be set up for automatic monitoring by the software attached to the ATUM. Once the level is set the software will maintain it for the duration of cutting. The water monitoring should be checked with the tape moving, as this may cause some small vibrations to the water level. Once the instrument and water monitoring is set up, the tape can be started and the ultramicrotome set to automatically cut. The system may then be left until the ultramicrotome reaches the end of its auto feed range and must be reset.

22.2.3 Mounting

When sectioning has been completed, the ATUM tape must be mounted for examination in the SEM. The tape is cut into strips and mounted on to conductive (doped) silicon discs using adhesive carbon tape; copper tape may be used to provide a path to ground (Figure 22.2a). For this, a mounting platform is supplied to aid cutting of the tape to the correct size (Figure 22.2b). The tape should be handled with forceps, laid flat with no wrinkling and care should be taken to avoid touching the sections, particularly with tools. It is vitally important at this stage that the tape strips are mounted in a known direction to maintain the order of the sections. As the tape itself is not conductive, care must be taken to ensure the sample is grounded to prevent charging in the electron microscope. For backscatter electron detection the entire disc may be carbon coated (to approximately 10 nm) after section mounting (Figure 22.2c) (Hayworth *et al.*, 2014; Kasthuri *et al.*, 2015). Depending on the image acquisition software used, it may be necessary to add fiducial markers to

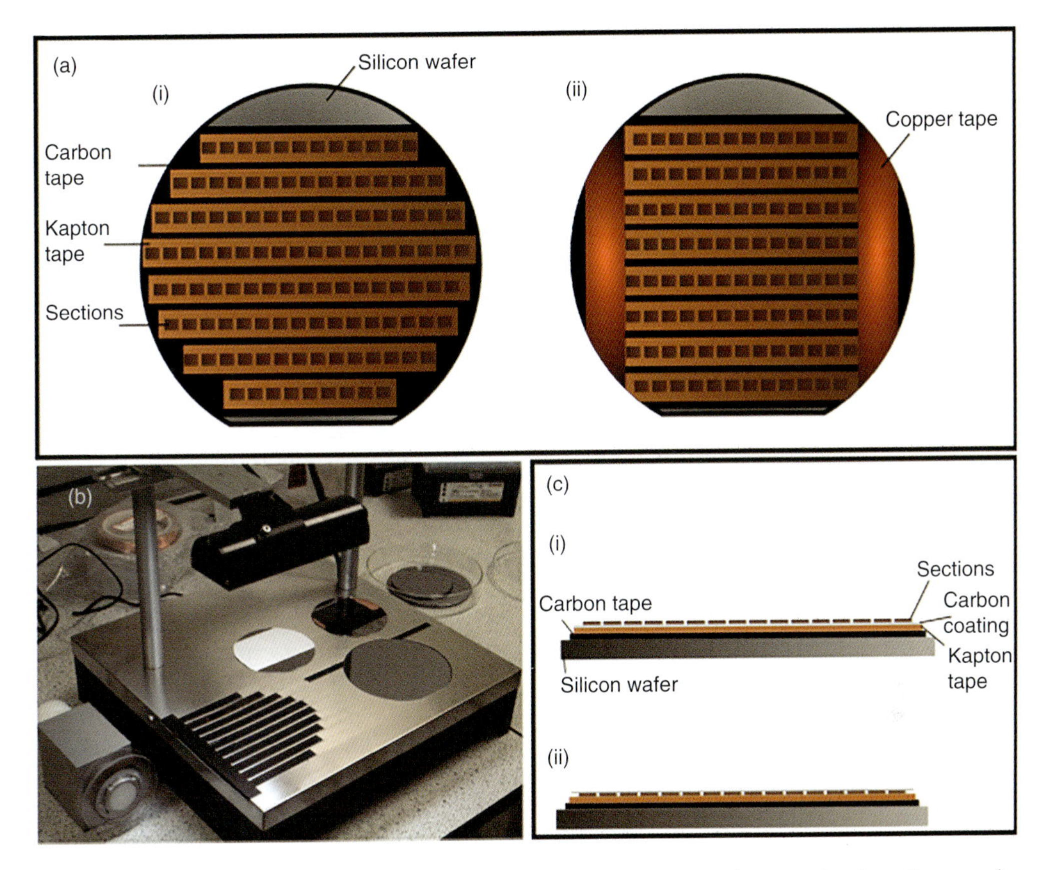

Figure 22.2 Mounting of ATUM sections. (a)(i) Sections are mounted on to circular silicon wafers using conductive carbon tape. (ii) Copper tape may be used to improve conductivity. (b) A workstation is supplied to make the mounting of sections easier. (c) Side view of the wafer showing the layers in the mounting of sections for (i) secondary electron and (ii) backscatter detection.

the ends of the tape strips, for which copper finder TEM grids have been used (Hayworth *et al.*, 2014). Mounting is also usually the most labour-intensive and time-consuming part of ATUM sample preparation.

22.2.4 Imaging

Efficient imaging of large numbers of ATUM sections requires automated microscopy approaches, as image acquisition must be targeted so the same area can be collected across potentially many thousands of sections without requiring an operator to manually select it. In addition, the time taken for image acquisition and the storage required for high-resolution images must also be considered. Several software and hardware approaches have been developed to address these issues on different microscope platforms. A free software package called WaferMapper was developed by Hayworth *et al.* for Zeiss Sigma and Merlin SEMs (Carl Zeiss Microscopy GmbH, Germany) that are fitted with Fibics scan generators (Fibics Inc., Canada) and was designed to work specifically with ATUM wafers. The software maps the complete wafer and acquires overviews of each section. Multiple wafers can be mapped and the software will present to the user an overview stack consisting of all of the mapped wafers. The user is then asked to choose and verify a target region and montage across the section stack. The software then acquires high-resolution images for the target region on each section on the wafer. As each wafer is re-loaded for high-resolution imaging the software uses fiducial markers to align the saved maps to the wafer. The software is open-source and freely available from the Google code repository (https://wafermapper.googlecode.com). It is written primarily in MatLab®, with communication between the software and the microscope and scan generator APIs handled by two C wrappers. Therefore it is potentially highly adaptable to different projects and microscope hardware platforms (Hayworth *et al.*, 2014).

Commercial software packages are also available for the control of ATUM wafer imaging. Also on the Zeiss platform, the Atlas 5 Array Tomography software (Carl Zeiss Microscopy) allows the mapping of wafers followed by semi-automatic selection of section boundaries and regions of interest (ROIs) within these boundaries. Once ROIs have been selected, multiresolution image acquisition protocols can be used to image the sections automatically. Atlas 5 also has the ability to use 2D and 3D stacks from light, confocal or X-ray microscopes for correlation and guiding the users to potentially interesting sites for SEM imaging (Eberle *et al.*, 2014).

On the JEOL platform, SEM Supporter (System in Frontier Inc., Japan) is also capable of imaging ATUM wafers. Similarly to the Zeiss system, SEM Supporter allows low-mag mapping of the array wafer, followed by the assignment of ROIs for imaging and automatic image acquisition. Light microscope montages of the array may be imported into the software for mapping and the positioning of ROIs refined by acquisition of low-magnification FESEM images, which may be saved into combined montages. ROIs may be defined on one section and positioning points used to copy the ROI to subsequent sections. ROIs may also define single image or montage regions. Focusing may be carried out manually or by autofocus at each ROI or at pre-defined intervals. A focus-mapping feature is also available for automatic focus adjustment as the stage is moved. The software aligns images 'on the fly' during image acquisition and manual adjustments may also be made. A montage image of a mammalian cochlea section collected by SEM Supporter and a higher resolution image of the same region can be seen in Figure 22.3b and c.

Speed of image acquisition is affected by the scan speed and the detector response time, meaning that the choice of signal used may affect the acquisition time, and for this reason secondary electron detection is often preferred (Hayworth *et al.*, 2014; Kasthuri *et al.*, 2015). As faster and more sensitive detectors are developed this will continue to improve, but currently an estimate for the imaging of a 1 mm × 1 mm block of tissue with a section thickness of 25 nm and an *x, y* resolution of 20 nm would be about 3 years (Knott and Genoud, 2013). An alternative approach to this problem is the use of a multibeam SEM, which combines many electron beams and detectors in a single column. An instrument with 61 beams arranged in a hexagonal array has been developed by Zeiss and shown to be capable of high-throughput imaging on a variety of different samples (Eberle *et al.*, 2015).

22.2.5 Alignment and Segmentation

The large number of sections that can be produced automatically by the ATUM mean that, depending on the number of sections, the size of the imaging area and the resolution required, the size of acquired image stacks could be in the terabyte or even petabyte

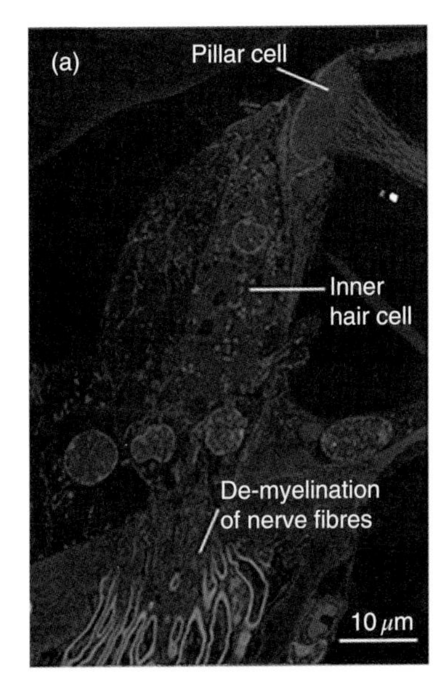

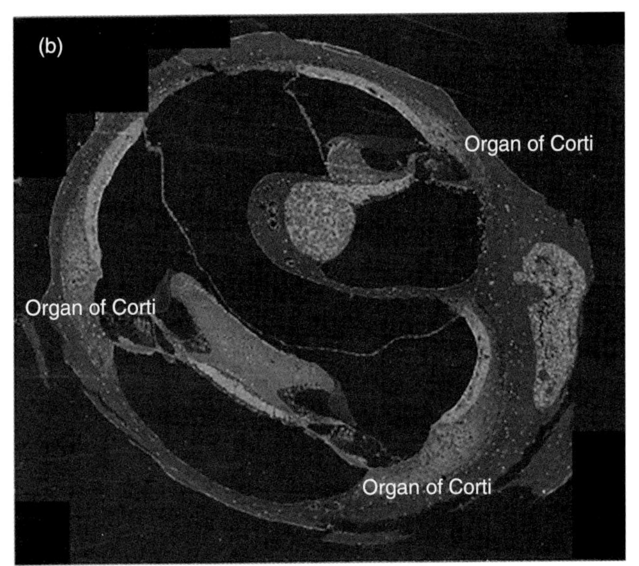

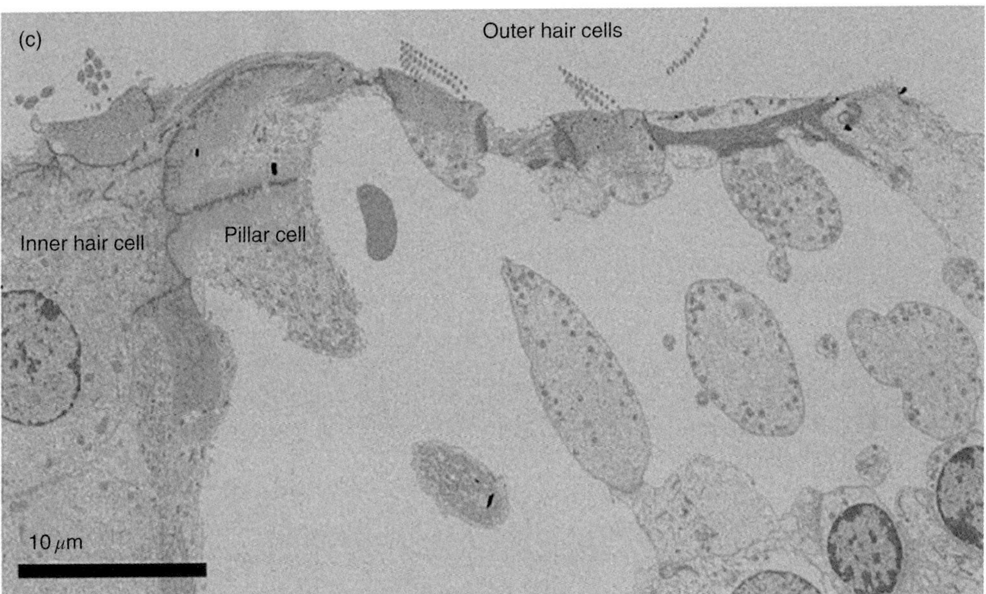

Figure 22.3 Backscatter imaging of serial and ATUM sections. Images of the organ of Corti and surrounding structures of the mammalian cochlea. (a) Backscatter image of an inner hair cell and surrounding structures from the mouse cochlea, stained with *en bloc* osmium tetroxide and uranyl acetate. No post-section staining has been applied. Details of inner hair cell structure, the structure of the supporting pillar cells and the de-myelination of nerve fibres as they pass through the basement membrane can be seen. (b) Montage image from ATUM section showing the mouse cochlea. Three cross-sections of the organ of Corti are labelled. (c) Higher magnification image from an ATUM section showing the tops and partial cell bodies of the three rows of outer hair cells. The top of a supporting pillar cell and inner hair cell are also shown.

range. Imaging data collected from the ATUM must be first aligned to produce an aligned image stack. Stitching and alignment can be carried out by several software packages, both freeware and commercial, including ImageJ or Fiji, etomo (part of IMOD), Amira (FEI Visualization Sciences Group, France) and Imaris (Bitplane, Switzerland) among others, although alignment of such large datasets may often require specialised software and research in this area is ongoing (Kaynig *et al.*, 2015). Once sections are aligned the volume may be segmented for 3D reconstruction. Although manual segmentation is possible using a variety of tools, the size of ATUM volumes means that the time and labour required to carry it out are often prohibitive. Therefore, computer-assisted or automated segmentation approaches must be explored. At Harvard, the laboratories of Prof. Jeffrey Lichtman together with the Visual Computing Group under Prof. Hanspeter Pfister have released several tools to aid with volume segmentation. A program for computer-assisted manual space-filling segmentation and annotation named 'VAST' (http://openconnecto.me/Kasthurietal2014/Code/VAST/) is capable of loading image stacks, which are hosted online, eliminating the requirement for all of the image data to be stored locally. The group also developed a suite of tools for automatic segmentation of neurons named RhoANA (http://www.rhoana.org) (Kasthuri *et al.*, 2015; Kaynig *et al.*, 2015), which can then be proofread using the tool Mojo or the web-based equivalent Dojo (Haehn *et al.*, 2014). Other automated/semi-automated segmentation packages are also available, including ilastik (http://ilastik.org) for automated segmentation (Sommer *et al.*, 2011; Kreshuk *et al.*, 2011) and semi-automated packages such as Ssecrett and NeuroTrace (http://www.neurotrace.org), a combined package for semi-automated segmentation and the handling of large data through a server–client paradigm (Jeong e*t al.*, 2010). A review of some of the available packages was compiled in 2012 by Helmstaedter and Mitra (2012).

The development of the ATUM system attempts to combine the best assets of different approaches to SEM-based serial section microscopy, the automation and ease of sample preparation available in SBF-SEM and FIB-SEM and the preservation of libraries of serial sections that may be reexamined at any time, like those produced by traditional array tomography. Although the technique is new, it has already produced impressive results, including the reconstruction of a region of neocortex (Kasthuri *et al.*, 2015) and insights into the 3D structure of endoplasmic reticulum (Terasaki *et al.*, 2013). These results also show the applicability of the technique on different resolution scales. Although much of the work so far using the ATUM has been concentrated on neural connectomics, as the system is further developed and disseminated it will be useful in solving many biological problems that require three-dimensional structural information. In particular, the potential for ATUM sections to cover vast length scales, compared to those practically achievable by other techniques, will make accessible important 3D information that has previously only been examined by lower resolution techniques. The ATUM can also be used for light microscopy, using glass or other tape types. Immunohistochemistry can also be performed on ATUM sections as it can be on other resin-embedded samples, and antibodies may even be eluted and the sections re-labelled with different antibodies (Micheva and Smith, 2007; Webster, Bentley and Kearney, 2015). Therefore, correlative microscopy should also be possible using ATUM section libraries.

Continuing development of the ATUM system involves the investigation of many aspects of ATUM usage, including sample preparation, the environmental conditions for sectioning, tape used, mounting methods and imaging modalities for ATUM sections. Information on this ongoing development can be found via the commercial developers RMC, who have set up an early adopters of technology exchange where early users of the ATUM system can discuss their discoveries (http://atumtome.com/exchange/). The relative simplicity of the ATUM system and its reliance on existing and well-established techniques in electron microscopy make it a valuable addition to the expanding toolbox of 3D electron microscopy techniques.

ACKNOWLEDGEMENTS

I would like to thank Fiona Winning (King's College London) for advice on this chapter and Andy Yarwood (JEOL UK) for imaging and information on the JEOL SEM supporter system. Work contained in this chapter was supported by the BBSRC and Action on Hearing Loss.

REFERENCES

Eberle, A. L., Selchow, O., Thaler, M., Zeidler, D. and Kirmse, R. (2014). Mission (im)possible - mapping the brain becomes a reality. *Microscopy* 64, 45-55.

Eberle, A.L., Mikula, S., Schalek, R., Lichtman, J., Tate, M.L. and Zeidler, D. (2015) High-resolution, high-throughput imaging with a multibeam scanning electron microscope. *J. Microsc.*, 259, 114–120.

Haehn, D., Knowles-Barley, S., Roberts, M., Beyer, J., Kasthuri, N., Lichtman, J. and Pfister, H. (2014) Design and evaluation of interactive proofreading tools for connectomics. *IEEE Transactions on Visualization and Computer Graphics*, 20, 2466–2475.

Hayworth, K.J., Morgan, J.L., Schalek, R., Berger, D.R., Hildebrand, D.G. and Lichtman, J.W. (2014) Imaging ATUM ultrathin section libraries with WaferMapper: A multi-scale approach to EM reconstruction of neural circuits. *Front Neural Circuits*, 8, 68.

Helmstaedter, M. and Mitra, P.P. (2012) Computational methods and challenges for large-scale circuit mapping. *Curr. Opin. Neurobiol.*, 22, 162–169.

Horstmann, H., Korber, C., Satzler, K., Aydin, D. and Kuner, T. (2012) Serial section scanning electron microscopy (S3EM) on silicon wafers for ultra-structural volume imaging of cells and tissues. *PLoS One*, 7, e35172.

Jeong, W.K., Beyer, J., Hadwiger, M., Blue, R., Law, C., Vazquez-Reina, A., Reid, R. C., Lichtman, J. and Pfister, H. (2010) Ssecrett and NeuroTrace: Interactive visualization and analysis tools for large-scale neuroscience data sets. *IEEE Comput. Graph Appl.*, 30, 58–70.

Kasthuri, N., Hayworth, K.J., Berger, D.R., Schalek, R.L., Conchello, J.A., Knowles-Barley, S., Lee, D., Vazquez-Reina, A., Kaynig, V., Jones, T.R. *et al.* (2015) Saturated reconstruction of a volume of neocortex. *Cell*, 162, 648–661.

Kaynig, V., Vazquez-Reina, A., Knowles-Barley, S., Roberts, M., Jones, T.R., Kasthuri, N., Miller, E., Lichtman, J. and Pfister, H. (2015) Large-scale automatic reconstruction of neuronal processes from electron microscopy images. *Med. Image Anal.*, 22, 77–88.

Knott, G. and Genoud, C. (2013) Is EM dead? *J. Cell Sci.*, 126, 4545–4552.

Kreshuk, A., Straehle, C.N., Sommer, C., Koethe, U., Cantoni, M., Knott, G. and Hamprecht, F.A. (2011) Automated detection and segmentation of synaptic contacts in nearly isotropic serial electron microscopy images. *PLoS One*, 6, e24899.

Micheva, K.D. and Smith, S.J. (2007) Array tomography: A new tool for imaging the molecular architecture and ultrastructure of neural circuits. *Neuron*, 55, 25–36.

Mikula, S. and Denk, W. (2015) High-resolution whole-brain staining for electron microscopic circuit reconstruction. *Nat. Methods*, 12, 541–546.

Schalek, R., Kasthuri, N., Hayworth, K., Berger, D., Tapia, J., Morgan, J., Turaga, S., Fagerholm, E., Seung, H. and Lichtman, J. (2011) Development of high-throughput, high-resolution 3D reconstruction of large-volume biological tissue using automated tape collection ultramicrotomy and scanning electron microscopy. *Microscopy and Microanalysis*, 17, 966–967.

Schalek, R., Wilson, A., Lichtman, J., Josh, M., Kasthuri, N., Berger, D., Seung, S., Anger, P., Hayworth, K. and Aderhold, D. (2012) ATUM-based SEM for high-speed large-volume biological reconstructions. *Microscopy and Microanalysis*, 18, 572–573.

Sommer, C., Strahle, C., Kothe, U. and Hamprecht, F.A. (2011) ilastik: Interactive learning and segmentation toolkit, in *Eighth IEEE International Symposiumm on Biomedical Imaging (IBSI) Proceedings*, pp. 230–233.

Tapia, J.C., Kasthuri, N., Hayworth, K.J., Schalek, R., Lichtman, J.W., Smith, S.J. and Buchanan, J. (2012) High-contrast en bloc staining of neuronal tissue for field emission scanning electron microscopy. *Nat. Protoc.*, 7, 193–206.

Terasaki, M., Shemesh, T., Kasthuri, N., Klemm, R.W., Schalek, R., Hayworth, Kenneth J., Hand, Arthur R., Yankova, M., Huber, G., Lichtman, J.W. *et al.* (2013) Stacked endoplasmic reticulum sheets are connected by helicoidal membrane motifs. *Cell*, 154, 285–296.

Webster, P., Bentley, D. and Kearney, J. (2015) The ATUMtome for automated serial sectioning and 3-D imaging. *Microscopy and Analysis*, 29, 19–23.

23

SBEM Techniques

Christel Genoud
Friedrich Miescher Institute for Biomedical Research, Basel, Switzerland

23.1 INTRODUCTION

The technique of serial block-face scanning electron microscopy (SBEM) using a microtome inside a scanning electron microscopy (SEM) is based on an automated ultramicrotome developed and designed by Denk and others (Denk and Horstmann, 2004) who were inspired by the work of Leighton (1981). The company GATAN (GATAN, Pleasanton, USA) is selling the commercial version under the name 3View®. Thermo Fisher Scientific is also selling a microtome inside an SEM in a version called Volumescope, where a deconvolution based on the beam energy is developed (https://www.fei.com/products/sem/teneo-vs-sem-for-life-sciences/). The aim of this technique is to automatically image the ultrastructure of biological samples at high resolution (<30 nm) throughout large volumes (>100 × 100 × 100 μm^3) inside the vacuum chamber of an SEM. During the same period, two other techniques were developed to retrieve a volume of tissue at an ultrastructural level (Kubota, 2015; Titze and Genoud, 2016). The first is focused ion beam scanning electron microscopy, FIBSEM (Knott *et al.*, 2008), well known and used in material science for a long time (Giannuzzi *et al.*, 2004; Orloff, 2009). It is limited to a smaller volume than SBEM. The second is Automatic Tape UltraMicrotomy (ATUM) requiring sample handling and registration procedures (Tapia *et al.*, 2012b; Hayworth *et al.*, 2014; Kasthuri *et al.*, 2015). With the present development of the correlative light and electron microscopy techniques (Watanabe *et al.*, 2011; Kukulski *et al.*, 2012; Müller-Reichert and Verkade, 2012; Peddie *et al.*, 2014), the fact that SBEM can image large volume makes it a complement to light microscopy techniques as the *in vivo* 2-photons (Briggman, Helmstaedter, and Denk, 2011; Helmstaedter *et al.*, 2013; Hoggarth *et al.*, 2015) or the super-resolution light microscopy (Kopek *et al.*, 2012). Electron microscopy allows the observation of structures below the resolution of the light techniques and is able to image all elements in a volume and not only specific structures tagged with fluorophores. The importance of SBEM is recognized

Biological Field Emission Scanning Electron Microscopy, First Edition.
Edited by Roland A. Fleck and Bruno M. Humbel.

in various areas of life sciences, particularly in neuroscience (Briggman and Bock, 2012; Helmstaedter, 2013; Kubota, 2015). An exhaustive list of applications can be found in the review of Peddie and Collinson (2014). The development of the multi-beam SEM (Eberle *et al.*, 2015) is a milestone, making it possible to envisage the acquisition of volumes as large as a mouse brain (Mikula, Binding and Denk, 2012; Perkel, 2014; Mikula and Denk, 2015).

23.2 SERIAL SECTIONS WITH TEM

The development of transmission electron microscopy (TEM) (Ruska and Knoll, 1932) offered resolution greater than that of light and thus allowed the ultrastructure of biological samples to be routinely viewed (Knott and Genoud, 2013). The technique of TEM employs the transmission of electrons through an ultrathin specimen with diffraction of the electrons as they pass through the specimen, contributing to the formation of an image. For ultrastructural visualization of biological systems it has been necessary for the development of multiple sample preparation techniques to maintain the ultrastructure of the biological sample whilst making them suitable for observation in a TEM (Marton, 1976). Samples have to be resistant to high vacuum and have to diffract electrons to generate an image when exposed to the electron beam. Biological samples for electron microscopy are fixed to keep them as close as possible to their living state by chemical fixation or cryo-fixation. Following fixation, heavy metals are then introduced to the samples to stain the biological elements present (Glauert and Lewis, 1988; Griffiths, 1993; Hayat, 2000). Finally, samples are embedded in plastic resins that are liquid at a given temperature but polymerized and hardened with heat or with UV light. The result consists of blocks of resin in which pieces of biological materials stained with heavy metals are embedded (Robards and Wilson, 1993).

The need for retrieving 3D information from sections in the TEM has been identified (Porter and Blum, 1953; Williams and Kallman, 1955). This need stemmed from the understanding that essential information lay in the third dimension of a tissue and that this information was averaged out when an electron beam passed through a 3D structure and formed a 2D image (Miranda *et al.*, 2015). Indeed, any object inserted in a TEM has to be thin enough for the electron beam to go through it. In the case of a small biological specimen like viruses, isolated proteins and mycoplasmae, they are thin enough to be stained and still can let an electron beam go through to retrieve information. Before the development of the volume electron techniques (Ohno *et al.*, 2014), the only two ways to image tissue samples was to cut thin slices out of the block with an ultramicrotome or to perform tomography on thicker sections (Crowther *et al.*, 1970; Hoppe and Grill, 1977; Frank, 2006; Gan and Jensen, 2012). The sections (20–300 nm) are collected on grids, which are then introduced in a TEM (Bozzolla and Russel, 1992). The image formation is obtained by the scattering of the electrons going through the thin section by the areas stained with heavy metals. As scattered electrons are filtered out in the microscope column, only electrons that went through samples are retained and are collected on a screen or on a charge-coupled device (CCD) chip of a camera or a direct detection device (DDD) in order to create a 2D projection image (Figure 23.1). The image obtained is a two-dimensional projection of a three-dimensional object. It leads to a loss of information. In Figure 23.1, case A and B cannot be distinguished in the two-dimensional plan as they both scatter the electrons the same way. In order to reduce the inherent loss of information due to the projection, the first possibility is to cut thinner sections (Sjostrand, 1958; Stevens *et al.*, 1980) in order to create more accurate two-dimensional projections that can be recombined in silico to reconstruct a three-dimensional model of the original object. The second option that has been developed

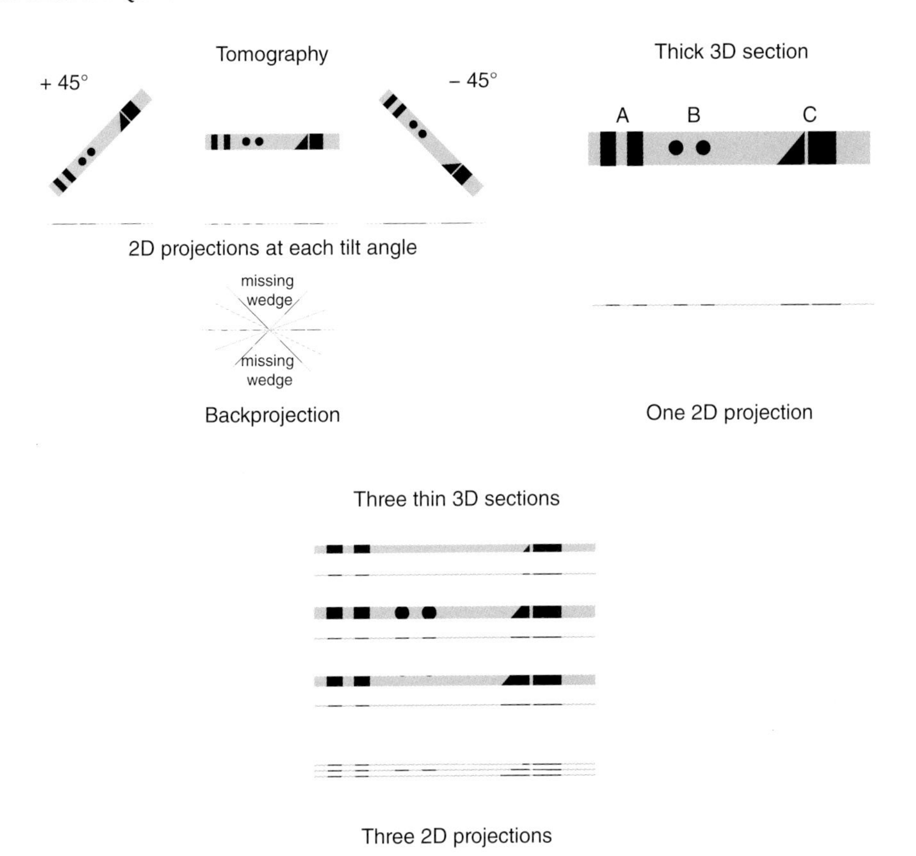

Figure 23.1 A formation of the image in a TEM. The electron beam goes through the thickness of the sample and is scattered by heavy metal in the section. The resulting image in 2D is a projection. On the right side, the option of serial sectioning is illustrated where volume information is retrieved from the alignment of the successive images obtaines. On the left side, the tomography option is illustrated where the volume information is retrieved following a back-projection issued from the 2D projections at different sample tilt angles.

is the tomography. In this case, rather than physically cutting the sample thinner, the thick section is tilted in the beam axes, creating as many two-dimensional projections as selected tilt angles. The two-dimensional images are then back-projected using software in order to reconstruct a three-dimensional volume through the thick section.

This technique has a limit: the more tilted is the thick section, the thicker is the volume of interaction with the electron beam. At a certain point, electrons can no longer go through the sample. This means that the tilt is limited and that information at acute angles are not retrieved, leading to loss of information in the three-dimensional reconstruction, called the missing wedge (Frank, 2006). By tilting in two axes, this can be reduced but never disappears. The penetration of the electron beam through the sample depends on its energy. One of the technical developments for TEM is to increase the beam energy. Therefore, TEM are available with different beam sources able to generate a beam going from 80 to 1000 kV. The higher the extracting voltage, the faster are the electrons, the deeper in the specimen they are able to penetrate. While an 80–120 kV TEM is sufficient to observe thin sections of resin-embedded samples and small specimens, higher voltages are recommended for applications like cryo-microscopy (Schröder, 2015) and tomography. In the case of cryo-microscopy, the contrast is based on a phase-shift through unstained samples. As the samples are highly beam-sensitive, there is a need to work in a low-dose mode, meaning few electrons going through the sample but at high speed (Schröder, 2015).

Out of the two options, serial sections and tomography, the field of neuroscience has always been keen in having serial sections. It is indeed the method of choice to follow tiny processes rather than large volumes and visualize the connections between neuronal cells on large fields of view (Harris, 1999; Knott *et al.*, 2002; Bumbarger *et al.*, 2013). Therefore, techniques to retrieve serial sections have been developed. The challenge is to cut hundreds of sections in series from a block without losing any section, to collect them on grids by keeping the order and then to image each section of each grid at the same position to obtain a series of images at the same location that represent a volume of tissue that is imaged (White *et al.*, 1986; Bumbarger *et al.*, 2013). As it is tedious work and the steps where you can lose material are numerous, the technique has always been limited to the size of the volume that can be obtained. It is important to notice that the collection of sections is difficult but the imaging is also not trivial. In order to achieve a large field of view in the TEM, it is necessary to make acquisitions in tiles and then to use a performing algorithm to align the tiles in one plane and then between the planes in Z (Bock *et al.*, 2011). Due to the beam interactions, each field of view is distorted in a different way and registration is a significant step demanding a lot of computer power (Cardona *et al.*, 2010; Kaynig *et al.*, 2010, 2013). One of the first successes in this approach has been the manual serial sectioning of *Caenorhabditis elegans* in order to reconstruct its nervous system together with its connections (White *et al.*, 1986; Varshney *et al.*, 2011). This technique is now successfully used to reconstruct neuronal circuits (Bock *et al.*, 2011; Bumbarger *et al.*, 2013; Takemura *et al.*, 2013; Ohyama *et al.*, 2015).

23.3 DEVELOPMENT OF ENVIRONMENTAL SEM

Scanning electron microscopes (Ardenne, 1938) use the same principles as TEM to generate an electron source and ultimately a signal. They use an electron beam of low energy compared to TEM (0.5–30 kV for SEM versus 80–1000 kV for TEM) that is shaped like a small spot. The small spot is driven by a scanning generator and scans the surface of a sample. Scanning parameters can be modulated to determine the pixel size, the number of pixels, the dwell time and the scanning pattern. The SEM do not require high beam energy as the beam is not going through the sample (one exception is the STEM in SEM, which is not discussed here) but is scanning the surface and images are formed by the detection of different signals coming from the interaction of the beam with the sample surface. It is widely used in material science fields to characterize the topology of a surface or the elemental composition of the surface of a sample. In life science, SEMs are commonly used to look at the topology of biological specimens. However, this technique has a major constraint: as the beam hits the sample and does not pass through the sample, the energy of the beam needs to be dissipated. If it is not, a phenomenon called charging occurs (Robinson, 1975). This means that electrons are trapped around the point where the beam is scanning: it is both damaging to the sample and creates artefacts in the imaging as it distorts the incoming beam while blocking or deviating the secondary electrons before they are detected (Titze and Denk, 2013).

The charging issue has been addressed by various techniques. The first approach is to coat the sample with a thin layer of conductive metal in order to prevent charging and create contrast. If the coating of the sample is not possible, a current can be directly applied on the sample but it only works if the sample is relatively conductive. Ultimately, it is possible to inject a gas into the SEM chamber to prevent charging. The major technique developed by the microscope manufacturers is the introduction of a gas in the chamber of the SEM without compromising the vacuum in the column (Danilatos, 1993, 2013a, 2013b). The manufacturers are now selling microscopes with a differential aperture, allowing a low pressure of gas (typically water or nitrogen) in the chamber (until 1 atm) while keeping the column at a high vacuum. It is called a low vacuum mode, environmental SEM (ESEM) or variable pressure SEM (VP-SEM). The incoming beam is scattered and creates a less sharp spot when it is going through the chamber until it hits the sample (skirting effect) so a compromise between the gas pressure in the chamber, the beam energy and the distance that the beam has to travel before hitting the sample has to be found (Danilatos, 1988). It also means that the detected electrons are scattered before reaching the detector (Danilatos, 2012, 2013a, 2013b). The use of the ESEM mode forces operators to find compromises as it introduces a new limit in the reachable resolution of the system. The more gas there is in the chamber, the greater is the scattering of the electrons, the higher the level of noise and the more limited the resolution. As biological samples in resin are not conductive, the use of ESEM has been a critical step in the development of SBEM (Denk and Horstmann, 2004; Titze and Denk, 2013). However, as soon as the potential of imaging blocks in the SEM has been identified (Denk and Horstmann, 2004), solutions to avoid low vacuum and optimize resolution have been investigated (Titze and Denk, 2013; Wanner, Genoud and Friedrich, 2016; Wanner *et al.*, 2016; Bouwer *et al.*, 2017). Low vacuum is indeed impairing the resolution, the signal-to-noise ratio and the speed of detection due to the scattering. The more the sample is charging, the more gas pressure must be used to dissipate charges. In order to keep a detectable signal, the electron dose must be increased by increasing the beam energy, the pixel dwell time or the beam current. Increasing any of these parameters increases the electron dose per area on the sample surface (Goldstein *et al.*, 1992). Increasing the beam voltage is the worst case as the beam will penetrate deeper into the block, thus impairing the resolution (Knott *et al.*, 2008). Increasing the dwell time and beam current should not be a problem as long as the sample can sustain the applied electron dose without damage (Wanner, Genoud and Friedrich, 2016; Wanner *et al.*, 2016). Embedded samples are highly sensitive to electron dose. By increasing any of the parameters cited above to get a better signal, you are getting closer to the limit where the sample will be damaged. At this point, scanning the surface is creating artefacts by altering the resin at the surface of the sample. Furthermore, the damages are not restrained to the surface but propagate deeper into the block, compromising the ability to cut this layer out of the block surface (Bouwer *et al.*, 2017). The drastic limitations triggered by the usage of low vacuum are impairing the use of SEM for its best performances and now many solutions are being investigated to be able to image resin blocks in high vacuum conditions. In high vacuum, there is no scattering in the chamber and a low voltage, high speed, high beam current can be used. Furthermore, it opens the possibility to use all the detectors available on the SEM while the low vacuum mode is excluding the in-chamber detectors (Everhart-Thornley detector).

23.3.1 Detection Modes in SEM Compatible with Resin Blocks

The detection modes compatible with the observation of biological samples embedded in plastic have until now focused on the two main signals generated in the SEM: secondary electrons (SE) as well as backscattered electrons (BSE) (Reimer, 1998). Both types of detection give a good signal (Denk and Horstmann, 2004; Titze and Denk, 2013). Detection of BSE has been more common so far as backscattered electron detectors (BSED) are performing well in both low vacuum mode as well as in high vacuum mode, generating good contrast between heavy metal labelled membranes and the surrounding resin (Denk and Horstmann, 2004; Helmstaedter, Briggman and Denk, 2008). However, it is also possible to detect the secondary electrons via the in-chamber Everhart-Thornley detector or via an in-lens detector for secondary electrons when the conditions allow the instrument to be operated at high vacuum (Tapia *et al.*, 2012a; Titze and Denk, 2013). As shown later in Table 23.2, the optimal voltage is in the range of 1.5–7 kV. It minimizes the damages on the block and charging artefacts whilst generating a signal high enough for the BSED detector. It is also at an optimal setting to restrain the beam penetration depth in the block and retrieves a signal from the surface (Seiter *et al.*, 2014). As a tissue in the resin is a combination of areas generating high levels of BSE (stained areas) and areas where the beam is penetrating deep without obstacles (empty resin), the resolution in *Z* is strongly determined by the beam energy. The higher the beam energy, the deeper is the penetration of electrons, the larger is the volume of interaction in the sample and the lower is the resolution as the BSE generating the signal for a given scan point is coming from a large volume around the scan point. It is in this case not possible to distinguish whether the signal is coming from the surface of the sample or from a layer at the surface. The higher the beam energy, the thicker is the layer and the lower is the resolution. At a low beam energy, the volume of interaction is smaller and the resolution is better as the electrons generating the signal for a given scan point are coming from a smaller volume around the scan point (Bouwer *et al.*, 2017).

23.3.2 Insertion of a Microtome in the SEM Chamber

In 1981, Leighton suggested combining a scanning electron microscope with a miniature ultramicrotome installed in the vacuum chamber of the scanning electron microscope. He built a prototype (US Patent No. 4377958, Remotely Operated Microtome) but the technique was not sufficiently developed at this time to offer automation. In 2004, Denk and Horstmann published the first paper describing the development of the SBEM. For their approach, they used a FEGSEM equipped with a low vacuum mode or environmental mode (QUANTA FEG 200, Thermo Fisher Scientific, Eindhoven, The Netherlands). They removed the door closing the chamber and manufactured a new door on which a microtome is attached on a moving stage. Compared to the classical ultramicrotome, the major difference is that the knife moves back and forth and the sample goes up each time for a determined step as small as 5 nm. By this way, the block face is always aligned with the electron beam at the same working distance to keep the surface of the block continuously in the focal plane. In order to extend the field of view, the stage is controlled by two piezo motors, allowing the precise movement of the stage to build mosaics. Software is written to allow automatic cycles of cutting/imaging. In order to optimize the detection, a homemade detector has been built with better sensibility and faster speed than the BSED on the market at this time (Denk and Horstmann, 2004).

GATAN, a company specialized in the developments of accessories for TEM and SEM commercialized the invention of Denk under the name 3View®. They kept the design of a microtome attached to a door closing the chamber of a SEM. While Denk used the scanning coil of the microscope itself, GATAN used their own solution, the DIGISCAN, a digital beam controller. At the level of the software, it is controlled by their existing software called Digital Micrograph (DM). Furthermore, they also commercialized a detector specific for

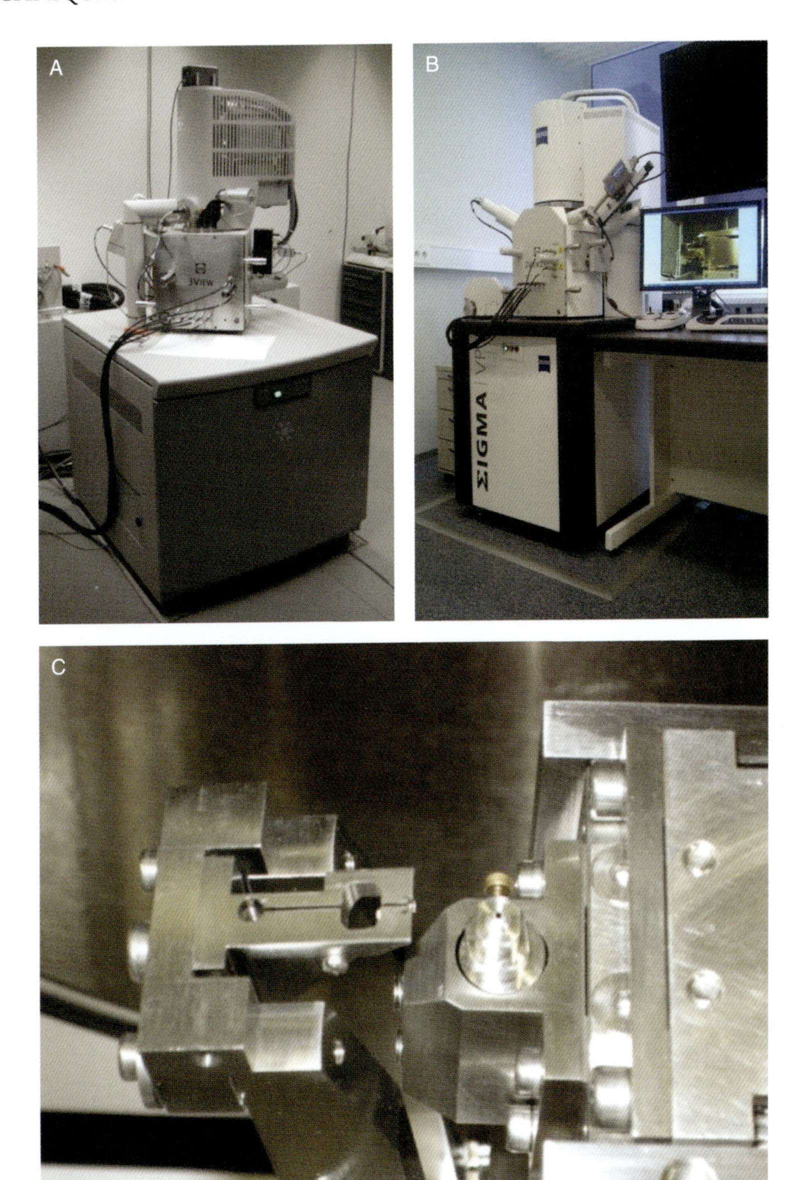

Figure 23.2 A. A 3View mounted on a FEI Quanta 200. B. A 3View mounted on a ZEISS SIGMA VP. C. The sample is located in the sample holder and the knife holder is in the 'clear position'.

3View®. The commercial version 3View® can be implemented on different types of SEM from different manufacturers (Figure 23.2).

The microscope manufacturer Thermo Fisher Scientific has developed its own solution called TENEO VS (VolumeScope). It is also a microtome using a diamond knife that is inserted in the chamber of a SEM microscope. In this approach, the microtome is not attached to the door but is sitting on the stage of the SEM. The detector is a BSED detector specially developed for this application by Thermo Fisher Scientific. Furthermore, another feature has been added: it is possible to modulate the beam energy while doing an acquisition. The surface of the block is imaged multiple times at different beam energies, generating a succession of images containing information coming from different depths of the block. By a deconvolution, a 3D reconstruction of a layer of tissue below the block surface is obtained. After imaging, this layer can be physically cut and the process can be reiterated (https://www.fei.com/products/sem/teneo-vs-sem-for-life-sciences/).

23.3.3 Sample Preparation

The samples must address a number of key variables. The resin block containing the tissue must be suitable for ultramicrotomy to allow for thin sectioning. However, this must be performed repeatedly and in conjunction with imaging. Thus, the resin/ sample must both resist the beam sufficiently and maintain cutting characteristics necessary so it can be cut as thin as needed. The sample also needs to have good contrast (between features) at a reasonable speed of acquisition time; therefore, biological tissues also require to be heavily stained to provide strong backscattered contrast between the resin and the tissue (Tapia *et al.*, 2012a). Originally, the first paper of Denk and Horstmann (2004) described a sample preparation based on horseradish peroxidase staining of the membranes to underline the plasma membranes and follow the neurites of the cells. In order to generalize the use of this in SEM microtome, other staining procedures have been considered (Briggman and Denk, 2006; Tapia *et al.*, 2012a). In TEM it is routine for sections to be contrasted with uranyl acetate and lead citrate on the grid after having been cut. This post-staining brings most of the contrast to the images (Robards and Wilson, 1993). However, such protocols are not feasible for SBEM as the contrast cannot be introduced into the sample within the block and it would be impractical to sequentially stain an exposed top surface of a block, return it to the SEM image, cut and repeat. However, TEM tissue preparation protocols also exist which employ an en bloc staining step (e.g. with uranyl acetate, osmium tetroxide, tannic acid). This en bloc approach has been adapted for SBEM. As a first approach, a protocol with two steps in osmium (reduced osmium followed by normal osmium) followed by immersion in uranyl acetate provided improved quality of the images compared to a protocol with one step in the osmium (Knott *et al.*, 2002). This general staining allowed observation of cytoplasmic details.

A significant development in the field of the 3View® has been the adaptation of the protocol of T.J. Deerinck in the laboratory of M. Ellisman for the 3View® (Deerinck *et al.*, 2010; Wilke *et al.*, 2013). The protocol has been proven to be highly suitable for 3View® in two major ways: first, it gives a high contrast that is perfect to optimize the settings of the SEM for cutting and image acquisition. Second, it makes the sample more conductive, reducing the problems of charging. As there are now a lot of applications where heavy staining of blocks is required, numerous variations of this kind of protocol have been developed and published (Table 23.1). All the variations have the basic pattern of:

- Two steps for osmium with or without a step in thiocarbohydrazide or tannic acid in between (OTO) (Seligman, Wasserkrug and Hanker, 1966; Walton, 1979; Caceci and Frankum, 1987).
- En bloc staining in uranyl acetate.
- A further step in a variant of lead preparation/copper (Walton, 1979; Tapia *et al.*, 2012a).
- Embedding in a version of 'hard epoxy resin'.

Table 23.1 Comparison of the protocols published for SBFSEM image acquisition

NCMIR	Briggman *et al.*	Tapia *et al.*	Mikula *et al.*	Starborg et al.	Xiao et al.
Deerinck *et al.* (2010), Wilke *et al.* (2013)	Briggman, Helmstaedter and Denk (2011)	Tapia *et al.* (2012a)	Mikula, Binding and Denk (2012)	Starborg *et al.* (2013)	Xiao*et al.* (2013)
FIX (2.5% glut, 2%PAF in cacodylate buffer 0.1M, pH 7.4)	FIX (2% glut, 4% sucrose in cacodylate buffer 0.1M, pH 7.2)	FIX (2.5% glut, 4%PAF in PBS, pH 7.4)	FIX (10% acrolein in cacodylate buffer 0.1M, pH 7.4)	FIX (2.5% glut in cacodylate buffer 0.1M, pH 7.4)	FIX (2.5% glut, 2%PAF in cacodylate buffer 0.1M, pH 7.4)
Wash (4% sucrose in cacodylate, pH 7.2)	Wash (4% sucrose in cacodylate, pH 7.2)	Wash (cacodylate, pH 7.4)	Wash (cacodylate, pH 7.4) 8 h	Wash (cacodylate, pH 7.4)	Wash (cacodylate, pH 7.4)
Reduced osmium 1%	Reduced osmium 1%	Osmium 2% in water	Periodic acid 90 mM	Reduced osmium 2%	Reduced osmium 1%
Wash	Wash	Wash	Wash	Wash	Wash
Thiocarbohydrazide 1%	Thiocarbohydrazide 1%	Thiocarbohydrazide 1%	Thiocarbohydrazide 1%	Tannic acid 1%	
Wash	Wash	Wash	Wash	Wash	
Osmium 2% in water	Osmium 2% in water	Reduced osmium 1% OR osmium+imidazole	Osmium 2% in water	Reduced osmium 1% OR osmium+ imidazole	Osmium 1% in water
Wash	Wash	Wash	Wash	Wash	Wash
Uranyl acetate 1%	Uranyl acetate 1%			Uranyl acetate 1%	Uranyl acetate 1%
Wash	Wash				Wash
Lead aspartate	Lead aspartate	Lead aspartate OR copper sulfate/lead citrate			
Wash	Wash	Wash	Wash	Wash	
Dehydration- ethanol and aceton	Dehydration – propylene oxide	Dehydration ethanol	Dehydration acetone	Dehydration- propylene oxide	Dehydration ethanol
Embbeding	Embbeding	Embbeding	Embbeding	Agar100Hard	Embbeding
Durcupan Hard	Epon Hard	Embed 812 Hard	Quetol/Spurr		Durcupan/ Embed812

Such protocols are now used in all kinds of biological sample preparations to get optimal conditions with the 3View® and other FEGSEM techniques (Tapia *et al.*, 2012). While this protocol is optimal for conventional samples of small sizes, it does not allow an even penetration of the staining for larger samples that are of interest for SBEM. Therefore, the group of Denk is developing the en bloc staining of entire mouse brains (Mikula, Binding and Denk, 2012; Mikula and Denk, 2015) while a protocol for mouse brain biopsies has been published allowing staining without gradient artefacts (Hua, Laserstein and Helmstaedter, 2015). The addition of a step in hydroquinone has been reported to improve the signal in the secondary electron mode (Togo, *et al.*, 2014). A protocol based on the platinum blue has been described for the visualization of purified humans chromosomes (Yusuf *et al.*, 2014).

23.3.4 Pre-embedding Immunolabelling

Concerning the immunolabelling, post-embedding techniques are of no use in the SBEM techniques. It is the reason why the researchers using SBEM are developing pre-embedding techniques of labelling, for example by using horseradish peroxidase (HRP) tagged molecules and staining with 3,3′-Diaminobenzidine tetrahydrochloride (DAB) staining (Jokitalo *et al.*, 2001; Puhka *et al.*, 2012) based on successful TEM studies using this technique (Stinchcombe *et al.*, 1995), injection of biocytine in neurons (Lang *et al.*, 2011) or anti-GFP immunolabelling with biotinylated secondary antibodies (Knott *et al.*, 2009; Blumer *et al.*, 2015). The use of immunolabelling for 3View® is still a challenge for two reasons: the penetration of the labelling and the preservation of the tissue around it. As you cut deep into the tissue with 3View®, you need staining that penetrates in all depths of the tissue. At the same time, you need to preserve the ultrastructure of the tissue, as it is necessary to adapt the protocols and find a compromise between the permeabilization allowing penetration of the labelling and preservation of the tissue. The necessity to develop an analogue of GFP for electron microscopy is becoming crucial. Two constructs exist. The first is called miniSOG (for mini singlet oxygen generator), a fluorescent flavoprotein engineered from *Arabidopsis* phototropin 2 (Shu *et al.*, 2011). MiniSOG is fluorescent under a light microscope and photoconversion of tissue expressing miniSOG generates the polymerization of DAB into an electron-dense reaction product enabling correlated light and EM even after aldehyde fixation without compromising the tissue quality by permeabilization (Perkins, 2014). The second construct is called APEX or APEX2, a monomeric peroxidase reporter derived from soybean or pea dimeric ascorbate peroxidase (Martell *et al.*, 2012; Lam *et al.*, 2015). This monomeric 28 kDa peroxidase is not fluorescent and only needs to be in contact with DAB and peroxide to generate an electron-dense signal (no illumination like miniSOG).

23.3.5 Sample Mounting

In order to be stable in SBEM, samples need to be mounted on special stubs that fit in the sample holder of the microtome. The stubs are done in such a way that the sample can be glued at the top of it with superglue or with epoxy resin (Denk and Horstmann, 2004; Starborg *et al.*, 2013). As this stage has a limited range in *X* and *Y* of approximately 800 micrometres, the centre of the sample should be aligned with the beam path.

As the angle of the diamond knife is not adjustable and the surface will be parallel to the stub surface in order to keep in focus all over the surface, it is important to orientate your region of interest at the time of the mounting on the stub. It will be not possible, as in the classical ultramicrotome, to adjust the angle of cutting on the 3View®.

Once the sample is glued on the stub, it can be fixed on a normal ultramicrotome in order to trim it properly. Usually, two sides are parallel to each other and are clean. The other two sides can be trimmed to define a square, a rectangle or a trapezoid. The two parallel sides that will be placed parallel to the knife edge can be trimmed with a slope of 45–50 degrees instead of being straight. The trimming knife can be used to obtain a surface of the block that is smooth and clean (Denk and Horstmann, 2004), as close as possible to the region of interest that needs to be cut.

In order to target a region of interest in the block on the stub, a light microscope can be used to look inside the block and visualize the location in *X*, *Y* and *Z* of the region of interest. This information can help to target the region of interest inside the SEM as it is not possible to look at the depth of the block once it is in the SEM, as the signal comes only from the surface.

23.3.6 Image Acquisition

For placement of the sample, the chamber is open and a binocular is accommodated on the door edge in order to visualize the block and the diamond knife. The first approach is manual and based on the same principles as for the classical ultramicrotomes. The reflection of a LED placed on the microtome helps to evaluate the distance between the knife and the sample surface. Once the manual approach is finished, an approach with the diamond knife can be done in order to have visual control of the first cuts and to be sure that the block face is cut before closing the chamber. When cutting has begun, the knife is clamped at its start position and the chamber is closed and pumped.

From this point, all settings are done via the software. The region to be imaged needs to be determined. Depending on the resolution required by the experiments, the settings of the scanning can be set (Table 23.2) and the number of regions of interest can be determined.

Once the electron dose sustained by your sample in order to minimize the damages and get the best resolution is set, the dwell time suitable for your experience can be determined and the total time for your acquisition estimated. If it is too long or if the signal-to-noise ratio is not good enough, it is necessary to increase the beam kV and beam current until satisfactory settings are obtained. The higher the beam voltage, the lower the *Z* resolution, as the beam is penetrating deeper into the block (Hennig and Denk, 2007; Bouwer *et al.*, 2017). This means that the sections will have to be thicker in order to avoid damage when cutting the next section and in order not to oversample your region of interest in *Z*. if the microscope allows work to be done in low vacuum, it is possible to also work on this parameter in order to decrease charging artefacts and work at a higher kV, which will produce a better signal-to-noise ratio but more scattering noise.

Once the parameters are determined, the ultimate step is to perform focus and stigmatism before beginning the acquisitions of the data or set autofocus and autostigmatism.

Images are taken sequentially and stored in the same computer. Acquisition can be stopped at any time to refine the parameters, determine new region of interests or readjust focus and stigmatism. An autofocus/autostigmatism option has been developed and allows the choice as to when and where the adjustments can be automatically done.

Table 23.2 Comparison of the parameters of acquisition of 3View data acquisitions published as of January 2015. The missing data in publication correspond to empty fields. Some publications where no details were given are omitted

Reference	Year	Microscope	keV	Spot size or beam current	Chamber pressure (Torr)	Dwell time (μs)	XY resolution (nm)	Z cut (nm)
Armer *et al.*	2010	FEI Quanta FEG 200	4	3	0.3	5	24	50
Briggman *et al.*	2011	FEI Quanta FEG 200	2.5		0.37		16.5	23
Chen *et al.*	2013	FEI Quanta FEG 200	2.5		0.25	10		30
Denk *et al.*	2004	FEI Quanta FEG 200	7.5	3	Low vac	30	6,7–26,7	30–100
Gillies *et al.*	2014	ZEISS Sigma VP	3	60 mm aperture size	Low vac	1	4.51	70
Hoppa *et al.*	2012	FEI Quanta FEG 200	4	3	0.5	5	12	50
Jurrus *et al.*	2009	FEI Quanta FEG 250	2-2.5	2-3			26	50
Lang *et al.*	2011	FEI Quanta FEG 250					25	
Motskin *et al.*	2011	FEI Quanta FEG 600	4	3		5	10–20	50
Müller-Reichert *et al.*	2010	FEI Quanta FEG 600					50	50
Mun *et al.*	2011	FEI Quanta FEG 600	5	3	0.5	80	50	50
Mustafi *et al.*	2011	FEI Quanta FEG 200	3	3	0.23	3	15	100
Oberti *et al.*	2015	FEI Quanta FEG 200	7	3				
Pfeifer *et al.*	2015	ZEISS Sigma VP	1.5	30 mm aperture size	High vac		55.7/24.3/5.4	50/50/25
Pinheiro *et al.*	2012	FEI Quanta FEG 200	3	3	0.2	5	12	50
Puhka *et al.*	2012	FEI Quanta FEG 250	2	3	0.23			50
Rezakhariha *et al.*	2011	FEI Quanta FEG 200	3	3	0.29	5	11	50
Rouquette *et al.*	2009	FEI Quanta FEG 200	5	3	0.25	8	20	50
Salloum *et al.*	2014	ZEISS Sigma VP	2	30 mm aperture size	High vac	1	10–11	75–80
Urwyler *et al.*	2015	ZEISS Merlin	1.9	77 pA	High vac		2	40–100
Wernitznig *et al.*	2015	FEI Quanta FEG 600	4				10	70
Wilke *et al.*	2013	FEI Quanta FEG 250			High vac – 0.1	5–8	5.2–6.3	50
Wilke *et al.*	2014	FEI Quanta FEG 250			High vac			
Xiao *et al.*	2013	FEI Quanta FEG 200	4.5	3	0.42	8	6	50
Young *et al.*	2014	FEI Quanta FEG 250	4			10	10	100
Zankel *et al.*	2014	FEI Quanta FEG 600	7		Low vac			30

23.3.7 Post-processing and Analysis of Data

The images can be saved in different formats (*.tif, *.jpg, *.dm4, *.dm3) or converted from one to the other. Depending on the required analysis afterwards, images need to be post-processed. In a general way, it is useful to perform a correction of the contrast in order to have even contrast through all images. As acquisition may take weeks, a slight change in contrast over time can be observed. Another post-processing step is the registration of the images. It is mainly a rigid translation operation to correct for some minor drift occurring during the acquisition of large datasets over weeks. The software commonly used for tomography as etomo and IMOD can be successfully used for registration as well as manual segmentation as long as the dataset size does not exceed the capacity of the software. Registration is crucial for any segmentation needed afterwards, manual or automatic, as it restores the continuity of the structures from one section to the other and allows the identification of the smaller structure in 3D. TrakEM2 in the software FIJI allows such an approach (Cardona *et al.*, 2012) as well as commercial software such as Amira® or Arivis®. If the ultimate analysis involves the tracing of structures over all the volume, Denk and colleagues have developed a software called Knossos (Helmstaedter, Briggman and Denk, 2008, 2011; Briggman, Helmstaedter and Denk, 2011), making the manual tracing more efficient. This software allows large datasets to be visualized and navigated while tracing the skeleton of neurons in 3D. The software Py-Knossos can also be used (Wanner, Genoud and Friedrich, 2016; Wanner *et al.*, 2016).

The strategy for tracing is dependent on the type of images obtained and the analysis that will be performed on the reconstructions (Tsai *et al.*, 2014). So far, few published algorithms have been reported to segment successfully organelles in SBEM datasets (Perez *et al.*, 2014) or neuronal segments (Helmstaedter, 2015; Wernitznig *et al.*, 2016).

23.4 RESULTS OBTAINED WITH THIS TECHNIQUE

Since the publication of the first article describing the methods (Denk and Horstmann, 2004), numerous articles have been published solving biological questions (see Peddie and Collinson, 2014).

23.4.1 Neuroscience

The field of connectomic is highly interested in SBEM (Helmstaedter, 2013; Ohno *et al.*, 2014; Perkel, 2014; Hoggarth *et al.*, 2015; Wanner *et al.*, 2016). The group of Denk have published numerous papers studying the correlation between calcium imaging and electron microscopy of the connectivity in the mouse retina (Briggman and Kristan, 2006, 2008; Helmstaedter, Briggman and Denk, 2008, 2011; Briggman, Helmstaedter and Denk, 2011; Helmstaedter *et al.*, 2013). By using tiles, they reach visualization of volumes as large as 60 x 350 x 300 cubic micrometres. 3View® has been used to visualize connectivity of calyces of Held (Xiao *et al.*, 2013), as well as connectivity in the hippocampus (Wilke *et al.*, 2013).

Many other fields of neuroscience are benefiting from 3View®. Quantitative data on modifications of synapse densities in the mouse brain have been obtained by applying stereology approaches (Peretti *et al.*, 2015). The connectivity of fusiforms cells in mammalian dorsal cochlear nucleus has been precisely assessed (Salloum *et al.*, 2014) as well as the analysis of retinal terminals in subnuclei of mouse visual thalamus (Yamasaki *et al.*, 2014).

The characterization of synapse alterations in a mouse model of Alzheimer disease has been quantified (Wilke *et al.*, 2014). The 3D reconstruction of spiny dendrite based on 3View® stacks has allowed automatic spine detection to be optimized *in vivo* (Blumer *et al.*, 2015). Combined with a correlative light and electron microscopic approach based on near-infrared branding (Bishop *et al.*, 2011), the synaptogenesis has been studied in *Drosophilae* larvae (Urwyler *et al.*, 2015). The identification of tetrad synapses in the visual circuit of the locust *Schistocercia gregaria* has been possible due to the combination of a large field of view and high resolution (Wernitznig *et al.*, 2015). Events like demyelination have been visualized and characterized (Yamasaki *et al.*, 2014).

23.4.2 Cell Biology

3View® is now used to study the cell ultrastructure (Hughes *et al.*, 2014), such as, for example, the structure of the endoplasmic reticulum in cultured cells (Puhka *et al.*, 2012), the shape and size of mitochondria (Zhuravleva *et al.*, 2012) and the cell compartments involved in autophagy (Biazik et al., 2015). Focusing on the nuclear structure, the distribution of chromatin and a specific staining for chromosomes based on platinum blue have been visualized by SBEM (Rouquette *et al.*, 2009; Pinheiro *et al.*, 2012; Yusuf *et al.*, 2014) or nanoparticle sequestration (Motskin *et al.*, 2011).

23.4.3 Organs and Tissues

SBEM has been used to study the organization of extracellular matrix: collagen fibres in the mouse tail (Starborg *et al.*, 2013), tendinopathies of the human Achilles tendon (Pingel *et al.*, 2014), matrix assembly in the developing chick cornea (Young *et al.*, 2014) and extracellular matrix around mouse muscle (Gillies *et al.*, 2014). It has allowed the 3D reconstruction of the limbal stem cells niche in the human eye (Dziasko *et al.*, 2014). The study of cardiac muscle ultrastructure (Pinali *et al.*, 2013, 2015; Pinali and Kitmitto, 2014) has been developed. The 3D ultrastructure of the enteroendocrine cells of the mouse has been elucidated in a correlative approach with confocal microscopy (Bohorquez *et al.*, 2014). The visualization of melanocyte (Mun *et al.*, 2011) in the mouse skin has been possible. Quantitative data on the ultrastructural architecture of mouse pancreatic islet (Pfeifer *et al.*, 2015) as well as the study of the exocrine cells in the pancreas (Hoppa *et al.*, 2012) have been done. An entire kidney glomerulus has been imaged (Arkill *et al.*, 2014). It has helped to characterize phagocytosis events in the rat retina (Mustafi *et al.*, 2011; Mustafi, Kikano and Palczewski, 2014) as well as cone development in the mouse retina (Bushong *et al.*, 2014). Hair cell death in the ear has been better understood (Anttonen *et al.*, 2014).

23.4.4 Organism

At the level of organisms, SBEM has been used to determine cell death and tissue regeneration in planarian (Pellettieri *et al.*, 2010). It has been used to better understand zebrafish larvae blood vessels anastomosis (Armer *et al.*, 2009) as well as the structure of *C. elegans* (Müller-Reichert *et al.*, 2010). The 3D reconstruction of many individual yeasts in different strains of mutant *Schizosaccharomyces pombe* has shown the formation/disappearance of aggregates (Oberti *et al.*, 2015). *Trypanosoma brucei* are organisms that can be reconstructed with SBEM (Demmel *et al.*, 2014).

23.4.5 Material Science

SBEM is now used in material science to determine the 3D structure of materials (Chen *et al.*, 2013; Zankel, Wagner and Poelt, 2014) as long as the studied material can be cut by a diamond knife. In these applications, as the sample has completely different properties compared to embedded biological samples, microtome can cut as thin as 2.5 nm (Thompson *et al.*, 2013) and all modes of detection can be combined to get information from the material cut. The interest of 3View® remains for material scientists the big field of view, allowing visualization of rare structure in the material observed.

23.5 CONCLUSION

SBEM is a powerful technique. For the time being, its development is driven by the neuroscience field and the requirements of the connectomic community (Briggman and Bock, 2012; Ohno *et al.*, 2014; Titze and Genoud, 2016), which is looking for large fields of view at a resolution allowing every neurite to be followed and synapses to be identified. However, the number of publications using this technique in all domains is increasing year after year (Peddie and Collinson, 2014) and it is now a well-accepted instrument used to reveal the 3D ultrastructure of numerous different samples. One of the consequences of spreading this technique is the need for reliable image analysis methods and software able to make sense out of the amount of data generated. It requires new efforts at the level of data storage, annotation and visualization (Perkel, 2014) in order to retrieve meaningful quantitative data out of the images.

REFERENCES

Anttonen, T., Belevich, I., Kirjavainen, A., Laos, M., Brakebusch, C., Jokitalo, E. and Pirvola, U. (2014) How to bury the dead: Elimination of apoptotic hair cells from the hearing organ of the mouse. *J. Assoc. Res. Otolaryngol.*, 15 (6), 975–992.

Ardenne, M. (1938) Das Elektronen-Rastermikroskop. Theoretische Grundlagen. *Z. Phys.*, 109, 553–579.

Arkill, K.P., Qvortrup, K., Starborg, T., Mantell, J.M., Knupp, C., Michel, C.C., Harper, S.J., Salmon, A.H., Squire, J.M., Bates, D.O. and Neal, C.R. (2014) Resolution of the three dimensional structure of components of the glomerular filtration barrier. *BMC Nephrol.*, 15, 24.

Armer, H.E., Mariggi, G., Png, K.M., Genoud, C., Monteith, A.G., Bushby, A.J., Gerhardt, H. and Collinson, L.M. (2009) Imaging transient blood vessel fusion events in zebrafish by correlative volume electron microscopy. *PLoS One*, 4 (11), e7716.

Biazik, J., Vihinen, H., Anwar, T., Jokitalo, E. and Eskelinen, E.L. (2015) The versatile electron microscope: An ultrastructural overview of autophagy. *Methods*, March, 75, 44–53.

Bishop, D., Nikic, I., Brinkoetter, M., Knecht, S., Potz, S., Kerschensteiner, M. and Misgeld, T. (2011) Near-infrared branding efficiently correlates light and electron microscopy. *Nat. Methods*, 8 (7), 568–570.

Blumer, C., Vivien, C., Genoud, C., Perez-Alvarez, A., Wiegert, J.S., Vetter, T. and Oertner, T.G. (2015) Automated analysis of spine dynamics on live CA1 pyramidal cells. *Med. Image Anal.*, 19 (1), 87–97.

Bock, D.D., Lee, W.C., Kerlin, A.M., Andermann, M.L., Hood, G., Wetzel, A.W., Yurgenson, S., Soucy, E.R., Kim, H.S. and Reid, R.C. (2011) Network anatomy and *in vivo* physiology of visual cortical neurons. *Nature*, 471 (7337), 177–182.

Bohorquez, D.V., Samsa, L.A., Roholt, A., Medicetty, S., Chandra, R. and Liddle, R.A. (2014) An enteroendocrine cell–enteric glia connection revealed by 3D electron microscopy. *PLoS One*, 9 (2), e89881.

Bouwer, J.C., Deerinck, T.J., Bushong, E., Astakhov, V., Ramachandra, R., Peltier, S.T. and Ellisman, M.H. (2017) Deceleration of probe beam by stage bias potential improves resolution of serial block-face scanning electron microscopic images. *Adv. Struct. Chem. Imaging*, 2 (1), 11.

Bozzolla, J.J. and Russel, L. (1992) *Electron Microscopy; Principles and Techniques for Biologists*, Sudbury, Massachussets.

Briggman, K.L. and Bock, D.D. (2012) Volume electron microscopy for neuronal circuit reconstruction. *Curr. Opin. Neurobiol.*, 22 (1), 154–161.

Briggman, K.L. and Denk, W. (2006) Towards neural circuit reconstruction with volume electron microscopy techniques. *Curr. Opin. Neurobiol.*, 16 (5), 562–570.

Briggman, K.L., Helmstaedter, M. and Denk, W. (2011) Wiring specificity in the direction-selectivity circuit of the retina. *Nature*, 471 (7337), 183–188.

Briggman, K.L. and Kristan, W.B., Jr (2006) Imaging dedicated and multifunctional neural circuits generating distinct behaviors. *J. Neurosci.*, 26 (42), 10925–10933.

Briggman, K.L. and Kristan, W.B. (2008) Multifunctional pattern-generating circuits. *Annu. Rev. Neurosci.*, 31, 271–294.

Bumbarger, D.J., Riebesell, M., Rodelsperger, C. and Sommer, R.J. (2013) System-wide rewiring underlies behavioral differences in predatory and bacterial-feeding nematodes. *Cell*, 152 (1–2), 109–119.

Bushong, E.A., Johnson, D.D., Kim, K.Y., Terada, M., Hatori, M., Peltier, S.T., Panda, S., Merkle, A. and Ellisman, M.H. (2014) X-ray microscopy as an approach to increasing accuracy and efficiency of serial block-face imaging for correlated light and electron microscopy of biological specimens. *Microscopy and Microanalysis: The official journal of the Microscopy Society of America, Microbeam Analysis Society and the Microscopy Society of Canada*, 2014, 1–8.

Caceci, T. and Frankum, K.E. (1987) Measurement of increased uptake of osmium in skin and in a gelatin model 'tissue' treated with tannic acid. *J. Microsc.*, 147 (Pt 1), 109–114.

Cardona, A., Saalfeld, S., Preibisch, S., Schmid, B., Cheng, A., Pulokas, J., Tomancak, P. and Hartenstein, V. (2010) An integrated micro- and macroarchitectural analysis of the *Drosophila* brain by computer-assisted serial section electron microscopy. *PLoS Biol.*, 8 (10).

Cardona, A., Saalfeld, S., Schindelin, J., Arganda-Carreras, I., Preibisch, S., Longair, M., Tomancak, P., Hartenstein, V. and Douglas, R.J. (2012) TrakEM2 software for neural circuit reconstruction. *PLoS One*, 7 (6), e38011.

Chen, B., Guizar-Sicairos, M., Xiong, G., Shemilt, L., Diaz, A., Nutter, J., Burdet, N., Huo, S., Mancuso, J., Monteith, A., Vergeer, F., Burgess, A. and Robinson, I. (2013) Three-dimensional structure analysis and percolation properties of a barrier marine coating. *Sci. Rep.*, 3, 1177.

Crowther, R.A., Amos, L.A., Finch, J.T., De Rosier, D.J. and Klug, A. (1970) Three dimensional reconstructions of spherical viruses by Fourier synthesis from electron micrographs. *Nature*, 226 (5244), 421–425.

Danilatos, G.D. (1988) *Foundations of Environmental Scanning Electron Microscopy*, Academic Press.

Danilatos, G.D. (1993) Bibliography of environmental scanning electron microscopy. *Microsc. Res. Tech.*, 25 (5–6), 529–534.

Danilatos, G.D. (2012) Backscattered electron detection in environmental SEM. *J. Microsc.*, 245 (2), 171–185.

Danilatos, G.D. (2013a) Implications of the figure of merit in environmental SEM. *Micron*, 44, 143–149.

Danilatos, G.D. (2013b) Electron scattering cross-section measurements in ESEM. *Micron*, 45, 1–16.

Deerinck, T., Bushong, E., Lev-Ram, V., Shu, X., Tsien, R. and Ellisman, M. (2010) Enhancing serial block-face scanning electron microscopy to enable high resolution 3-D nanohistology of cells and tissues. *Microsc. Microanal.*, 16, 1138–1139.

Demmel, L., Schmidt, K., Lucast, L., Havlicek, K., Zankel, A., Koestler, T., Reithofer, V., de Camilli, P. and Warren, G. (2014) The endocytic activity of the flagellar pocket in *Trypanosoma brucei* is regulated by an adjacent phosphatidylinositol phosphate kinase. *J. Cell Sci.*, 127 (Pt 10), 2351–2364.

Denk, W. and Horstmann, H. (2004) Serial block-face scanning electron microscopy to reconstruct three-dimensional tissue nanostructure. *PLoS Biol.*, 2(11), e329.

Dziasko, M.A., Armer, H.E., Levis, H.J., Shortt, A.J., Tuft, S. and Daniels, J.T. (2014) Localisation of epithelial cells capable of holoclone formation *in vitro* and direct interaction with stromal cells in the native human limbal crypt. *PLoS One*, 9 (4), e94283.

Eberle, A.L., Mikula, S., Schalek, R., Lichtman, J.W., Tate, M.L. and Zeidler, D. (2015) High-resolution, high-throughput imaging with a multibeam scanning electron microscope. *J. Microsc.*, August, 259 (2), 114–120.

Frank, J. (2006) *Electron Tomography*, Springer, Albany, NY.

Gan, L. and Jensen, G.J. (2012) Electron tomography of cells. *Q. Rev. Biophys.*, 45 (1), 27–56.

Giannuzzi, L. A. (2004) *Introduction to Focused Ion Beams: Instrumentation, Theory, Techniques and Practice*, Springer Press.

Gillies, A. R., Bushong, E.A., Deerinck, T.J., Ellisman, M.H. and Lieber, R.L. (2014) Three-dimensional reconstruction of skeletal muscle extracellular matrix ultrastructure. *Microsc. Microanal.*, 20 (6), 1835–1840.

Glauert, A.M. and Lewis, P.R. (1988) *Biological Specimen Preparation for Transmission Electron Microscopy*, Princeton University Press.

Goldstein, J., Newbury, D.E., Echlin, P., Joy D.C., Alton, D., Romig Jr, A.D., Lyman, C.E., Fiori, C., Lifshin, E. (1992) *Scanning Electron Microscopy and X-Ray Microanalysis: A Text for Biologists, Material Scientists and Geologists*, Plenum Press, New York.

Griffiths, G. (1993) *Fine Structure Immunocytochemistry*, Springer-Verlag.

Harris, K.M. (1999) Structure, development, and plasticity of dendritic spines. *Curr. Opin. Neurobiol.*, 9(3), 343–348.

Hayat, M.A. (2000) *Principles and Techniques of Electron Microscopy Biological Applications*, Cambridge University Press.

Hayworth, K.J., Morgan, J.L., Schalek, R., Berger, D.R., Hildebrand, D.G. and Lichtman, J.W. (2014) Imaging ATUM ultrathin section libraries with WaferMapper: A multi-scale approach to EM reconstruction of neural circuits. *Front Neural Circuits*, June 27, 8, 68.

Helmstaedter, M. (2013) Cellular-resolution connectomics: Challenges of dense neural circuit reconstruction. *Nat. Methods*, 10 (6), 501–507.

Helmstaedter, M. (2015) The mutual inspirations of machine learning and neuroscience. *Neuron*, 86 (1), 25–28.

Helmstaedter, M., Briggman, K.L. and Denk, W. (2008) 3D structural imaging of the brain with photons and electrons. *Curr. Opin. Neurobiol.*, 18 (6), 633–641.

Helmstaedter, M., Briggman, K.L. and Denk, W. (2011) High-accuracy neurite reconstruction for high-throughput neuroanatomy. *Nat. Neurosci.*, 14 (8), 1081–1088.

Helmstaedter, M., Briggman, K.L., Turaga, S.C., Jain, V., Seung, H.S. and Denk, W. (2013) Connectomic reconstruction of the inner plexiform layer in the mouse retina. *Nature*, 500 (7461), 168–174.

Hennig, P. and Denk, W. (2007) Point-spread functions for backscattered imaging in the scanning electron microscope. *Journal of Applied Physics*, 102 (123101).

Hoggarth, A., McLaughlin, A.J., Ronellenfitch, K., Trenholm, S., Vasandani, R., Sethuramanujam, S., Schwab, D., Briggman, K.L. and Awatramani, G.B. (2015) Specific wiring of distinct amacrine cells in the directionally selective retinal circuit permits independent coding of direction and size. *Neuron*, 86 (1), 276–291.

Hoppa, M.B., E. Jones, E., J. Karanauskaite, J.,R. Ramracheya, R.,M. Braun, M.,S. C. Collins, S.C.,Q. Zhang, Q., Clark, A., Eliasson, L., Genoud, C., Macdonald, P.E., Monteith, A.G., Barg, S., Galvanovskis, J. and Rorsman, P. (2012) Multivesicular exocytosis in rat pancreatic beta cells. *Diabetologia*, 55 (4), 1001–1012.

Hoppe, W. and Grill, B. (1977) Prospects of three-dimensional high resolution electron microscopy of non-periodic structures. *Ultramicroscopy*, 2 (2–3), 153–168.

Hua, Y., Laserstein, P. and Helmstaedter, M. (2015) Large-volume en-bloc staining for electron microscopy-based connectomics. *Nat. Commun.*, 6, 7923.

Hughes, L., Hawes, C., Monteith, S. and Vaughan, S. (2014) Serial block face scanning electron microscopy – the future of cell ultrastructure imaging. *Protoplasma*, 251 (2), 395–401.

Jokitalo, E., Cabrera-Poch, N., Warren, G. and Shima, D.T. (2001) Golgi clusters and vesicles mediate mitotic inheritance independently of the endoplasmic reticulum. *J. Cell Biol.*, 154 (2), 317–330.

Jurrus, E., Hardy, M., Tasdizen, T., Fletcher, P.T., Koshevoy, P., Chien, C.B. *et al.* (2009) Axon tracking in serial block-face scanning electron microscopy. *Medical Image Analysis*, 13 (1), 180–188.

Kasthuri, N., Hayworth, K.J., Berger, D.R., Schalek, R.L., Conchello, J.A. Knowles-Barley, S., Lee, D., Vazquez-Reina, A., Kaynig, V., Jones, T.R., Roberts, M., Morgan, J.L., Tapia, J.C., Seung, H.S., Roncal, W.G., Vogelstein, J.T., Burns, R., Sussman, D.L., Priebe, C.E., Pfister, H. and Lichtman, J.W. (2015) Saturated reconstruction of a volume of Neocortex. *Cell*, 162 (3), 648–661.

Kaynig, V., Fischer, B., Muller, E. and Buhmann, J.M. (2010) Fully automatic stitching and distortion correction of transmission electron microscope images. *J. Struct. Biol.*, 171 (2), 163–173.

Kaynig, V., Vazquez-Reina, A., Knowles-Barley, S., Roberts, M., Jones, T.R., Kasthuri, N. *et al.* (2015) Large-scale automatic reconstruction of neuronal processes from electron microscopy images. *Medical Image Analysis*, 22 (1), 77–88.

Knott, G. and Genoud, C. (2013) Is EM dead? *J. Cell Sci.*, 126 (Pt 20), 4545–4552.

Knott, G.W., Quairiaux, C., Genoud, C. and Welker, E. (2002) Formation of dendritic spines with GABAergic synapses induced by whisker stimulation in adult mice. *Neuron*, 34 (2), 265–273.

Knott, G., Marchman, H., Wall, D. and Lich, B. (2008) Serial section scanning electron microscopy of adult brain tissue using focused ion beam milling. *J. Neurosci.*, 28 (12), 2959–2964.

Knott, G.W., Holtmaat, A., Trachtenberg, J.T., Svoboda, K. and Welker, E. (2009) A protocol for preparing GFP-labeled neurons previously imaged *in vivo* and in slice preparations for light and electron microscopic analysis. *Nat. Protoc.*, 4 (8), 1145–1156.

Kopek, B.G., Shtengel, G., Xu, C.S., Clayton, D.A. and Hess, H.F. (2012) Correlative 3D superresolution fluorescence and electron microscopy reveal the relationship of mitochondrial nucleoids to membranes. *Proc. Natl Acad. Sci. USA*, 109 (16), 6136–6141.

Kubota, Y. (2015) New developments in electron microscopy for serial image acquisition of neuronal profiles. *Microscopy (Oxford)*, 64 (1), 27–36.

Kukulski, W., Schorb, M., Kaksonen, M. and Briggs, J.A. (2012) Plasma membrane reshaping during endocytosis is revealed by time-resolved electron tomography. *Cell*, 150 (3), 508–520.

Lam, S.S., Martell, J.D., Kamer, K.J., Deerinck, T.J., Ellisman, M.H., Mootha, V.K. and Ting, A.Y. (2015) Directed evolution of APEX2 for electron microscopy and proximity labeling. *Nat. Methods*, 12 (1), 51–54.

Lang, S., Drouvelis, P., Tafaj, E., Bastian, P. and Sakmann, B. (2011) Fast extraction of neuron morphologies from large-scale SBFSEM image stacks. *J. Comput. Neurosci.*, 31 (3), 533–545.

Martell, J.D., Deerinck, T.J., Sancak, Y., Poulos, T.L., Mootha, V.K., Sosinsky, G.E., Ellisman, M.H. and Ting, A.Y. (2012) Engineered ascorbate peroxidase as a genetically encoded reporter for electron microscopy. *Nat. Biotechnol.*, 30 (11), 1143–1148.

Marton, L. (1976) Early application of electron microscopy to biology. *Ultramicroscopy*, 1 (4), 281–296.

Mikula, S., Binding, J. and Denk, W. (2012) Staining and embedding the whole mouse brain for electron microscopy. *Nat. Methods*, 9 (12), 1198–1201.

Mikula, S. and Denk, W. (2015) High-resolution whole-brain staining for electron microscopic circuit reconstruction. *Nat. Methods*, June, 12 (6), 541–546.

Miranda, K., Girard-Dias, W., Attias, M., de Souza, W. and Ramos, I. (2015) Three dimensional reconstruction by electron microscopy in the life sciences: An introduction for cell and tissue biologists. *Mol. Reprod. Dev.*, July–August, 82 (7–8), 530–547.

Motskin, M., Muller, K.H., Genoud, C., Monteith, A.G. and Skepper, J.N. (2011) The sequestration of hydroxyapatite nanoparticles by human monocyte-macrophages in a compartment that allows free diffusion with the extracellular environment. *Biomaterials*, 32 (35), 9470–9482.

Müller-Reichert, T. and Verkade, P. (2012) Introduction to correlative light and electron microscopy. *Methods Cell Biol.,* 111, xvii–xix.

Müller-Reichert, T., Mancuso, J., Lich, B. and McDonald, K. (2010) Three-dimensional reconstruction methods for *Caenorhabditis elegans* ultrastructure. *Methods Cell Biol.*, 96, 331–361.

Mun, J.Y., Jeong, S.Y., Kim, J.H., Han, S.S. and Kim, I.H. (2011) A low fluence Q-switched Nd:YAG laser modifies the 3D structure of melanocyte and ultrastructure of melanosome by subcellular-selective photothermolysis. *J. Electron Microsc. (Tokyo)*, 60 (1), 11–18.

Mustafi, D., Kikano, S. and Palczewski, K. (2014) Serial block face-scanning electron microscopy: A method to study retinal degenerative phenotypes. *Curr. Protoc. Mouse Biol.*, 4 (4), 197–204.

Mustafi, D., Kevany, B.M., Genoud, C., Okano, K., Cideciyan, A.V., Sumaroka, A., Roman, A.J., Jacobson, S.G., Engel, A., Adams, M.D. and Palczewski, K. (2011) Defective photoreceptor phagocytosis in a mouse model of enhanced S-cone syndrome causes progressive retinal degeneration. *FASEB J.*, 25 (9), 3157–3176.

Oberti, D., Biasini, A., Kirschmann, M.A., Genoud, C., Stunnenberg, R., Shimada, Y. and Buhler, M. (2015) Dicer and hsp104 function in a negative feedback loop to confer robustness to environmental stress. *Cell Rep.*, 10 (1), 47–61.

Ohno, N., Katoh, M., Saitoh, H., Saitoh, S. and Ohno, S. (2014) Three-dimensional volume imaging with electron microscopy toward connectome. *Microscopy (Oxford)*, 2014, 1–10.

Ohyama, T., Schneider-Mizell, C.M., Fetter, R.D., Aleman, J.V., Franconville, R., Rivera-Alba, M., Mensh, B.D., Branson, K.M., Simpson, J.H., Truman, J.W., Cardona, A. and Zlatic, M. (2015) A multilevel multimodal circuit enhances action selection in *Drosophila. Nature*, 520 (7549), 633–639.

Orloff, J. (ed.) (2009) *Handbook of Charged Particle Optics*, 2nd edn, CRC Press, Boca Raton, 665 pp.

Peddie, C.J. and Collinson, L.M. (2014) Exploring the third dimension: Volume electron microscopy comes of age. *Micron*, 61, 9–19.

Peddie, C.J., Blight, K., Wilson, E., Melia, C., Marrison, J., Carzaniga, R., Domart, M.C., O'Toole, P., Larijani, B. and Collinson, L.M. (2014) Correlative and integrated light and electron microscopy of in-resin GFP fluorescence, used to localise diacylglycerol in mammalian cells. *Ultramicroscopy*, 143, 3–14.

Pellettieri, J., Fitzgerald, P., Watanabe, S., Mancuso, J., Green, D.R. and Sanchez Alvarado, A. (2010) Cell death and tissue remodeling in planarian regeneration. *Dev. Biol.*, 338 (1), 76–85.

Peretti, D., Bastide, A., Radford, H., Verity, N., Molloy, C., Martin, M.G., Moreno, J.A., Steinert, J.R., Smith, T., Dinsdale, D., Willis, A.E. and Mallucci, G.R. (2015) RBM3 mediates structural plasticity and protective effects of cooling in neurodegeneration. *Nature*, February 12, 518 (7538), 236–239.

Perez, A.J., Seyedhosseini, M., Deerinck, T.J., Bushong, E.A., Panda, S., Tasdizen, T. and Ellisman, M.H. (2014) A workflow for the automatic segmentation of organelles in electron microscopy image stacks. *Front Neuroanat.*, 8, 126.

Perkel, J.M. (2014) Mapping Neural Connections. *Biotechniques*, 57, 230–236.

Perkins, G.A. (2014) The use of miniSOG in the localization of mitochondrial proteins. *Methods Enzymol.*, 547, 165–179.

Pfeifer, C.R., Shomorony, A. Aronova, M.A., Zhang, G., Cai, T., Xu, H., Notkins, A.L. and Leapman, R.D. (2015) Quantitative analysis of mouse pancreatic islet architecture by serial block-face SEM. *J. Struct. Biol.*, 189 (1), 44–52.

Pinali, C. and Kitmitto, A. (2014) Serial block face scanning electron microscopy for the study of cardiac muscle ultrastructure at nanoscale resolutions. *J. Mol. Cell Cardiol.*, November, 76, 1–11.

Pinali, C., Bennett, H., Davenport, J.B., Trafford, A.W. and Kitmitto, A. (2013) Three-dimensional reconstruction of cardiac sarcoplasmic reticulum reveals a continuous network linking transverse-tubules: this organization is perturbed in heart failure. *Circ. Res.*, 113 (11), 1219–1230.

Pinali, C., Bennett, H.J., Davenport, J.B., Caldwell, J.L., Starborg, T., Trafford, A.W. and Kitmitto, A (2015) Three-dimensional structure of the intercalated disc reveals plicate domain and gap junction remodeling in heart failure. *Biophys. J.*, 108 (3), 498–507.

Pingel, J., Lu, Y., Starborg, T., Fredberg, U., Langberg, H., Nedergaard, A., Weis, M., Eyre, D., Kjaer, M. and Kadler, K.E. (2014) 3-D ultrastructure and collagen composition of healthy and overloaded human tendon: Evidence of tenocyte and matrix buckling. *J. Anat.*, 224 (5), 548–555.

Pinheiro, I., Margueron, R., Shukeir, N., Eisold, M., Fritzsch, C., Richter, F.M., Mittler, G., Genoud, C., Goyama, S., Kurokawa, M., Son, J., Reinberg, D., Lachner, M. and Jenuwein, T. (2012) Prdm3 and Prdm16 are H3K9me1 methyltransferases required for mammalian heterochromatin integrity. *Cell*, 150 (5), 948–960.

Porter, K.R. and Blum, J. (1953) A Study in Microtomy for electron microscopy.*Anatomical Record*, 117 (4), 685–710.

Puhka, M., Joensuu, M., Vihinen, H., Belevich, I. and Jokitalo, E. (2012) Progressive sheet-to-tubule transformation is a general mechanism for endoplasmic reticulum partitioning in dividing mammalian cells. *Mol. Biol. Cell*, 23 (13), 2424–2432.

Reimer, L. (1998) *Scanning Electron Microscopy: Physics of Image Formation and Microanalysis*, Springer.

Rezakhaniha, R., Fonck, E., Genoud, C. and Stergiopulos, N. (2011) Role of elastin anisotropy in structural strain energy functions of arterial tissue. *Biomechanics and Modeling in Mechanobiology*, 10 (4), 599–611.

Robards, A.W. and Wilson, A.J. (1993) *Procedures in Electron Microscopy*, John Wiley & Sons.

Robinson, V.N. (1975) The elimination of charging artefacts in the scanning electron microscope. *Journal of Physics E : Scientific Instruments*, 8 (8).

Rouquette, J., Genoud, C., Vazquez-Nin, G.H., Kraus, B., Cremer, T. and Fakan, S. (2009) Revealing the high-resolution three-dimensional network of chromatin and interchromatin space: A novel electron-microscopic approach to reconstructing nuclear architecture. *Chromosome Res.*, 17 (6), 801–810.

Ruska, E. and Knoll, M. (1932) Das Elektronenmikroskop. *Z. Phys.*, 78, 318–339.

Salloum, R.H., Chen, G., Velet, L., Manzoor, N.F., Elkin, R., Kidd, G.J., Coughlin, J., Yurosko, C., Bou-Anak, S., Azadi, S., Gohlsch, S., Schneider, H. and Kaltenbach, J.A. (2014) Mapping and morphometric analysis of synapses and spines on fusiform cells in the dorsal cochlear nucleus. *Front Syst. Neurosci.*, 8, 167.

Schröder, R.R. (2015) Advances in electron microscopy: A qualitative view of instrumentation development for macromolecular imaging and tomography. *Arch. Biochem. Biophys.*, 581, 25–38.

Seiter, J., Muller, E., Blank, H., Gehrke, H., Marko, D. and Gerthsen, D. (2014) Backscattered electron SEM imaging of cells and determination of the information depth. *J. Microsc.*, 254 (2), 75–83.

Seligman, A.M., Wasserkrug, H.L. and Hanker, J.S. (1966) A new staining method (OTO) for enhancing contrast of lipid – containing membranes and droplets in osmium tetroxide – fixed tissue with osmiophilic thiocarbohydrazide(TCH). *J. Cell Biol.*, 30 (2), 424–432.

Shu, X., Lev-Ram, V., Deerinck, T.J., Qi, Y., Ramko, E.B., Davidson, M.W., Jin, Y., Ellisman, M.H. and Tsien, R.Y. (2011) A genetically encoded tag for correlated light and electron microscopy of intact cells, tissues, and organisms. *PLoS Biol.*, 9 (4), e1001041.

Sjostrand, F.S. (1958) Ultrastructure of retinal rod synapses of the guinea pig eye as revealed by three-dimensional reconstructions from serial sections. *J. Ultrastruct. Res.*, 2 (1), 122–170.

Starborg, T., Kalson, N.S., Lu, Y., Mironov, A., Cootes, T.F., Holmes, D.F. and Kadler, K.E. (2013) Using transmission electron microscopy and 3View to determine collagen fibril size and three-dimensional organization. *Nat. Protoc.*, 8 (7), 1433–1448.

Stevens, J.K., Davis, T.L., Friedman, N. and Sterling, P. (1980) A systematic approach to reconstructing microcircuitry by electron microscopy of serial sections. *Brain Res.*, 2 (3), 265–293.

Stinchcombe, J.C., Nomoto, H., Cutler, D.F. and Hopkins, C.R. (1995) Anterograde and retrograde traffic between the rough endoplasmic reticulum and the Golgi complex. *J. Cell Biol.*, 131 (6 Pt 1), 1387–1401.

Takemura, S.Y., Bharioke, A., Lu, Z., Nern, A., Vitaladevuni, S., Rivlin, P.K., Katz, W.T., Olbris, D.J., Plaza, S.M., Winston, P., Zhao, T., Horne, J.A., Fetter, R.D., Takemura, S., Blazek, K., Chang, L.A., Ogundeyi, O., Saunders, M.A., Shapiro, V., Sigmund, C., Rubin, G.M., Scheffer, L.K.,

Meinertzhagen, I.A. and Chklovskii, D.B. (2013) A visual motion detection circuit suggested by *Drosophila* connectomics. *Nature*, 500 (7461), 175–181.

Tapia, J.C., Kasthuri, N., Hayworth, K.J., Schalek, R., Lichtman, J.W., Smith, S.J. and Buchanan, J. (2012a). High-contrast en bloc staining of neuronal tissue for field emission scanning electron microscopy. *Nat. Protoc.*, 7 (2), 193–206.

Tapia, J.C., Wylie, J.D., Kasthuri, N., Hayworth, K.J., Schalek, R., Berger, D.R., Guatimosim, C., Seung, H.S. and Lichtman, J.W. (2012b). Pervasive synaptic branch removal in the mammalian neuromuscular system at birth. *Neuron*, 74 (5), 816–829.

Thompson, G.E., Hashimoto, T., Zhong, X.L., Curioni, M., Zhou, X.R., Skeldon, P., Withers, P.J., Carr, J.A. and Monteith, A.G. (2013) Revealing the three dimensional internal structure of aluminium alloys. *Surface and Interface Analysis*, 45 (10), 1536–1542.

Titze, B. and Denk, W. (2013) Automated in-chamber specimen coating for serial block-face electron microscopy. *J. Microsc.*, 250 (2), 101–110.

Titze, B. and Genoud, C. (2016) Volume scanning electron microscopy for imaging biological ultrastructure. *Biol. Cell*, 108 (11), 307–323.

Togo, A., Ohta, K., Higashi, R. and Nakamura, K. (2014) En bloc staining with hydroquinone treatment for block face imaging. *Microscopy (Oxford)*, 63 (Suppl. 1), i34–i35.

Tsai, W.T., Hassan, A., Sarkar, P., Correa, J., Metlagel, Z., Jorgens, D.M. and Auer M. (2014) From voxels to knowledge: A practical guide to the segmentation of complex electron microscopy 3D-data. *J. Vis. Exp.*, (90), e51673.

Urwyler, O., Izadifar, A., Dascenco, D., Petrovic, M., He, H., Ayaz, D., Kremer, A., Lippens, S., Baatsen, P., Guerin, C.J. and Schmucker, D. (2015) Investigating CNS synaptogenesis at single-synapse resolution by combining reverse genetics with correlative light and electron microscopy. *Development*, 142 (2), 394–405.

Varshney, L.R., Chen, B.L., Paniagua, E., Hall, D.H. and Chklovskii, D.B. (2011). Structural properties of the *Caenorhabditis elegans* neuronal network. *PLoS Comput. Biol.*, 7 (2), e1001066.

Walton, J. (1979) Lead aspartate, an en bloc contrast stain particularly useful for ultrastructural enzymology. *J. Histochem. Cytochem.*, 27 (10), 1337–1342.

Wanner, A.A., Genoud, C. and Friedrich, R.W. (2016) 3-dimensional electron microscopic imaging of the zebrafish olfactory bulb and dense reconstruction of neurons. *Sci. Data*, 3, 160100.

Wanner, A.A., Genoud, C., Masudi, T., Siksou, L. and Friedrich, R.W. (2016) Dense EM-based reconstruction of the interglomerular projectome in the zebrafish olfactory bulb. *Nat. Neurosci.*, 19 (6), 816–825.

Watanabe, S., Punge, A., Hollopeter, G., Willig, K.I., Hobson, R.J., Davis, M.W., Hell, S.W. and Jorgensen, E.M. (2011) Protein localization in electron micrographs using fluorescence nanoscopy. *Nat. Methods*, 8 (1), 80–84.

Wernitznig, S., Rind, F.C., Polt, P., Zankel, A., Pritz, E., Kolb, D., Bock, E. and Leitinger, G. (2015) Synaptic connections of first-stage visual neurons in the locust *Schistocerca gregaria* extend evolution of tetrad synapses back 200 million years. *J. Comp. Neurol.*, 523 (2), 298–312.

Wernitznig, S., Sele, M., Urschler, M., Zankel, A., Polt, P., Rind, F.C. and Leitinger, G. (2016) Optimizing the 3D-reconstruction technique for serial block-face scanning electron microscopy. *J. Neurosci. Methods*, 264, 16–24.

White, J. G., Southgate, E., Thomson, J.N. and Brenner, S. (1986) The structure of the nervous system of the nematode *Caenorhabditis elegans*. *Philos. Trans. R. Soc. Lond. B Biol. Sci.*, 314 (1165), 1–340.

Wilke, S.A., Antonios, J.K., Bushong, E.A., Badkoobehi, A., Malek, E., Hwang, M., Terada, M., Ellisman, M.H. and Ghosh, A. (2013) Deconstructing complexity: Serial block-face electron microscopic analysis of the hippocampal mossy fiber synapse. *J. Neurosci.*, 33 (2), 507–522.

Wilke, S.A., Raam, T., Antonios, J.K., Bushong, E.A., Koo, E.H., Ellisman, M.H. and A. Ghosh, A. (2014) Specific disruption of hippocampal mossy fiber synapses in a mouse model of familial Alzheimer's disease. *PLoS One*, 9 (1), e84349.

Williams, R.C. and Kallman, F. (1955). Interpretations of electron micrographs of single and serial sections. *Journal of Biophysical and Biochemical Cytology*, 1 (4), 301.

Xiao, L., Michalski, N., Kronander, E., Gjoni, E., Genoud, C., Knott, G. and Schneggenburger, R. (2013) BMP signaling specifies the development of a large and fast CNS synapse. *Nat Neurosci.*, 16 (7), 856–864.

Yamasaki, R., Lu, H., Butovsky, O., Ohno, N., Rietsch, A.M., Cialic, R., Wu, P.M., Doykan, C.E., Lin, J., Cotleur, A.C., Kidd, G., Zorlu, M.M., Sun, N., Hu, W., Liu, L., Lee, J.C., Taylor, S.E., Uehlein, L., Dixon, D., Gu, J., Floruta, C.M., Zhu, M., Charo, I.F., Weiner, H.L., and Ransohoff, R.M. (2014) Differential roles of microglia and monocytes in the inflamed central nervous system. *J. Exp. Med.*, 211 (8), 1533–1549.

Young, R. D., Knupp, C., Pinali, C., Png, K.M., Ralphs, J.R., Bushby, A.J., Starborg, T., Kadler, K.E. and Quantock, A.J. (2014) Three-dimensional aspects of matrix assembly by cells in the developing cornea. *Proc. Natl Acad. Sci. USA*, 111 (2), 687–692.

Yusuf, M., Chen, B., Hashimoto, T., Estandarte, A.K., Thompson, G. and Robinson, I. (2014) Staining and embedding of human chromosomes for 3-D serial block-face scanning electron microscopy. *Biotechniques*, 57 (6), 302–307.

Zankel, A., Wagner, J. and Poelt, P. (2014) Serial sectioning methods for 3D investigations in materials science. *Micron*, 62, 66–78.

Zhuravleva, E., Gut, H., Hynx, D., Marcellin, D., Bleck, C.K., Genoud, C., Cron, P., Keusch, J.J., Dummler, B., Esposti, M.D. and Hemmings, B.A. (2012) Acyl coenzyme A thioesterase Them5/Acot15 is involved in cardiolipin remodeling and fatty liver development. *Mol. Cell Biol.*, 32 (14), 2685–2697.

24

FIB-SEM for Biomaterials

Lucille A. Giannuzzi
L.A. Giannuzzi & Associates LLC, Fort Myers, FL, USA
EXpressLO LLC, Lehigh Acres, FL, USA

24.1 INTRODUCTION AND FIB BASICS

The combined focused ion beam microscope–scanning electron microscope (FIB-SEM) instrument has been used for many biomaterials applications over the years. In this chapter, we emphasize its use for non-cryo-applications (handled elsewhere in this book) and show examples for the study of dental implants and soft biomaterials that are made vacuum compatible (e.g., via staining and plastic embedding or critical point drying). Details on FIB theory, instrumentation, and techniques may be found elsewhere (Giannuzzi and Stevie 2005), but some key points are described below.

FIB milling techniques were first developed on a single column Ga^+ ion beam instrument, that is, an FIB column mounted vertically on a vacuum chamber and stage arrangement similar to an SEM instrument as shown in the schematic diagram in Figure 24.1a. It is important to understand how the FIB itself works prior to the addition of an SEM column. At the very least, FIB instrumentation consists of a focused ion beam column operating within a vacuum chamber, a multi-axis sample stage, a secondary electron (SE) detector, and a gas injector system (GIS). Optional items shown in Figure 24.1a include a secondary ion (SI) detector, an electron flood gun (used to mitigate sample charging artifacts), additional GISs, analytical detectors, in situ probes, manipulators, or other specialty stages (e.g., cooling, heating, straining, electrical, etc.). In an FIB-SEM instrument, an SEM column and imaging or analytical detectors optimized for electron-induced interactions are then also added (not shown). Thus, the FIB-SEM can get quite crowded and an instrument loaded with detectors and capabilities may humorously be referred to as a "Christmas tree." As an aside, it should also be noted that it becomes more difficult to optimize the FIB-SEM for a single application as more and more capabilities are added. An FIB-SEM operator should

Biological Field Emission Scanning Electron Microscopy, First Edition.
Edited by Roland A. Fleck and Bruno M. Humbel.

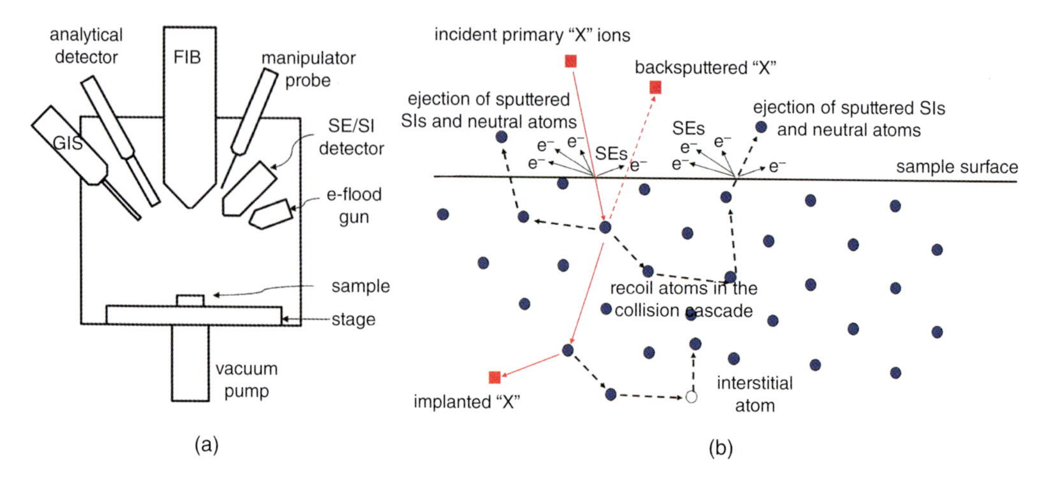

Figure 24.1 (a) Schematic diagram of an FIB chamber. (b) Schematic diagram of ion–solid interactions.

become familiar with its chamber geometry to avoid stage motion sample collisions with detectors and pole pieces.

An understanding of ion–solid interactions is necessary to optimize the FIB usage of the FIB-SEM. In particular, the FIB itself is useful as a microscope and characterization via imaging or analytics, and, specifically, for specimen preparation techniques such as milling (i.e., sectioning or slicing) and ion beam assisted deposition (Giannuzzi and Stevie 2005). The basic interactions that occur in a solid are shown in the schematic diagram in Figure 24.1b. These ion–solid interactions vary with ion species, ion energy, incidence angle, and target mass and crystallinity (Giannuzzi and Stevie 2005). The beauty of focused ion beam processing is the use of FIB imaging via the collection of SEs or SIs for site-specific placement of the ion beam for milling or deposition (i.e., site-specific sample preparation). Momentum transfer between the energetic ions and the target atoms create recoil atom motion within the collision cascade analogous to a cue ball hitting a rack of billiard balls. The energetic recoil atoms can in turn create additional recoil atom motion. If a moving recoil atom near a free surface has energy greater than its surface binding energy, it can escape from the surface as a sputtered ion or sputtered neutral atom. Hence, the ion beam is used to site specifically sputter atoms (or mill/remove/section) the target. For specimen preparation, the objective is to mill or remove all but the region of interest.

Particle or ion beam assisted chemical vapor deposition is used to mark regions of interest and protect the region underneath the deposit from being milled away. As shown in Figure 24.2, ion deposition is executed by inserting a gas injector needle to within ~100 μm of the sample surface and opening a valve that sprays an organometallic, hydrocarbon, carbonyl, or orthosilicate-based gas on to the sample (Giannuzzi and Stevie 2005). In Figure 24.2, a gas that may be used for Pt deposition is shown. The direct energetic ion beam and secondary electrons created from the ion–solid interactions during ion beam rastering cracks or decomposes the absorbed organometallic gas, leaving behind a metal-rich conductive layer (or insulting deposition layer depending on the specific gas used). The decomposed gas particles and gas not used in the process are removed via vacuum pumping. For specimen preparation, either a Pt-rich or C-rich deposition layer is most often used. Note that for either case of ion beam or electron beam deposition, the deposit is not pure and will consist of a composition that approximates the starting gas composition (Langford *et al.* 2007). Note also that the composition of an ion-induced deposition will include that ion.

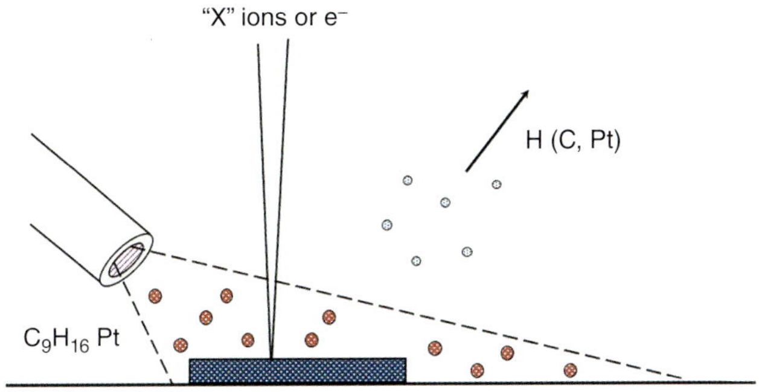

"X" ions or e^-

H (C, Pt)

C_9H_{16} Pt

Figure 24.2 Schematic diagram of particle–beam-induced deposition.

The Ga^+ FIB uses a liquid metal ion source to generate ions (Giannuzzi and Stevie 2005). The FIB column then accelerates these ions at energies from a few hundred eV up to 30–50 keV. While SEM columns typically use magnetic lenses for focusing electrons, the much larger ion mass requires the use of electrostatic lenses and scan coils to focus and move or raster the beam. The beam size is controlled by physical apertures that limit the beam current (i.e., diameter). At 30 keV, the Ga^+ ion beam may vary from ~1 pA (~2–5 nm) up to 50–100 nA (>1 μm). Several other ion sources have been developed over the years allowing for applications over a range of beam currents (Smith *et al.* 2014). Since techniques for using these additional ion sources and columns refer to methods developed with Ga^+, the discussion of techniques and methods below centers on Ga^+ FIB-based instruments.

The sputter yield varies with target, ion species, ion energy, and incidence angle. The sputter yield for a given ion does not vary monotonically with atom number and to a first approximation scales with material density that is related to d-shell atomic filling. The ion energy, ion type, and ion range will determine what sort of damage, if any, is transferred to the target surface. A detailed discussion on sputtering and ion range physics is beyond the scope of this chapter and is found elsewhere (Giannuzzi and Stevie 2005). The reader should appreciate that FIB methods to date have been developed with an understanding of ion–solid interactions to optimize both quality and speed. Ga^+ and other heavy ions are stopped within a few tens of nanometers from the sample surface and therefore will not influence any of the analyses discussed below.

As mentioned above, both SEs and SIs may be generated during ion–solid interactions. While some use the terminology "FIB image" or scanning ion image (SIM) to refer to an image acquired using ions as the primary beam, this does not indicate which signal was used to collect the image. Thus, when referring to images acquired with the FIB, it is useful to differentiate between the ion-induced SE signal (iSE) or the ion-induced SI signal (iSI). This nomenclature also discriminates from an electron-induced SE signal (eSE) acquired with the SEM. The iSE or iSI signal may be collected with the same detectors used in the SEM such as the Everhart-Thornley detector that are biased accordingly. Alternatively, the FIB-induced signals may be collected with additional detectors optimized for FIB signal and geometry using either a continuous dynode electron multiplier (CDEM) or multi-channel plates (Giannuzzi and Stevie 2005).

Topographic contrast for FIB iSE imaging is similar to SEM eSE imaging. That is, the iSE signal is lowest for surfaces at 0° incidence angle (i.e., for a surface normal to the beam direction) and increases to a maximum as the incidence angle increases (Ohya *et al.* 2009). However, iSE varies considerably from eSE contrast for both differences in materials (i.e., atomic number or Z contrast) as well as crystallographic orientation (i.e., channeling contrast). Indeed, the iSE signal can vary considerably across the periodic table and may be linked to contributions from d-shell filling, which influences both ion stopping power and material sputter yields (Giannuzzi and Utlaut 2011). For crystalline materials, the iSE channeling contrast depends on the ion range defined by the non-channeled fraction and critical angle while the eSE channeling contrast is related to diffraction phenomenon (Giannuzzi and Michael 2013).

Finally, the FIB column requires replenishment of consumables, which adds considerably to its cost of ownership. The Ga^+ ion source will deplete after ~1000 hours or so of usage. During FIB operation, both beam acceptance apertures and beam defining apertures will be milled away, causing degradation to aperture shape requiring replacement. Gas material for GIS units will deplete and require replacement. In addition, CDEM or multi-channel plates will degrade with usage and also require replacement. All these consumables last for ~1000–2000 ion beam usage hours at a cost of ~$10k. Any FIB instrument would generally require at least one complete consumable change out per year. For an FIB system that runs continuously 24 hours per day, 365 days per year, the consumable cost can easily exceed $40k. Note, that this operational cost is in addition to a vendor or third party service contract necessary to maintain the FIB (or FIB-SEM), which can add ~$25–$75k per year depending on the instrument vintage. While the FIB-SEM can be expensive to operate and maintain, it routinely provides indispensable information for academic and industrial researchers in ways not previously attained.

24.2 GEOMETRY FOR 2D SECTIONING AND IMAGING

Understanding FIB-SEM applications basics is best understood starting with the FIB itself, shown in the schematic diagram in Figure 24.3. A protective layer is deposited marking the region of interest and then the FIB is used to sputter away material a few micrometers in front of the deposit. Once this initial rectangular trench is opened up using a large beam (e.g., > 10 nA), the FIB is used to mill the cross-sectioned trench wall using successively

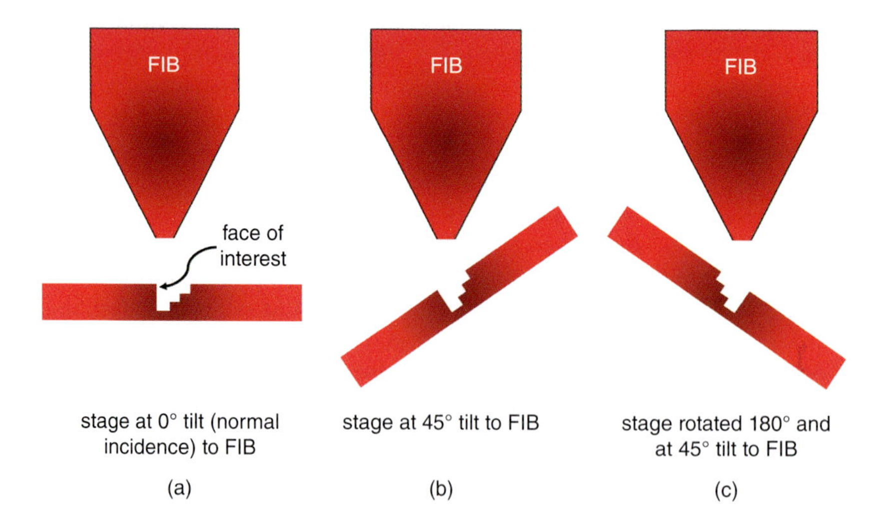

Figure 24.3 Schematic diagram of FIB to specimen geometry relationships. (a) Conventional FIB milling geometry, (b) FIB imaging geometry on a single beam FIB, and (c) FIB imaging geometry on an FIB-SEM.

smaller and smaller beam currents (e.g., 5 nA, 1 nA, 500 pA, 100 pA), to create a polished face that is parallel or nearly parallel to the ion beam direction. Note that the cross-section face created is along the direction of the ion beam (e.g., perpendicular to the original sample surface). The polished face should always be FIB milled parallel to the ion beam direction to avoid topological FIB milling artifacts. Thus, the plane of the face of interest can always be changed by mounting the sample appropriately or by changing the stage tilt and/or using a pre-tilted sample holder. Note that a stage tilt is required to FIB image the FIB milled cross-sectional face of interest, as shown in Figure 24.3a and b. On an FIB-SEM, a stage tilt and stage rotation may be necessary to FIB image the FIB milled cross-section, as shown in Figure 24.3c.

The added benefit to the FIB-SEM is that both the ion and electron beams can be positioned into coincidence such that the SEM can directly view and end-point FIB milling operations without stage tilting or rotation, as shown in Figure 24.4. The combination of the SEM with an FIB also allows additional imaging capabilities via the collection of eSE or backscattered electron (BSE) signals and analytical methods using the SEM, such as X-ray energy dispersive spectrometry (XEDS). A schematic diagram showing the instrumentation set-up is shown in Figure 24.4. Most commercial vendors of FIB-SEM instrumentation use the configuration shown in Figure 24.4a, where the SEM is mounted on the specimen chamber vertically and the FIB is positioned at an angle with respect to the SEM. An alternative configuration is shown in the "FIB-centric" platform in Figure 24.4b, where the FIB is mounted vertically and the SEM is angled on the chamber. Either of these configurations in Figure 24.4 may be used for the applications described below. Note that in either configuration, the SEM has direct line of sight access for imaging the FIB milled cross-sectioned surface created. In either of these FIB-SEM geometries, SEM images are obtained from a tilted specimen face and must be corrected for foreshortening in the *y*-direction. The foreshortening angle between the FIB and SEM varies among vendors from ~35° to 38°. Note that the FIB or SEM column may also operate independently at any stage tilt or stage height (within the geometrical constraints of the chamber configuration) to optimize its individual performance. In most cases, the FIB-SEM is operated to work together with the sample height adjusted to a geometrically defined coincident point such that any ion processing can be viewed by SEM imaging without moving or tilting the specimen as in Figure 24.4. When using an FIB-SEM, if FIB imaging of the FIB milled cross-section is of interest, then

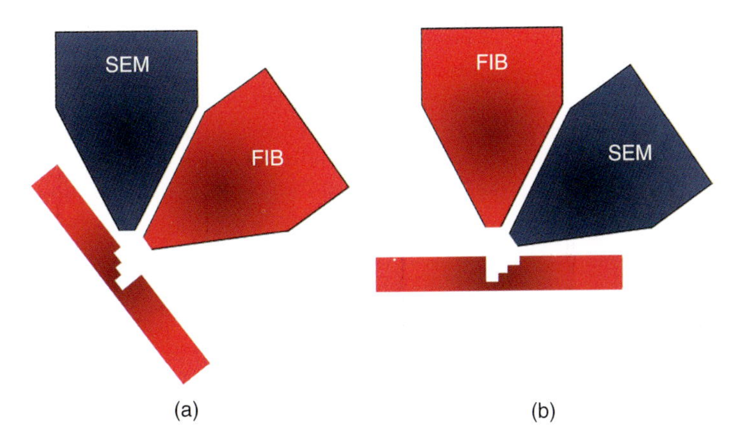

Figure 24.4 Schematic diagram of FIB-SEM instrumentation. (a) The SEM is vertical and the FIB is angled with respect to the SEM. (b) The FIB is vertical and the SEM is angled with respect to the FIB.

the specimen stage must be rotated and tilted as in Figure 24.3c to avoid collision with the SEM column. (Note that most vendors limit or prevent stage tilting in the "negative" direction to reduce collisions.)

Computer control of the FIB-SEM instrument is now routinely available and allows for unattended FIB milling and specimen preparation, and SEM imaging or acquisition of analytical information, stage motion, and precise placement of either the ion beam or the electron beam using image recognition of fiducial marks. In addition, auto focus and tilt correction routines may be implemented depending on the imaging or FIB-SEM operating conditions utilized.

For TEM specimen preparation, trenches on either side of the region of interest is FIB milled away. The specimen is undercut and lifted out using either an in situ or ex situ probe. FIB polishing can be performed either prior to or after lift-out, depending on the exact methods used. Entire chapters have been written on TEM specimen preparation and the reader is referred elsewhere for details (Giannuzzi and Stevie 2005; Mayer *et al.* 2007; Giannuzzi *et al.* 2015).

Depending on the specimen type, a typical 2D cross-section of ~10 μm wide and 5 μm deep may take ~15–30 minutes to prepare with a Ga^+ FIB, while a transmission electron microscopy specimen of the same size may take ~20–60 minutes to prepare. Once feature sizes approach ~50 μm or more, the Ga^+ FIB becomes time and cost prohibitive and the advantage of a larger mass Xe^+ ion plasma FIB source that can generate microamperes of beam current becomes more suitable (Smith *et al.* 2014).

24.3 GEOMETRY FOR 3D SECTIONING AND IMAGING

3D FIB tomography experiments were performed on single beam FIB instruments, whereby an FIB image was obtained from an FIB milled face and repeated to create a stack of images that were reconstructed to form a 3D structural model of the specimen (Dunn and Hull 1999). These first 3D tomograms were obtained manually with the user controlling all of the FIB slicing and FIB imaging operations. 3D FIB tomography became more popular using the FIB-SEM combined with computer automation of the FIB milling and SEM imaging steps. Now scripts or instrument programming can acquire multi-imaging signals as well as analytical X-ray or crystallography information, creating 3D cubes of information on microstructure, chemistry, phase, and crystal structure (Uchic *et al.* 2007; Cantoni and Holzer 2014; Kotula *et al.* 2014).

Most 3D FIB tomography acquisition sample set-up routines use a geometry like that shown in Figure 24.5. A protective deposition layer (e.g., Pt, C, or other) is used to define the volume for tomography. The FIB is then used to mill out a large U-shaped region around the volume of interest. This U-shaped trench allows for sputtered material to escape during subsequent FIB slicing so that redeposited material does not block imaging.

The time to acquire a set of 2D slices stacked together to form a 3D may take hours to days depending on the voxel size (i.e., slice thickness and imaging resolution per slice size) and number of slices. If it takes ~1 minute to mill the FIB slice and ~1 minute to acquire the SEM image, then 100 slices will take ~200 minutes of instrument acquisition time (not counting set-up time and off-line analysis time). While researchers have been working on smart milling and image acquisition techniques to reduce acquisition time and cost associated with 3D FIB-SEM tomography (Narayan *et al.* 2014), one can see how this acquisition time and cost can increase significantly as the volume of interest with improved resolution increases.

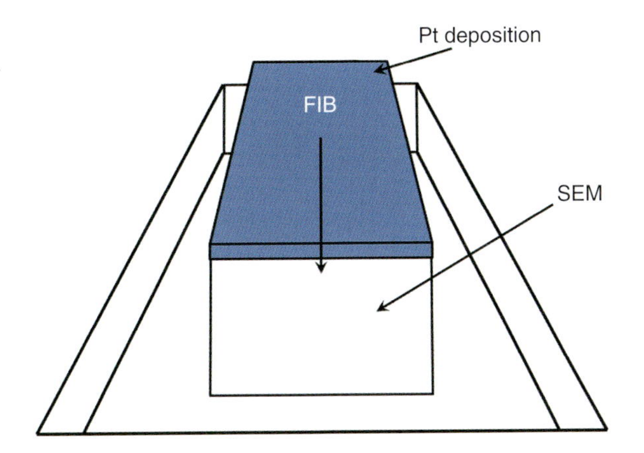

Figure 24.5 Schematic diagram of the set-up geometry for 3D FIB tomography acquisition.

24.4 APPLICATIONS OF 2D SECTIONING AND IMAGING WITH FIB-SEM

FIB-SEM applications are shown with bone on a dental Nobel Biocare TiUnite implant, which was surgically removed due to implant mobility upon stage 2 uncovering of the implant (Giannuzzi *et al.* 2005, 2007). A Quanta 3D FIB-SEM (FEI Company, Eindhoven, The Netherlands) was used for the following work. A BSE SEM image of a site-specific FIB section from the top of an implant thread is shown in Figure 24.6a. Note that this is a fairly large section for a Ga^+ FIB and it took a couple of hours to complete. Only ~50 μm of the center portion of the entire section was FIB polished for detailed analysis. The cross-section shows three distinct contrast layers, with the Ti the brightest at the bottom of the layer, the coating a bit lighter and on top of the Ti, and the bone the darkest contrast layer at the top of the section. The bone growth into the porous coating structure is demonstrated with the iSE FIB image in Figure 24.6b. Note that the iSE FIB image in Figure 24.6b shows different contrast formation mechanisms compared to the BSE image in Figure 24.6a (Ohya *et al.*

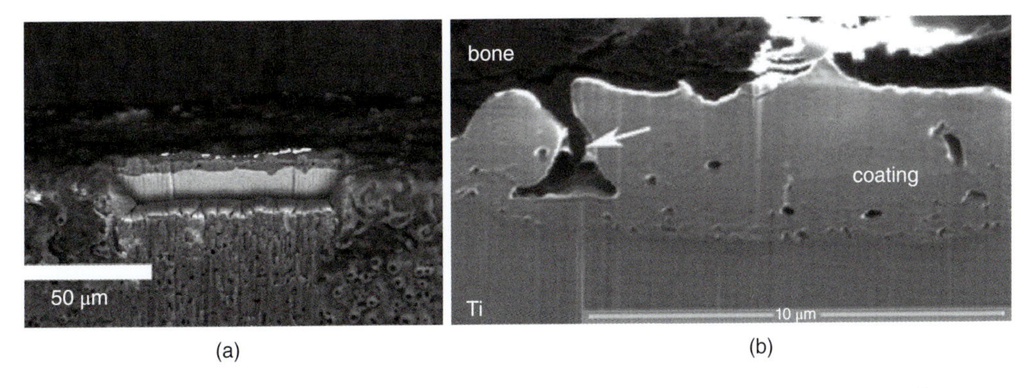

Figure 24.6 (a) BSE SEM image of an FIB milled section of bone on a dental implant. (b) iSE FIB image from the FIB milled section in (a).

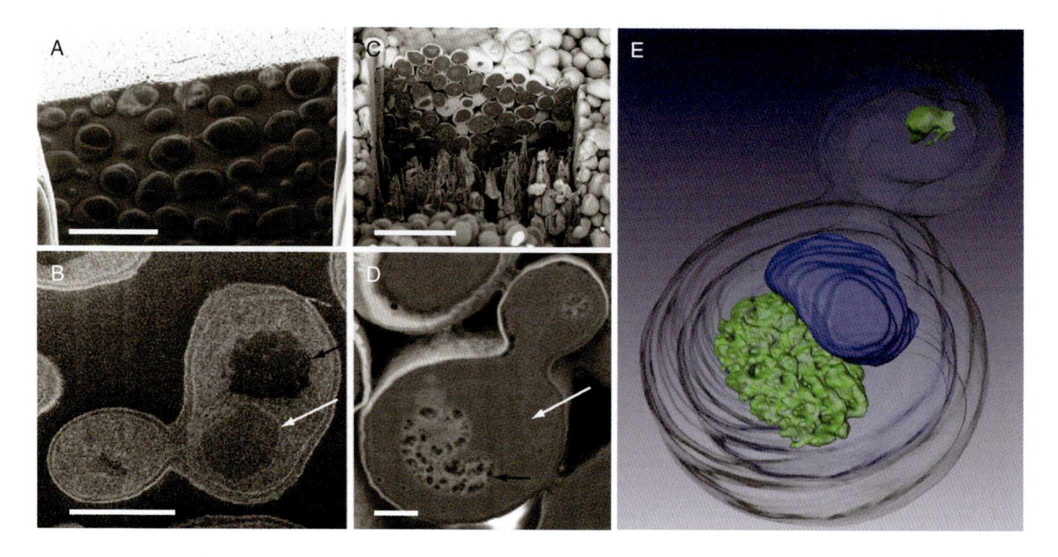

Figure 24.7 FIB-SEM analysis of yeast cells. SEM images of FIB sections of plastic-embedded (A and B) and critical point-dried (C and D) yeast cells. (E) 3D visualization of critical point-dried yeast cells showing the cell wall (gray envelope), the vacuolar region (green), and the nucleus (blue) of an individual, budding yeast cell. Scale bars: (A and C) 10 μm, (B) 2 μm, (D) 0.5 μm. Reprinted with permission from Heymann *et al.* (2006) .

2009; Giannuzzi and Utlaut 2011). The bone growth is observed to be in intimate contact only along certain portions of the coating. Porosity is observed within the coating and the bone. The arrow in Figure 24.6b denotes a region of bone growing into a pore. Note that the bone did not completely grow to fill the open pore space.

A second example of FIB-SEM cross-sectioning and imaging is shown using yeast cells that were prepared to make them vacuum compatible (Heymann *et al.* 2006). Figure 24.7A, and B shows FIB-SEM results from chemically fixed and resin-embedded cells (from an FEI Strata 400 DualBeam instrument) and Figure 24.7C and D shows 2D FIB-SEM results from critical point-dried cells (from an FEI Nova 600 NanoLab DualBeam instrument). SEM images at low magnification (Figure 24.7A and C) and high magnification (Figure 24.7B and D) are shown. The long white arrows point to the location of the nuclear membrane in the budding yeast cells shown (Figures 24.7B and D) and the short black arrow in Figure 24.7B points to the locations of the vacuoles. Figure 24.7E shows a 3D reconstruction of the yeast cells, which will be described in more detail below.

24.5 3D FIB-SEM TOMOGRAPHY

An FEI Quanta 3D FIB-SEM was used for the following work. FIB-SEM tomography is accomplished by creating an FIB section as in Figures 24.5 and 24.6 above, and then continuing to FIB slice to create fresh faces for subsequent imaging. Each new face is imaged and the process may be repeated for hundreds of slices (or more). This process may take anywhere from a couple of hours to many days depending on the size of each section and the number of sections. Figure 24.8 shows seven consecutive BSE SEM images using an FIB slice of 100 nm between images. This is a subset of a total of 68 images. Looking only at the

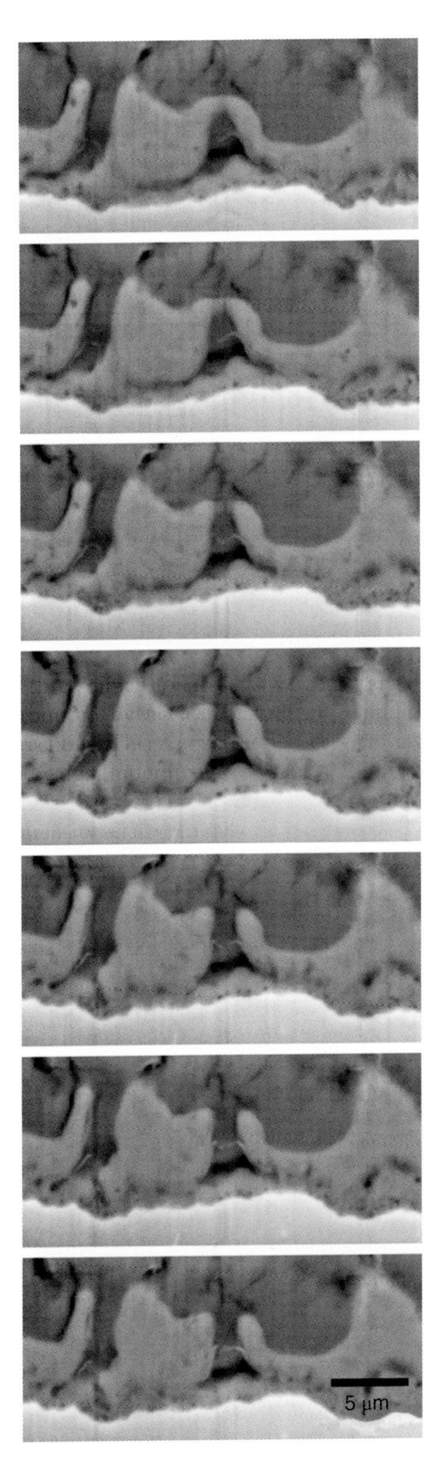

Figure 24.8 Seven consecutive BSE SEM images using an FIB slice of 100 nm between images.

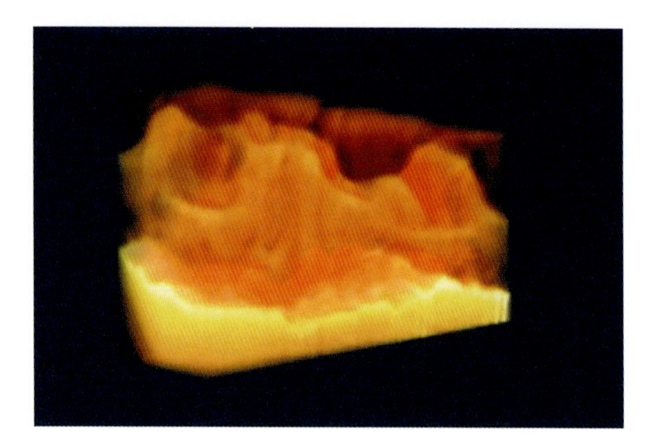

Figure 24.9 3D model of reconstructed stack of 68 BSE SEM images.

topmost image in Figure 24.8, it seems that bone appears somehow underneath a piece of coating. These series of images show that, consistent with Figure 24.6b, bone growth occurs into and then laterally within a pore. This understanding of the mechanism of bone growth into pores is only realized using a 3D data set. Figure 24.9 shows a 3D reconstruction from the 68 stack of images obtained from this full data set.

The 3D reconstruction shown in Figure 24.7E was achieved in a similar fashion. Each successive FIB milling step removed a slice of ~100 nm in thickness. 100 slices/images were obtained and reconstructed to show the 3D structure of prominent cellular components such as the outer cell membrane, the vacuolar region, and the nucleus (Figure 24.7E). The total time taken to generate this data set was about 2 hours. This does not include the hours necessary to perform the offline data processing and reconstruction of the image stack.

24.6 3D FIB-SEM TOMOGRAPHY WITH MULTI-SIGNAL SEM ACQUISITION

Multi-signal SEM tomography is advantageous because it allows the collection of either secondary electron (SE) or backscattered electron (BSE) signals. The SE signal is best suited for imaging topography while the BSE signal is ideal for identifying multi-phase samples (Goldstein *et al.* 2003). Since an FIB milled face is smooth, the SE signal is most useful for imaging internal topography such as voids or porosity. Since FIB milling is a destructive process, it is necessary to collect as much information as possible from each FIB milled face. In addition, to save time and increase throughput, it is advantageous to collect multiple signals from a single electron beam acquisition process. An SEM configured with unique detector chains allows for individual signal optimization. Therefore, the acquisition of multiple signals from the same FIB slice allows for complementary and optimized characterization from the same region of interest.

3D analysis using FIB-SEM tomography was performed on an Auriga FIB-SEM (Carl Zeiss, Oberkochen, Germany) CrossBeam instrument (Giannuzzi 2012a,b). The data set presented below consisted of the simultaneous acquisition of in-lens BSE (using the EsB detector) and in-lens SE signals with a single electron beam scan after each FIB slice. The

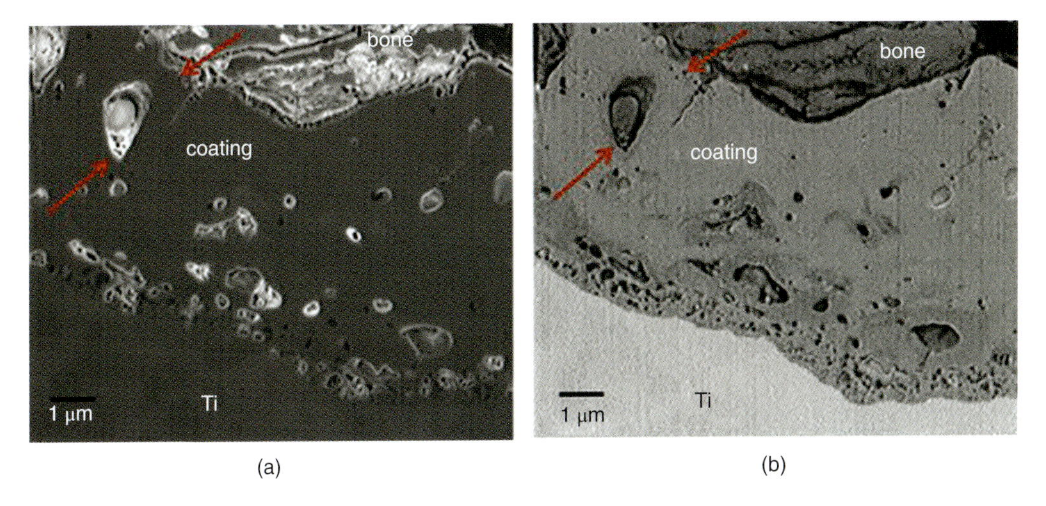

Figure 24.10 SEM images from the same FIB slice showing simultaneous acquisition of (a) an in-lens SE image and (b) an in-lens BSE image.

imaging pixel size was 21.8 nm (x, y) with pixel dimensions of 1024 × 768 and 216 slices were FIB milled at a z dimension of 20 nm per slice using a beam current of 1 nA. The SEM imaging was acquired using an accelerating voltage of 3 keV with the EsB grid set to 1500 V. The total volume sampled was ~21 μm × 21 μm × 4.3 μm and it took about 4 hours to acquire the data set. All images were acquired with tilt correction compensation. The images were aligned, rendered, segmented, and quantified using the Avizo Fire software package.

Figure 24.10 shows (a) an in-lens SE image and (b) an in-lens BSE image from the same FIB slice of the data set. The arrows in both images point to similar features on the same slice. The BSE image shows the obvious presence of three different phases: (i) the Ti substrate indicated by the lightest signal at the bottom of the image, (ii) the ceramic coating indicated by the intermediate grayscale signal located in the middle of the image, and (iii) the bone indicated by the darkest signal conforming to the rough coating surface at the top of the image. The in-lens SE image shows a bright signal outlining the edges of pores in the ceramic coating as well as pores and cracks along the bone/ceramic interface and within the bone itself. Cracking at the bone/ceramic interface indicates that the osseointegration does not always result in intimate contact with the bone and ceramic coating as previously noted above. Note that, in some cases, the porosity in the SE image does not show extensive bright edges. Pores in the BSE image indicate larger dimensions compared to the SE image. In addition, pores can be seen to appear in slices from the BSE images before they appear in the SE images. This occurs because BSEs originate from a larger volume both spatially (in x and y) and depth of volume (in z) from the sample compared to SEs. Thus, quantification of porosity may provide an overestimate of the size and distribution using BSE images. The lower arrow indicates a pocket of bone that has grown in three dimensions into a volcano-shaped pore and hence in this single 2D section appears to exist below the bone/ceramic interface. 3D volume reconstructions of the SE and BSE images are shown in Figure 24.11a and b respectively. The differences in the contrast signal and emphasis in the volume reconstructions are dramatic. This illustrates the importance of 3D analysis of features to truly understand their connectivity and spatial relationships throughout the volume.

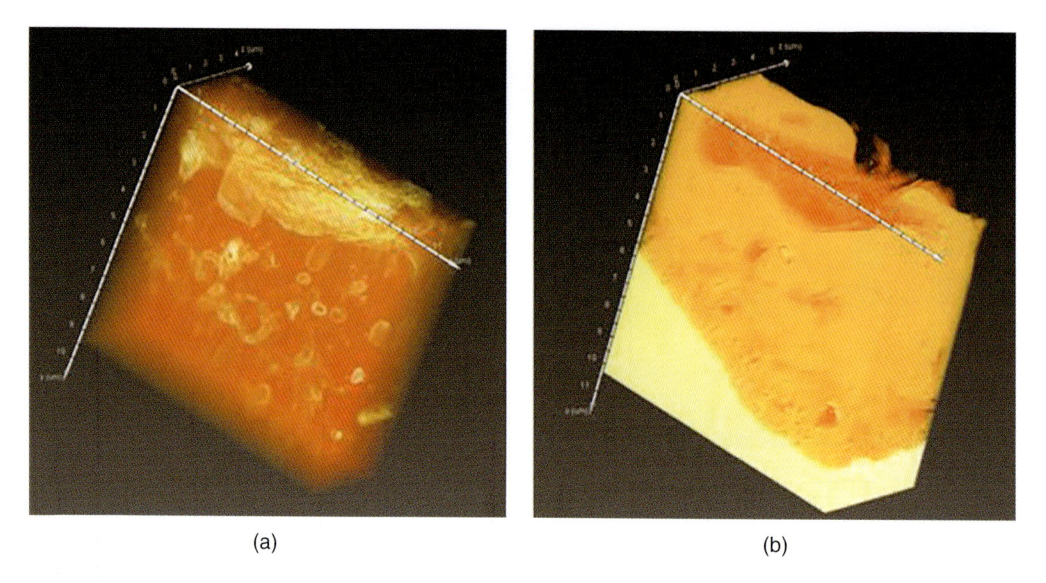

Figure 24.11 3D volume reconstructions from the image data sets of (a) the in-lens SE images and (b) the in-lens BSE images.

24.7 TEM SPECIMEN PREPARATION WITH FIB-SEM

The Quanta 3D FIB-SEM instrument were used in the following work. FIB milling for TEM specimen preparation is one of the most popular uses of FIB-SEM. These specimen preparation methods were first applied to dental implants by Giannuzzi et al. (2005, 2007), and since then used in other publications for TEM studies on dental implants (Jarmar *et al.* 2008a,b; Grandfield *et al.* 2010). FIB techniques allow for site-specific sectioning of hard and soft materials within the same section devoid of mechanical damage often incurred in ultramicrotomed prepared specimens (Giannuzzi *et al.* 2007).

Figure 24.12 shows (a) a BSE SEM image of a FIB thinned specimen showing the bone/coating/Ti substrate interfaces and corresponding TEM images of (b) the bone/coating interface and (c) the coating/Ti substrate interface. Higher magnification TEM images of the coating and the bone are shown in Figure 24.12a and b respectively. There are numerous microstructural features observed from the TEM images that are not discernable using SEM imaging. An amorphous phase is observed between the crystalline bone structure and the coating (see Figure 24.12b). The thickness of this apparent organic amorphous phase varies along the bone/coating interface. The bone was observed in its "natural" state, that is, no staining was used to observe the college fibrils yielding the obvious crystalline diffraction contrast (see Figures 24.13b and 24.14b). The Ti substrate contains a large number of dislocations. Nanoporosity and nanocrystallinity is observed within the coating, particularly at the coating/Ti interface. The coating (Figure 24.13a) is comprised of both amorphous TiO_x ($aTiO_x$) as well as nanocrystalline anatase TiO_2. It was previously shown that interdiffusion of Ca, P, and O occurs at the bone/coating interface, indicating that chemical bonding as well as mechanical bonding occurs at this interface (Giannuzzi *et al.* 2007).

TEM specimens can likewise be prepared for chemically fixed and embedded yeast cells, as shown in Figure 24.14. The cells were negatively stained and decorated with 15 nm gold

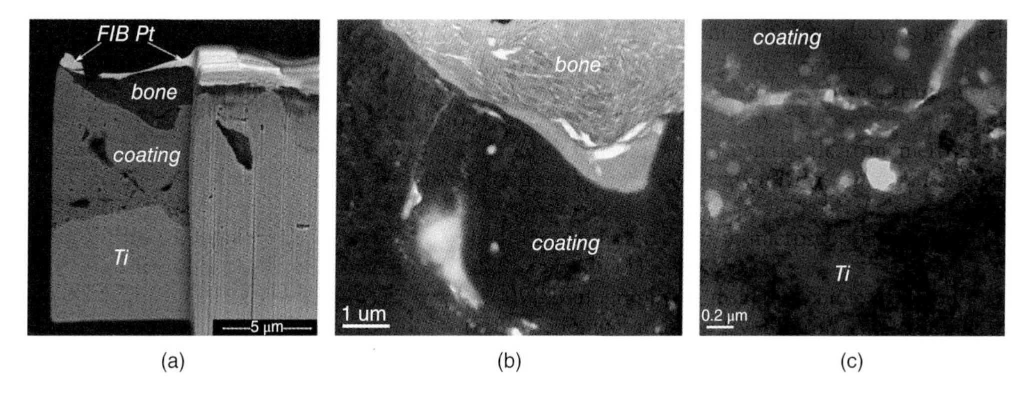

Figure 24.12 (a) BSE SEM image of a FIB milled TEM specimen. (b) TEM image of the bone/coating interface. (c) TEM image of the coating/Ti substrate interface.

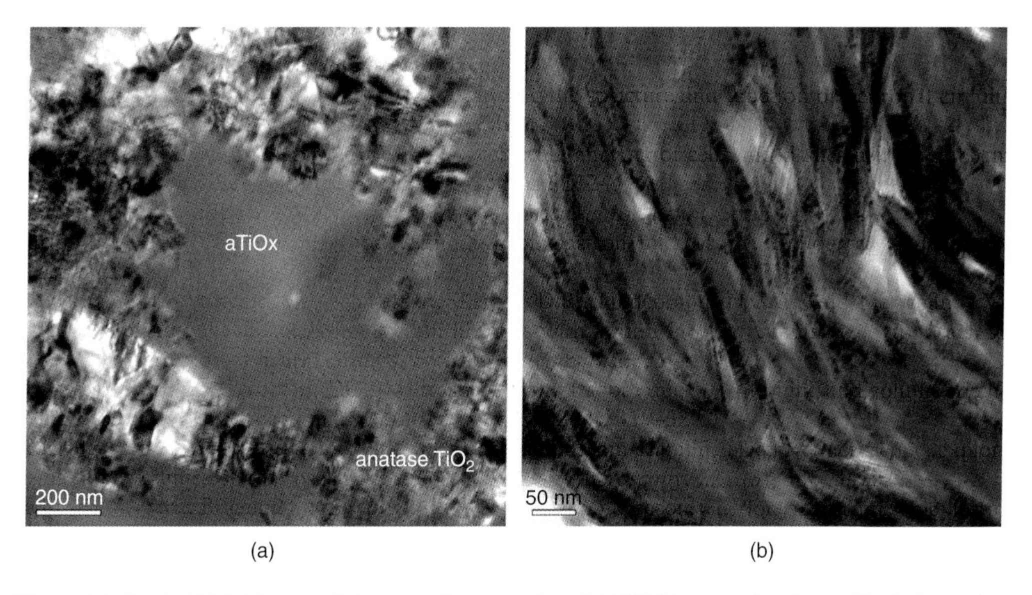

Figure 24.13 (a) TEM image of the two-phase coating. (b) TEM image of collagen fibrils formed in the bone.

particles used as fiducial markers for tomographic experiments (Williams and Carter 2009). A plastic section (about 12 µm × ~16 µm × ~1 µm) was FIB milled into a block of resin containing chemically fixed yeast as cells and lifted out using an in situ probe (Figure 24.14A). The section was transferred to a half-grid (indicated by the arrow) (Figure 24.14B and C). This specimen was then further FIB milled to a thickness of ~200 nm (Figure 24.14D and E). TEM tomography (Williams and Carter 2009) was performed and a tomographic slice obtained from a 3D volume reconstruction of a region in the section is shown in Figure 24.14E. Each of the black dots, indicated by the white arrow, approximately corresponds to the staining of a single ribosome in the cytoplasm. The TEM tomography was performed on an FEI Tecnai F30 microscope operating at 300 keV.

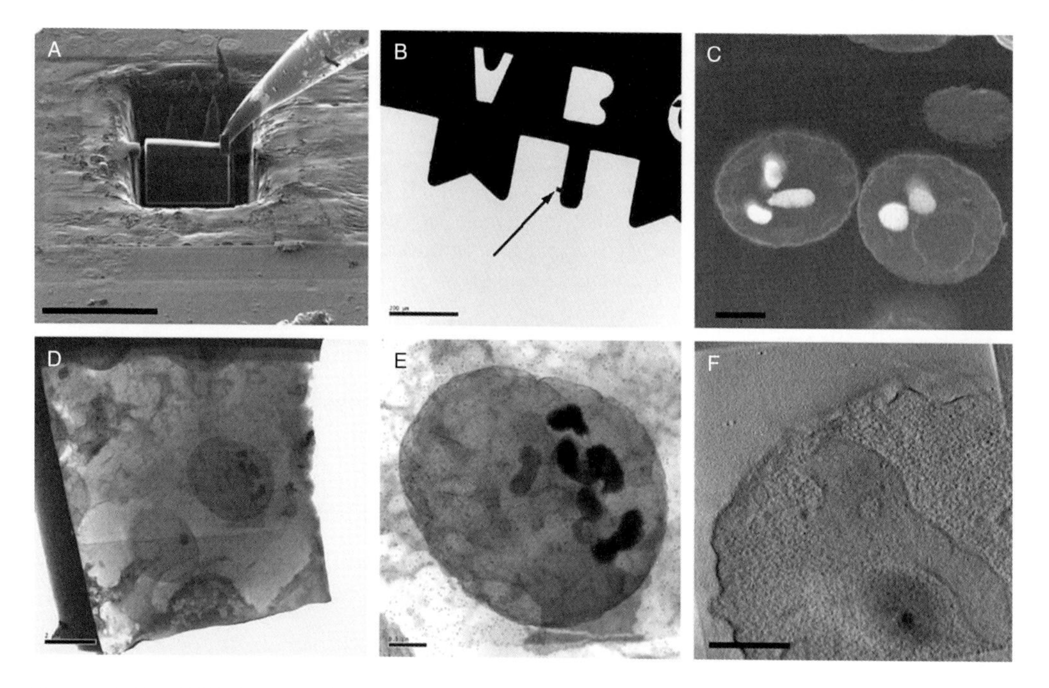

Figure 24.14 SEM (A–C) and TEM (B, D–F) images showing FIB lift-out of chemically fixed and embedded yeast cells. (C) SEM of the thick FIB lift-out. (D and E) TEM images of FIB thinned yeast cells. (F) A tomographic slice obtained from a 3D volume reconstruction of a region in the section. Scale bars: (A) 20 μm, (B) 200 μm, (C) 1 μm, (D) 2 μm, (E and F) 0.5 μm. Reprinted with permission from Heymann *et al.* (2006).

24.8 SUMMARY

As described, FIB-SEM for 2D and 3D specimen preparation and analysis is indispensable for site-specific biomaterials characterization of morphology and microstructure and composition. FIB-SEM techniques are excellent high resolution methods. Going forward, however, these methods begin to exhibit limitations when combining both high resolution with large volume analysis. Extending FIB-SEM to larger volumes becomes both data and time intensive. The advent of plasma FIB columns enables larger volumes to be sectioned but the serial beam scanning required for SEM imaging is a bottleneck in the process. Thus, techniques and methods are needed that combine large volume analysis with high resolution and speed. Continued progress with smart imaging acquisition techniques (e.g., image at high resolution only when a feature of interest comes into view) (Narayan *et al.* 2014) or perhaps the application of multi-imaging beams (Mohammadi-Gheidari and Kruit 2010) may be utilized in the future.

ACKNOWLEDGEMENTS

Nicholas J. Giannuzzi, DDS (Miller Place, NY) and Mario J. Capuano, DDS (Long Island Oral and Maxillofacial Surgery, Selden, NY) provided samples and helpful discussions. Daniel Phifer, Ritch Gursky, and Lee Pullan from FEI Company (now ThermoFisher Scientific) helped with this work. Jeff Marshman (Carl Zeiss) helped with the Auriga FIB-SEM set-up. Thanks also to Mike Marsh (Marsh Imaging and Visualization (now with Object Research Systems)) and Patrick Barthelemy (FEI Company (now ThermoFisher Scientific)) for help with the 3D segmenting and analysis. This work was originally funded by FEI Company (now ThermoFisher Scientific) and Carl Zeiss.

REFERENCES

Marco Cantoni and Lorenz Holzer, Advances in 3D focused ion beam Tomography, *MRS Bulletin*, 39, 354–360 (2014). DOI: 10.1557/mrs.2014.54.

D.N. Dunn, and R. Hull, Reconstruction of three-dimensional chemistry and geometry using focused ion beam microscopy, *Appl. Phys. Lett.*, 75, 3414–3416 (1999). https://dx.doi.org/10.1063/1.125311.

Lucille A. Giannuzzi, Multi-signal FIB/SEM tomography, in *Proceedings of SPIE 8378, Scanning Microscopies 2012: Advanced Microscopy Technologies for Defense, Homeland Security, Forensic, Life, Environmental, and Industrial Sciences*, 83780P (May 1, 2012a). DOI: 10.1117/12.919821.

Lucille A. Giannuzzi, Optimizing morphology and structure with multi-signal FIB/SEM tomography, *MRS Proceedings*, 1421, mrsf11-1421-pp07-03 (2012b). DOI: 10.1557/opl.2012.432.

L.A. Giannuzzi and J.R. Michael, Comparison of channeling contrast between ion and electron images, *Microsc. Microanal.*, 19, 344–349 (2013).

L.A. Giannuzzi and F.A. Stevie (eds), *Introduction to Focused Ion Beams: Instrumentation, Theory, Techniques and Practice*, Springer, New York (2005).

L.A. Giannuzzi and M. Utlaut, Non-monotonic material contrast in scanning ion and scanning electron images, *Ultramicroscopy*, 111, 1564–1573 (2011).

Lucille A. Giannuzzi, Daniel Phifer, Nicholas J. Giannuzzi, and Mario J. Capuano, Two-dimensional and 3-dimensional analysis of bone/dental implant interfaces with the use of focused ion beam and electron microscopy, *J. Oral Maxillofac. Surg.*, 65, 737–747 (2007).

Lucille A. Giannuzzi, Nicholas J. Giannuzzi, and Mario J. Capuano, FIB, SEM, and TEM of bone/dental implant interfaces, *Microsc. Microanal.*, 11 (Suppl. 2), 998–999 (2005).

Lucille A. Giannuzzi, Zhiyang Yu, Denise Yin, Martin P. Harmer, Qiang Xu, Noel S. Smith, Lisa Chan, Jon Hiller, Dustin Hess, and Trevor Clark, Theory and new applications of ex situ lift out, *Microsc. Microanal.*, 21, 1034–1048 (2015). DOI: 10.1017/S1431927615013720.

Joseph Goldstein, Dale E. Newbury, David C. Joy, Charles E. Lyman, Patrick Echlin, Eric Lifshin, Linda Sawyer, and J.R. Michael, *Scanning ElectronMicroscopy and X-rayMicroanalysis*, 3rd edn, Springer (2003).

K. Grandfield, E.A. McNally, A. Palmquist, G.A. Botton, P. Thomsen, and H. Engqvist, Visualizing biointerfaces in three dimensions: Electron tomography of the bone–hydroxyapatite interface, *J. R. Soc. Interface*, October 6, 7 (51), 1497–1501 (2010).

Jurgen A.W. Heymann, Mike Hayles, Ingo Gestmann, Lucille A. Giannuzzi, Ben Lich, and Sriram Subramaniam, Site-specific 3D imaging of cells and tissues with a dual beam microscope, *J. Struct. Biol.*, 155 (1), 63–73 (2006).

Tobias Jarmar, Anders Palmquist, Rickard Brånemark, Leif Hermansson, Håkån Engqvist, and Peter Thomsen, Technique for preparation and characterization in cross-section of oral titanium implant surfaces using focused ion beam and transmission electron microscopy, *Journal of Biomedical Materials Research Part A*, 1003–1009 (2008a).

Tobias Jarmar, Anders Palmquist, Rickard Brånemark, Leif Hermansson, Håkån Engqvist, and Peter Thomsen, Characterization of the surface properties of commercially available dental implants using scanning electron microscopy, focused ion beam, and high-resolution transmission electron microscopy, *Clinical Implant Dentistry and Related Research*, 10 (1), 11–22 (2008b).

Paul G. Kotula, Gregory S. Rohrer, and Michael P. Marsh, Focused ion beam and scanning electron microscopy for 3D materials characterization, *MRS Bulletin*, 39, 361–365, (2014). DOI: 10.1557/mrs.2014.55.

R.M. Langford, T.-X. Wang, and D. Ozkaya, Reducing the resistivity of electron and ion beam assisted deposited Pt, *Microelectronic Engineering*, 84, 784–788 (2007).

Joachim Mayer, Lucille A. Giannuzzi, Takeo Kamino, and Joseph Michael, TEM sample preparation and FIB-induced damage, *MRS Bulletin*, 32, 400–407 (2007).

Mohammadi-Gheidari and P. Kruit, Electron optics of multi-beam scanning electron microscope, *Nuclear Instruments and Methods in Physics Research Section A: Accelerators, Spectrometers, Detectors and Associated Equipment*, 645, 60–67 (2010). DOI: 10.1016/j.nima.2010.12.090.

Kedar Narayan, Cindy M. Danielson, Ken Lagarec, Bradley C. Lowekamp, Phil Coffman, Alexandre Laquerre, Michael W. Phaneuf, Thomas J. Hope, and Sriram Subramaniam, Multi-resolution correlative focused ion beam scanning electron microscopy: Applications to cell biology, *Journal of Structural Biology*, 185 (Issue 3), 278–284 (2014). https://doi.org/10.1016/j.jsb.2013.11.008.

K. Ohya, T. Yamanaka, K. Inai, and T. Ishitani, Comparison of secondary electron emission in helium ion microscope with gallium ion and electron microscopes, *Nuclear Instruments and Methods in Physics Research B*, 267 (4), 584–589 (2009).

Noel S. Smith, John A. Notte, and Adam V. Steele, Advances in source technology for focused ion beam instruments, *MRS Bulletin*, 39, 329–335 (2014). DOI: 10.1557/mrs.2014.53.

Michael D. Uchic, Lorenz Holzer, Beverley J. Inkson, Edward L. Principe, and Paul Munroe, Three-dimensional microstructural characterization using focused ion beam tomography, *MRS Bulletin*, 32, 408–416 (2007).

David B. Williams and C. Barry Carter, *Transmission Electron Microscopy: A Textbook for Materials Science*, 2nd edn, Springer (2009).

25

New Opportunities for FIB/SEM EDX in Nanomedicine: Cancerogenesis Research

Damjana Drobne[1], Sara Novak[1], Andreja Erman[2] and Goran Dražić[3]

[1]*Department of Biology, Biotechnical Faculty, University of Ljubljana, Ljubljana, Slovenia*
[2]*Institute of Cell Biology, Faculty of Medicine, University of Ljubljana, Ljubljana, Slovenia*
[3]*National Institute of Chemistry, Laboratory for Materials Chemistry, Ljubljana, Slovenia*

25.1 INTRODUCTION

Engineered nanoparticles (NP) show great promise in nanomedicine. They are large enough to avoid rapid elimination through the kidney and small enough to penetrate the vasculature of the tumour tissues. Because tumour sites have a leaky, immature vasculature with wider fenestrations than normal mature blood vessels, nanomaterials could selectively accumulate in tumour tissues, while they cannot penetrate the intact endothelial barrier (Cho *et al.*, 2008). Nanomedicines and their payload can enter the cells by endocytosis and accumulate there (Maeda, 2001; Cherukuri, 2010). In this way, it is possible to improve the transport of anticancer compounds to tumours, the targeting and detection of cancer cells, as well as the study of cancerogenesis and the efficiency of anticancer treatment.

A successful application of nanomaterials in biomedicine highly depends on understanding the body distribution of NPs, their retention in the body, cell recognition and cellular uptake. An assortment of methods is available for this task. Ostrowski *et al.* (2015) made a comprehensive review on the different imaging techniques available for localizing inorganic as well as organic nanoparticles in tissues, cells and subcellular compartments. These authors have pointed out the importance of observing particle distribution and tissue responses. They concluded that there is no single technique available for this purpose.

Biological Field Emission Scanning Electron Microscopy, First Edition.
Edited by Roland A. Fleck and Bruno M. Humbel.

With each technique applied, a compromise has to be made between the optical resolution and the specificity and sensitivity of NP detection.

The FIB/SEM approach offers a good compromise between spatial resolution and the possibility of imaging the relatively large surfaces of tissue quickly and efficiently (Kizilyaprak, Daraspe and Humbel, 2014). The SEM and FIB systems integrated into a single imaging device constitute a powerful scientific instrument for the study of biological specimens and soft materials. Although the resolution of SEM is not as good as in TEM, the analysed volume can be much larger. An important reason for imaging large volumes is low nanoparticle concentrations *in vivo* and their irregular tissue distribution (Kempen *et al.*, 2013).

The FIB/SEM approach is increasingly used in investigating the interactions of cells with nanomaterials (Table 25.1). Internalized nanomaterials were assessed either by imaging the sectioned surface using backscattered electrons (BSEs) or by energy-dispersive X-ray spectroscopy (EDX). Most authors have studied individual cells from *in vitro* systems.

In the work presented here, we have studied the early stages of cancerogenesis of the mouse urinary bladder epithelium (called urothelium), with the additional aim to show the applicability of FIB/SEM in distinguishing between the cancer cells and the normal cells. The cancer cells from cell line MB49 used in our study were first incubated with magnetic $CoFe_2O_4$ NPs and then injected in the mouse urinary bladder. We performed FIB/SEM EDX analyses on pieces of the isolated bladder epithelium. We will furthermore discuss the advantages of the method over other available techniques in investigating cancerogenesis and cancer treatment.

25.2 MATERIALS AND METHODS

We have applied the FIB/SEM EDX method to detect selective cell internalization of NPs and to observe cells with internalized NPs at tissue level. We have observed an internalization of magnetic $CoFe_2O_4$ NPs (synthesized by the Group for Nano and Biotechnological Applications, Faculty of Electrical Engineering, University of Ljubljana, Ljubljana, Slovenia) in cancer cells from cell line MB49.

The cells were incubated with NPs for 24 hours and then injected via transurethral catheterization into the *in vitro* urinary bladder of C57BL/6JOlaHsd mice. After 1 hour, the urinary bladders were taken from the mice and fixed in a mixture of 4% formaldehyde and 2% glutaraldehyde in 0.2 M cacodylate buffer (pH 7.4) at 4 °C for 3–4 hours. The tissue samples were rinsed in 0.2 M cacodylate buffer, post-fixed in 1% osmium tetroxide (in the same buffer) for 1 hour at 4 °C and dehydrated in progressive series of acetone. Finally, the specimens were critical point dried, fixed on brass holders with silver paint (high purity silver paint, SPI), sputter-coated with gold and analysed with an FIB/SEM EDX microscope (FEI Helios NanoLab 650 from Hillsboro, Oregon, USA, equipped with Oxford Instruments EDXS system with X-max 50 mm^2 SDD detector and Inca software, Abingdon, Oxfordshire, UK).

Based on the morphological characteristics of cancer cells, we have selected an area with cancer cells and confirmed the presence of Co and Fe inside them with EDX. We selected one of the cancer cells with internalized NPs for FIB milling and SEM imaging.

First we deposit a few nm layer of Pt to protect the sample from ion beam damages. The rough milling conditions to open a trench employed ion currents of 0.43 nA at 30 keV. Lower beam currents of 0.2 nA to 100 pA were used to polish the cross-section. The beam size in the case of rough milling was approximately 50 nm in diameter. For polishing, it ranged from 10 nm of diameter or higher onwards. The secondary electron detectors were

Table 25.1 Literature review on FIB/SEM EDX application for investigating interactions between biological samples and nanoparticles (NPs)

Type of investigated biological sample by FIB/SEM technique	Investigated nanomaterials	Composition study	Sample preparation	Reference
In vitro cells (hMCS cells)	Ag-NPs	EDX	SEM mode-dry sample	Greulich *et al.* (2011)
In vitro cells (SH-SY5YY cells)	Fe_3O_4-NPs	EDX	SEM mode-dry sample	Riggio *et al.* (2012)
In vitro cells (SH-SY5Y cells)	Fe_3O_4-NPs	EDX	SEM mode-dry sample	Calatayud *et al.* (2013)
In vitro cells (THP-1 cells)	Au-NPs	EDX	SEM mode-dry sample	Garcia *et al.* (2013)
In vitro cells (HT 29 cells)	Pt-NPs	BSEs	SEM mode-dry sample	Pelka *et al.* (2009)
In vitro cells (human dendritic cells)	Au-NPs	BSEs	SEM mode-dry sample	Tomic *et al.* (2014)
In vitro cells (HepG2 cells)	Nonoporous substrate	/	SEM mode-dry sample	Bruggemann (2013)
In vitro cells (NIH3T3)	Nanostructured substrate	/	TEM mode-resin embedded	Wierzbicki *et al.* (2013)
Tissue (lung epithelium)	CNTs	/	TEM mode-resin embedded	Kobler *et al.* (2014)
Tissue (bladder epithelium)	Au-NPs	EDX	SEM mode-dry sample	Current chapter

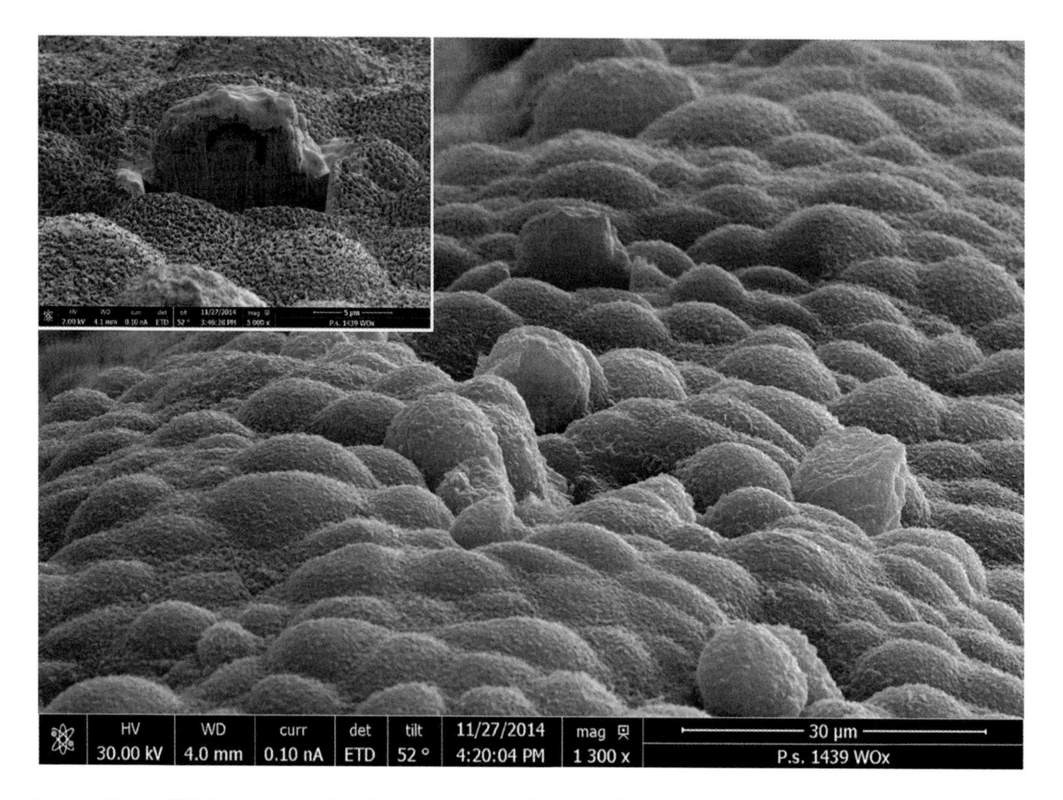

Figure 25.1 SEM micrograph of mouse urothelium with FIB milling operation on selected cell (details in upper left corner). FIB milling operation conducted on epithelial tissue prepared for conventional scanning electron microscopy (fixed and dried).

the Everhart Thornely Detector (ETD) and the secondary electron–secondary ion imaging detector (ICE). The SEM imaging was performed by means of the FEG electron column available in the same system, with a subnanometer resolution at 30 and also at 1 keV, due to a monochromator. A 100 or 200 pA electron current was used for imaging at 52 degree tilt and 4 mm working distance. Tilt and working distance were optimized for ion cutting. Images were taken at 1536×1103 pixels resolution and 300 µs per pixel dwell time. EDXS analyses were performed at 15 keV for a better acquisition rate and elemental mapping and at 5 keV, as a compromise between characteristic X-ray lines excitation and penetration depth for point analysis. Using 5 keV, we were able to detect Fe and Co L lines with a spatial resolution better than 100 nm.

We concluded that FIB/SEM EDX enables *in situ* sample manipulation followed by an *ad libidum* milling of the selected region (aproximately $10 \times 10 \times 10$ µm^3) for subsurface investigation, which was of significant importance when the internalized nanomaterials discriminated cancer from normal cells.

In parallel, cancer cells were prepared for TEM investigation. In this case, samples were fixed for 3 hours at 4 °C in a mixture of 4% formaldehyde and 2% glutaraldehyde in 0.2 M cacodylate buffer (pH 7.4). Overnight rinsing in 0.33 M saccharose in 0.2 M cacodylate buffer was followed by post-fixation with 1% OsO_4 for 1 hour. After dehydration in graded series of ethanol, tissue samples were embedded in Epon resin (Serva Electrophoresis, Heidelberg, Germany). Ultrathin sections were contrasted with uranyl acetate and lead

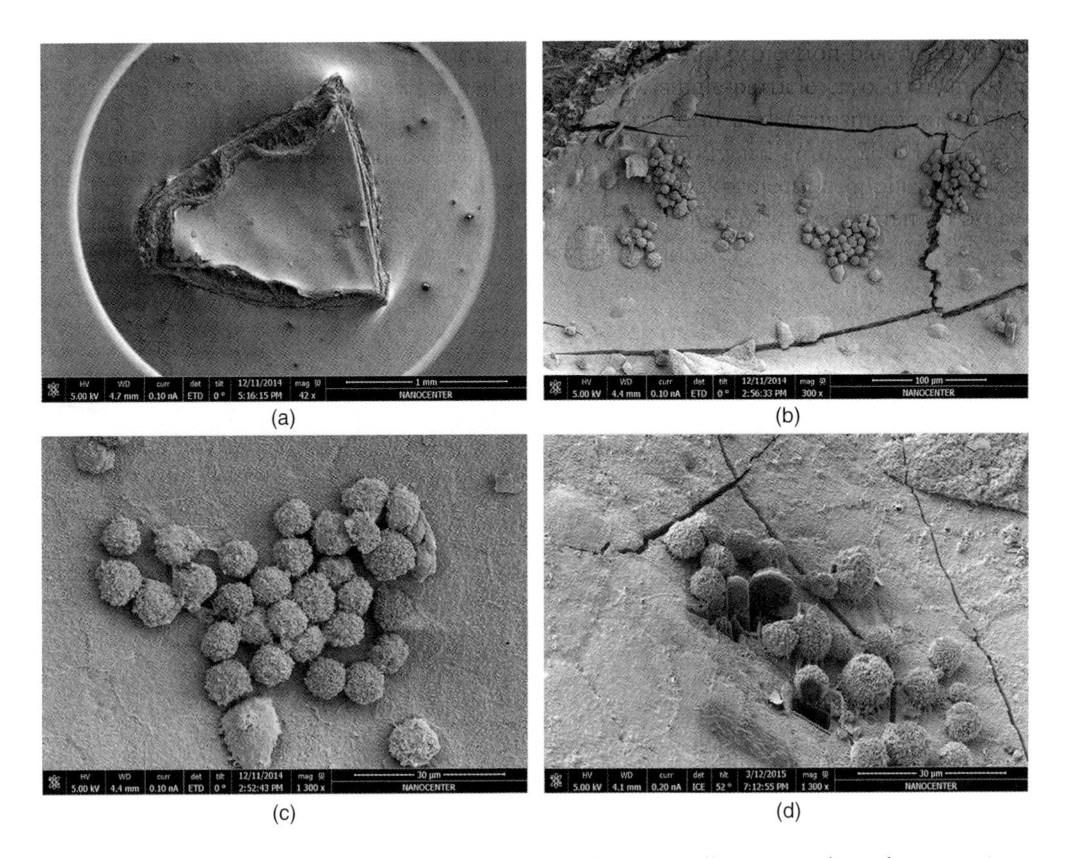

Figure 25.2 SEM micrograph of mouse urothelium with MB49 cells on its surface of mouse urinary bladder tissue piece with MB49 cells on lumenal urothelial surface. (a) Investigated tissue sample is in a millimetre size range. (b,c and d) SEM inspection allowed selection of the region for FIB milling, which was approximately 10 μm long.

citrate and examined with a transmission electron microscope (JEOL 100 CX, Tokyo, Japan).

25.3 RESULTS

The early stages of the carcinogenesis of the mouse urinary bladder epithelium were investigated with FIB/SEM EDX. The MB49 cancer cells were injected into the mouse urinary bladder and 1 hour after the injection the urinary bladder was isolated and prepared for inspection with TEM and FIB/SEM EDX. Since normal and cancer cells differ in the presence of endosomes filled with $CoFe_2O_4$ NPs, we expected the cells filled with NPs to be tumour cells.

The advantage of using FIB/SEM is in the possibility of selecting *ad libitum* any region on the sample surface to be FIB milled (Figures 25.1 and 25.2). The sample prepared for FIB/SEM could be of a millimetre size range (Figure 25.2a).

In this study, we were interested in the subcellular presence of endosomes filled with $CoFe_2O_4$ NPs; therefore the milling operation was conducted on some selected cells and the cut (milling) was up to 10 μm in size (Figures 25.3 and 25.4). The milling of a single

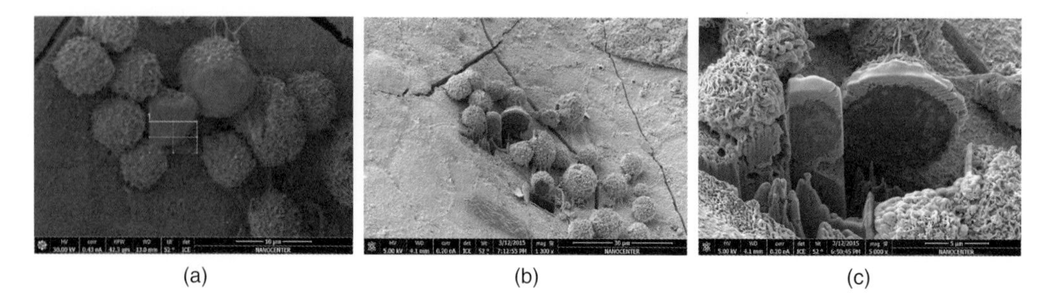

Figure 25.3 (a) Region selected for FIB milling and (b and c) outcome of FIB milling operation.

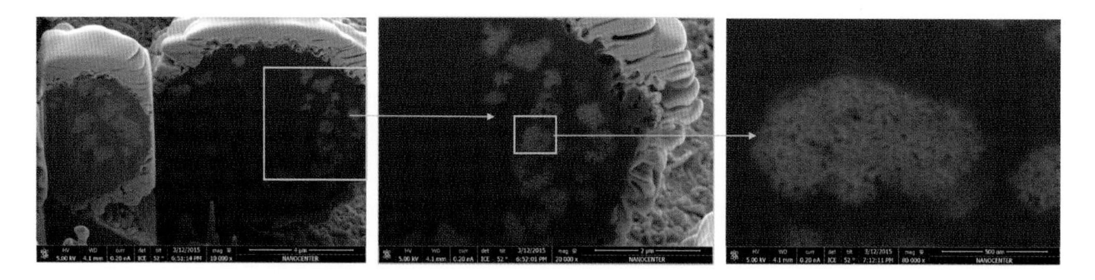

Figure 25.4 Detailed SEM imaging of FIB milled region.

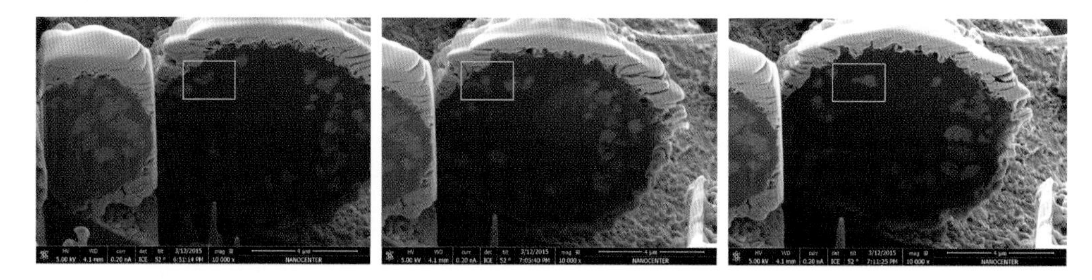

Figure 25.5 SEM micrographs show results of successive FIB milling steps, revealing internal distribution of structure with different compositional contrasts. In frame: endosomes filled with $CoFe_2O_4$ NPs.

cell was in the direction from the apical part of the cell toward its basal part. The milling and polishing took a few minutes. Before milling, the platinum layer (1–2 μm thick) was deposited on the surface exposed toward the ion beam in order to protect the surface from mechanical damage (Figures 25.3 and 25.4).

The fixative we have used in our study for the FIB/SEM investigation preserved as many cellular elements as possible, including molecules in the cytosol; therefore the FIB milled interior appeared homogeneous with a minimal topographic contrast (Figure 25.4). The inspection of the FIB milled cell interior therefore indicated only the presence of structures with different compositional contrast using an ICE detector and were presumably metal filled vesicles (Figure 25.5). The successive FIB milling of the same region allowed an investigation of the intracellular distribution of endosomes filled with $CoFe_2O_4$ NPs.

The size, shape and the location of these vesicles were the same as revealed by the TEM investigation (Figure 25.6a). They were found only in cancer cells (the uppermost cells,

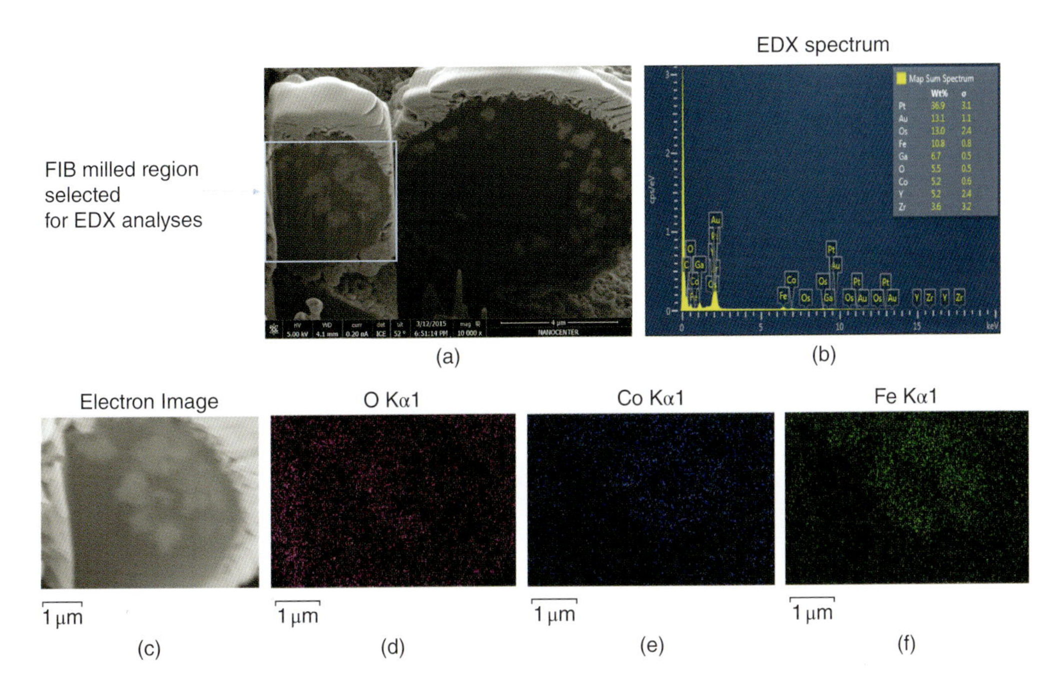

Figure 25.6 (a and c) EDX analyses of FIB milled cell interior. EDX spectrum of investigated region shows presence of some most abundant elements. EDX maps (d to f) show distribution of two elements belonging to $CoFe_2O_4$ NPs, (e) Co and (f) Fe.

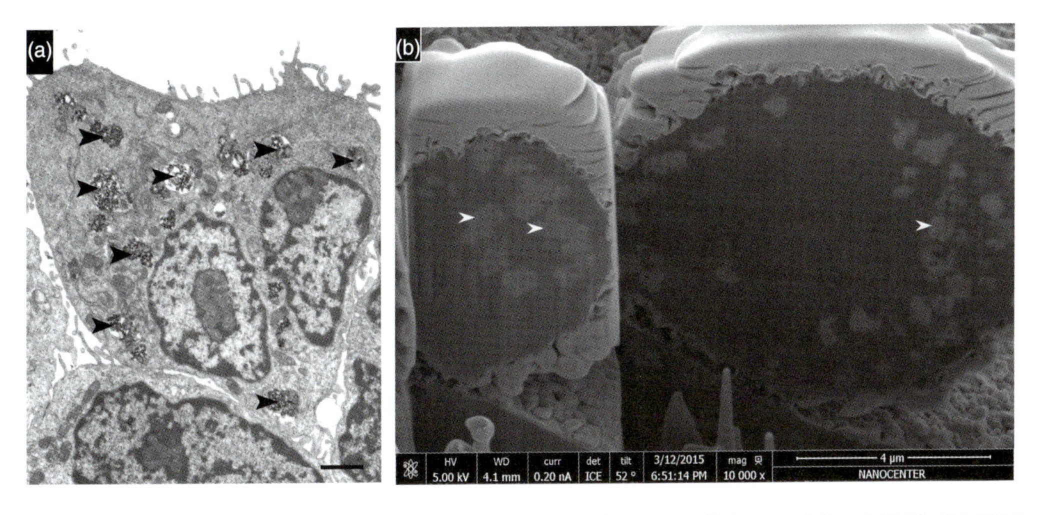

Figure 25.7 (a) TEM micrograph shows ultrastructure of cancer cell from cell line MB49. (b) SEM micrograph of FIB shows exposed interior of MB49 cell. Arrowheads indicate endosomes filled with $CoFe_2O_4$ NPs.

lying on the urothelial surface), but not in the normal urothelial cells, which were part of the epithelial tissue (beneath) (Figures 25.6 and 25.8). The chemical composition of these intracellular elements was obtained by EDX analyses, confirming the presence and overlapping of cobalt and iron (Figure 25.7e and f). Both elements were not present in other cell regions, as seen on the EDX maps.

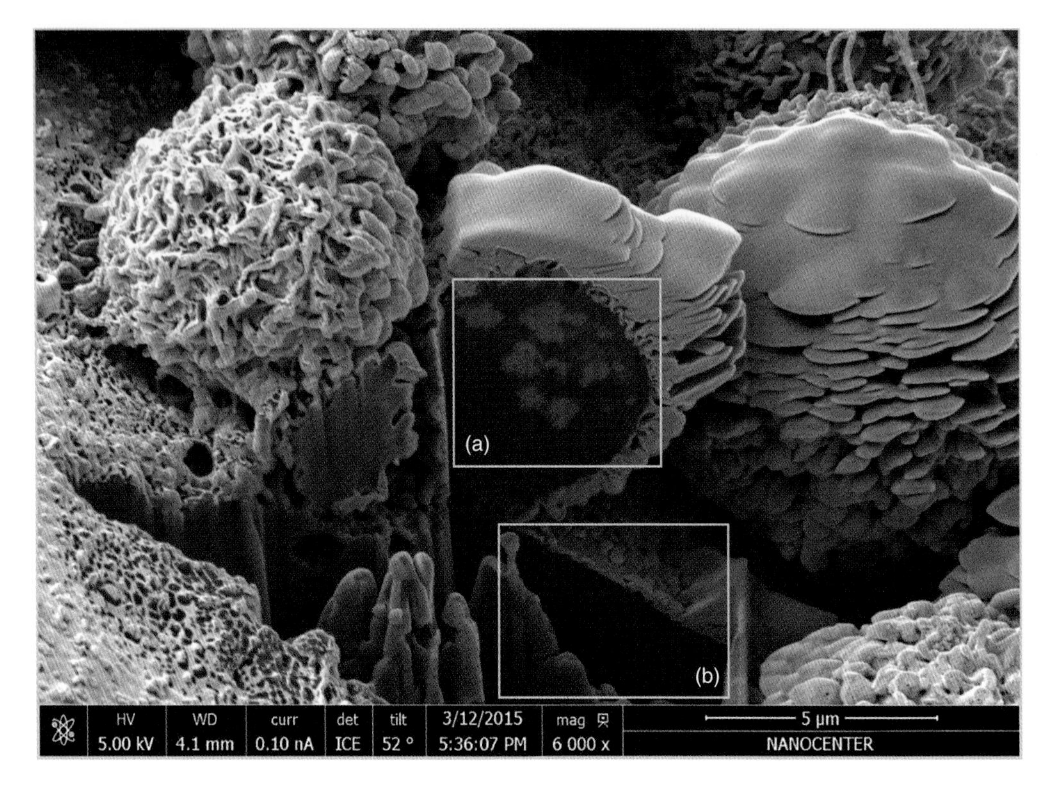

Figure 25.8 Cell region of cancer cell which contains (a) endosomes filled with $CoFe_2O_4$ NPs and (b) cell region of normal urothelial cell without endosomes filled with $CoFe_2O_4$ NPs.

25.4 DISCUSSION

This study shows successful FIB/SEM EDX analyses of *ex vivo* tissue samples combined with cells treated with nanoparticles and prepared with a conventional sample preparation procedure for SEM. A similar approach has already been applied before by other investigators, but only on cells from cell cultures interacting with nanomaterials (Table 25.1).

The tissue exposed to nanomaterials and the investigation with FIB/SEM EDX was reported by Købler *et al.* (2014), but they have investigated the resin-embedded tissue (mice lung), which has been intratracheally installed with carbon nanotubes (CNTs). The prime aim of using FIB/SEM in their study was to locate the region of interest for the FIB milling. These authors suggest the use of FIB/SEM as a complementary tool to TEM. In our study, tissue samples were dried and the aim of the FIB/SEM application was to expose the cell interior for the EDX analyses.

Our results confirm that FIB/SEM EDX provides a good compromise between spatial resolution and the possibility of quickly imaging relatively large volumes of tissue. EDX enables chemical analyses of exposed cells, which is of high importance when the efficiency of internalization and the retention of nanodiagnostics or nanotherapeutics are studied. The investigation of large volumes of nanoparticle treated samples is of the outmost importance, because nanomaterials *in vivo*, either intentionally administered (diagnostic or treating agents) or unintentionally introduced (pollutants), are expected to be unevenly distributed and in low amounts.

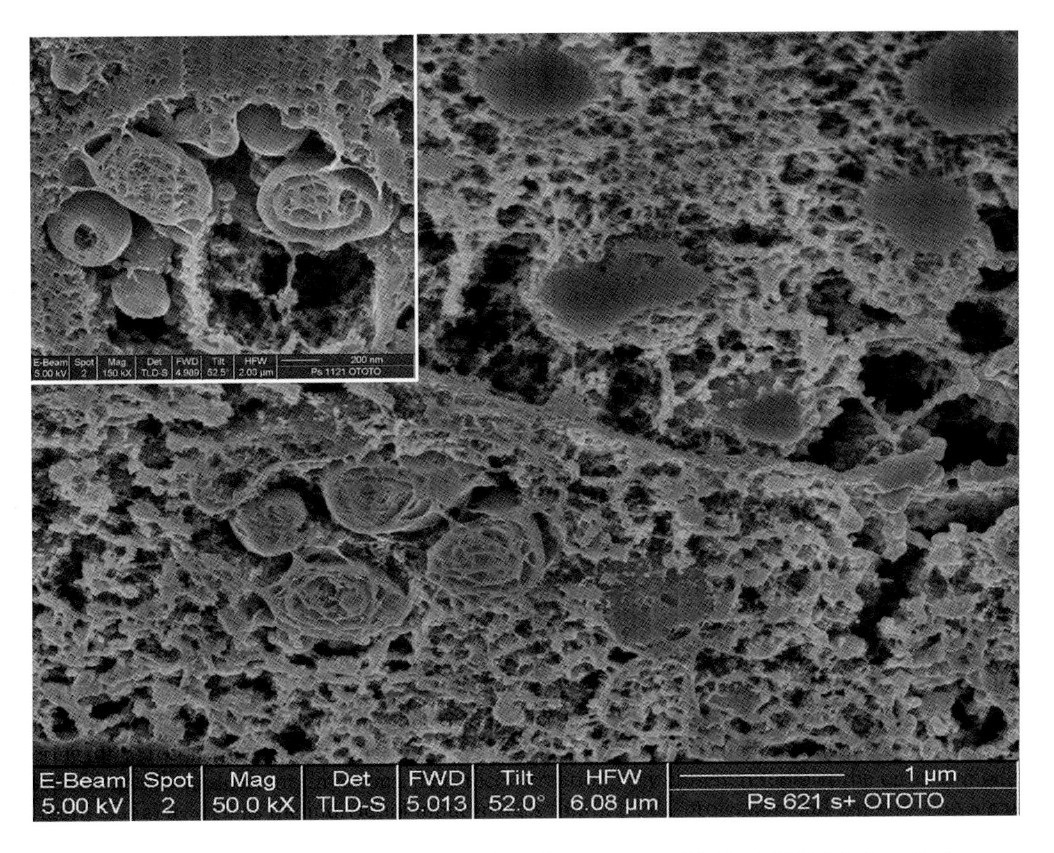

Figure 25.9 FIB exposed cell interior. Photo by courtesy of F. Tatti, FEI, Modena, Italy.

In the past, the preparation of samples for site-specific milling and imaging by FIB/SEM has typically adopted the embedding techniques used for the TEM samples. The reason for this lies in the envisaged potential of FIB/SEM for the TEM lamella preparation or for the automated FIB sectioning for 3D tomography. In some instances, FIB/SEM was proven to be applicable for the structural analyses of specific intracellular components like lamellar structures or lipid bodies (Figure 25.9).

Recently, FIB/SEM EDX has shown high potential in nanomedicine as an efficient tool for cell and tissue imaging. Even if its resolution cannot reach that of TEM, the possibility to image samples of millimetre scale and choose the spots for FIB milling is of great importance. Namely, all biological tissues are three dimensional and span a range of length scales from nanometres to hundreds of millimetres. These are not ideally suited to the current three-dimensional characterization techniques such as transmission electron tomography. The advantage of using FIB/SEM for *in situ* sample manipulation followed by *ad libidum* milling of a selected region for subsurface investigation is outstanding.

The only drawback of FIB/SEM is that it is an abrasive method, so there is no way of retrieving an interesting field to obtain a higher resolution image (Scott, 2011).

In view of all the aforementioned facts, with the FIB/SEM EDX analysis we were able to detect cancer cells in the urinary bladder to confirm that they are cancer cells and to distinguish them from normal urothelial cells, which is very difficult to realize by conventional TEM. Additionally, with this type of analysis we were able to study the early stages of carcinogenesis *ex vivo*, when cancer cells start to adhere to urothelial cells and make first contact with them before they start to proliferate. This field of carcinogenesis research is still poorly examined.

We conclude that the FIB milling and the SEM imaging of tissue *ex vivo* can provide evidence on the cellular internalization of nanomaterials used in biomedicine. We demonstrate the utility of the focused ion beam scanning electron microscopy combined with the energy dispersive X-ray spectrometry for 3D morphological and elemental correlative analysis of subcellular features. Although recent advances in super-resolution light microscopy techniques and traditional transmission electron microscopy methods can provide cellular imaging at a wide range of length scales, a simultaneous 3D morphological and elemental imaging of cellular features at nanometre scale can only be achieved with techniques such as focused ion beam scanning electron microscopy with the energy dispersive X-ray spectrometry capability (Scott, 2011).

In the future, the development of FIB/SEM could go in the direction of better exploration of cellular ultrastructural research and toward quantitative elemental analyses.

ACKNOWLEDGEMENT

A part of the work was conducted at the Centre of Excellence in Nanoscience and Nanotechnology, Ljubljana, Slovenia.

REFERENCES

Bruggemann, D. (2013) Nanoporous aluminium oxide membranes as cell interfaces. *Journal of Nanomaterials*, 2013, Article ID460870, 18 pp.

Calatayud, M.P., Riggio, C., Raffa, V., Sanz, B., Torres, T.E., Ibarra, M.R., Hoskins, C., Cuschieri, A., Wang, L., Pinkernelle, J., Keilhofff, G. and Goya, G.F. (2013) Neuronal cells loaded with PEI-coated Fe_3O_4 nanoparticles for magnetically guided nerve regeneration. *Journal of Materials Chemistry B*, 1, 3607–3616.

Cherukuri, P.C.S. (2010) Use of nanoparticles for targeted, noninvasive thermal destruction of malignant cells, in *Cancer Nanotechnology, Methods and Protocols* (eds S.R. Grobmyer and B.M. Moudgil), Humana Press, London, pp. 359–373.

Cho, K.J., Wang, X., Nie, S.M., Chen, Z. and Shin, D.M. (2008) Therapeutic nanoparticles for drug delivery in cancer. *Clinical Cancer Research*, 14, 1310–1316.

Garcia, C. P., Sumbayev, V., Gilliland, D., Yasinska, I.M., Gibbs, B.F., Mehn, D., Calzolai, L. and Rossi, F. (2013) Microscopic analysis of the interaction of gold nanoparticles with cells of the innate immune system, *Scientific Reports*, 3.

Greulich, C., Diendorf, J., Simon, T., Eggeler, G., Epple, M. and Koller, M. (2011) Uptake and intracellular distribution of silver nanoparticles in human mesenchymal stem cells. *Acta Biomaterialia*, 7, 347–354.

Kempen, P.J., Thakor, A.S., Zavaleta, C., Gambhir, S.S. and Sinclair, R. (2013) A scanning transmission electron microscopy approach to analyzing large volumes of tissue to detect nanoparticles. *Microscopy and Microanalysis*, 19, 1290–1297.

Kizilyaprak, C., Daraspe, J. and Humbel, B.M. (2014) Focused ion beam scanning electron microscopy in biology. *Journal of Microscopy*, 254, 109–114.

Kobler, C., Saber, A.T., Jacobsen, N.R., Wallin, H., Vogel, U., Qvortrup, K. and Molhave, K. (2014) FIB-SEM imaging of carbon nanotubes in mouse lung tissue. *Analytical and Bioanalytical Chemistry*, 406, 3863–3873.

Maeda, H. (2001) The enhanced permeability and retention (EPR) effect in tumor vasculature: The key role of tumor-selective macromolecular drug targeting. *Advances in Enzyme Regulation*, 41, 189–207.

Ostrowski, A., Nordmeyer, D., Boreham, A., Holzhausen, C., Mundhenk, L., Graf, C., Meinke, M.C., Vogt, A., Hadam, S., Lademann, J., Ruhl, E., Alexiev, U. and Gruber, A.D. (2015) Overview about the localization of nanoparticles in tissue and cellular context by different imaging techniques. *Beilstein Journal of Nanotechnology*, 6, 263–280.

Pelka, J., Gehrke, H., Esselen, M., Turk, M., Crone, M., Brase, S., Muller, T., Blank, H., Send, W., Zibat, V., Brenner, P., Schneider, R., Gerthsen, D. and Marko, D. (2009) Cellular uptake of platinum nanoparticles in human colon carcinoma cells and their impact on cellular redox systems and DNA integrity. *Chemical Research in Toxicology*, 22, 649–659.

Riggio, C., Calatayud, M.P., Hoskins, C., Pinkernelle, J., Sanz, B., Torres, T.E., Ibarra, M.R., Wang, L.J., Keilhoff, G., Goya, G.F., Raffa, V. and Cuschieri, A. (2012) Poly-l-lysine-coated magnetic nanoparticles as intracellular actuators for neural guidance. *International Journal of Nanomedicine*, 7, 3155–3166.

Scott, K. (2011) 3D elemental and structural analysis of biological specimens using electrons and ions. *Journal of Microscopy*, 242, 86–93.

Tomic, S., Dokic, J., Vasilijic, S., Ogrinc, N., Rudolf, R., Pelicon, P., Vucevic, D., Milosavljevic, P., Jankovic, S., Anzel, I., Rajkovic, J., Rupnik, M.S., Friedrich, B. and Colic, M. (2014) Size-dependent effects of gold nanoparticles uptake on maturation and antitumor functions of human dendritic cells *in vitro*. *PloS One*, 9.

Wierzbicki, R., Kobler, C., Jensen, M.R.B., Lopacinska, J., Schmidt, M.S., Skolimowski, M., Abeille, F., Qvortrup, K. and Molhave, K. (2013) Mapping the complex Morphology of Cell Interactions with Nanowire Substrates Using FIB-SEM. *PloS One*, 8.

26

FIB-SEM Tomography of Biological Samples: Explore the Life in 3D

Caroline Kizilyaprak[1]**, Damien De Bellis**[1]**, Willy Blanchard**[1]**, Jean Daraspe**[1] **and Bruno M. Humbel**[1,2]

[1]*Electron Microscopy Facility, University of Lausanne, Lausanne, Switzerland*
[2]*Imaging Section, Okinawa Institute of Science and Technology, Onna-son, Okinawa, Japan*

26.1 INTRODUCTION

Since the first electron micrograph of cells using transmission electron microscopy in 1945 (Porter *et al.*, 1945), electron microcopy (EM) methods have continually improved our appreciation of complex cellular organization. The ultimate challenge of biological EM is to image life in its most natural shape and investigation in three-dimensions (3D) of biological samples appears essential. 3D information in a transmission electron microscope can be obtained using either serial ultrathin sections (Sjöstrand, 1958; Ware and Loperesti, 1975; Stevens *et al.*, 1980) or tilt-series TEM tomography (Hoppe, 1981; Frank 1992). Using TEM tomography, semi-thin sections from 100 to 500 nm are placed in a high-voltage TEM respectively between 100 and 400 keV and imaged at increasing angles around the center of the sample. Sample volumes can be reconstructed from the resulting image sequences (for reviews see Baumeister *et al.*, 1999; McIntosh *et al.*, 2005; Subramaniam, 2005). TEM tomography is a very powerful technique for obtaining high-resolution structural data of macromolecules, organelles, and small cells but may not be applicable when larger volumes need to be analyzed; this is mainly because the section thickness is limited. For specimens with a greater section thickness, chromatic aberration produces a drastic degradation in resolution, particularly at high tilt angles, where the sample thickness effectively doubles at 60° tilt and almost triples at 70° tilt. To reduce degradation due to chromatic aberration, very high-voltage electron microscopes in the range of 1–3 MeV can be used to image sections

Biological Field Emission Scanning Electron Microscopy, First Edition.
Edited by Roland A. Fleck and Bruno M. Humbel.

up to 3 μm in thickness without dramatic resolution loss (Wilson *et al.*, 1992; Martone *et al.*, 2000). Another method, named scanning transmission electron microscopy (STEM) tomography, offers interesting advantages to generate 3D structural information with nanometer to subnanometer resolution in thick specimens (>500 nm). STEM tomography has attracted attention in the field of material sciences but unfortunately still remains an unfamiliar territory for biologists. The first application of STEM for 3D imaging of very thick (>1 μm) stained sections was reported by Beorchia *et al.* (1993). Since then, few examples of STEM imaging on stained sections have been published. In 2007, Yakushevska and colleagues used STEM tomography in a high angular annular dark field (HAADF) mode and showed that HAADF-STEM tomography is a powerful tool to study stained biological specimens. Compared to TEM tomography, HAADF-STEM tomography yields better contrast and a better signal-to-noise ratio using the same electron dose (Yakushevska *et al.*, 2007). Finally, serial sectioning TEM can be used to obtain 3D data of much thicker volumes than is possible with TEM tomography. Serial sectioning is a mature method and has much contributed to a better understanding of the local 3D organization of biological samples. For example, a whole nematode *Caenorhabditis elegant* has been reconstructed using this method (White *et al.,* 1986). This study is still considered a seminal effort because serial sectioning is a labor-intensive and time-consuming method that requires intensive operator involvement and skills in cutting sections, gathering data, and reconstructing volumes. The number of sections that need to be handled can be reduced by combining tomography with serial sectioning (Sotto *et al.*, 1994), but again the reconstruction of an entire eukaryotic cell can be very challenging (Noske *et al.*, 2008).

Recent advances in instrumentation and specimen preparation resulted in alternative methods to perform serial sectioning that simplify and accelerate the analysis of large volumes. Todays scanning electron microscopes (SEM) achieve ~ 1 nm spot diameters at landing energies low enough that only the top few nanometers of the section surface contribute to form the image (Erdman and Bell, 2012). In addition, the development of the automatic tape-collecting UltraMicrotome (ATUM) enables serially sectioned cells or tissue to be collected automatically from the knife's water trough in the ultramicrotome (Kasthuri *et al.*, 2007). Serial sections are automatically collected on a partially submerged conveyor belt-like conductive tape. The tape with the serially cut ultrathin sections is mounted on silicon wafers in order to be imaged in an SEM equipped with either an "Array Tomography/ATLAS" (Carl Zeiss Microscopy GmbH) or "wafer mapping", which is the original software developed by the Lichtman laboratory. These systems capture serial electron micrographs of the same region of interest in the serial ultrathin sections one by one automatically after manual positioning of the sections (Hayworth *et al.*, 2014). This method, is now commercially available (ATUMtome, Boeckeler Instruments, Tucson, USA) and represents the first step towards automation of SEM serial-sections. This technique, used as an alternative to TEM serial-sections, appears particularly well suited to the analysis of large surface areas. In addition, the sections can be stored, re-imaged multiples times, and with on-section heavy-metal stained improved signal-to-noise ratios (Schalek *et al.*, 2011). Despite multiple advantages present with this method, studies using these approaches are still few, certainly due to a significant and time-consuming technical input.

Over the last 20 years, development of two "destructive" methods has led to a dramatic increase in 3D EM studies: (1) Serial Block-face–SEM (SBF-SEM) (Denk and Horstmann, 2004) and (2) Focused Ion Beam–SEM (FIB-SEM) (Heymann *et al.*, 2006; Gestmann *et al.*, 2004; Young *et al.*, 1993). Once again, the advances in field emission scanning electron microscopy (SEM) to enable low-voltage backscattered electron imaging (Erdman and Bell, 2012) has led to new 3D techniques based on block-face imaging to explore the cell volume. The block-face scanning approach captures electrons backscattered from below the block's surface of a resin embedded sample. Only the top few nanometers of the surface are imaged with sufficient resolution and depth discrimination contributing to form the image. To obtain 3D data, the block-face needs to be repeatedly imaged, with the top slice removed between image acquisitions. One possibility is to remove the top slice with an ultramicrotome in the chamber of a SEM (serial block-face SEM: SBF-SEM). This method was introduced 30 years ago by Leighton, who constructed a microtome for cutting sections inside the microscope chamber (Leighton, 1981). In 2004, Denk and Horstmann showed that with a custom-designed microtome in an SEM, 3D ultrastructural data can be obtained at the resolution and with a volume sufficient to follow local neural circuits (Denk and Horstmann, 2004). SBF-SEM has been driven by neurobiology as an alternative to the tedious serial sectioning and traditional TEM imaging needed to resolve the conflicting requirements of imaging complex neural networks and achieving high resolution (Friedrich *et al.*, 2013; Kasthuri *et al.*, 2007; Tapia *et al.*, 2012). For most biological specimens, the lateral resolution of SBF-SEM can reach $\sim$ 5 nm and z resolution of SBF-SEM is about 25 nm, limited by different factors: the minimal section thickness that can be cut with an oscillating diamond-knife, the hardness of the resin block (depending of the resin used and of the amount of heavy metals in the block), and the width of the block-face (which has to be less than 1 mm to reach 25 nm of slice thickness). In general, the harder the sample, the thinner the section can be (decreasing the lifetime of the diamond-knife). The removal of material in smaller increments is possible with focused ion beam (FIB) milling. The maximum size of specimen that can be examined by SBF-SEM is currently limited by the physical constraints of the ultramicrotome and the biological specimen properties to around 800 μm^2, but the total volume that can be investigated covers a wide range from 1000 μm^3 (Schwartz *et al.*, 2012) to immense volumes exceeding 1 000 000 μm^3 (Briggman *et al.*, 2011; Helmstaedter *et al.*, 2013). The variation in volumes is related to differences in biological samples and in instrumentation. For example, non-brain specimens are known to be particularly sensitive to accumulation of local negative charge, leading to sectioning artifacts and low image quality. To overcome this drawback, Tietze and Denk (2013) have shown that in-chamber coating of SBF-SEM samples successfully eliminated the charging phenomenon, increasing the signal-to-noise ratio. In a recent study, Nguyen and colleagues used conductive resins by adding the carbon black filler, Ketjen black, to resins commonly used for electron microscopic observations of biological specimens. When these conductive resins were used, the charging effect of samples was significantly improved. Interestingly, they showed that carbon black appears as dark granular aggregates, distributed outside the tissues, but was also found in vessels. These results suggest that using conductive resins is a simple option for SEM imaging of biological samples prone to charging (Nguyen *et al.*, 2016).

The judicious combination of focused ion beam (FIB) milling and block-face imaging in the same SEM constitutes one of the more encouraging methods to obtain volume information of a wide variety of biological samples (Knott *et al.*, 2008; De Winter *et al.*, 2009; Heymann *et al.*, 2009; Merchan-Perez *et al.*, 2009; Villinger *et al.*, 2012; Beckwith *et al.*, 2015; Blazquez-Llorca *et al.*, 2015; Bosch *et al.*, 2015; Cretoiu *et al.*, 2015). This method is commonly named FIB-SEM tomography. During FIB-SEM tomography, the FIB generates a beam of focused gallium ions to remove by ion milling a thin layer of a block's surface as thin as 5 nm. Subsequent milling and imaging of the block-face creates an image stack where isotropic voxels can be as small as 5 nm × 5 nm × 5 nm (Wei *et al.*, 2012; Narayan *et al.*, 2014). FIB-SEM currently offers the highest voxel resolution, approaching the resolution of TEM tomography, which makes it the reference volume EM method for high-resolution imaging of a site-specific area. The main benefits of FIB-SEM tomography compared to

SBF-SEM are a significant improvement in axial resolution, maximally ~ 3 nm, and the ability to target a small region of interest without discarding the rest of the block-face (Hekking *et al.*, 2009). The initial development of FIB-SEM instruments was driven by computer chip failure analysis, repair, and circuit modification in semiconductor technology (for an overview see Giannuzzi and Stevie, 2005). The first evidence of successful application of FIB-SEM tomography for biological sample was published in 1993 (Young *et al.*, 1993).

In this chapter, we will concentrate on volume investigation using FIB-SEM tomography. Before discussing volume characteristics and biological results, we would like to address the issue of samples preparation, which is essential for all 3D reconstruction efforts. Biological samples have to be optimized for FIB-SEM imaging conditions in order to increase volume resolution. Therefore, at least three parameters have to be taken into consideration.

1. *For SEM block-face imaging, backscattered detectors are used to collect electrons and generate images.* Contrast in the electron micrographs depends on the accumulation of heavy, electron-dense atoms (heavy metals) on the structures of interest. The different protocols currently available in the literature will be discussed and compared.
2. *Sample milling using FIB can induce surface damages.* During SEM imaging, the area of interest is exposed to a high electron dose. The stability of biological specimens is a main issue for 3D analysis. To ensure the dimensional integrity of the final volume of the cell, it is essential to assess the properties of resins used to embed biological specimens and to determine the resin formulation that has the best stability in the electron beam. The most commonly used resins to perform FIB-SEM tomography will be described.
3. *Using FIB-SEM to perform tomography requires observation of some geometric rules.* The ion column is mounted with a specific angle, depending on the instrument used, to the electron column. The conventional geometry for FIB-SEM tomography is milling normal to the ion beam and imaging at this dictated angle, which can cause problems such as a gradient in brightness from the top to the bottom of the block-face along with a decrease of signal with depth and shadowing from the sidewalls (De Winter *et al.*, 2009). To overcome these problems strategies for milling and imaging have been developed that will be presented in this chapter.

Finally, to highlight the power of the FIB-SEM tomography method, two sets of data from mouse liver samples will be presented. In both volumes mitochondria were segmented in order to appreciate the very complex mitochondrial network at the resolution of the electron microscopy comprising almost a whole liver cell.

26.2 FOCUS ON SAMPLE PREPARATION

Before biological tissue can be imaged using FIB-SEM tomography, the sample must be prepared by fixation, stained and embedded in resin. These steps are determining for the quality of the two-dimensional image acquisition in SEM and subsequent three-dimensional reconstruction of features of interest. Image acquisition time is the main limiting factor for large volumes. Each image is generated as a result of the electron beam scanning the block surface. A balance needs to be found between acquisition time, resolution, and an adequate signal to generate images with the suitable contrast. To enhance the signal-to-noise ratio and to reduce the acquisition time, high contrast is needed. Contrast in electron micrographs depends on the accumulation of heavy, electron dense atoms (heavy metals) on the structures of interest. In addition, a high metal ion concentration can reduce the charging effect during electron image acquisition. New ways to prepare samples in order to obtain higher contrast are explored and many labs are working on ways to improve tissue staining.

26.2.1 Chemical Fixation

The most common FIB-SEM approaches are using variations in the conventional chemical fixation protocols that include an aldehyde primary fixation in combination with osmium tetroxide and uranyl acetate *en bloc* staining or include thiocarbohydrazide-osmium tetroxide treatment (Tanaka and Mitsushima, 1984), modified osmium with potassium ferricyanide (De Bruijn, 1973), and/or tannic acid (Jimenez *et al.*, 2009) for enhanced contrast (Table 26.1).

Imaging of resin-embedded biological tissue in the FIB-SEM indicated that standard TEM embedding protocols using glutaraldehyde primary fixation followed by osmium tetroxide post-fixation and *en bloc* staining with uranyl acetate produced suitable contrast for a good backscattered electron (BSE) signal and they are currently largely accepted by the community (see Table 26.1(3) later). However, in certain cases, following OsO_4 post-fixation some electron dense granules of variable sizes can occur in the resin block (Hendriks and Eestermans, 1982) (see Figure 26.1). The exact cause of these precipitates is not known and the presence of reduced osmium in the specimen seems to be a prerequisite to its prevention

Table 26.1 Variation in chemical fixation protocols described in the literature to perform FIB-SEM tomography of biological samples

1. Reduced osmium with potassium ferro- or ferri-cyanide	**2. Extended osmification with osmium-thiocarbohydrazide**	**3. Osmium tetroxide**
Murphy *et al.* (2011)	Murphy *et al.* (2011)	Leser *et al.* (2009)
Hekking *et al.* (2009)	Leser *et al.* (2010)	Heymann *et al.* (2006)
Heymann *et al.* (2009)	Leser *et al.* (2009)	Murphy *et al.* (2010)
Cretoiu *et al.* (2015)	Felts *et al.* (2010)	
Medeiros *et al.* (2012)		
De Winter *et al.* (2009)		
Paredes-Santos *et al.* (2012)		
+ tannic acid		**+ uranyl acetate**
Bushby *et al.* (2011)		Murphy *et al.* (2011)
Armer *et al.* (2009)		Leser *et al.* (2010)
Jimenez *et al.*, (2010)		Leser *et al.* (2009)
		Merchan-Perez *et al.* (2009)
		Bosch et al. (2015)
		Blazquez-Llorca *et al.* (2015)
		Narayan *et al.* (2014)
		Hayworth *et al.* (2015)
		Steinmann *et al.* (2013)
+ uranyl acetate		**+ tannic acid + uranyl acetate**
Beckwith *et al.* (2015)		Wierzbicki *et al.* (2013)
Hayworth *et al.* (2015)		
+ OsO_4 + uranyl acetate		
Knott *et al.* (2008)		
Knott *et al.* (2011)		

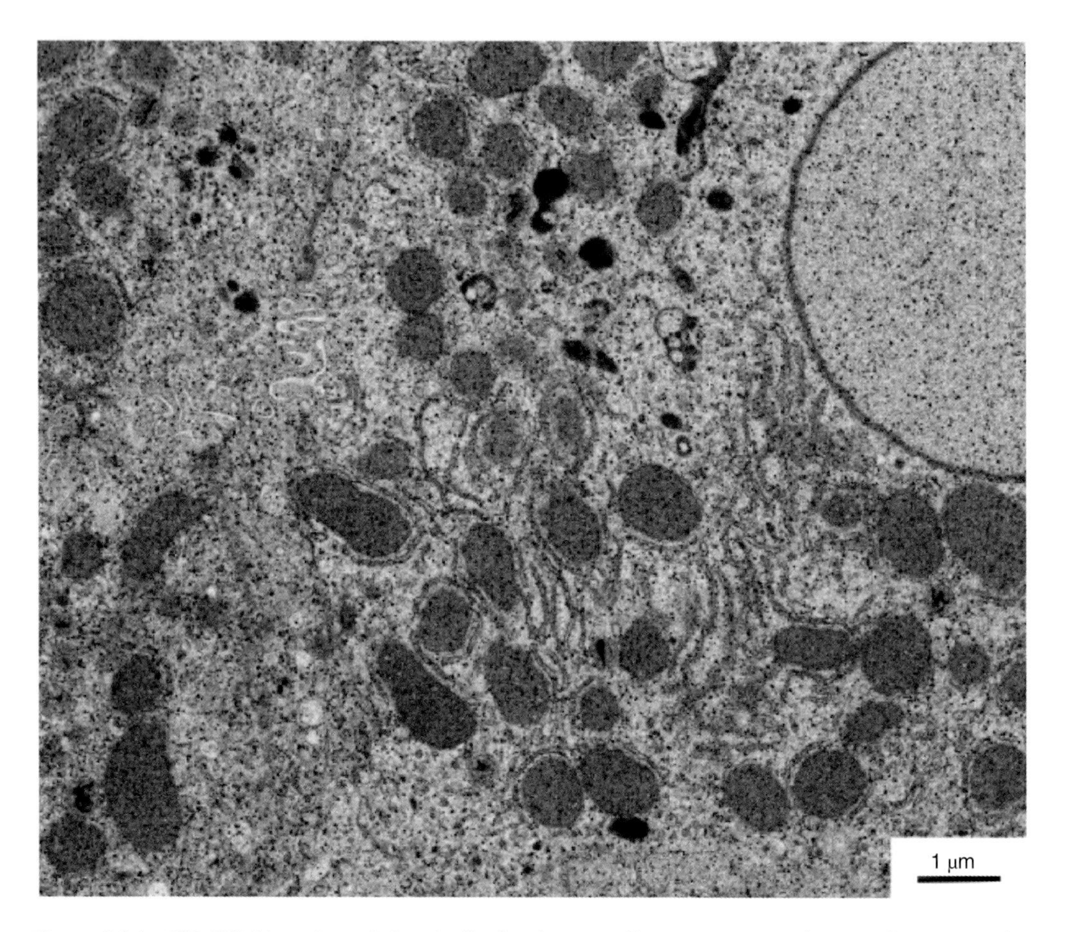

Figure 26.1 FIB-SEM imaging of chemically fixed mouse liver containing electron dense granules. Samples were fixed using 2.5% glutaraldehyde followed by post-fixation in 2% OsO_4 dehydrated in acetone and embedded in Hard Plus resin. The milled cross-section was imaged at a low acceleration voltage (2 keV) in backscatter electron mode using the through-the-lens detector.

(Hayat, 1981). Reduced osmium using potassium ferro- or ferri-cyanide (Table 26.1(1)) is also one of the methods of choice to enhance *en-bloc* staining of a biological specimen. Osmium tetroxide in combination with ferrocyanide has also been used for enhanced staining of a wide variety of cellular components. De Bruijn, in 1968, first demonstrated that the electron density of glycogen can be enhanced selectively by adding 0.05 M $K_3Fe^{3+}(CN)_6$ to 1% OsO_4 during post-fixation (De Bruijn, 1968; De Bruijn, 1973). The electron density of lipid droplets and membrane is also increased.

Osmium staining can be dramatically increased by treatment with thiocarbohydrazide and additional osmification (OTO method). The extended osmification enhances specially the contrast of the lipid components of the cells. The OTO staining was originally developed by Seligman *et al.* in 1966 and is currently appreciated for block-face imaging to enhance contrast and reduce charging (Seligman *et al.*, 1966). However, extensive use of osmium can have a destructive effect on cell morphology (Behrman, 1983; Maupin-Szamier and Pollard, 1978.). With the OTO method, the increased contrast is confined to structures that were osmophilic to begin with, and no new non-stained components of the tissue will become visible. In addition, Hayworth *et al.* report that samples prepared with extended osmification tend to crack during sectioning, presumably because the increased density of heavy metals makes them brittle (Hayworth *et al.*, 2015).

Primary glutaraldehyde fixation seems to be an essential prerequisite for sample preparation in order to perform FIB-SEM investigations. However, there are some indications that sequential double fixation using glutaraldehyde followed by OsO_4 has undesirable effects on certain specimens, especially on isolated cells. One possible explanation for these effects is that during pre-fixation the specimen is subjected to the detrimental effects of glutaraldehyde, such as shrinkage and lipid extraction before application of osmium (Hayat, 1981). To prevent post-fixation relocation of membrane and lipids, which have considerable mobility even after aldehyde fixation, the use of a glutaraldehyde plus osmium mixture as the primary fixation is recommended (Schwarz, 1973). Fixation with the mixture is known to result in sharp membrane definition and especially good preservation of nucleoprotein-, lipid-, and polysaccharide-containing structures. However, glycogen is defined poorly. We decided to take the advantages of the mixture fixation and to improve the glycogen staining by adding potassium ferricyanide to the mixture. For numerous varieties of samples (animal tissue as liver, muscle, retina; plant tissue; and for cell-culture cells), we recommend using a mixture containing 2.5% glutaraldehyde and 2% OsO_4 modified with 1.5% $K_3Fe^{3+}(CN)_6$ in phosphate buffer for 2 hours at room temperature. To reduce charging effects during FIB-SEM tomography, chemical fixation was followed by *en bloc* staining with 2% aqueous uranyl acetate for 1 hour. Under these conditions we obtained nicely preserved and well-stained chemically fixed liver samples for SEM observation (Figure 26.2A). However, from time to time, the penetration of the mixture is not sufficient and a gradient of contrast appears through the block of tissue (this effect has never been observed in isolated cells). In this case, only the outer layers of cells in the tissue block are well fixed and well stained. If a gradient occurs it is recommended to use only the peripheral layers of these samples for electron microscopy. Nevertheless, this method is useful for enhancing the contrast of various subcellular components such as endoplasmic reticulum, mitochondria, glycogen granules, Golgi apparatus, and peroxisome with urate oxidase crystals (see Figure 26.2A respectively for sidebars 1, 2, 3, 4, and 5).

At present, the large majority of FIB-SEM investigations are based on chemical fixation with large variations of the fixation protocols. Currently, image stacks with a voxel size of 5 nm can be reasonably well achieved using FIB-SEM tomography. At this resolution care has to be taken that the resolution of the sample preparation, that is, the quality of the preservation of the ultrastructure, does not fall behind. Therefore, further efforts have to be made to prepare samples as close as possible to the native state with a protocol that is also compatible for FIB-SEM investigations. Cryo-immobilization of samples using high-pressure freezing (HPF) following by freeze-substitution (FS) is the method of choice to investigate subcellular architecture in the absence of artifacts that may be introduced by conventional chemical fixation (Humbel and Schwarz, 1989).

26.2.2 High-Pressure Freezing and Freeze-Substitution

Cryo-microscopy using cryo-immobilized samples offers, in principle, the unique opportunities to investigate subcellular architecture in the absence of artifacts that are introduced by chemical fixatives and heavy metal stains. Though cryo-FIB-SEM tomography can be done (Schertel *et al.*, 2013) for exploring subcellular architecture at high resolution in the range of nanometers, there is still the need of using resin-embedded samples and room temperature FIB-SEM tomography. Freeze-substitution and resin embedding, after cryo-fixation, constitutes a very interesting way to combine best ultrastructural preservation, good contrast, and

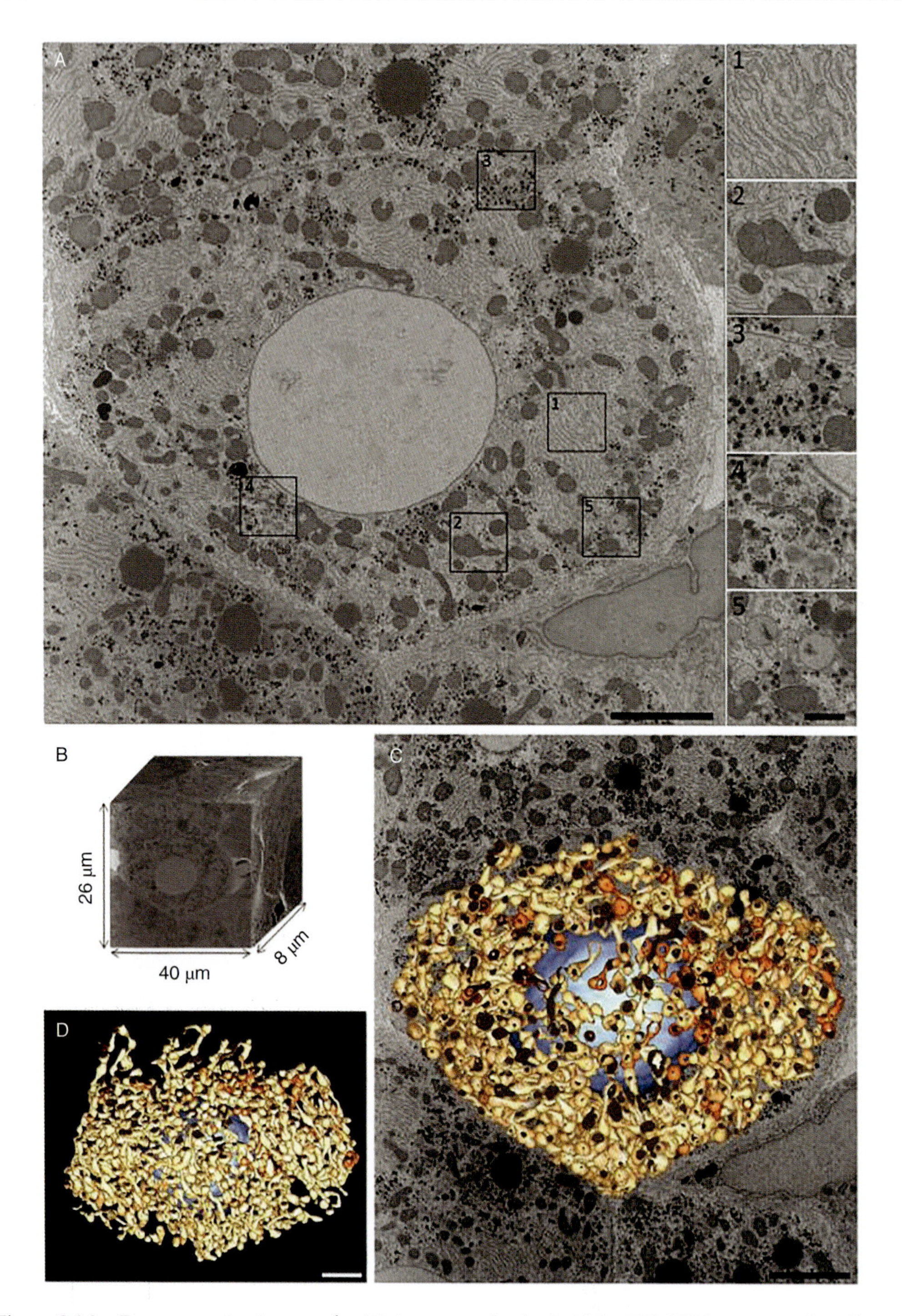

Figure 26.2 Representative image of a 3D image stack obtained by FIB-SEM tomography. Chemically fixed liver is imaged in an FIB-SEM at an accelerating voltage of 2 keV using a backscattered electron with the through-the-lens detector. (A) Electron micrograph of a liver cell from mouse, chemically fixed with a mixture of 2.5% glutaraldehyde and 2% reduced OsO_4 with 1.5% $K_3Fe^{3+}(CN)_6$. Using this sample preparation the endoplasmic reticulum (1), the mitochondria (2), the glycogen (3), the Golgi apparatus (4), and peroxisomes are nicely contrasted. (B) The same sample is used to perform FIB-SEM tomography, where a stack of 800 images with a field of view of 40 μm × 26 μm is generated with a slice thickness of 10 nm. (C) In the stack of images the mitochondria are segmented in gold and the nucleus is segmented in blue. The segmentation is done manually using IMOD (Kremer *et al.*, 1996) (University of Colorado, Boulder, Colorado, USA; http://bio3d.colorado.edu/imod). (D) Overview of the complex mitochondria network. The stack has an isotropic voxel size of 10 nm.

Table 26.2 Variation in freeze-substitution protocols described in the literature for FIB-SEM tomography on biological samples cryo-fixed by HPF

1. OsO_4 + uranyl acetate + glutaraldehyde	Murphy *et al.* (2011)
2. OsO_4 + uranyl acetate + H_2O	Villinger *et al.* (2012) Wei *et al.* (2012) Hayworth *et al.* (2015)
3. Chemical pre-fixation following by HPF/FS in OsO_4 + uranyl acetate	Remis *et al.* (2014)

good conductivity of the sample. Actually, during the FS process a variety of chemical agents and heavy metals can be added to the organic solvent in order to increase the signal-to-noise ratio and decrease the charging effects during FIB-SEM investigation.

Up to now, only a few studies used HPF and FS to prepare samples for FIB-SEM tomography (Table 26.2) and few groups used cryo-methods as their standard preparation to investigate biological samples.

Cryo-fixation and FS are considered as the gold standards for electron microscopic preparation of whole cells and tissue. In certain cases, this is the only way to visualize delicate structures that are difficult or impossible to visualize using conventional chemical fixation. One particular delicate structure is that of bacterial biofilms. Using HPF and FS, Remis *et al.* discovered a previously unknown extension of cell-to-cell interactions within a biofilm of *M. xanthus*. It is obvious that such small membrane appendages are notoriously difficult to preserve faithfully with conventional methods (Remis *et al.*, 2014).

In contrast to their expectation, in their study Murphy and colleagues testing 24 different preparation protocols, including HPF and FS methods (Murphy *et al.*, 2011), did not find substantial differences in the preservation of the cellular structures and contrast in their sample of T cells. Because no improvements using cryo-fixation protocols were observed they continued with chemically fixed cells for the study.

We used mouse liver to compare chemically fixed and HPF/FS prepared samples (see Figure 26.3). Preliminary comparisons of the images allowed us to visualize some interesting

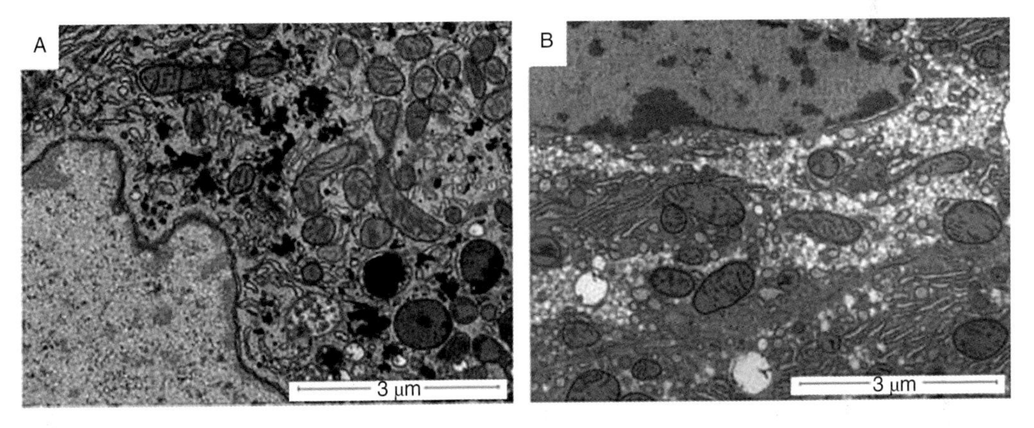

Figure 26.3 FIB SEM imaging of mouse liver at 2 keV using the backscattered electron mode with the in-column detector. (A) The sample was chemically fixed in a mixture of 2.5% glutaraldehyde and 2% OsO_4 reduced with 1.5% $K_3Fe^{3+}(CN)_6$, dehydrated in acetone, and embedded in Hard Plus resin. (B) Corresponding sample prepared using the high-pressure freezing and freeze-substitution method. The cryo-immobilized liver samples were freeze-substituted in acetone containing 2% OsO_4 and 2.5% H_2O and then embedded in Hard Plus resin.

differences in terms of sample contrast and ultrastructural preservation. However, further efforts need to be made to better understand contrast formation in SEM, especially when diverse combinations of heavy metals and chemicals are used in the substitution medium. For TEM investigations, FS is usually done in acetone or methanol containing OsO_4, uranyl acetate with or without water, and glutaraldehyde (Table 26.2). However, we do not know how these chemicals influence the BSE signal. Interestingly, comparison of TEM and BSE-SEM images of cryo-fixed, freeze-substituted *S. cerevisiease* showed very similar images except for the cell walls, which had a more pronounced BSE contrast (Wei *et al.*, 2012). This result suggests that the contrast in TEM and SEM may be different and therefore a systematic study of the role of the different ingredients in the substitution medium, fixatives, and contrasting agents on the contribution to the final image contrast is needed.

For the studies described in this chapter we cryo-fixed mouse liver with Compact 2 (Engineering Office M. Wohlwend GmbH, Sennwald, Switzerland) and freeze-substituted at −90 °C for 24 h in anhydrous acetone supplemented with 2% OsO_4 and 2.5% H_2O or 3% glutaraldehyde. Samples were then warmed (5 °C/h) to −60 °C in an automated freeze-substitution device (AFS2, Leica Microsystems, Vienna, Austria). After 8 h the temperature was raised to −30 °C (5 °C/h) and the samples were kept at this temperature for 8 h before rinsing 3 times in acetone for 20 min. The samples were then infiltrated with gradually increasing concentrations of Hard plus Resin (EMS, Hatfield, PA, USA) in anhydrous acetone (1:3, 1:2, 2:3 volume ratio and finally pure resin) for 2.5 h each step while raising the temperature from –30 °C to 25 °C (11 °C/h).

26.2.3 Resin Embedding

During FIB-SEM tomography, the sample is exposed to a high electron and ion dose. While the ions are only used to remove a thin layer of the resin, the effects of the electron dose on the sample are of greater importance. The electron irradiation effects are known to reduce the intrinsic resolution of the samples. In addition, they can locally deform the structure and show charging effects that deteriorate the final image in such a way that the final 3-D reconstruction becomes challenging with additional complications for automated segmentation.

In the literature, four different resin formulations are described for FIB-SEM investigation. These resins are commonly used to perform electron microscopy (see Table 26.3). For FIB-SEM tomography, the influence of the Ga ion beam on the resin may be of less importance but the resin qualities like density can have a strong influence on the milling speed and the evenness of the polished surface. Crevasses, called curtaining, are the most prominent artifacts of the milling process and are a result of large density differences between cellular components. Further, our knowledge and calculations for the ion impact are mainly restricted to silicon.

Therefore we decided to study the Ga ion beam interaction on five different resin formulations. For a comprehensive information on epoxy and methacrylate resins (see Lee and Neville, 1967 and Newman and Hobot, 2012):

1. Hard-Plus resin-812 (EMS # 14115) composed of 50 g of Hard-Plus resin-812, 50 g of hardener Hard-Plus, and 2.5 ml of accelerator. All the components are available ready to use in the Hard-Plus resin-812 kit.

Table 26.3 List of resin formulations used to perform FIB-SEM tomography on biological samples

Araldite or Durcupan resin (Durcupan is a trademark of Sigma Aldrich GmbH)	Knott *et al.* (2008); Armer *et al.* (2009); Merchan-Perez *et al.* (2009); Bushby *et al.* (2011); Knott *et al.* (2011); Steinmann *et al.* (2013); Blazquez-Llorca *et al.* (2015); Bosch, *et al.* (2015); Hayworth *et al.* (2015)
Epon resin or replacement of Epon 812: **- Embed 812 (EMS)** **- Eponate 12 (Shell Chemical)** **- Poly/Bed 812 (Polysciences Inc)** **- Agar 100 resin kit (Agar scientific)**	Heymann *et al.* (2009); Leser *et al.* (2009); Felts *et al.* (2010); Leser *et al.* (2010); Murphy *et al.* (2010); Murphy *et al.* (2011); Medeiros *et al.* (2012); Paredes-Santos *et al.* (2012); Wei *et al.* (2012); Hayworth *et al.* (2015); Wanner *et al.*, (2013)
Araldite + Epon resin	Remis *et al.* (2014)
Methacrylate HM20	Schneider *et al.* (2011)

2. Durcupan ≪modified≫ (ACM, Sigma # 44610) (used without dibutyl phthalate) composed of 49% v/v Durcupan A/M resin, 49% v/v 964 hardener, and 2% v/v 964 accelerator.
3. Durcupan/Epon resin (Matsko and Mueller, 2004) is composed of 49% w/w Durcupan/Epon stock solution, 49% w/w dodecenyl succinic anhydride (DDSA), and 2% w/w DMP-30. The Durcupan/Epon stock solution consisted of 43% w/w epoxy embedding medium (Sigma # 45345) and 57% w/w Durcupan A/M resin (Sigma # 44611).
4. Epon ≪modified≫ (Sigma # 45359) (used without dodecenyl succinic anhydride (DDSA)) composed of 52% v/v epoxy embedding medium, 46% v/v methyl nadic anhydride (MNA), and 2% v/v 2,4,6-tri(dimethylaminomethyl) phenol (DMP-30).
5. HM20 (EMS # 14340) composed of 82.5% w/w monomer E, 17% w/w cross-linker D, and 0.5% w/w initiator C.

In each resin block the same rectangular pattern (2 μm × 4 μm) was milled using three different currents, 40 pA, 80 pA, and 230 pA, at a constant ion dose per pA (=16.25 pC/μm^2 pA). The volume of the milled pattern was calculated for each resin formulation and, using the volume of resin removed and the ion dose, the volume per dose was calculated to calibrate the specific milling rate for each resin formulation (Table 26.4) (Kizilyaprak *et al.*, (2015)).

Generally, the milling and patterning protocols of FIB instruments use the specific milling rate of silicon. Therefore, for biological applications a calibration of the specific milling rate for the different resins is highly beneficial. The Hard Plus resin was the most resistant, with

Table 26.4 The calculated volume per dose for each resin formulation.

Resin	Volume per dose (μm^3/nC)
Hard Plus resin	0.37
Epon “modified”	0.88
Durcupan “modified”	0.51
Epon/Durcupan	0.54
HM20	0.73

the smallest removed volume per dose (0.37 μm^3/nC). The Epon "modified" was the least resistant, with a volume per dose more than two times higher than the volume per dose of the Hard Plus resin.

To confirm the calibration, the same blocks were milled using the specific milling rates measured by our experiments. The results confirmed the correctness of the calibration (Figure 26.4). During the milling tests another interesting observation could be made.

Figure 26.4 Milling of resin block using a specific volume per dose calibration calculated for each resin formulation (adapted from Kizilyaprak *et al.*, (2015)). A defined rectangular pattern was milled: $x = 2$ µm, $y = 4$ µm, and $z = 0.2$ µm, 0.4 µm, 0.8 µm, and 1.6 µm using the calibrated milling rates. Images of the corresponding milling patterns in the hard plus resin (A), in the Epon "modified" (B), in the Durcupan "modified" (C), in the Epon/Durcupan (D), and in the Lowicryl HM20 (E) resin. Note the holes in the base of Epon "modified" and Lowicryl HM20.

The base of the milled cuboids of the most vulnerable resins, Epon ≪modified≫ and HM20, had holes, looking like "Emmental" cheese. The other resins analyzed, Hard Plus, Durcupan ≪modified≫, and Epon/Durcupan mixture had a clean, undamaged base. The holes might be created by a melting effect induced by the Ga ion irradiation. In the FIB machine, the samples are irradiated with positively charged Ga ions produced by a liquid metal ion source accelerated to the energy of 30 keV. The ion collisions initiating sputter removal can cause severe damage to the remaining bulk of material (Rubanov and Munroe, 2004; Drobne et al., 2007). To minimize the damage created during FIB milling, the sample can be protected with a platinum layer, reducing the sample heating caused by FIB irradiation. In our study, we found that a thick carbon layer (0.7–1 μm) was more efficient to reduce the ion beam induced curtaining effect caused by incident ions during FIB-SEM tomography. Most likely the carbon deposition is more homogeneous than the platinum deposition (Kizilyaprak *et al.*, (2015)).

26.3 FOCUS ON THE GEOMETRY OF THE SAMPLE IN RELATION TO THE ION AND ELECTRON BEAM

To be investigated in 3D, the sample has to be mounted in the dual beam microscope. The two beams are built into the instrument with a specific angle depending on the manufacturer and coincide at their focal points. To perform FIB-SEM tomography, the geometry of the samples has to be optimized regarding the geometry of the instrument used.

26.3.1 Geometry of the Instrument

There are five manufacturers for FIB-SEM instruments. Depending on the manufacturer, the angle between the ion and the electron column is different, for example, 52° for FEI, 54° Zeiss and JEOL, 55° for Tescan, and 90° for Hitachi (Figure 26.5). Hence, when normal to the ion beam the freshly milled block-face is imaged at this specific angle (respectively 52°,

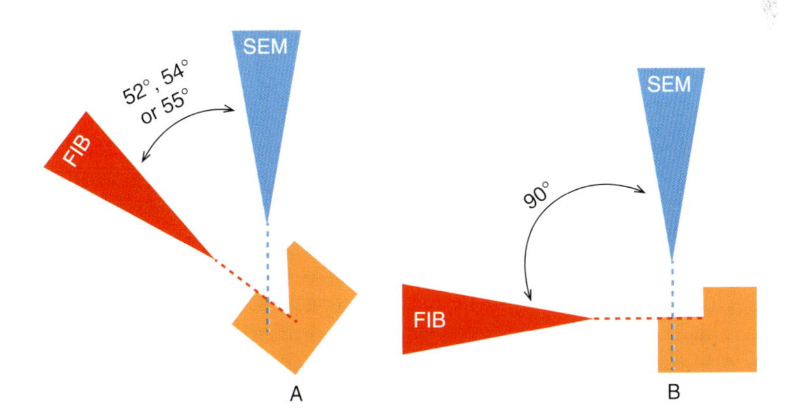

Figure 26.5 Geometry of the FIB-SEM instruments available on the market. (A) The ion beam is inclined with a specific angle to the electron beam. The angle is 54° (Carl Zeiss Microscopy GmbH, Oberkochen, Germany, and JEOL Ltd, Tokyo, Japan), 52° (FEI Company, Eindhoven, The Netherlands), or 50° (Tescan, Brno-Kohoutovice, Česká Republika). (B) The ion beam is normal to the electron beam (NX9000; Hitachi Technologies Corporation, Naka, Japan).

54°, or 55°) to the electron beam, shorting the images compared to the real image by the cosine of the inclination angle. This can easily be corrected by stretching the *y* axis with a factor, the cosine of the angle. To avoid this stretching, Hitachi decided to mount the ion column at 90° to the electron beam. Previously, Hitachi has developed FIB-SEM instruments dedicated for TEM lamella preparation and patterning with an ion beam vertically mounted and an electron beam inclined with a specific angle of 54° and 58° respectively for the NB500 and NX2000 instruments. This geometry is not optimized for FIB-SEM tomography. With the NX9000 the ion beam is mounted normal to the electron beam and this geometry is particularly suitable for biological samples. The freshly milled block-face can be imaged directly normal to the electron beam.

We use a FIB-SEM Helios NanoLab 650 (FEI, Eindhoven, The Netherlands) with a build-in ion beam column inclined at 52° to the electron beam. In order to improve the image quality (contrast and resolution) the geometry of the sample is optimized.

26.3.2 Geometry of the Resin Block

Traditionally, the sample to be fibbed is tilted so that the ion beam is normal to the surface. Cutting and imaging is done at this inclination during the whole collection of data. This geometry can impede electrons of deeper regions of the fosse created by the ion beam to reach the detectors, resulting in a brightness gradient from the top of the section surface to the bottom. De Winter *et al.* showed that this effect can be avoided when the sample block is kept at a 0° tilt angle and the ion beam is cutting oblique to the block surface (De Winter *et al.*, 2009). This geometry, however, may have some implications on the cutting process. The shadowing effect can also be avoided by trimming the resin block to expose the complete surface of the volume of interest (Knott *et al.*, 2008). In both cases, the sample has to be kept at the working distances where the ion beam and the electron beam coincide, in our instrument at 4 mm, and the images are shortened due to the inclination angle (see above). To increase the resolution of the images it would be advantageous to get closer to the pole piece during imaging and to image the surface normal to the electron beam.

To have optimum geometry for the milling and for imaging, we use a strategy adapted from Dr Jiao (FEI, personal communication) (Table 26.5: 4): milling at 90° to the block surface and imaging normal to the imaging face. The resin embedded sample is glued on a stub with an electrically conductive resin in such way that the osmicated sample overhangs the stub. The stub is installed in the ultramicrotome and the block-face (which will be normal to the ion beam: milling face) is trimmed. The sample is rotated by 90° in order to trim the block-face that is normal to the electron beam (imaging face). The two block-faces are perfectly smoothened with the region of interest at the edge of the resin block. Then the stub is mounted on the FIB-SEM stage using a pre-tilted 45° aluminum SEM mount. To remove a layer from the imaging face, the sample is tilted at 7° where the milling face is normal to the ion beam. The block is tilted to 45° to be imaged normal to the electron beam (Kizilyaprak *et al.*, (2015)). This strategy results in higher image resolution, as milling and imaging can be done under optimum conditions. For milling, we use an ion beam with low currents in the range of 800 pA at an acceleration voltage of 30 keV at a working distance of 4 mm. Imaging is done at 1–2 mm working distance with a 2 keV accceleration voltage, also with low currents in the range of 800 pA, using the immersion mode of the lens and collecting the BSEs with the through-the-lens and the in-column detectors.

Table 26.5 List of sample geometry used for FIB-SEM investigations at a specific angle of 52°

Sample geometry	References
 1. *Adapted from Villinger et al. (2012):* 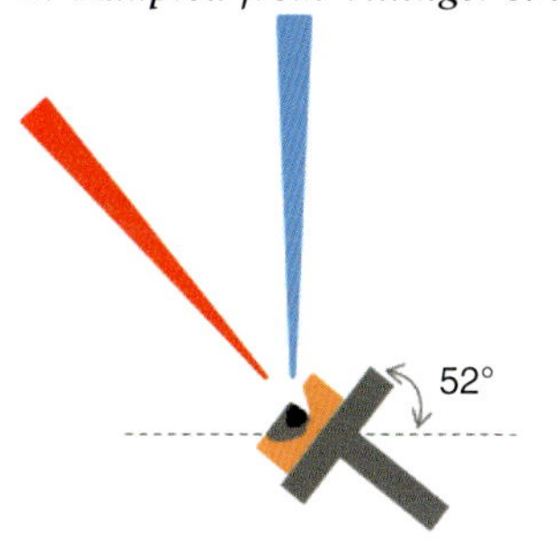	Villinger *et al.* (2012); Heymann *et al.* (2006, 2009); Bushby *et al.* (2011); Earl *et al.* (2010)
2. *Adapted from Beckwith et al. (2015):* 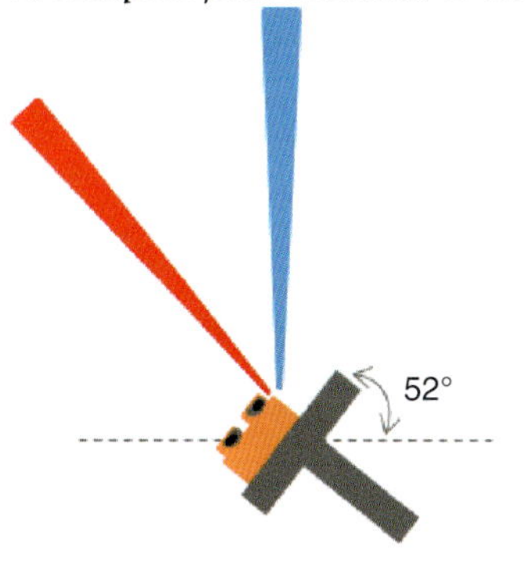	Beckwith *et al.* (2015)
3. *Adapted from De Winter et al. (2009):* 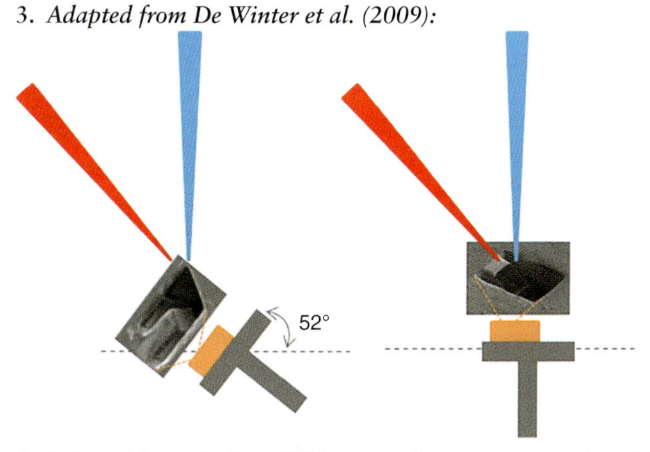	De Winter *et al.* (2009); Hekking *et al.* (2009); Medeiros *et al.* (2012); Paredes-Santos *et al.* (2012); Wierzbicki *et al.* (2013); Jimenez *et al.* (2010); Scott, (2011)
4. *Adapted from Dr. Jiao (FEI, personal communication) and Kizilyaprak et al. (2015):* 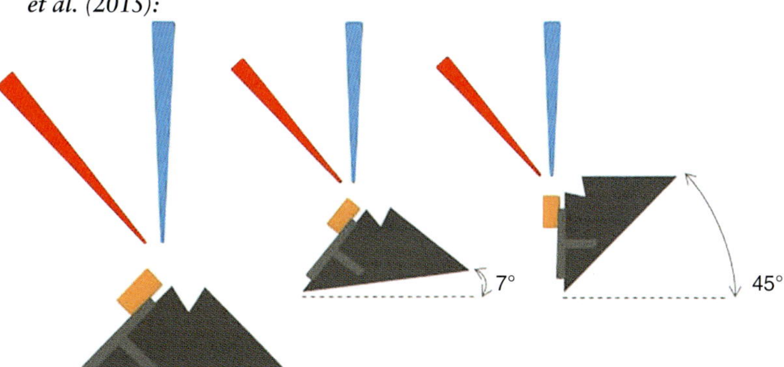	Kizilyaprak *et al.* (2015)

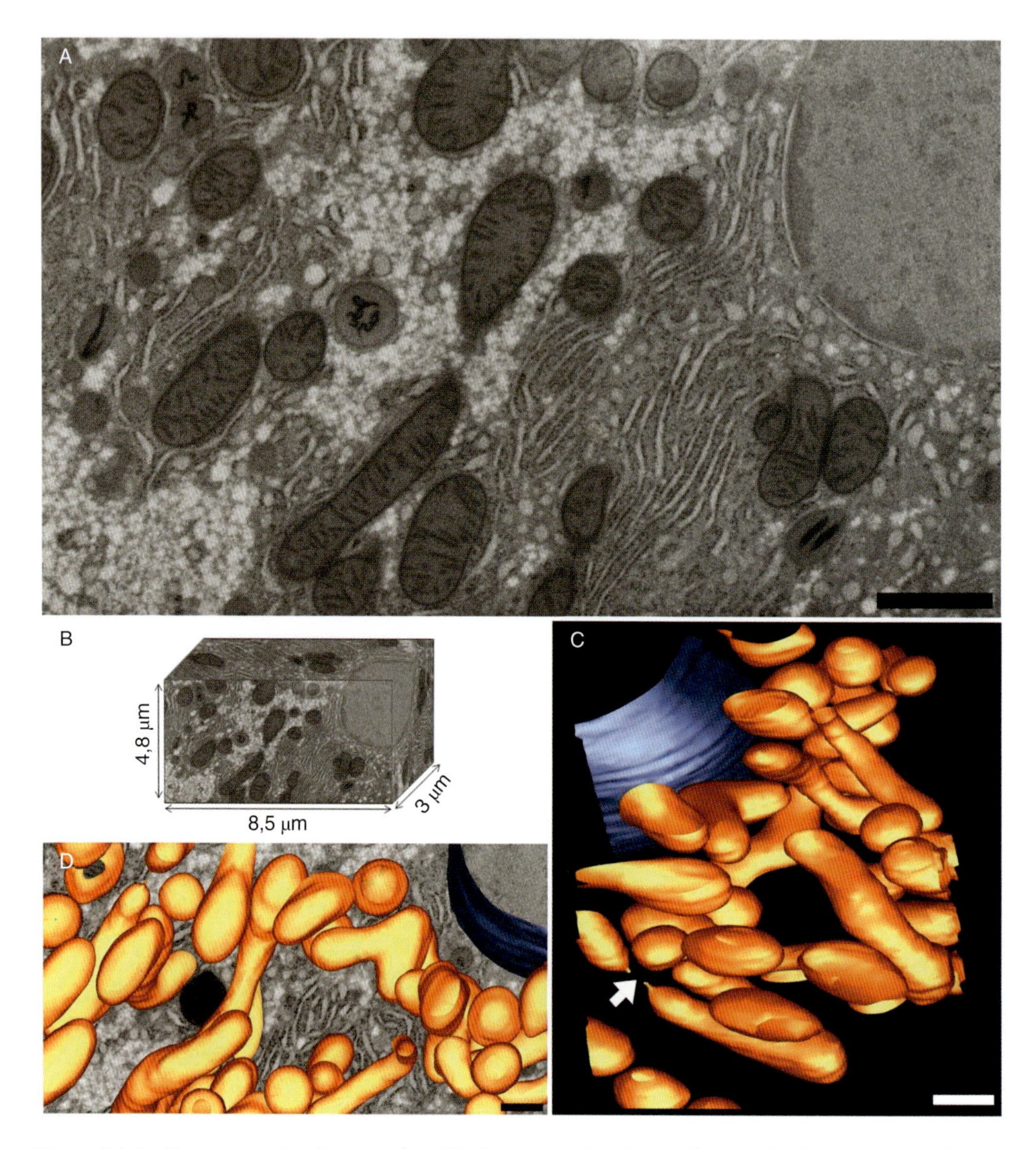

Figure 26.6 Representative image of a 3D image stack obtained by FIB-SEM tomography. (A) High-pressure cryo-immobilized and freeze-substituted liver sample is imaged at an accelerating voltage of 2 keV using back-scattered electron in the in-column detector. Scanning electron micrograph of a subregion in a hepatic cell from mouse liver after HPF and FS in acetone containing 2% OsO_4 and 3% glutaraldehyde, embedded in Hard Plus Epon following the protocol of substitution described in paragraph: "High pressure freezing and freeze substitution". Note that with this condition of preparation the glycogen granules appear white, the mitochondria are nicely preserved and contrasted, and fine structures as ribosomes are well stained, and appear as black dots along the endoplasmic reticulum. (B) The same sample is used to perform FIB-SEM tomography, a stack of 300 images with a field of view of 8.5 μm × 4.8 μm is generated with a slice thickness of 10 nm. (C) In the stack of images the mitochondria are segmented in gold and the nucleus is segmented in blue. Note that with these conditions of sample preparation, some rare events can be seen as the fission or fusion process between two distinct mitochondria (with arrow). The segmentation is done manually using IMOD. (D) Overview of the complex mitochondria network. The stack has a pixel size of 2.9 nm.

26.4 BEAUTY OF THE FIB-SEM INVESTIGATION

26.4.1 Example of the Mitochondrial Network in Mouse Liver

In TEM images mitochondria seem to be isolated organelles, resembling the size and shape of those of *E. coli* (Figures 26.2A and 26.6A). This is also the way they are sketched in textbooks. Looking, however, in three dimensions, it becomes obvious that mitochondria form an extensive and very active network. They are continuously moving, dividing, and fusing. They exist in many different shapes including small balls and long branching rods and the shapes, size, and number are constantly changing. FIB-SEM tomography is a very powerful method to investigate the mitochondria network in 3D. This approach allows easy study of mitochondrial interactions with, for example, the ER or lipid droplets, and quantifying gross changes of the mitochondrial network caused by diseases, for example, non-alcoholic steatohepatitis (Jayakumar *et al.*, 2011).

In addition, FIB-SEM tomography allows investigation of alteration in mitochondrial morphology, which are not only indicative of the organelles function but in many cases regulate the ability of mitochondria to carry out their numerous roles. In fact, mitochondria are known to generate ATP but they are central to several vital cellular functions. Mitochondria are implicated in programmed cell death, biosynthesis of haem complexes, calcium signaling, oxidation of fatty acids, and signal transduction in the innate immune response (Scott and Youle, 2010). Many of these roles are related to the ability of mitochondria to undergo the highly coordinated processes of fission (division of a single organelle into two or more independent structures) or fusion (the opposing reaction). These actions occur simultaneously and continuously in many cell types and the balance between them regulates the overall morphology of mitochondria within a given cell. Here, we demonstrate our preliminary results of a mitochondrial network in a liver cell.

FIB-SEM tomography was performed on chemically fixed liver samples (Figure 26.2). The total volume analyzed was 40 µm × 26 µm × 8 µm with a voxel size of 10 nm. In the final volume, mitochondria were segmented manually and the complex network has been depicted. Using these conditions of sample preparation and image resolution, in single images we can recognize the individual cigar-shaped morphology of mitochondria as they are expected from the thin section view in TEM, but in 3D a highly complex mitochondria network throughout the cells becomes evident. Knowing the artifacts of chemical fixation we decided to use cryo-fixation and freeze-substitution. Indeed, the overall morphology of the liver sample was thus much better preserved. Remarkably, we even found a mitochondrion that is either in the process of fusion or fission (Figure 26.6C). As the long extension looked more like it had ruptured we therefore prefer to think that the mitochondria were caught shortly after division. This result clearly shows that cellular processes can be trapped in a time lapse manner and new, unexpected biological dynamics can be studied at the electron microscopy resolution.

26.5 CONCLUSION

Elucidating the three-dimensional (3D) spatial distribution of organelles within cells and tissues is essential for investigating numerous cellular processes. Tomography in the transmission electron microscope (TEM tomography) is the method of choice for 3D imaging of cellular structures down to 3 nm resolution (Baumeister *et al.*, 1999). However, TEM tomography is typically limited to 500 nm thick sections, making the reconstruction of an entire eukaryotic cell technically very challenging (Noske *et al.*, 2008). There is a need for a technology that can be used for rapid 3D imaging of large mammalian cells to provide information at nanometer resolution. In the last two decades, block-face SEM imaging of resin embedded tissues using serial sectioning of the resin block directly inside the SEM chamber was developed and improved. Using a focused ion beam (FIB) or a build-in ultramicrotome (serial block-face, SBF-SEM) scientists were able to collect large volumes of 3D EM information at resolution that could address many important biological questions.

In this chapter, we presented some examples of studies using FIB-SEM tomography. Typically, any stained and embedded resin samples prepared for TEM examination can be used for FIB-SEM tomography. However, consideration has to be given to artifacts and surface damages induced by FIB milling and imaging. Ion beam artifacts originate mainly from curtaining (non-planar milling of the surface) and redeposition of material sputtered from the sample surface (Drobne *et al.*, 2007). Different protocols of fixation widely used for FIB-SEM tomography were compared in order to find the best compromise to improve the signal-to-noise ratio, preserve the ultrastructure, and reduce charging effects of biological samples. The different resin formulations and the resin block geometry strategies were explored. The milling rate was measured for each resin formulation and the damages caused by the ion impact on the resin were analyzed. The geometry of the sample was optimized to improve the imaging conditions using detection of the backscattered electrons. The best condition of sample preservation is used to highlight the complex mitochondrial networking in liver cell at unprecedented resolution, making FIB-SEM tomography the most promising technology at the moment to study organelles' interaction with a resolution below 5 nm in all three dimensions.

REFERENCES

Armer, H.E., Mariggi, G., *et al.* (2009) Imaging transient blood vessel fusion events in zebrafish by correlative volume electron microscopy. *PLoS One*, 4 (11), e7716.

Baumeister, W., Grimm, R., *et al.* (1999) Electron tomography of molecules and cells. *Trends in Cell Biology*, 9 (2), 81–85.

Beckwith, M.S., Beckwith, K.S., *et al.* (2015) Seeing a *Mycobacterium*-infected cell in nanoscale 3D: Correlative imaging by light microscopy and FIB/SEM tomography. *PLoS One*, 10 (9), e0134644.

Behrman, E. (1983) The chemistry of osmium tetroxide fixation. *The Science and Biology of Specimen Preparation for Microscopy and Microanalysis*, 1–5.

Bell, D. and Erdman, N. (2012) *Low Voltage Electron Microscopy – Principles and Applications*, John Wiley & Sons.

Beorchia, A., Heliot, L., *et al.* (1993) Applications of medium-voltage STEM for the 3-D study of organelles within very thick sections. *Journal of Microscopy*, 170 (Pt 3), 247–258.

Blazquez-Llorca, L., Hummel, E., *et al.* (2015) Correlation of two-photon *in vivo* imaging and FIB/SEM microscopy. *Journal of Microscopy*, 259 (2), 129–136.

Bokhari, S.H. and Sauer, J.R. (2005) A parallel graph decomposition algorithm for DNA sequencing with nanopores. *Bioinformatics*, 21 (7), 889–896.

Bosch, C., Martinez, A., *et al.* (2015) FIB/SEM technology and high-throughput 3D reconstruction of dendritic spines and synapses in GFP-labeled adult-generated neurons. *Frontiers in Neuroanatomy*, 9, 60.

Briggman, K. L., Helmstaedter, M., *et al.* (2011) Wiring specificity in the direction-selectivity circuit of the retina. *Nature* 471(7337): 183.

Bushby, A.J., P'Ng K, M., *et al.* (2011) Imaging three-dimensional tissue architectures by focused ion beam scanning electron microscopy. *Nature Protocols*, 6 (6), 845–858.

Cretoiu, D., Gherghiceanu, M., *et al.* (2015) FIB-SEM tomography of human skin telocytes and their extracellular vesicles. *Journal of Cellular and Molecular Medicine*, 19 (4), 714–722.

De Bruijn, W. (1968) A modified OsO_4-(double) fixation procedure which selectively contrasts glycogen, in *Proceedings of 4th European Region Conference on EM*, vol. II, p. 65.

De Bruijn, W. (1973) Glycogen, its chemistry and morphologic appearance in the electron microscope: I. A modified OsO4 fixative which selectively contrasts glycogen. *Journal of Ultrastructure Research* 42(1-2): 29–50.

Denk, W. and Horstmann, H. (2004) Serial block-face scanning electron microscopy to reconstruct three-dimensional tissue nanostructure. *PLoS Biology*, 2 (11), e329.

De Winter, D.A., Schneijdenberg, C.T., *et al.* (2009) Tomography of insulating biological and geological materials using focused ion beam (FIB) sectioning and low-kV BSE imaging. *Journal of Microscopy*, 233 (3), 372–383.

Drobne, D., Milani, M., *et al.* (2007) Surface damage induced by FIB milling and imaging of biological samples is controllable. *Microscopy Research and Technique*, 70 (10), 895–903.

Earl, J., Leary, R., *et al.* (2010) Characterization of dentine structure in three dimensions using fib-sem. *Journal of Microscopy* 240(1): 1–5.

Felts, R.L., Narayan, K., *et al.* (2010) 3D visualization of HIV transfer at the virological synapse between dendritic cells and T cells. *Proceedings of the National Academy of Sciences of the United States of America*, 107 (30), 13336–13341.

Frank, J. (1992). *Electron Tomography*, Springer.

Friedrich, R.W., Genoud, C., *et al.* (2013) Analyzing the structure and function of neuronal circuits in zebrafish. *Frontiers in Neural Circuits*, 7, 71.

Gestmann, I., Hayles, M., *et al.* (2004) Site-specific 3D imaging of cells and tissues using dualbeam technology. *Microscopy and Microanalysis* 10(S02): 1124.

Giannuzzi, L. and Stevie, F. (2005) *Introduction to Focused Ion Beam*, Springer.

Hayat, M.A., (1981) Principles and techniques of electron microscopy. *Biological applications*, Edward Arnold.

Hayworth, K.J., Morgan, J.L., Schalek, R., Berger, D.R., Hildebrand, D.G.C., and Lichtman, J.W. (2014) Imaging ATUM ultrathin section libraries with WaferMapper: A multi-scale approach to EM reconstruction of neural circuits. *FrontNeural Circuits*, 8, 1–18.

Hayworth, K.J., Xu, C.S., *et al.* (2015) Ultrastructurally smooth thick partitioning and volume stitching for large-scale connectomics. *Nature Methods*, 12 (4), 319–322.

Hekking, L.H., Lebbink, M.N., *et al.* (2009) Focused ion beam-scanning electron microscope: exploring large volumes of atherosclerotic tissue. *Journal of Microscopy*, 235 (3), 336–347.

Helmstaedter, M., Briggman, K.L., *et al.* (2013) Connectomic reconstruction of the inner plexiform layer in the mouse retina. *Nature* 500(7461): 168.

Hendriks, H. R. and I. L. Eestermans (1982) Electron dense granules and the role of buffers: artefacts from fixation with glutaraldehyde and osmium tetroxide. *Journal of Microscopy* 126(2): 161–168.

Heymann, J.A., Hayles, M., *et al.* (2006) Site-specific 3D imaging of cells and tissues with a dual beam microscope. *Journal of Structural Biology*, 155 (1), 63–73.

Heymann, J.A., Shi, D., *et al.* (2009) 3D imaging of mammalian cells with ion-abrasion scanning electron microscopy. *Journal of Structural Biology*, 166 (1), 1–7.

Hoppe, W. (1981) Three-dimensional electron microscopy. *Annual Review of Biophysics and Bioengineering*, 10 (1), 563–592.

Hoving, S., Brunstein, F., *et al.* (2005) Synergistic antitumor response of interleukin 2 with melphalan in isolated limb perfusion in soft tissue sarcoma-bearing rats. *Cancer Research*, 65 (10), 4300–4308.

Humbel, B. and Schwarz, H. (1989) Freeze-substitution for immunochemistry in *Immuno-gold Labeling in Cell Biology*, CRC Press, Boca Raton, FL, pp. 115–134.

Jayakumar, S., Guillot, S., *et al.* (2011) Ultrastructural findings in human nonalcoholic steatohepatitis. *Expert review of gastroenterology & hepatology* 5(2): 141–145.

Jimenez, N., Vocking, K., *et al.* (2009) Tannic acid-mediated osmium impregnation after freeze-substitution: A strategy to enhance membrane contrast for electron tomography. *Journal of Structural Biology*, 166 (1), 103–106.

Jimenez, N., Van Donselaar, E., *et al.* (2010) Gridded Aclar: preparation methods and use for correlative light and electron microscopy of cell monolayers, by TEM and FIBSEM. *Journal of Microscopy* 237(2): 208–220.

Kasthuri, N., Hayworth, K., *et al.* (2007) New technique for ultra-thin serial brain section imaging using scanning electron microscopy. *Microscopy and Microanalysis*, 13, 26–27.

Kizilyaprak, C., Longo, G., *et al.* (2015) Investigation of resins suitable for the preparation of biological sample for 3-D electron microscopy. *Journal of structural biology* 189(2): 135–146.

Kleefstra, T., Smidt, M., *et al.* (2005) Disruption of the gene Euchromatin Histone Methyl Transferase1 (Eu-HMTase1) is associated with the 9q34 subtelomeric deletion syndrome. *Journal of Medical Genetics*, 42 (4), 299–306.

Knott, G., Marchman, H., *et al.* (2008) Serial section scanning electron microscopy of adult brain tissue using focused ion beam milling. *The Journal of Neuroscience: the official Journal of the Society for Neuroscience*, 28 (12), 2959–2964.

Knott, G., Rosset, S., *et al.* (2011) Focussed ion beam milling and scanning electron microscopy of brain tissue. *Journal of Visualized Experiments: JoVE*, (53), e2588.

Kremer, J.R., Mastronarde, D.N., *et al.* (1996) Computer visualization of three-dimensional image data using IMOD. *Journal of Structural Biology*, 116 (1), 71–76.

Lee, H. and K. Neville (1967) Book Review-Handbook of Epoxy Resins. *Industrial & Engineering Chemistry* 59(9): 16–17.

Leser, V., Drobne, D., *et al.* (2009) Comparison of different preparation methods of biological samples for FIB milling and SEM investigation. *Journal of Microscopy*, 233 (2), 309–319.

Leser, V., Milani, M., *et al.* (2010) Focused ion beam (FIB)/scanning electron microscopy (SEM) in tissue structural research. *Protoplasma*, 246 (1–4), 41–48.

Leighton, S.B. (1981) SEM images of block faces, cut by a miniature microtome within the SEM – a technical note. *Scanning Electron Microsc.*, 73–76.

Martone, M.E., Deerinck, T.J., *et al.* (2000) Correlated 3D light and electron microscopy: Use of high voltage electron microscopy and electron tomography for imaging large biologicas Structures. *Journal of Histotechnology*, 23 (3), 261–270.

Matsko, N. and Mueller, M. (2004) AFM of biological material embedded in epoxy resin. *Journal of Structural Biology*, 146 (3), 334–343.

Maupin-Szamier, P. and T. D. Pollard (1978) Actin filament destruction by osmium tetroxide. *The Journal of cell biology* 77(3): 837–852.

McIntosh, R., Nicastro, D., *et al.* (2005) New views of cells in 3D: An introduction to electron tomography. *Trends in Cell Biology*, 15 (1), 43–51.

Medeiros, L.C., De Souza, W., *et al.* (2012) Visualizing the 3D architecture of multiple erythrocytes infected with *Plasmodium* at nanoscale by focused ion beam-scanning electron microscopy. *PLoS One*, 7 (3), e33445.

Merchan-Perez, A., Rodriguez, J.R., *et al.* (2009) Counting synapses using FIB/SEM microscopy: A true revolution for ultrastructural volume reconstruction. *Frontiers in Neuroanatomy*, 3, 18.

Murphy, G.E., Lowekamp, B.C., *et al.* (2010) Ion-abrasion scanning electron microscopy reveals distorted liver mitochondrial morphology in murine methylmalonic acidemia. *Journal of Structural Biology*, 171 (2), 125–132.

Murphy, G.E., Narayan, K., *et al.* (2011) Correlative 3D imaging of whole mammalian cells with light and electron microscopy. *Journal of Structural Biology*, 176 (3), 268–278.

Narayan, K., Danielson, C.M., *et al.* (2014) Multi-resolution correlative focused ion beam scanning electron microscopy: Applications to cell biology. *Journal of Structural Biology*, 185 (3), 278–284.

Newman, G. R. and Hobot, J. A. (2012) *Resin microscopy and on-section immunocytochemistry*, Springer Science & Business Media.

Nguyen, H. B., Thai, T.Q., *et al.* (2016) Conductive resins improve charging and resolution of acquired images in electron microscopic volume imaging. *Scientific reports* 6: 23721.

Noske, A.B., Costin, A.J., *et al.* (2008) Expedited approaches to whole cell electron tomography and organelle mark-up in situ in high-pressure frozen pancreatic islets. *Journal of Structural Biology*, 161 (3), 298–313.

Paredes-Santos, T.C., de Souza, W., *et al.* (2012) Dynamics and 3D organization of secretory organelles of *Toxoplasma gondii*. *Journal of Structural Biology*, 177 (2), 420–430.

Porter, K.R., Claude, A., *et al.* (1945) A study of tissue culture cells by electron microscopy: Methods and preliminary observations. *The Journal of Experimental Medicine*, 81 (3), 233–246.

Remis, J.P., Wei, D., *et al.* (2014) Bacterial social networks: Structure and composition of *Myxococcus xanthus* outer membrane vesicle chains. *Environmental Microbiology*, 16 (2), 598–610.

Rubanov, S. and Munroe, P.R. (2004) FIB-induced damage in silicon. *Journal of Microscopy*, 214 (Pt 3), 213–221.

Schalek, R., Kasthuri, N., *et al.* (2011) Development of High-Throughput, High-Resolution 3D Reconstruction of Large-Volume Biological Tissue Using Automated Tape Collection Ultramicrotomy and Scanning Electron Microscopy. *Microscopy and Microanalysis* 17(SupplementS2): 966–967.

Schertel, A., Snaidero, N., *et al.* (2013) Cryo FIB-SEM: Volume imaging of cellular ultrastructure in native frozen specimens. *Journal of Structural Biology*, 184 (2), 355–360.

Schneider, P., Meier, M., *et al.* (2011) Serial FIB/SEM imaging for quantitative 3D assessment of the osteocyte lacuno-canalicular network. *Bone*, 49 (2), 304–311.

Schwarz, H. (1973) *Immunelektronenmikroskopische Untersuchungen über Zelloberflächenstrukturen aggregierender Amöben von Dictyostelium discoideum*, Fachbereich Biologie, Eberhard-Karis Universitaet Tuebingen

Schwartz, C. L., Heumann, J.M., *et al.* (2012) A detailed, hierarchical study of Giardia lamblia's ventral disc reveals novel microtubule-associated protein complexes. *PLoS One* 7(9): e43783.

Scott, I. and Youle R.J. (2010) Mitochondrial fission and fusion. *Essays in biochemistry* 47: 85–98.

Scott, K. (2011) 3D elemental and structural analysis of biological specimens using electrons and ions. *Journal of Microscopy* 242(1): 86–93.

Seligman, A.M., Wasserkrug, H.L., *et al.* (1966) A new staining method (OTO) for enhancing contrast of lipid – containing membranes and droplets in osmium tetroxide – fixed tissue with osmiophilic thiocarbohydrazide (TCH). *The Journal of Cell Biology*, 30 (2), 424–432.

Sjöstrand, F.S. (1958) Ultrastructure of retinal rod synapses of the guinea pig eye as revealed by three-dimensional reconstructions from serial sections. *J. Ultrastruct. Res.*, 2 (1), 122–170.

Soto, G. E., Young, S.J., *et al.* (1994) Serial section electron tomography: A method for three-dimensional reconstruction of large structures. *NeuroImage*, 1 (3), 230–243.

Steinmann, U., Borkowski, J., *et al.* (2013) Transmigration of polymorphnuclear neutrophils and monocytes through the human blood-cerebrospinal fluid barrier after bacterial infection *in vitro*. *Journal of Neuroinflammation*, 10: 31.

Stevens, J. K., Davis, T.L., *et al.* (1980) A systematic approach to reconstructing microcircuitry by electron microscopy of serial sections. *Brain Research Reviews*, 2 (1), 265–293.

Subramaniam, S. (2005) Bridging the imaging gap: Visualizing subcellular architecture with electron tomography. *Current Opinion in Microbiology*, 8 (3), 316–322.

Tanaka, K. and Mitsushima, A. (1984) A preparation method for observing intracellular structures by scanning electron microscopy. *Journal of Microscopy*, 133 (Pt 2), 213–222.

Tapia, J.C., Kasthuri, N., Hayworth, K., Schalek, R., Lichtman, J.W., Smith, S.J., and Buchanan, J. (2012) High contrast en bloc staining of neuronal tissue for field emission scanning electron microscopy. *Nature Protocols*, 7, 193–206.

Titze, B. and Denk, W. (2013) Automated in-chamber specimen coating for serial block-face electron microscopy. *J. Microsc.*, 250, 101–110.

Villinger, C., Gregorius, H., *et al.* (2012) FIB/SEM tomography with TEM-like resolution for 3D imaging of high-pressure frozen cells. *Histochemistry and Cell Biology*, 138 (4), 549–556.

Wanner, G., Schafer, T., *et al.* (2013) 3-D analysis of dictyosomes and multivesicular bodies in the green alga Micrasterias denticulata by FIB/SEM tomography. *Journal of structural biology* 184(2): 203–211.

Ware, R.W. and Lopresti, V. (1975) Three-dimensional reconstruction from serial sections, in *International Review of Cytology* (eds G.H. Bourne, J.F. Danielli, and K.W. Jeon), vol. 40, Academic Press, pp. 325–440.

Wei, D., Jacobs, S., *et al.* (2012) High-resolution three-dimensional reconstruction of a whole yeast cell using focused-ion beam scanning electron microscopy. *BioTechniques*, 53 (1), 41–48.

White, J.G., Southgate, E., Thomson, J.N., and Brenner, S. (1986) The structure of the nervous system of the nematode Caenorhabditis elegans. *Phil. Trans. R. Soc. Lond. A*, 314 (1), 340.

Wierzbicki, R., Kobler, C., *et al.* (2013) Mapping the complex morphology of cell interactions with nanowire substrates using FIB-SEM. *PLoS One*, 8 (1), e53307.

Wildiers, H., Highley, M.S., *et al.* (2003) Pharmacology of anticancer drugs in the elderly population. *Clinical Pharmacokinetics*, 42 (14), 1213–1242.

Wilson, C.J., D. Mastronarde, D.N., *et al.* (1992) Measurement of neuronal surface area using high-voltage electron microscope tomography. *NeuroImage*, 1 (1), 11–22.

Yakushevska, A.E., Lebbink, M.N., *et al.* (2007) STEM tomography in cell biology, *Journal of Structural Biology*, 159 (3), 381–391.

Young, R.J., Dingle, T., *et al.* (1993) An application of scanned focused ion-beam milling to studies on the internal morphology of small Arthropods. *Journal of Microscopy (Oxford)*, 172, 81–88.

27

Three-Dimensional Field-Emission Scanning Electron Microscopy as a Tool for Structural Biology

J.D. Woodward[1] **and R.A. Wepf**[2,3]

[1]*Department of Integrative Biomedical Sciences in the division of Medical Biochemistry & Structural Biology and Structural Biology Research Unit, University of Cape Town, Cape Town, South Africa*
[2]*ETH ScopeM, Swiss Federal Institute of Technology, ETH-Hönggerberg Zürich, Switzerland*
[3]*UQ, CMM, University of Queensland, Brisbane, Australia*

27.1 INTRODUCTION

In order to fully understand the processes leading to cellular function, biologists will require dynamic three-dimensional structural data at atomic detail of the myriad of molecular interactions taking place within the undisturbed living cell (e.g. Baumeister, 2005). Certainly no single (existing) imaging technique can come anywhere close to satisfying this requirement. Instead it is likely that various different imaging modes will need to be superimposed to provide complementary information (Figure 27.1). X-ray crystallography, biomolecular NMR, modelling and, more recently, single-particle cryo-electron microscopy (Bartesaghi *et al.*, 2015) can define the atomic resolution structure of purified proteins and protein complexes. At the opposite end of the resolution spectrum, confocal microscopy, computed tomography and serial section reconstruction can image organelles, whole cells and even tissues and organs (Figure 27.1).

Between these two extremes, there is a region referred to as 'the resolution gap', which spans the atomic (~0.1 nm) and cellular resolution (~100 nm) ranges (Figure 27.1). In this region, technologies with the capacity to produce three-dimensional data at

Biological Field Emission Scanning Electron Microscopy, First Edition.
Edited by Roland A. Fleck and Bruno M. Humbel.

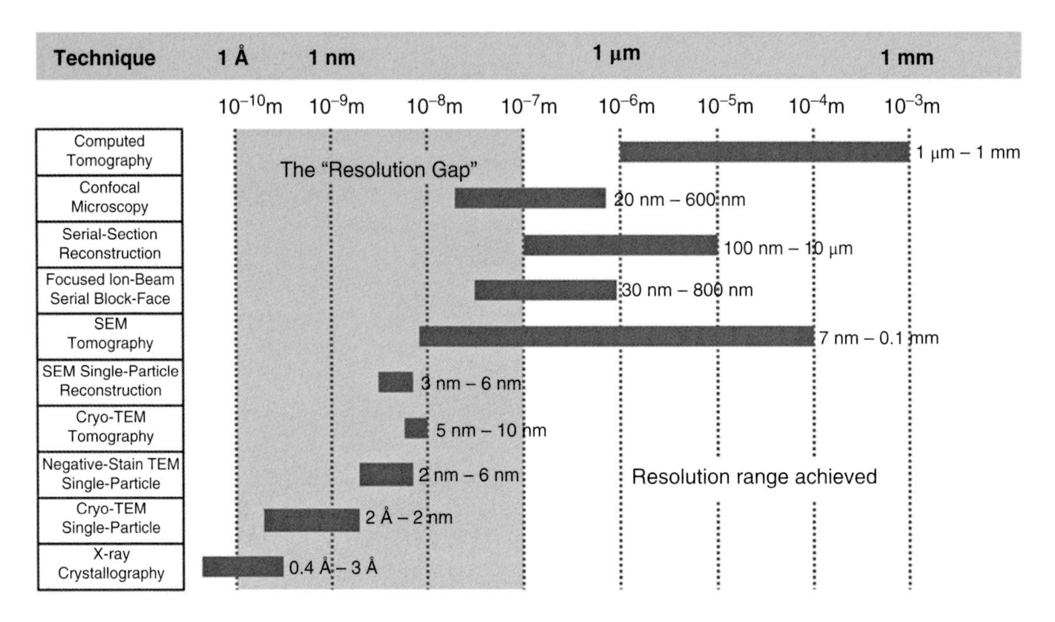

Figure 27.1 Imaging techniques as a function of resolution. Dark grey bars indicate the approximate resolution ranges that can be attained by each technique (Perkins, Sun and Frey, 2009; de Boer, Hoogenboom and Giepmans, 2015). The 'Resolution Gap' (light grey) spans the resolution required for accurate determination of the locations of atoms and that required to define cellular structure (Frank, 2006).

sufficient resolution to identify the *in situ* structural signatures of individual proteins in a close-to-lifelike condition will allow the integration of atomic and cellular resolution information (Medalia *et al.*, 2002). Pseudo-atomic resolution data of macromolecular machines, organelles or even whole cells can then be generated by docking atomic resolution data into lower-resolution maps (e.g. Bohm *et al.*, 2000). The cryo-field emission scanning electron microscope fulfils the requirements for resolution and structural preservation and has historically been an important tool for the structural characterisation of the *in situ* macromolecular organisation of the cell (e.g. Goldberg *et al.*, 1997).

Humans can intuitively recognize scanning electron micrographs as images of three-dimensional objects and interpret them qualitatively, but computational means are required to extract quantitative three-dimensional information for structural purposes (Tafti *et al.*, 2015). This has been achieved in a number of ways, with various requirements for type and quality of input micrographs and limitations on the output three-dimensional data (Figure 27.2). Current approaches use either (1) the surface normal versus intensity information of FE/SEM images, for example shape from shading or SEM single-image topography (Beil and Carlson, 1991) and four quadrant BSE detector 3D (Berger, 2007), or (2) by calculating depth information by triangulation after identifying homologous points between tilt pairs (Piazzesi, 1973) ('SEM stereo pair' or SEM photogrammetry), or (3) physically ablating the specimen in fine layers and imaging the resulting exposed face (Denk and Horstmann, 2004), focused ion beam serial block-face reconstruction (Figure 27.2).

SEM stereo pair and SEM photogrammetry are successful techniques for visualizing the topography of beam resistant samples (e.g. Stampfl *et al.*, 1996; Scherer and Koledenik, 2001; Cornille *et al.*, 2003). These algorithms require high-dose images of textured samples produced at low magnification. However, under the conditions necessary for

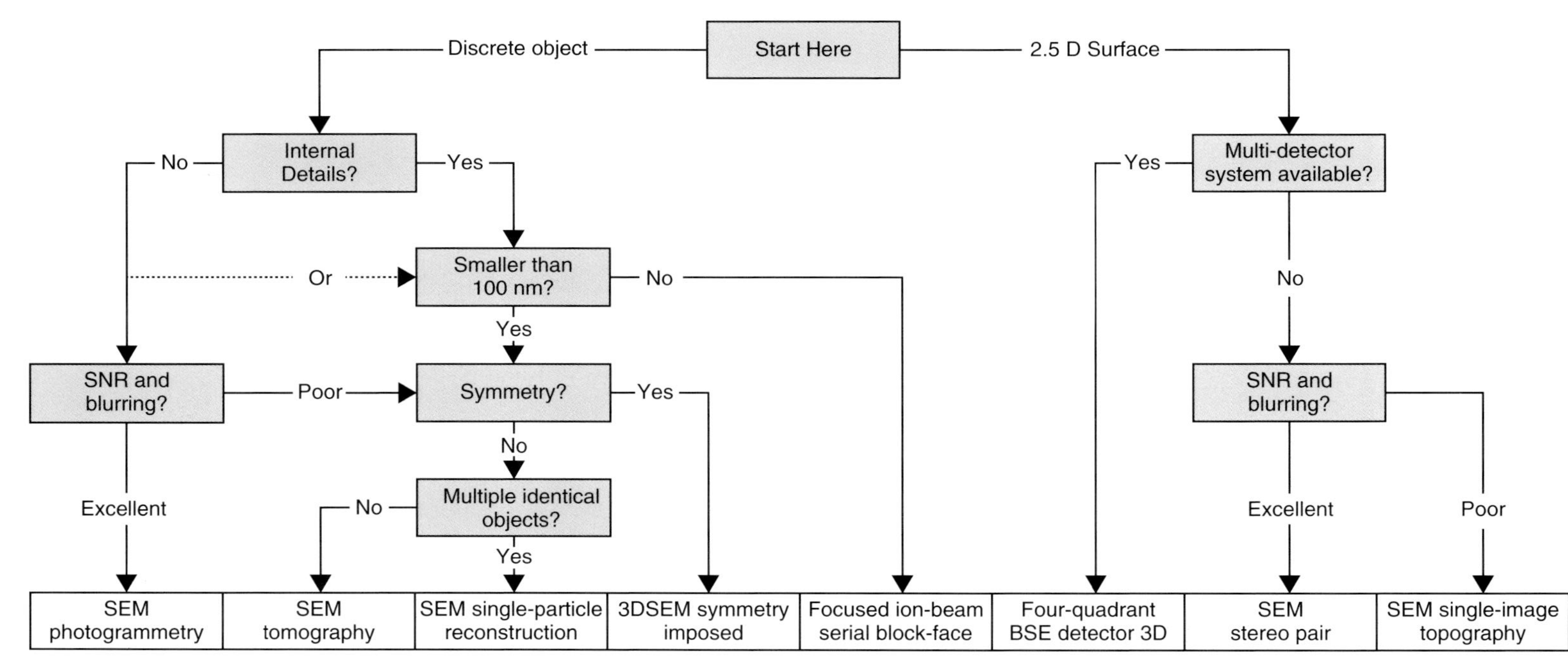

Figure 27.2 Three-dimensional scanning electron microscopy technique decision diagram. The only options for full three-dimensional reconstruction (discrete object) of macromolecular biological samples (poor signal-to-noise ratio and blurring) currently available are backprojection-based reconstruction (SEM tomography, SEM single-particle reconstruction and 3DSEM symmetry imposed reconstruction).

macromolecular structural determination certain artefacts become apparent: blurring obscures detail near the resolution limit of the microscope and the signal-to-noise ratio may be poor, a consequence of imaging at low dose. Regions with low electron-escape probability, edge effects or charging may drastically change the appearance of homologous points between micrographs making reconstruction challenging.

An alternative approach is to use backprojection-based reconstruction, which overcomes these problems to produce successful full three-dimensional reconstructions of even smooth samples in the presence of charging, poor signal-to-noise ratio and blurring (Woodward and Sewell, 2010). We first applied backprojection to in-lens secondary electron field emission images of helical macromolecular assemblies prepared using negative stain and cryo-metal shadowing (Woodward, Wepf and Sewell, 2009). Later it was demonstrated that the technique could also be applied to asymmetric samples imaged by secondary electron FE/SEM tomography (Woodward and Sewell, 2010; Lück *et al.*, 2010; Facron and Field, 2015; Okyay *et al.*, 2016) as well as icosahedral single particles (Miller *et al.*, 2011).

27.2 THEORY

27.2.1 Backprojection

The inverse Radon transform can be used to recover an unknown three-dimensional function from a series of parallel line integrals (Radon, 1917). In practice, this means that an object can be reconstructed from two-dimensional projection images given a sufficient number of different angular views. An illustration of the process with a two-dimensional example is given in Figure 27.3.

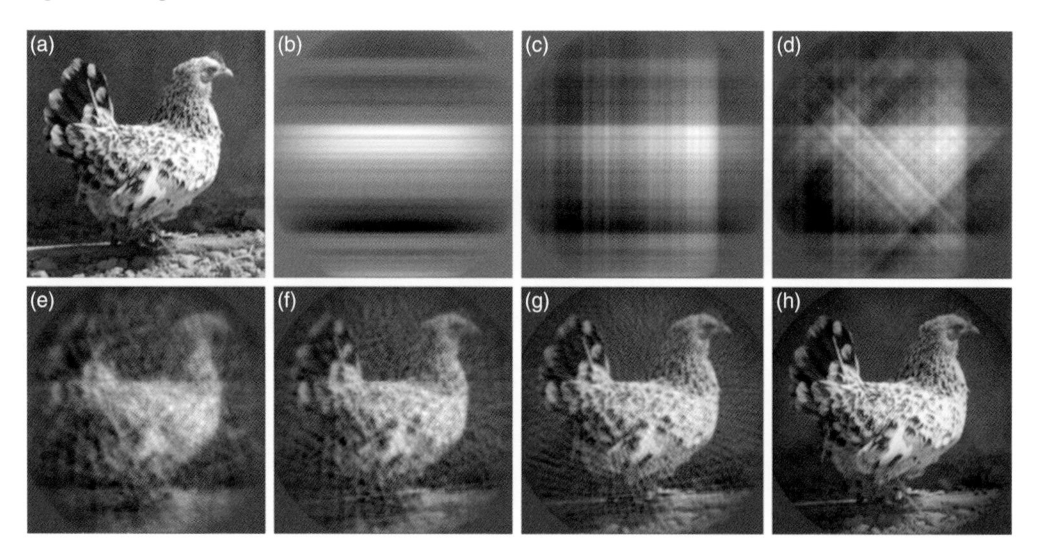

Figure 27.3 The inverse Radon transform in two-dimensions. (a) A two-dimensional density distribution, in this case a photograph of a chicken, can be 'projected' in any direction by summing the pixel densities in that direction and forming a one-dimensional intensity profile. (b) The photograph has been projected horizontally and the resulting filtered backprojection (Harauz and Van Heel, 1986) reconstruction is shown. (c) Reconstruction from two projections at 90° to one another. As the number of projections is increased to (d) four, (e) 15, (f) 30, (g) 60 and (h) 180, the resolution of the reconstruction improves.

This approach forms the mathematical underpinning for all projection-based reconstruction techniques such as X-ray computed tomography, single-particle cryo (transmission) electron microscopy, helical- and icosahedral reconstruction and (transmission) electron tomography. It also forms the basis of our approach (Woodward *et al.*, 2009) for reconstructing FESEM/SEM data in three dimensions. How backprojection applies to images captured in the SEM is not immediately obvious because the SEM image formation process does not superficially appear to generate 'density projections'.

27.2.2 Image Formation in the SEM

In the scanning electron microscope, a focused incident electron beam is scanned in a raster across the surface of a sample while, depending on the detectors in operation, various signals are recorded as a function of beam position (Pawley, 1997). An image is generated by assigning a greyscale value proportional to the signal flux at the corresponding sample position to each pixel. For our purposes, the secondary electron signal is the most practical because it shows a high signal-to-noise ratio, as the secondary electron coefficient is relatively high and insensitive to atomic number (Goldstein *et al.*, 1992, pp. 108–109). This imaging modality also accurately reproduces the high-resolution topography of the sample (Cazaux, 2004).

Secondary electrons are formed when a sample is ionized through inelastic collisions with energetic primary and/or scattered backscattered electrons. In the scanning electron microscope, they are scattered along the path of the incident electron beam (Wolff, 1954; Goldstein *et al.*, 1992, p. 111) with energies generally below 50 eV (Pawley, 1997). However, only those that form sufficiently close to the sample surface can escape, get detected and contribute to image formation. The probability that a secondary electron of a given energy within the sample will be detected depends on a number of factors related to sample composition, the local geometry of the sample and, most importantly, the distance between its point of origin and the surface of the sample (Figure 27.4a). We can therefore define a function: 'the secondary electron escape probability distribution' (Woodward and Sewell, 2010) by combining all of the factors that influence the number of secondary electrons detected (Figure 27.4b).

Because the mean-free path of the incident electron beam is far greater than the secondary electron escape distance (Goldstein *et al.*, 2012, p. 66), the number of secondary electrons detected is approximately proportional to the line integral through the accessible secondary electron escape probability distribution in the beam direction (Figure 27.4b). At moderately high magnification, the incident electron beam can be assumed to be parallel (Hemmleb and Schubert, 1997), meaning that scanning electron microscopy images approximate beam-accessible and orthographic projections of this distribution.

27.2.3 SEM Backprojection

The process of image formation can be simulated *in silico* by generating a voxel model of an arbitrary shape (in this case a beetle) (Figure 27.5a) that consists of an approximation to the secondary-electron escape probability distribution. In cross-section (Figure 27.5b and c) this appears as a narrow white outline that decreases rapidly in intensity with distance from the surface. Projecting this volume on to a two-dimensional plane illustrates

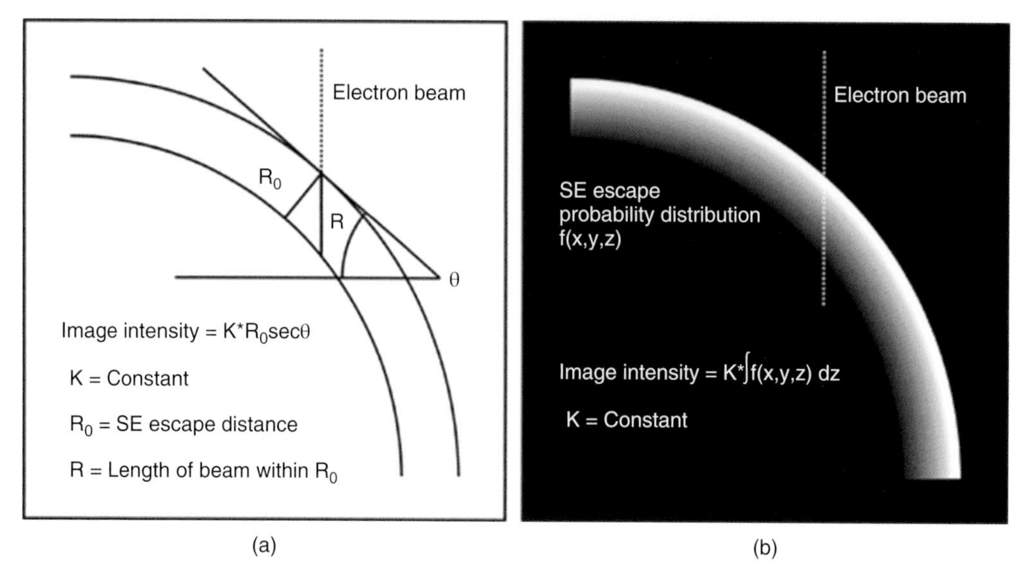

Figure 27.4 SEM 'projections'. (a) A typical approximation used to describe the secondary electron image formation process occurring in a scanning electron microscope. Only those secondary electrons arising within R_0 of the surface of the sample can escape. Assuming an equal number of electrons are formed along the length of the incident electron beam, the number of secondary electrons detected is proportional to R, which is equal to the secant of the slope of the sample. (b) A better approximation is given by taking a line integral through the electron probability distribution $f(x, y, z)$ shown here in cross-section. Once the incident electron beam enters the 'black area', no secondary electrons can be detected.

what a complete projection through the SE escape probability distribution would look like (Figure 27.5d).

Because the probability of secondary electron escape decreases exponentially with increasing sample depth (Seiler, 1967) and biological samples are usually thicker than ~10 nm, the majority of contrast arises from close to the specimen surface on the beam accessible side. This provides only half the information required to reconstruct the full secondary electron escape probability distribution. Each image therefore represents a partial projection (Woodward *et al.*, 2008) of the full three-dimensional density distribution (Figure 27.6). However, if a sufficient number of images from a range of different viewing directions can be correctly aligned, each partial central section is summed, leading to a complete recovery of the original distribution (Woodward *et al.*, 2008) (Figure 27.7).

The resulting volumetric reconstruction is limited to the beam accessible parts of the sample; therefore the angular range used in imaging should be as wide as possible. This is either achieved by physically tilting the sample, that is *Tomographic 3DSEM* (Figure 27.8a), by imaging symmetric samples, that is *Helical 3DSEM/Icosahedral 3DSEM*, or by capturing single images of multiple identical objects in different orientations, that is *Single Particle 3DSEM* (Figure 27.8b). Parts of the sample that are partially obscured due to the geometry of the sample and regions of low electron escape probability (such as concave surfaces) lead to low occupancy (decreased intensity) of these features in the final reconstruction.

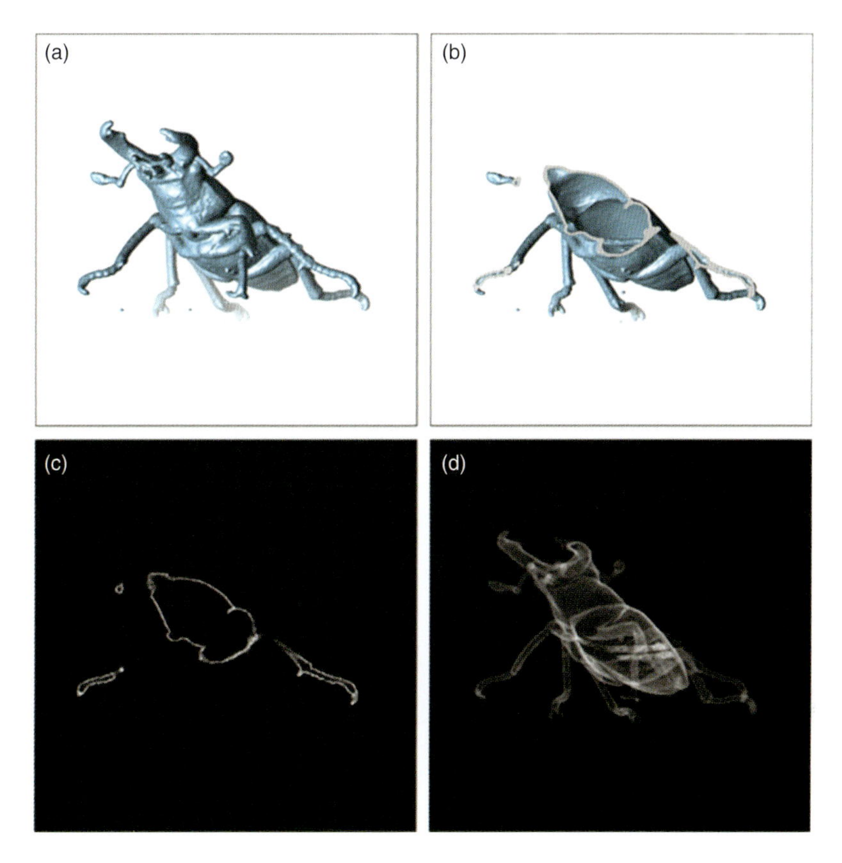

Figure 27.5 SEM image simulation. (a) Surface rendering using UCSF Chimera (Pettersen *et al.*, 2004) of a computer model of a beetle, generated from CT data (courtesy of J. Kastner, Wels College of Engineering) and modified by assigning each voxel a greyscale intensity that is proportional to the likelihood of secondary electron escape from that position. This can be seen in cross-section (b, c). (d) Projecting the volume on to a two-dimensional plane by summing the voxel intensities in one direction produces an image that in some respects resembles scanning electron microscope images, except that the beam inaccessible parts of the distribution have also been sampled.

27.3 TOMOGRAPHIC 3DSEM

The simplest way to collect a dataset suitable for three-dimensional reconstruction is to physically rotate the sample relative to the electron beam (Figure 27.8a) (Frank, 2006). The angular increment between images will depend on the resolution required and the electron dose that the sample can withstand, but is typically around 1°. This process is analogous to TEM tomography and can be efficiently achieved using a variety of different strategies, the aim of which is to produce a large number of in-focus images of the entire sample, at the same working distance and magnification but from different orientations. The relative orientations (Euler angles) of each image should be recorded from the stage goniometer as precisely as possible. No part of the image should be stationary; it is therefore often necessary to manually erase the background from the images (Woodward and Sewell, 2010), the most laborious part of the process.

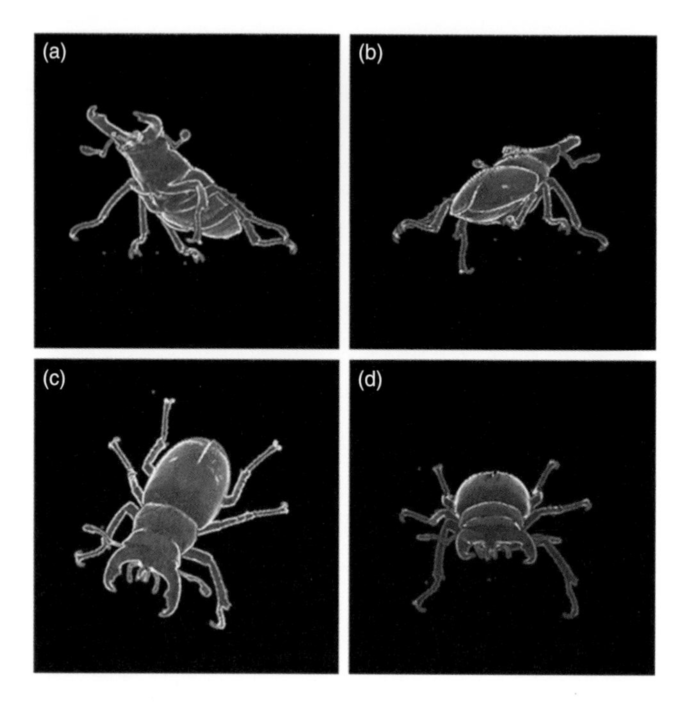

Figure 27.6 Simulated SEM images. (a) This view corresponds to Figure 27.5d, except that the sum of pixel intensities is calculated from one side and the summation process is halted when a voxel with an intensity of zero is encountered. These images closely approximated scanning electron microscope images and can be produced in any number of arbitrary directions, for example (b, c, d). Images were generated using Matlab 2013.

27.3.1 Tilt Strategies

Geometrically, single-axis tilt tomography is the simplest method of data collection: the sample stage should be able to tilt at least 60° along an axis perpendicular to the incident beam (Frank, 2008). In some cases it may be possible to tilt the stage from –60° to +60°, in which case the data collection strategy consists of tilting the stage and then translating it to re-centre the area of interest (Figure 27.9a). If, on the other hand, the stage is only able to tilt in a single direction, one solution is to collect images from 0° to 60° and then rotate the sample around the *z*-axis or remount the sample on the stage at 180° to its original position. The general shortcoming of the single-axis tilt strategy is that it leads to missing views of the sample and a substantially lower resolution in the direction of the 'missing wedge' (Frank, 2008) (Figure 27.9a).

One way of improving matters is to collect a second tilt series perpendicular to the first, which decreases this missing information. This strategy is referred to as a dual-axis tilt experiment (Mastronarde, 1997). This involves all of the steps required to collect a single-axis tilt dataset, a 90° stage rotation or sample remount (about the beam axis) and the collection of a second single-axis tilt dataset (Figure 27.9b). The optimal strategy for TEM tomography is to rotate the sample 180° along an axis that is exactly perpendicular to the beam direction (Frank, 2006), but for practical reasons this usually cannot be achieved. In the case of SEM tomography, this is often a viable strategy, except that the sample is rotated 360° (because both sides of the sample need to be imaged) (Figure 27.9d). The most practical

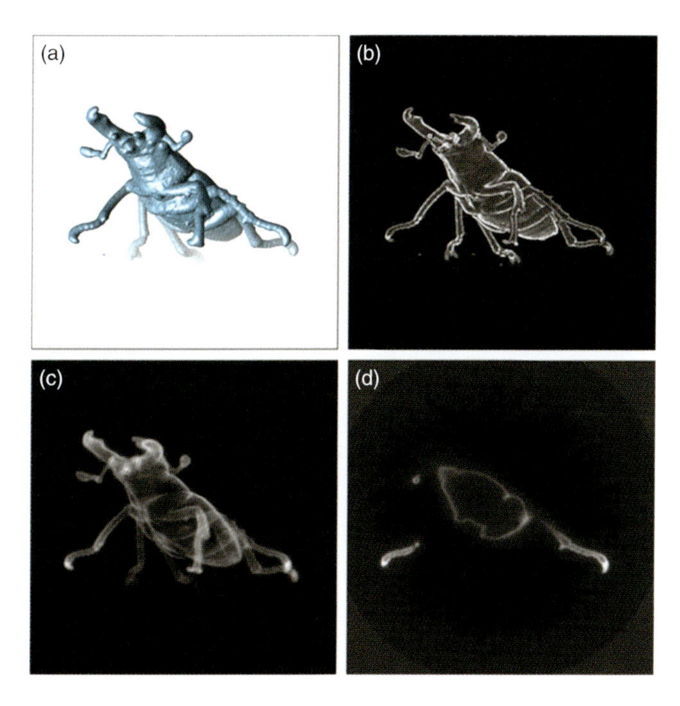

Figure 27.7 Recovery of full three-dimensional density from partial projections. (a) Surface rendering of a reconstruction, produced by backprojecting 200 simulated scanning electron microscope images from different directions. This reconstruction very closely resembles the original model (Figure 27.5a) as well as the individual input images (b). In (c, d), projection and cross-section it can be seen that the original density has been fully recovered. Some regions (such as the tips of the limbs) show an artefactual enhanced intensity relative to the original density (Figure 27.5c, d). This is because these regions are over-represented in the input images.

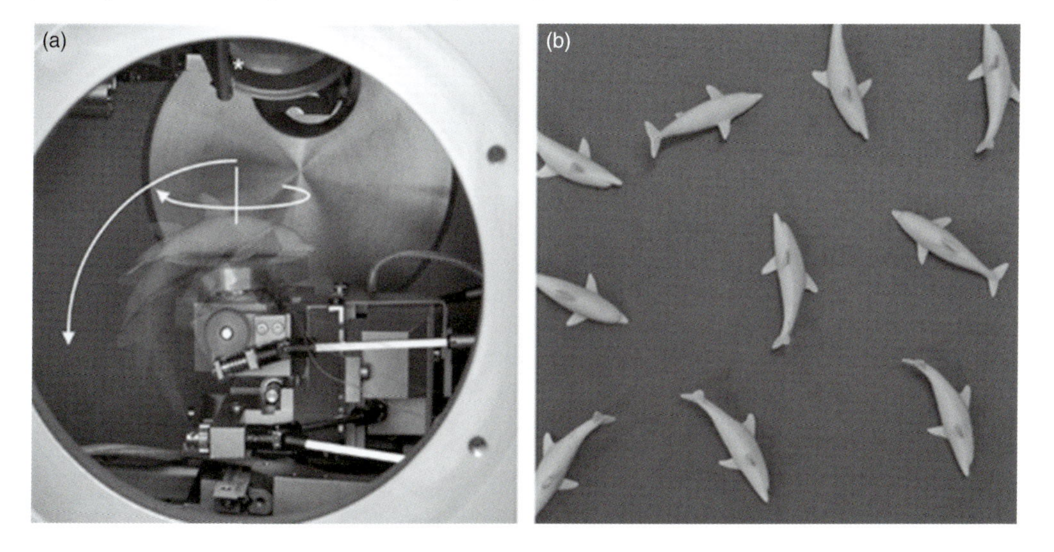

Figure 27.8 Data collection strategies. (a) Tomographic reconstruction: the stage is tilted and rotated relative to the primary electron beam (*) to produce a variety of different views of the sample. In this example, the stage can rotate 360° and tilt 90°. Depending on the stage used, a variety of different strategies for data collection can be employed. (b) Single-particle data collection strategy: single images comprised of identical objects lying in random orientations are collected.

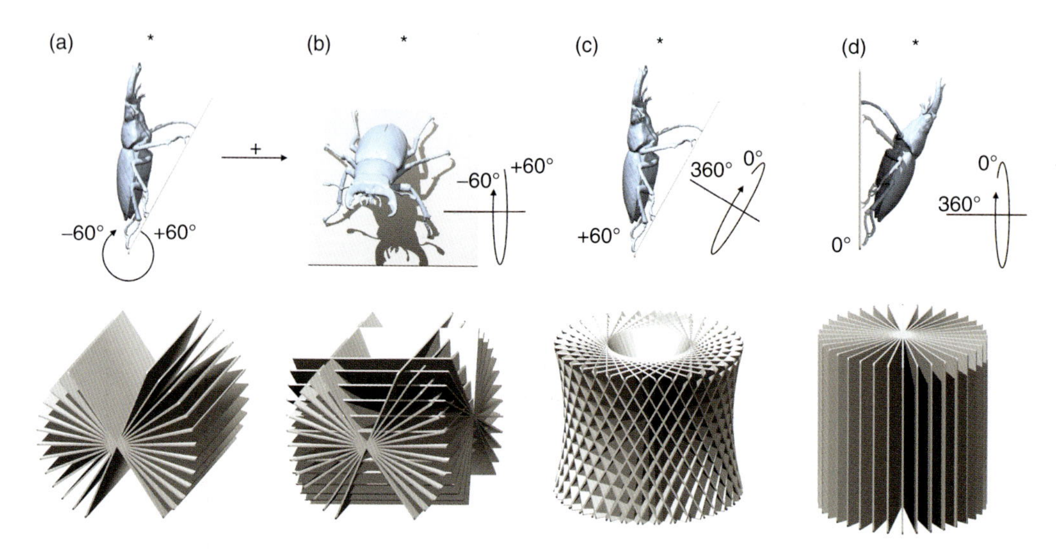

Figure 27.9 Scanning electron tomography tilt strategies. (a) Single-axis tilt tomography: the stage is tilted from +60° to −60° relative to the primary electron beam (*). 'The missing wedge' consists of missing views of the sample. (b) Dual-axis tilt tomography: a second dataset 90° from the first is collected, which partially compensates for the missing data. 'The missing wedge' is now 'the missing pyramid'. (c) Conical tilt tomography: 360° rotation about a 60° axis; 'the missing pyramid' is now 'the missing cone'. (d) Complete tomographic rotation: 360° rotation about a 90° axis; there are no missing views of the sample. Position of the primary electron beam (*).

general method is to rotate the specimen 360° around an axis that lies approximately 60°–70° from the beam direction (Woodward and Sewell, 2010) (Figures 27.9c and 27.10). This tilt geometry is called conical tilt tomography and minimizes the missing data.

27.3.2 Signal-to-Noise Ratio

When the incident electron beam irradiates a biological sample, ionization resulting from inelastic interactions with orbital electrons leads to the rearrangement of chemical bonds, formation of free radicals and vaporization and liquefaction of sample fragments (Figure 27.11). The degree of this radiation damage is proportional to the total dose received by the sample at a given beam energy and specimen temperature (Wade, 1984; Lamvik, 1991). This means that the total electron dose irradiating the sample must be less than that required to destroy the features of interest. This electron dose must be divided between all of the different angular views in the dataset. This means that for beam-sensitive samples there is a trade-off between the number of images that can be recorded and the signal-to-noise ratio of those images (Figure 27.11).

27.3.3 Resolution

The high-resolution limit of the final reconstruction is ultimately limited by the resolution of the individual input images. However, it could end up being substantially lower if an insufficient number of images are used for backprojection (Figure 27.3). Assuming an equally

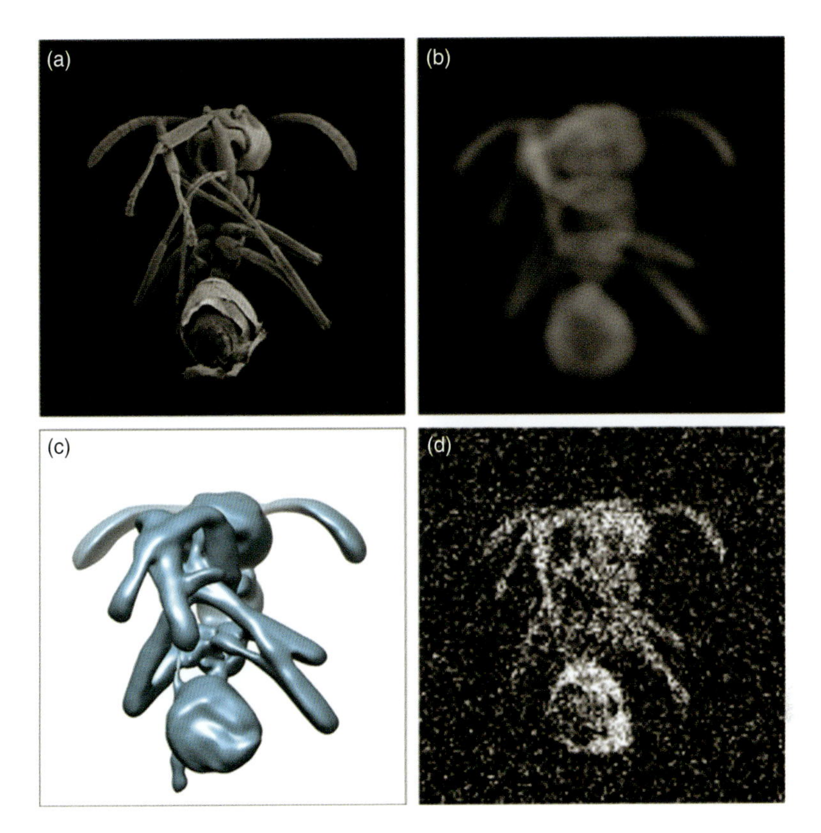

Figure 27.10 Ant reconstruction. (a) A series of 50 images were acquired over 360° (in 7.2° increments) using conical tilt tomography at an angle of 70°. The background was manually set to black (8-bit greyscale value of 0) using an image editor. Readings from the stage goniometer were converted into a list of Spider (Frank *et al.*, 1996) Euler angles. (b) Three-dimensional reconstruction of the images by real-space backprojection using the simultaneous iterative reconstruction technique (SIRT) implementation in Spider (BP RP). (c) Surface rendering of the reconstructed density using Chimera (Pettersen *et al.*, 2004). (d) Introduction of noise and blurring only slightly degrades the resulting reconstruction (Woodward and Sewell, 2010).

spaced angular distribution of images, the resolution of the reconstruction is given by the central section theorem (van Heel and Harauz, 1986) with one proviso: at least two images are required to generate a 'true' projection of the specimen (one from the 'front' and one from the 'back') (Woodward and Sewell, 2010) given by:

> Object of diameter (D) to a resolution (d) with (N) images is $N \approx 2\pi D/d$
> In the case of the beetle (Figure 15.7):
> $N = 200$ (images used)
> $D \approx 20$ mm (approximate 'diameter' of the beetle)
> A maximum resolution of $d \approx 0.6$ mm

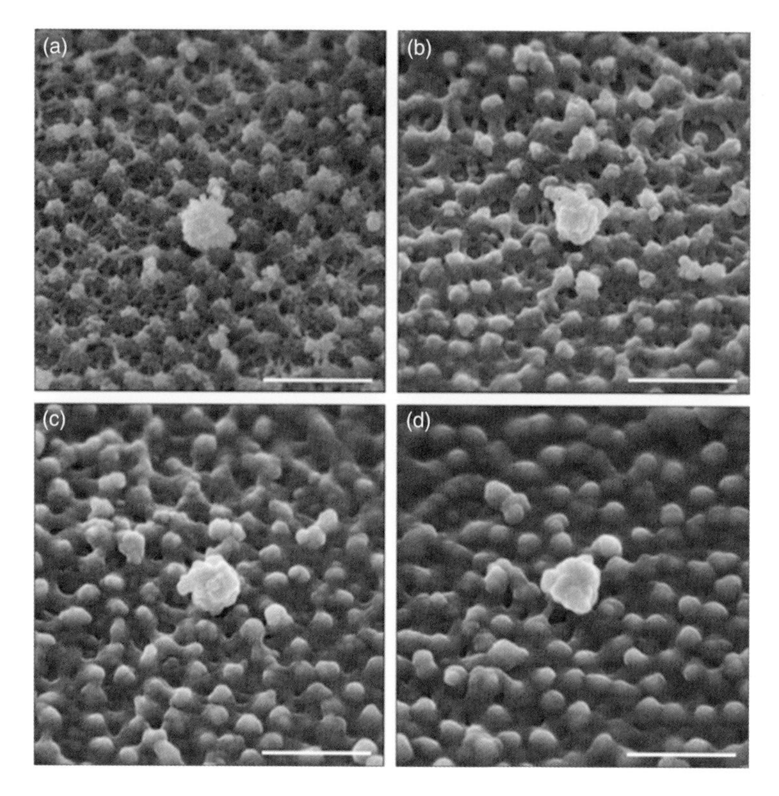

Figure 27.11 Sample melting as a result of electron damage. These images show a high electron-dose conical tilt tomography dataset of a metal-coated sample taken over 180° in 4.5° increments. (a) While the structural preservation and signal-to-noise ratio of the first image is excellent, by image 13 (b) the structures of interest are nearly entirely lost. Further liquefaction of the sample is evident in images 27 (c) and 40 (d). Scale bars: 500 nm. Images kindly supplied by Dr Martin Goldberg, Durham University.

27.4 SINGLE-PARTICLE RECONSTRUCTION

The technique of SEM single-particle analysis relies on allowing a macromolecular sample, consisting of multiple identical objects, to be imaged on a flat surface (Figure 27.8b). Negative staining or freeze-drying and metal-shadowing is used to enhance contrast, stabilize the sample and reduce charging (Stokroos *et al.*, 1998). If the particles lie in random orientations (Figure 27.12a) the requirement for multiple views of the (identical) object(s) is satisfied. Only the relative orientations of the particles need to be calculated and because of the excellent contrast of scanning electron microscope images (compared to transmission electron micrographs) this is trivial. If, on the other hand, the particles lie in a single preferential orientation, then two images can be captured: an untilted image and one tilted by approximately 60° (random conical tilt). Particles that are symmetrical, such as icosahedral viruses (Figure 27.12c) or other large macromolecular complexes such as helices (Figure 27.12e and g), require only a few images, as each view of the particle already contains multiple views of the asymmetric unit (Woodward and Wepf, 2014).

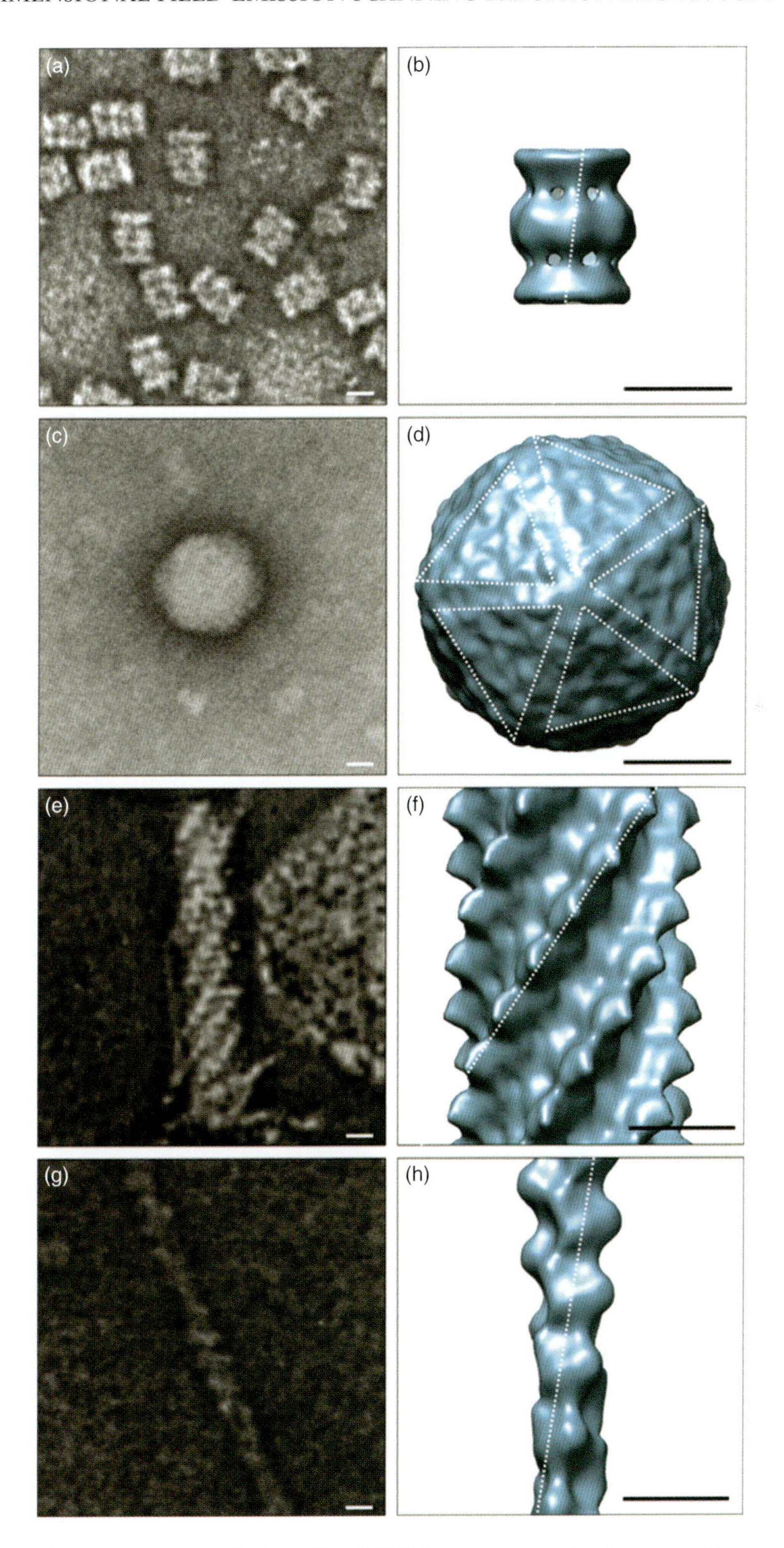

Figure 27.12 Contrast-inverted in-lens SE FESEM negative stain images with corresponding surface-fitted backprojection reconstructions showing handedness: (a, b) proteasome 20S core particle, (c, d) *Heterocapsa circularisquama* RNA virus, (e, f) bacteriophage T4 tail and (g, h) actin filament. Scale bars: 10 nm.

27.5 IN PRACTICE

27.5.1 Sample Preparation

Negative-stain samples for in-lens FESEM secondary electron imaging are prepared in the same way as those for TEM (Frank, 2006). Purified macromolecular complexes of at least ~200 kDa in size at a concentration of approximately 0.5 mg/ml (in buffer) are made up to 1:1, 1:10 and 1:100 dilutions. Approximately 2 μl of each is pipetted on to the surface of three glow-discharged carbon-coated copper grids. The sample is allowed to absorb for 30 s before being carefully blotted with a piece of filter paper. The grid is then placed on to two droplets of water and three droplets of 2% uranyl acetate or uranyl formate placed on Parafilm and blotted with filter paper between washings. A thin film of stain solution should be visible on the surface of the grids. The grids are allowed to air dry at room temperature for a few minutes.

Metal-shadowed macromolecular samples are prepared by pipetting ~2 μl of purified sample on to glow-discharged, carbon-coated grids and allowing the sample to absorb for two to three minutes before being rinsed in distilled water, blotted and plunged into liquid nitrogen. The sample is cryo-transferred into a freeze-drying metal-shadowing instrument and freeze-dried for two hours at a temperature of 193 K (–80 °C) and pressure of 1027 mbar before being shadowed with 0.8 nm of tantalum/tungsten (Ta/W) (Bachmann, Abermann and Zingsheim, 1969; Wepf *et al.*, 1994). Once the sample has been coated, only a slight loss of resolution occurs if the sample is allowed to reach room temperature before being transferred into the microscope for cryo-imaging (Woodward and Wepf, 2014).

The electron *dose* to the sample with probe current (I) and dwell time (t) at a pixel sampling (p) and $e = 1.60 \times 10^{-19}$ is Dose $= It/(ep^2)$

In the case of *Heterocapsa circularisquama* RNA virus negative stain images (Figure 15.12a):

Data size = 1280 × 960 pixels

Scan speed = 0.5 s per frame

Therefore dwell time = 0.5 s/(1280 × 960) = 4.1×10^{-7} s

Length of one pixel = 2.14 Å

Probe current = 100 pA = 1×10^{-10} A

A dose of 55 $e^-/Å^2$

27.5.2 Tomographic Reconstruction

In order to produce a three-dimensional reconstruction by backprojection, a series of images with associated angular and alignment information needs to be input into an appropriate software package. Angular information is taken from the stage goniometer and alignment parameters are usually calculated semimanually through the software interface. Several commercial and open source software packages are available, including ASTRA Toolbox (Palenstijn, Batenburg and Sijbers, 2011), EM3D (Ress *et al.*, 1999), IMOD (Kremer, Mastronarde and McIntosh, 1996), protomo (Winkler, 2007), PyTom (Hrabe *et al.*, 2012), RAPTOR (Amat *et al.*, 2008), TOM Toolbox (Nickell *et al.*, 2005), TomoJ (Messaoudii *et al.*, 2007) and Spider (Frank *et al.*, 1996). The software package EM2EM (Image Science, Germany) can be used to convert between Tiff and EM image formats.

27.5.3 Single-Particle Reconstruction

Individual particles need to be selected from micrographs manually using Boxer (Ludke, Baldwin and Chiu, 1999) or automatically using DoG Picker (Voss *et al.*, 2009), FindEM (Roseman, 2003), Signature (Chen and Grigorieff, 2007), SwarmPS (Woolford *et al.*, 2007) or TMaCS (Zhao, Brubaker and Rubinstein, 2013). Negative stain images must be contrast-inverted before particle selection because the signal arises from the stain surrounding the particle and not the particle itself, producing images of 'black' particles surrounded by 'white' stain (Woodward *et al.*, 2009). An iterative refinement approach is used to determine the relative orientations of the particles: the images are randomly backprojected and then the original images are assigned Euler angles by matching them to reprojections of the reconstruction. This process is iterated and the reconstruction gradually improves in quality. If handedness information is to be retained, only the front-facing side of the reconstruction should be reprojected (Woodward *et al.*, 2009). Several different software packages are available for performing single-particle reconstruction; these include Spider (Frank *et al.*, 1996) and Eman (Ludke *et al.*, 1999) as well as a host of specialized packages. The resulting volumes can be visualized in UCSF Chimera (Pettersen *et al.*, 2004).

27.6 APPLICATIONS

27.6.1 Handedness

Biological macromolecules are chiral (Pasteur, 1850) because they are comprised of constituent molecules with one or more asymmetric carbons (Van't Hoff, 1875). Opposite enantiomers cannot be superimposed in three dimensions, but they can be rotated relative to one another such that every atom occupies the same x and y coordinates as the corresponding atom in the mirror image molecule and differs only in its z coordinate. This means that (macro)molecules with opposite handedness are indistinguishable from one another in two-dimensional projection. TEM single-particle reconstructions at a resolution lower than 0.7 nm (where the helical striations of alpha helices can be resolved) are inherently ambiguous with regard to handedness and this needs to be resolved with additional experiments (Klug and Finch, 1968).

Handedness is assigned in a number of different ways: for instance, if one or more component proteins making up a complex have had their structures solved by X-ray crystallography, then docking these into a low-resolution map can identify the correct handedness of the overall complex. Assignment can also be based on homology with a structure of known quaternary structure, but this is less ideal because quaternary structure is less conserved than tertiary structure (Han *et al.*, 2009). Tilting (Klug and Finch, 1968; Belnap, Olson and Baker, 1997) or metal shadowing (in the case of helices) (Tyler and Branton, 1980) can be used to resolve the handedness of TEM reconstructions, but these experiments can be difficult, time-consuming, require specialist equipment and may lead to ambiguous results.

Because the secondary electron signal in the scanning electron microscope decreases exponentially with increasing distance from the specimen surface (Seiler, 1967), the majority of contrast arises from the beam-facing side of the sample. This eliminates the 180° ambiguity usually associated with projection images and allows direct determination of handedness: compare Figure 27.7b and c – is the beetle facing forwards or backwards? Furthermore, the excellent signal-to-noise ratio of secondary electron images means that handedness can be unambiguously assigned to helical samples even in unprocessed images (Figure 27.12e and g). In the case of non-helical complexes, single images are not sufficient to assign handedness (Figure 27.12a and c). In this case SEM single-particle reconstruction can be used as a tool to assign handedness (Miller *et al.*, 2011) (Figure 27.12b, d, f and h).

27.6.2 Bridging the Resolution Gap

Using backprojection 3DSEM, a resolution of ~3 nm can be achieved in negative stain with ~150 views of the asymmetric unit, while metal-shadowing delivers ~4 nm resolution with ~100 views (Woodward and Wepf, 2014). The molecular envelope is sufficiently well resolved in negative stain to unambiguously dock atomic-resolution data into the map (Figure 27.13). While the excellent signal-to-noise ratio of metal-shadowed data means that given only 30 low-dose views of the asymmetric unit, individual actin monomers can be resolved in the actin fibre (Woodward and Wepf, 2014). This means that given the right sample preparation strategy, with the capacity to image the intact macromolecular structure of the cell, atomic-resolution interpretation of *in situ* macromolecular complexes within the resulting three-dimensional map is theoretically feasible.

While FE/SEM is conventionally thought of as a surface-imaging technique, at high magnification, secondary electrons arising from a depth of ~10 nm make up a significant proportion of the final image (Woodward *et al.*, 2009). This contributes to apparent blurring and

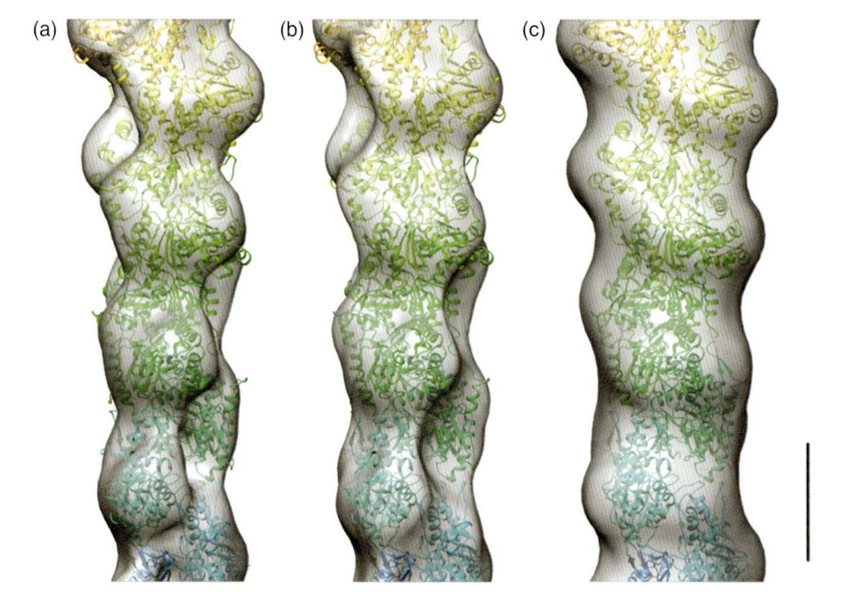

Figure 27.13 Accuracy of protein structure replication by negative stained- and metal-coated in-lens SE FESEM. (a–c) F-actin atomic model (Holmes *et al.*, 2003) docked into negative stain TEM, negative stain SEM and rotationally shadowed tungsten reconstructions at 2–3 nm resolution. In all three cases the reconstructed density follows the outer contour of the atomic model and allows the three-dimensional location and orientation of the individual actin monomers (42 kDa, 4–7 nm diameter) within the filament to be determined. Structural preservation is similar between TEM- and SEM-negative stain (a, b), but the ~1 nm thick layer of tungsten coating the surface of the metal shadowed F-actin obscures the finer structural details (c). Scale bar: 5 nm.

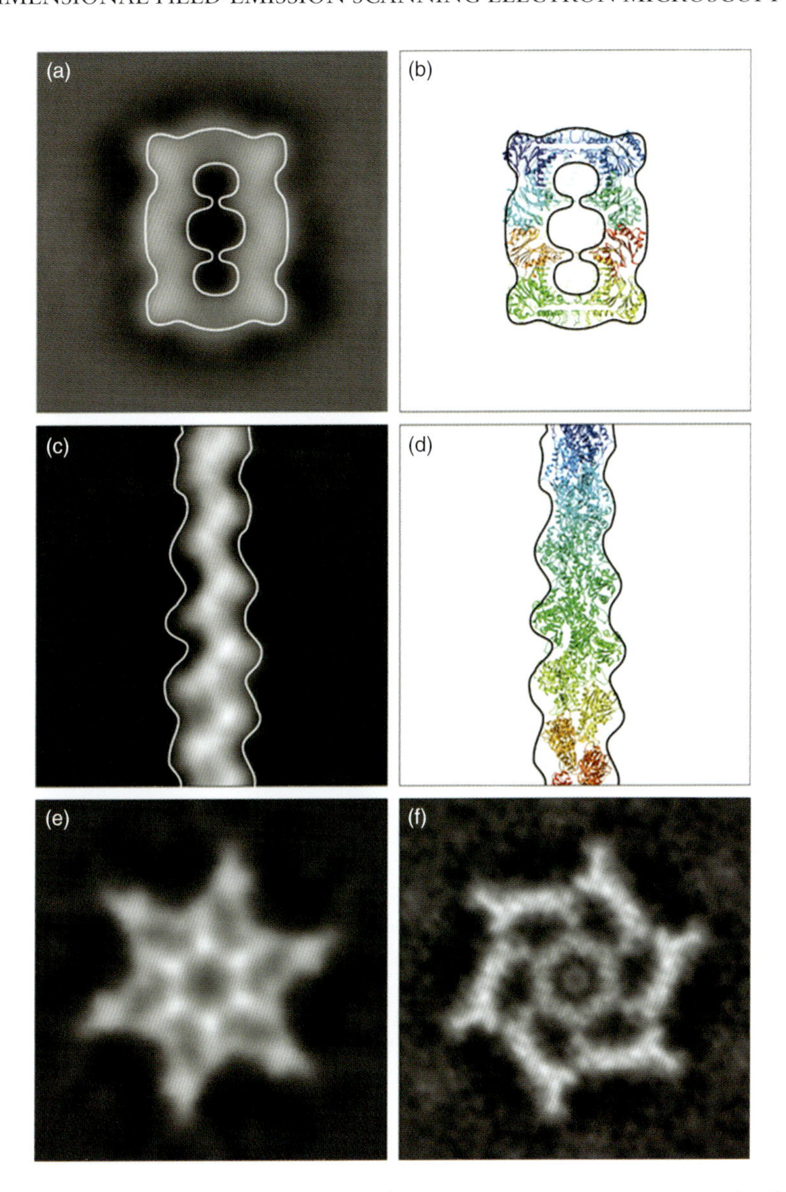

Figure 27.14 Negative stain macromolecular in-lens SE FESEM 3D reconstruction leading to internal details. (a) Cross-section through the centre of a proteasome core particle 3D FESEM reconstruction at ~3 nm resolution. The surface of the macromolecular complex is defined by thresholding (shown in white) to enclose the known volume of the complex. Clear internal cavities can be seen, which correspond to those seen in (b) the high-resolution cryo-TEM reconstruction of the *T. acidophilum* proteasome (PDB 3c91) (Rabl *et al.*, 2008). (c) Cross-section through the centre of the actin filament and (d) comparison with atomic resolution model (Holmes *et al.*, 2003). (e) Cross-section through a 3DFESEM bacteriophage T4 tail helical reconstruction, cavities between the tail proteins and those corresponding to the tail tube correspond to those seen in (f) a cryo-TEM reconstruction of the T4 tail (Kostyuchenko *et al.*, 2005; EMDB ID: 1126). Scale bar: 10 nm.

difficulties in the interpretation of FE/SEM macromolecular images, because components of the signal arising from the surface are summed with those from deeper within the sample. Backprojection-based three-dimensional reconstruction deconvolutes this information and provides clear internal details arising from stain that has penetrated into the interior of the protein complex (Figure 27.14). In contrast-inverted FE/SEM negative stain images, the protein component appears white; although the surrounding stain is black, stain-excluded regions corresponding to individual protein monomers are clearly visualized (Figure 27.14a, c and e).

REFERENCES

Amat, F., Moussavi, F., Comolli, L.R., Elidan, G., Downing, K.H. and Horowitz, M. (2008) Markov random field based automatic image alignment for electron tomography. *Journal of Structural Biology*, 161 (3), 260–275.

Bachmann, R., Abermann, R.A. and Zingsheim, H.P. (1969) Hochauflosende Gefrieratzung. *Histochimie*, 20, 133–142.

Bartesaghi, A., Merk, A., Banerjee, S., Matthies, D., Wu, X., Milne, J.L.S. and Subramaniam, S. (2015) 2.2 Å resolution cryo-EM structure of β-galactosidase in complex with a cell-permeant inhibitor. *Science*, 348, 1147–1151.

Baumeister, W. (2005) From proteomic inventory to architecture. *FEBS Letters*, 579, 933–937.

Beil, W. and Carlsen, I.C. (1991) Surface reconstruction from stereoscopy and 'shape form shading' in SEM images. *MVA*, 4, 271–285.

Belnap, D.M., Olson, N.H. and Baker, T.S. (1997) A method for establishing the handedness of biological macromolecules. *Journal of Structural Biology*, 120, 44–51.

Berger, D. (2007) Characterization of a 4-quadrant-large-angles BSE-detector for in-situ 3D quantitative analysis of the sample morphology in SEM. *Microscopy and Microanalysis*, S03, 74–75.

Bohm, J., Frangakis, A.S., Hegerl, R., Nickell, S., Typke, D. and Baumeister, W. (2000) Toward detecting and identifying macromolecules in a cellular context: Template matching applied to electron tomograms. *Proceedings of the National Academy of Sciences*, 97 (26), 14245–14250.

Cazaux, J. (2004) Recent developments and new strategies in scanning electron microscopy. *Journal of Microscopy*, 217, 16–35.

Chen, J.Z. and Grigorieff, N. (2007). SIGNATURE: A single-particle selection system for molecular electron microscopy. *Journal of Structural Biology*, 157, 168–173.

Cornille, N., Garcia, D., Sutton, M.A., McNeill, S.R. and Orteu, J.J. (2003) Automated 3-D reconstruction using a scanning electron microscope, in *SEM Conference on Experimental and Applied Mechanics*, Charlotte, North Carolina, 2–4 June 2003.

de Boer, P., Hoogenboom, J.P. and Giepmans, B.N.G. (2015) Correlated light and electron microscopy: Ultrastructure lights up! *Nature Methods*, 12, 503–513.

Denk, W. and Horstmann, H. (2004) Serial block-face scanning electron microscopy to reconstruct three-dimensional tissue nanostructure. *PLoS Biol.*, 2 (11), e329.

Fakron, O.M. and Field, D.P. (2015) 3D image reconstruction of fiber systems using electron tomography. *Ultramicroscopy*, 149, 21–25.

Frank, J. (2006) *Three-Dimensional Electron Microscopy of Macromolecular Assemblies*, Oxford University Press, New York.

Frank, J. (ed.) (2008) *Electron Tomography: Methods for Three-Dimensional Visualization of Structures in the Cell, Springer*, Albany, NY.

Frank, J., Radermacher, M., Penczek, P., Zhu, J., Li, Y., Ladjadj, M. and Leith, A. (1996) SPIDER and WEB: Processing and visualization of images in 3D electron microscopy and related fields. *Journal of Structural Biology*, 116, 190–199.

Goldberg, M.W., Wiese, C., Allen, T.D. and Wilson, K.L. (1997) Dimples, pores, star-rings, and thin rings on growing nuclear envelopes: Evidence for structural intermediates in nuclear pore complex assembly. *Journal of Cell Science*, 110, 409–420.

Goldstein, J.I. (ed.) (2012) *Practical Scanning Electron Microscopy: Electron and Ion Microprobe Analysis*, Springer Science & Business Media.

Goldstein, J.I., Newbury, D.E., Echlin, P., Joy, D.C., Fiori, C. and Lifshin, E. (eds) (1992) *Scanning Electron Microscopy and X-Ray Microanalysis*, 2nd edn, Plenum Press, New York.

Han, B.G., Dong, M., Liu, H., Camp, L., Geller, J., Singer, M., Hazen, T.C., Choi, M., Witkowska, H.E., Ball, D.A., Typke, D., Downing, K.H., Shatsky, M., Brenner, S.E., Chandonia, J.M., Biggin, M.D. and Glaeser, R.M. (2009) Survey of large protein complexes in *D. vulgaris* reveals great structural diversity. *Proc. Natl Acad. Sci. USA*, 106, 16580–16585.

Harauz, G. and van Heel, M. (1986) Exact filters for general geometry three dimensional reconstruction. *Proceedings of the IEEE Computer Vision and Pattern Recognition Conference*, 73, 146–156.

Hemmleb, M. and Schubert, M. (1997) Digital microphotogrammetry – determination of the topography of microstructures by scanning electron microscope, in *Conference Proceedings of the Second Turkish-German Joint Geodetic Days*, pp. 745–752.

Holmes, K.C., Angert, I., Kull, F.J., Jahn, W. and Schröder, R.R. (2003). Electron cryo-microscopy shows how strong binding of myosin to actin releases nucleotide. *Nature*, 425 (6956), 423–427.

Hrabe, T., Chen, Y., Pfeffer, S., Cuellar, L.K., Mangold, A.V. and Förster, F. (2012) PyTom: A python-based toolbox for localization of macromolecules in cryo-electron tomograms and subtomogram analysis. *Journal of Structural Biology*, 178 (2), 178–188.

Klug, A. and Finch, J.T. (1968) Structure of viruses of the papilloma-polyoma type: IV. Analysis of tilting experiments in the electron microscope. *Journal of Molecular Biology*, 31, 1–12.

Kostyuchenko, V.A., Chipman, P.R., Leiman, P.G., Arisaka, F., Mesyanzhinov, V.V. and Rossmann, M.G. (2005) The tail structure of bacteriophage T4 and its mechanism of contraction. *Nature, Structural & Molecular Biology*, 12 (9), 810–813.

Kremer, J.R., Mastronarde, D.N. and McIntosh, J.R. (1996) Computer visualization of three-dimensional image data using IMOD. *Journal of Structural Biology*, 116 (1), 71–76.

Lamvik, M.L. (1991) Radiation damage in dry and frozen hydrated organic material. *Journal of Microscopy*, 16, 171–181.

Lück, S., Sailer, M., Schmidt, V. and Walther, P. (2010) Three-dimensional analysis of intermediate filament networks using SEM tomography. *Journal of Microscopy*, 239, 1–16.

Ludtke, S.J., Baldwin, P.R. and Chiu, W. (1999) EMAN: Semiautomated software for high-resolution single-particle reconstructions. *Journal of Structural Biology*, 128 (1), 82–97.

Mastronarde, D.N. (1997) Dual-axis tomography: An approach with alignment methods that preserve resolution. *Journal of Structural Biology*, 120, 343–352.

Medalia, O., Weber, I., Frangakis, A.S., Nicastro, D., Gerisch, G. and Baumeister, W. (2002) Macromolecular architecture in eukaryotic cells visualized by cryoelectron tomography. *Science*, 298, 1209–1213.

Messaoudii, C., Boudier, T., Sanchez Sorzano, C.O. and Marco, S. (2007) TomoJ: Tomography software for three-dimensional reconstruction in transmission electron microscopy. *BMC Bioinformatics*, 8, 288.

Miller, J.L., Woodward, J.D., Chen, S., Jaffer, M., Weber, B., Nagasaki, K., Tomaru, Y., Wepf, R., Roseman, A., Varsani, A. and Sewell, B.T. (2011) Three-dimensional reconstruction of *Heterocapsa circularisquama* RNA virus by electron cryo-microscopy. *Journal of General Virology*, 92 (8), 1960–1970.

Nickell, S., Förster, F., Linaroudis, A., Net, W.D., Beck, F., Hegerl, R., Baumeister, W. and Plitzko, J.M. (2005) TOM software toolbox: Acquisition and analysis for electron tomography. *Journal of Structural Biology*, 149 (3), 227–234.

Okyay, G., Héripré, E., Reiss, T., Haghi-Ashtiani, P., Auger, T. and Enguehard, F. (2016) Soot aggregate complex morphology: 3D geometry reconstruction by SEM tomography applied on soot issued from propane combustion. *Journal of Aerosol Science*, 93, 63–79.

Palenstijn, W.J., Batenburg, K.J. and Sijbers, J. (2011) Performance improvements for iterative electron tomography reconstruction using graphics processing units (GPUs). *Journal of Structural Biology*, 176 (2), 250–253.

Pasteur, L. (1850) Recherches sur les propriétés spécifiques des deux acides qui composent l'acide racémique. *Ann. Chim. Phys.*, 3e sér., 28, 56–99.

Pawley, J. (1997) The development of field-emission scanning electron microscopy for imaging biological surfaces. *Scanning*, 19, 324–336.

Perkins, G., Sun, M.G. and Frey, T.G. (2009) Chapter 2: Correlated light and electron microscopy/electron tomography of mitochondria *in situ*. *Methods in Enzymology*, 456, 29–52.

Pettersen, E.F., Goddard, T.D., Huang, C.C., Couch, G.S., Greenblatt, D.M., Meng, E.C. and Ferrin, T.E. (2004) UCSF Chimera – a visualization system for exploratory research and analysis. *J. Comput. Chem*, 25: 1605-1612.

Piazzesi, G. (1973) Photogrammetry with the scanning electron microscope. *Journal of Physics E: Scientific Instruments*, 6, 392–396.

Rabl, J.I., Smith, D.M., Yu, Y., Chang, S.C., Goldberg, A.L. and Cheng, Y. (2008) Mechanism of gate opening in the 20S proteasome by the proteasomal ATPases. *Molecular Cell*, 30 (3), 360–368.

Radon, J. (1917) Über die Bestimmung von Funktionen durch ihre Integralwerte längs gewisser Mannigfaltigkeiten. Ber Verh Sächs Akad Wiss Leipzig, *Math. Nat. kl*, 69, 262–277.

Ress, D., Harlow, M.L., Schwarz, M., Marshall, R.M. and McMahan, U.J. (1999) Automatic acquisition of fiducial markers and alignment of images in tilt series for electron tomography. *Journal of Electron Microscopy*, 48 (3), 277–287.

Roseman, A.M. (2003) Particle finding in electron micrographs using a fast local correlation algorithm. *Ultramicroscopy*, 94 (3–4), 225–236.

Scherer, S. and Kolednik, O. (2001) A new system for automatic surface analysis in SEM. *Microscopy and Analysis*, 70, 15–17.

Seiler, H. (1967) Einige aktuelle Probleme der Sekundärelektronenemission. *Z. angew. Phys.*, 22, 249.

Stampfl, J., Scherer, S., Gruber, M. and Kolednik, O. (1996) Reconstruction of surface topographies by scanning electron microscopy for application in fracture research. *Applied Physics A*, 63, 341–346.

Stockroos, I., Kalicharan, D., Van Der Want, J.J.L. and Jongebloed, W.L. (1998) A comparative study of thin coatings of Au/Pd, Pt and Cr produced by magnetron sputtering for FE-SEM. *Journal of Microscopy*, 189, 79–89.

Tafti, A.P., Kirkpatrick, A.B., Alavi, Z., Owen, H.A. and Yu, Z. (2015) Recent advances in 3D SEM surface reconstruction. *Micron*, 78, 54–66.

Tyler, J.M. and Branton, D. (1980) Rotary shadowing of extended molecules dried from glycerol. *J. Ultrastruct. Res.*, 71 (2), 95–102.

van Heel, M. and Harauz, G. (1986) Resolution criteria for three dimensional reconstruction. *Optik*, 73, 119–122.

Vant' Hoff, J.H. (1875) *La Chimie dans l'Espace*, Rotterdam.

Voss, N.R., Yoshioka, C.K., Radermacher, M., Potter, C.S. and Carragher, B. (2009) DoG Picker and TiltPicker: Software tools to facilitate particle selection in single particle electron microscopy. *Journal of Structural Biology*, 166 (2), 205–213.

Wade, R.H. (1984) The temperature dependence of radiation damage in organic and biological material. *Ultramicroscopy*, 14, 265–270.

Wepf, R., Aebi, U., Bremer, A., Haider, M., Tittmann, P., Zach, J. and Gross, H. (1994) High resolution SEM of biological macromolecular complexes. *Proceedings of the Annual Meeting – Electron Microscopy*. San Francisco Press, CA, USA.

Winkler, H. (2007) 3D reconstruction and processing of volumetric data in cryo-electron tomography. *Journal of Structural Biology*, 157 (1), 126–137.

Wolff, P.A. (1954) Theory of secondary electron cascade in metals. *Physical Review*, 95 (1), 56–66.

Woodward, J.D. and Sewell, B.T. (2010) Tomography of asymmetric bulk specimens imaged by scanning electron microscopy. *Ultramicroscopy*, 110 (2), 170–175.

Woodward, J.D. and Wepf, R. (2014) Macromolecular 3D SEM reconstruction strategies: Signal to noise ratio and resolution. *Ultramicroscopy*, 144, 43–49.

Woodward, J.D., Wepf, R. and Sewell, B.T. (2009) Three-dimensional reconstruction of biological macromolecular complexes from in-lens scanning electron micrographs. *Journal of Microscopy*, 234 (3), 287–292.

Woodward, J.D., Weber, B., Scheffer, M., Benedik, M.J., Hoenger, A. and Sewell, B.T. (2008) Helical structure of unidirectionally shadowed metal replicas of cyanide hydratase from *Gloeocercospora sorghi*. *Journal of Structural Biology*, 161 (2), 111–119.

Woolford, D., Ericksson, G., Rothnagel, R., Muller, D., Landsberg, M.J., Pantelic, R.S., McDowall, A., Pailthorpe, B., Young, P.R., Hankamer, B. and Banks, J. (2007) SwarmPS: Rapid, semi-automated single particle selection software. *Journal of Structural Biology*, 157, 174–188.

Zhao, J., Brubaker, M.A. and Rubinstein, J.L. (2013) TMaCS: A hybrid template matching and classification system for automated particle selection. *Journal of Structural Biology*, 181 (3), 234–242.

28

Element Analysis in the FEGSEM: Application and Limitations for Biological Systems

Alice Warley[1] and Jeremy N. Skepper[2]

[1]*Centre for Ultrastructural Imaging, King's College London, London, UK and Visiting Professor Department of Histology and Cell Biology, Faculty of Medicine, University of Granada, Granada, Spain*

[2]*Cambridge Advanced Imaging Centre, University of Cambridge, Cambridge, UK*

28.1 INTRODUCTION

Since its commercial development in the 1980s energy dispersive spectroscopy (EDS, also known commonly as X-ray microanalysis) has become an established technique in the imaging lab for the identification and quantification of elemental content within biological specimens. The advantages of EDS compared to flame photometry or inductively coupled plasma mass spectrometry are the ability to localise elements to a specific cell within a tissue, or to a specific localisation within a cell, as well as its multielement capability. Whilst techniques such as nanoSIMS (Clode *et al.* 2007) and proton and synchrotron microprobe analysis (Ortega *et al.* 2009) have similar capabilities they remain more specialised whereas EDS is generally available in the majority of EM facilities.

Compared to transmission electron microscopy, which is used in analytical applications that require information at high, (nm) spatial resolution (Porter *et al.* 2007; Aronova and Leapman 2012; Brown and Hondow 2013), the SEM has the advantage of ease of use, so that the technique lends itself to rapid screening of samples. EDS in the SEM is well established for pathological diagnosis (Ingram *et al.* 1999; Dvorackova *et al.* 2015) (see also Chapter 14 by Gunning and Hauröder in this volume), biomedical research (Reynolds *et al.* 2004), cell physiology (Hall and Gupta 1982; Roomans 1999; Marshall and Xu 1998; Marshall *et al.* 2012), apoptosis (Fernandez-Segura *et al.* 1999b; Skepper *et al.* 1999; Alaminos

Biological Field Emission Scanning Electron Microscopy, First Edition.
Edited by Roland A. Fleck and Bruno M. Humbel.

et al. 2007), plant cell research (Echlin and Taylor 1986; Canny and Huang 1993; McCully *et al.* 2000) and environmental studies (Zapotoczny *et al.* 2007; Tylko *et al.* 2005). More recently, the burgeoning field of nanoparticle research has extended the field of applications for EDS in the SEM. In addition, the proposal to use EDS in conjunction with FIB milling for 3D element analysis (Scott and Ritchie 2009; Scott 2011) has opened up new possible areas for the investigation of biological specimens.

Continued refinement of field emission guns (see also Chapter 1 by Joy in Volume I) and the development of silicon drift detectors (Newbury 2006; Burgess *et al.* 2013) have together considerably improved both the resolution and the sensitivity of EDS in the SEM. However, the ease of use of modern microscopes and analytical software can make such investigations seem disarmingly simple but, for success, these applications require a good understanding of the principles and the limitations of the EDS technique. In this chapter, we will briefly discuss specimen preparation, which for biologists is the most important step in the analytical process. We will introduce the principles that underlie X-ray generation in the SEM and problems likely to be encountered in analysis, the processes for element identification and quantification of results will be outlined and, where appropriate, the advantages and disadvantages for using an FEGSEM will be emphasised. This chapter can only act as a brief overview. The reader must understand that there is no set procedure either for specimen preparation or for carrying out X-ray microanalysis. We have tried to point out the options available and to reference pertinent papers that the reader can follow in their investigations.

28.2 SPECIMEN PREPARATION

For structural studies using the FEGSEM the emphasis must be on preserving the structure of the sample so that it most closely resembles the tissue or cells in their living state. In studies using microanalysis there has to be a compromise between adequate structural preservation and the necessity for retention of the elements of interest. The options for specimen preparation are: aqueous or anhydrous chemical fixation followed by critical point drying, freeze-drying or resin-embedding; cryo-immobilisation followed by either freeze-drying or freeze-substitution and resin-embedding; or bulk samples can be frozen and prepared for analysis in the frozen hydrated state. The various different methods for freezing samples have been reviewed in Skepper (2000).

28.2.1 Chemical Fixation, Critical Point Drying, Freeze-Drying or Resin-Embedding

Chemical fixation produces very good ultrastructural preservation, but all ions are lost during the early stages of aldehyde fixation when cell membranes are permeabilised. In addition, heavy metals used in fixatives and stains (Os, Pb, Fe and U) may substitute for endogenous elements in the sample; for example, Ca and P in hydroxyapatite readily exchange with Os and U. Conditions of fixation can be altered to avoid extraction or exchange of elements; for example, samples containing hydroxyapatite may be fixed in aldehydes buffered to pH 8.0 or, if the chemistry is more important than structure, they can be fixed in cold methanol.

After choice of primary fixation the samples should be dehydrated without further treatment with osmium or uranyl acetate. Dehydration can be achieved either by critical point drying or by freeze-drying after cryo-immobilisation. If mineral inclusions are the sole interest, cryo-immobilised material can be air dried. Since the major loss or exchange of elements happens prior to these processes it is unlikely any further loss of elements will occur.

An alternative to critical point drying is resin-embedding and, since most commercial FEGSEMs come with the option of a STEM detector, the analysis of thin resin sections with resultant improved spatial resolution is becoming increasingly common (Hondow *et al.* 2011). If such an approach is intended, it should be realised that the action of floating a resin section on the water bath of a diamond knife may extract elements (Boothroyd 1964). For samples that are very difficult to section, such as mineralised bone or atheroma, large areas of the block face of resin-embedded samples can be polished relatively easily using a Histo-diamond knife. Material prepared in this way is compatible with serial block face imaging in the dual beam FIB FEGSEM (see Chapter 24 by Giannuzzi and Chapter 26 by Kizilyaprak *et al.*) or 3-View (see Chapter 23 by Genoud). Using these techniques image acquisition can be alternated with X-ray mapping.

28.2.2 Cryo-Immobilisation and Freeze-Drying

Cryo-immobilisation is the only method that can retain ions at their site of origin at sub-cellular resolution. Its major limitation is the increasing size of ice crystals formed as a function of depth in the sample during the freezing process. In specimens that have been cryo-immobilised by impact on to polished metal mirrors or by plunging into liquid cryogens cooled in liquid nitrogen, there is only a small region, typically no more than 10 μm in depth, with ice crystals less than 10–15 nm in size (Figure 28.1). High pressure freezing

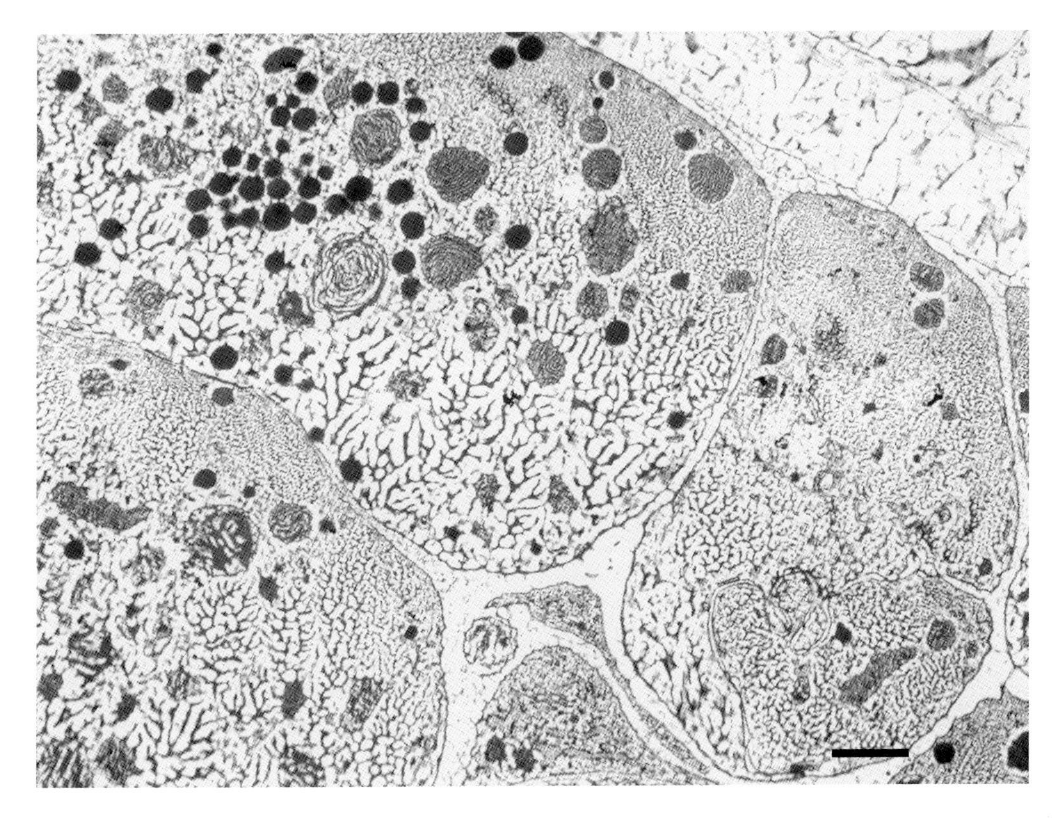

Figure 28.1 Atrial cardiac tissue cryoimmobilized by impact against a polished, cooled copper block. The smallest ice crystals are found in the surface regions of the cells closest to the site of impact. Marker = 1 μm. Reproduced from Skepper (2000) with permission.

increases the depth of well frozen material to perhaps 200 μm but, although hyperbaric freezing has been used successfully for the study of plant tissue (Frey *et al.* 1997), the only use of this method for the study of elemental distribution in mammalian cells that the authors are aware of is on erythrocytes (Zierold *et al.* 1991), which are atypical due to their lack of nucleus and high dry mass content. If this approach is envisaged for other types of cell or tissue the effects of high pressure freezing on element distribution will need to be determined. For EDS analysis of *bulk* specimens in the FEGSEM, it may not be necessary to strive for the best cryo-immobilisation since, even though a probe of a few nanometres in size can be formed, when that probe penetrates into soft biological tissue the resulting interaction volume, measured in μm, is generally much greater than the size of the individual ice crystals measured in nm. This is discussed further in Section 28.4.

Monolayers of cell cultures grown on either gold grids or Millicell inserts (Millicell-PCF®, Milipore, Bedford, MA) can be prepared quite simply by rapid washing in ice cold distilled water to remove overlying medium and cryo-immobilised in a liquid cryogen (Skepper *et al.* 1999; Fernandez-Segura *et al.* 1999a). If samples are to be analysed as sections in STEM mode, more care will be required with cryo-preparation and the procedures developed for TEM analysis will need to be followed. Cells in suspension or red blood cells can be concentrated by centrifugation and submicrolitre volumes mounted on pins and frozen in liquid propane or ethane (Fernandez-Segura *et al.* 1999a; Warley and Skepper 2000; Mauritz *et al.* 2011). Larger samples such as plant roots may be frozen by plunging into cryogen or by impact on to a metal mirror or with cryo-pliers (McCully *et al.* 2000). A specimen prepared in this way is shown in Figure 28.1. Once the sample is frozen it can be stored under liquid nitrogen or in the vapour phase of liquid nitrogen indefinitely.

Monolayers of cells are freeze-dried before analysis (Skepper *et al.* 1999; Fernandez-Segura *et al.* 1999a). Frozen samples mounted on pins can be transferred to a cryoultramicrotome and sections prepared and mounted on Formvar coated TEM grids or Formvar coated holes in Melinex coverslips before freeze-drying. Freeze-dried specimens are analysed in the FEGSEM using either secondary electron or backscattered electron detectors or STEM detectors to locate areas of interest.

As an alternative to freeze-drying, frozen bulk samples can be further prepared by freeze-substitution followed by resin-embedding, but there is a need to ensure that substitution does not result in element redistribution. Marshall's group developed freeze-substitution against acrolein in diethyl ether and showed that this did not cause redistribution of either diffusible elements or minerals (Marshall and Wright 1991; Marshall *et al.* 2007). Others have used substitution in the presence of precipitating agents such as oxalate (Thirion *et al.* 1997) or KF (Hardt and Plattner 1999) to maintain Ca at its intracellular sites during the substitution process. Once embedded such preparations can be used for block face imaging, as described earlier.

28.2.3 Preparation for Analysis in the Frozen-Hydrated State

Bulk frozen specimens can be transferred to the preparation chamber of a cryoFEGSEM, fractured, etched and coated before transfer to the microscope for analysis in the frozen-hydrated state. Alternatively, the pin with a sample of tissue can be planed in a cryoultramicrotome before transfer to the cryoFEGSEM and sufficient ice can be removed by sublimation to identify surface features such as plant cell vacuoles before analysis (McCully *et al.* 2000; Huang *et al.* 1994). More recently, Marshall's group (Marshall and Xu 1998; Marshall *et al.* 2012) have avoided the sublimation step, which can compromise quantitative analysis and used element imaging to provide structural information from the frozen, planed specimen surface. The advantage of analysis in the frozen-hydrated state is that specimens are analysed as near as possible to the *in vivo* situation, allowing more relevant quantitative information to be obtained (Section 28.7.2).

28.2.4 Choice of Coating Material

The traditional choice of coating material was carbon as it is almost featureless even when >20 nm are deposited. Its X-ray peak is also well below those of most elements with physiological significance. Carbon is deposited by evaporation so there is a danger of thermal damage to soft samples. The current generation of FEGSEMs have such high resolution that they require very thin layers of metal coating with the finest grain size since, if the coating layer is more than a few nanometres in thickness, it will obscure surface detail. This is less important for microanalysis but the ability to eliminate surface charging with a thin layer of metal is useful and metals such as aluminium, chromium, gold, iridium, platinum or gold/palladium can all be deposited by sputter coating as layers as thin as 1 nm. On analysis, gold, iridium and platinum all generate multiple X-ray peaks that overlie peaks of biological interest, but most analytical software allows these to be identified and separated from those generated by the biological material. Chromium and aluminium do not overlie peaks of biological interest, but are, however, both oxidising metals and as they oxidise [O] will increase. Aluminium has been the metal of choice for coating specimens destined for analysis in the frozen-hydrated state, where it helps to stop dissipation of the charge over the surface of the sample and enables beam penetration into the specimen (Marshall 1987).

28.3 PRODUCTION OF X-RAYS AND X-RAY DETECTION

28.3.1 Production of X-Rays

Element analysis in the EM takes advantage of the interactions of the incident electron beam with atoms in the specimen that result in the emission of X-rays. Rearrangement of orbital electrons after the ejection of inner shell electrons by the incident electron beam results in the production of characteristic X-rays, the energy of which depends on both the atomic number of the atom in which the transition is occurring and the electron shells involved; these characteristic X-rays are used for element identification. X-rays are also produced when the electron beam is slowed by interaction with atomic nuclei in the specimen, resulting in a continuous band of radiation beneath the characteristic X-ray peaks, known variously as continuum radiation, bremsstrahlung (braking radiation) or white radiation. This background radiation limits the sensitivity of the technique, which is determined by the ability to clearly determine the characteristic peak from the background counts.

In the SEM the operating voltage used for analysis directly affects X-ray production. Excitation of characteristic X-rays only occurs provided that incident electrons have sufficient energy to cause inner cell ionisations. For a given atom, the energy required to cause inner shell ionisations is known as the critical excitation potential, the value of which varies increasing with atomic number and with the orbital shell of the electrons involved. Maximum X-ray production occurs when the operating voltage is 2 to 3 times higher than the energy required for excitation of a given X-ray line. The ratio of beam energy to critical excitation potential is known as the overvoltage. At a given overvoltage the number of X-rays produced also depends on the current in the focused probe, increasing as the probe current increases. The probability that an X-ray will be emitted from the specimen is termed the fluorescent yield, the value of which also increases with atomic number and is greater for K X-ray lines than for L X-ray lines.

28.3.2 X-Ray Detection

Emitted X-rays are collected either according to their wavelength using the technique of wavelength dispersive spectroscopy (WDS) or their energy, using energy dispersive X-ray spectroscopy (EDS). A comparison of the two techniques is given in Table 28.1.

Despite its superior resolution, WDS has been the technique least used for biological applications. WDS has a very limited capability for the detection of elements since each detector crystal can only detect one element at a time (although up to four detectors may be fitted to a column). In addition, high electron doses are required due to the low detection efficiency and this can result in severe specimen damage. Nevertheless, the high sensitivity makes WDS a complementary technique for the study of elements present in low or trace concentrations and its high resolution may be required for applications that have a mineral component (Rasch *et al.* 2003) or where severe peak overlap occurs.

EDS using an Si(Li) detector has been most commonly used for the analysis of biological specimens. The main advantages are its ease of use and the reduced electron dose requirement due to both the multielement capability (all elements present are detected in a single analytical run) and better efficiency because of the increased solid angle of detection. Nevertheless, the resolution of these detectors (defined as full width half maximum height) is much poorer than that of WDS spectrometers due to broadening of peaks during the collection process.

Over the past decade there have been considerable improvements in EDS detector performance with the introduction of silicon drift detectors (SDD) (Newbury 2006; Burgess *et al.* 2013). With these the process of X-ray detection is the same as in the Si(Li) detector but their design makes them much less sensitive to noise, so that operation at liquid nitrogen temperatures is not required. Peltier cooling is sufficient, resulting in a more compact piece of equipment that allows clusters of detectors to be fitted to a single column (Newbury and Ritchie 2015). The resolution of the SDD does not deteriorate with increased surface area so the use of large surface area detectors and/or clustered detectors results in considerable improvements in detection efficiency. SDDs also have the advantages of shorter processing times compared to Si(Li) detectors and their resolution is not compromised at high count rates. These improvements allow the operator to manipulate beam current and live time to reduce either dosage and damage to the specimen or to improve sensitivity through higher

Table 28.1 Comparison of the performance of wavelength dispersive (WDS) and energy dispersive (EDS) X-ray spectrometers

		EDS	
	WDS	Si(Li)	SDD
Multielement	Limited	Yes	Yes
Detection efficiency	~0.5%	~5%[a]	> 5%[a]
Resolution (FWHM)	2–20 eV	130–180 eV[b]	Similar to Si(Li)[b] Does not deteriorate at high count rate
Required dose	High	Low	Low
Counting time	Long	Short	Short
Trace elements	Yes	No	Yes (Huang *et al.* 1994)
Spectrum artefacts		Yes	Yes

[a] Detection efficiency depends on the solid angle, which also depends on the specimen-to-detector distance and the surface area of the detector crystal/wafer.
[b] For Si(Li) detectors resolution deteriorates with increased count rates and with increasing detector surface area. SDDs do not suffer from these problems.

count rates. It has been shown that for materials science specimens SEM fitted with SDD can equal the performance of a WDS equipped electron probe microanalyser (Newbury and Ritchie 2015). However, it should be noted that such increased performance may not be achievable with biological specimens due to their inherently low count rates.

28.4 INTERACTION VOLUME

In an SEM electrons from the incident electron beam that penetrate the specimen are scattered by the mass of the specimen and lose energy until they are eventually stopped. During their passage through the specimen those electrons that have sufficient energy excite X-ray production from atoms within the depth of the specimen and, if these X-rays have sufficient energy, they will either interact with other atoms or exit the specimen surface and be detected. The sum of these processes results in an interaction volume (Figure 28.2), which has several consequences for analysis. These are:

(i) The volume from which X-rays are emitted is greater both in depth and width than the area on the surface defined by the focused electron beam so that a given X-ray peak cannot necessarily be localised to a specific surface feature but may originate some

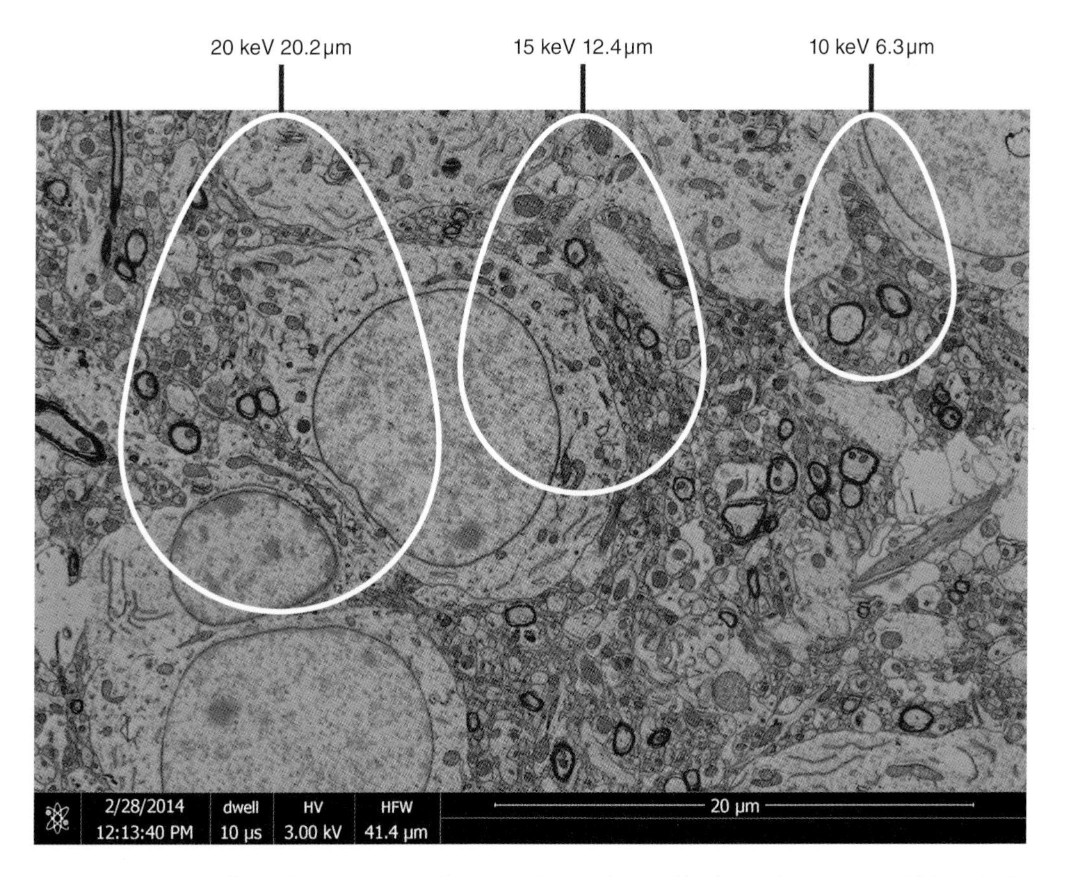

Figure 28.2 The effect of interaction volume on the analysis of biological specimens. Although the beam is focused on to a point on the specimen surface, electrons penetrate into the depth of the specimen and the beam broadens due to interaction of the electrons with the specimen mass so that X-rays are emitted from a region that is much greater than that defined by the focused electron probe. Penetration and broadening of the beam is greater when higher accelerating voltages are used.

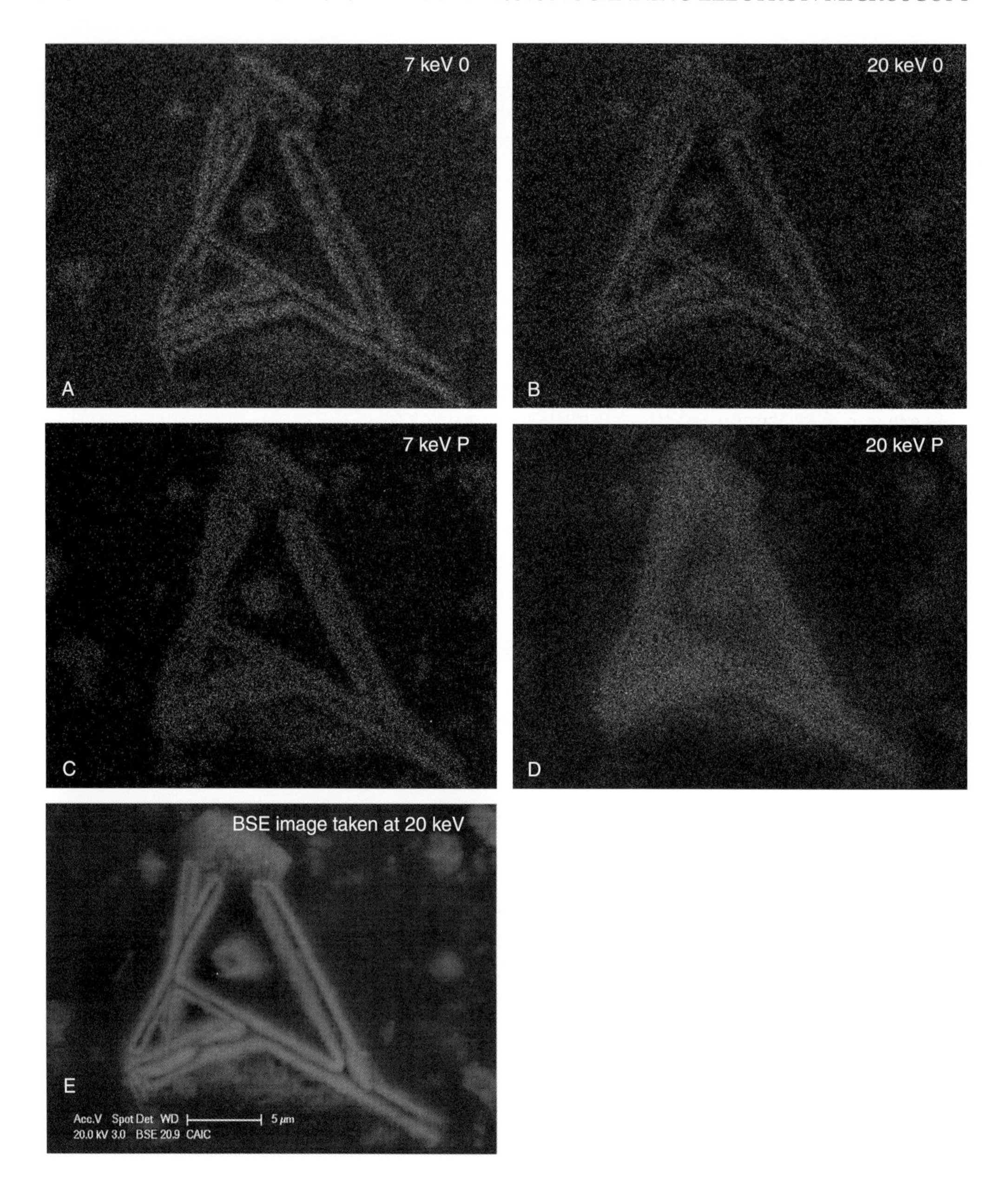

Figure 28.3 The effect of accelerating voltage on X-ray mapping of biological specimens. X-ray maps acquired at voltages of 7 and 20 keV of the K_α lines of O and P in a specimen comprised of crystals of hydroxyapatite embedded in resin and the corresponding BSE image. Maps acquired at 20 keV are blurred due to the spreading of the beam, resulting in the production of X-rays from points below the visible surface of the specimen.

distance away, that is the spatial resolution of the analysis is reduced. The emission of X-rays from a wide source below the specimen surface makes mapping difficult in low atomic number specimens (Figure 28.3).

(ii) The actual size of the interaction volume varies for different elements, being greatest for low atomic number elements that require less energy for X-ray production.

(iii) X-rays excited below the specimen surface may be lost due to absorption within the specimen, a process that may cause secondary X-ray emission; both of these affect the number of X-rays detected in the resultant spectrum and thus quantification.
(iv) The various interactions of incident electrons that produce the interaction volume lead to damage to the specimen, which is increased if a high beam current is used.

28.4.1 Size of Interaction Volume in Biological Specimens

The size of the interaction volume in a given specimen is determined by both the initial accelerating voltage and the composition of the specimen itself. Higher accelerating voltages result in increased penetration of the beam into the specimen and thus a larger interaction volume. Penetration of the beam into the specimen, and thus the interaction volume, is also greater in specimens of low atomic number, such as the majority of biological specimens. Knowledge about the extent of interaction volume is necessary to be able to interpret experimental data in a meaningful way.

The extent of beam penetration can be determined experimentally or calculated from theory. Experiments in one of the author's labs have shown penetration through a 5 μm thick epoxy resin placed on a gold coated Al stub generating peaks from the underlying gold coat and Al stub at operating voltages of 15 keV or 12 keV. At 10 keV the interaction volume remained within the thickness of the epoxy resin section and no peaks for gold and aluminium were generated.

For cryoimmobilized/freeze-dried material Fernandez-Segura *et al.*, using the Bethe range equation, calculated that beam penetration into cells would be 20 μm at 20 keV, but would be reduced to 6.3 μm at 10 keV; these calculations were validated experimentally (Fernandez-Segura *et al.* 1999a). Marshall *et al.*, using Monte Carlo calculations, obtained similar values, a depth of approximately 17 μm at 15 keV in freeze-dried material (Marshall *et al.* 2012). These results demonstrate that the use of high accelerating voltage (15 to 20 keV) on freeze-dried biological specimens will cause the beam to overpenetrate individual cells so that any spectra obtained will be a composite of either cells and interstitium or cells and their underlying substrate (see Figure 28.2). However, overpenetration has an advantage for specimens in which relatively large ice crystals have formed (Figure 28.1). The interaction volume (Figure 28.2) is much greater than the volume of individual ice crystals so that signals from both ions and the organic mass displaced by the growing ice crystal are included in the resulting spectrum. If lower voltages are used to minimise beam penetration then the interaction volume will be confined to the part of the specimen where ice crystal size is minimal so that again analytical results should not be affected by their presence.

Beam penetration is reduced in biological specimens examined in the frozen hydrated state. Both Echlin (2001) and Marshall *et al.* (2012) calculated depth penetration in fully frozen hydrated specimens. Echlin found values of 4 μm for 10 keV and Marshall 4 μm at 5 keV. This lower depth of penetration confines X-ray production to within cell boundaries, reducing some of the uncertainties found when analysing specimens in the freeze-dried state. Analysis of frozen-hydrated material is, however, not straightforward. Specialised equipment is required for cryotransfer and maintaining the specimen in the fully hydrated state (Section 28.2.3). In addition, increased continuum production from the hydrated matrix results in a lower sensitivity compared to freeze-dried material. Despite this, consideration of the spatial resolution problem led Echlin to propose that the analysis of frozen-hydrated materials at voltages of <5 keV had considerable promise for biological microanalysis (Echlin 2001), but this has not been achieved to date, probably due to practical difficulties both in specimen preparation and in conducting analysis at low voltage, as described in the following section.

28.5 EDS USING LOW VOLTAGES

Although operation at low beam energy improves spatial resolution, the procedure is not straightforward, principally because of low excitation of X-rays. Voltage instability, low beam current and the presence of surface coatings also affect the outcome.

28.5.1 Low Excitation of X-Rays

Operation at 5 keV is below the optimal overvoltage (Section 28.3.1) for excitation of the K X-ray lines of most of the biologically important elements, especially K and Ca, resulting in reduced X-ray production (Figure 28.4). For higher atomic number elements, for example constituents of nanoparticles, the critical excitation potential for K X-ray lines is not met so that only L or M X-ray lines are produced and peak assignment depends on these. Given that the yield of L and M X-rays is less efficient than for the corresponding K lines (Joy 1998) and that the spectra produced consist of families of lines that are close together in energy or overlapping, this leads to a situation in which automatic assignment of peaks becomes very difficult and misidentification of peaks probable (Newbury 2005; Statham 2006; Camus 2011) (also see Section 28.6). It is imperative that robust peak deconvolution routines are used when interpreting such spectra. Severe peak overlap is particularly problematic when X-ray maps are collected using low voltages. The routine of setting a window over the peak area of the element of interest cannot be used. In this situation spectral imaging mapping routines in which a full spectrum is acquired at each pixel point and stored are mandatory. The spectra can subsequently be subject to full deconvolution and quantitative maps acquired (Marshall and Xu 1998; Marshall *et al.* 2012; LeFurgey *et al.* 1992; Thompson 2013).

28.5.2 Voltage Instability

At low voltage operation very small fluctuations can have a large effect on X-ray production; for example, at 5 keV a change of only 100 eV in beam energy alters Ca K_α intensity by more than 20% (Statham 2006). Modern FEGSEMs offer the advantage of stable operation at low voltages compared to SEMs fitted with tungsten hairpin filaments where voltage instability precludes low voltage operation.

28.5.3 Low Beam Current

As operating voltage is decreased, spot size and current in the focused probe also decrease, resulting in an overall reduction of X-rays produced (Figure 28.4), the reduction in X-ray counts can be remedied by increasing the probe current (Figure 28.5). A key feature of

→

Figure 28.4 The effect of decreasing accelerating voltage on X-ray production from a specimen composed of sodium potassium tartrate crystals. (A) At 20 keV X-ray production from both Na and K atoms is equally efficient. (B and C) As voltage is decreased below the critical excitation potential for K K_α X-rays there is a decrease in production of these relative to the production of Na K_α X-rays. (D) At 4.5 keV, K K_α X-ray production is severely decreased largely due to 4 keV being close to the critical excitation potential for this X-ray line. Note, however, that as voltage is decreased the current in the focused probe is also decreased and this will decrease the production of both Na K_α and K K_α X-rays.

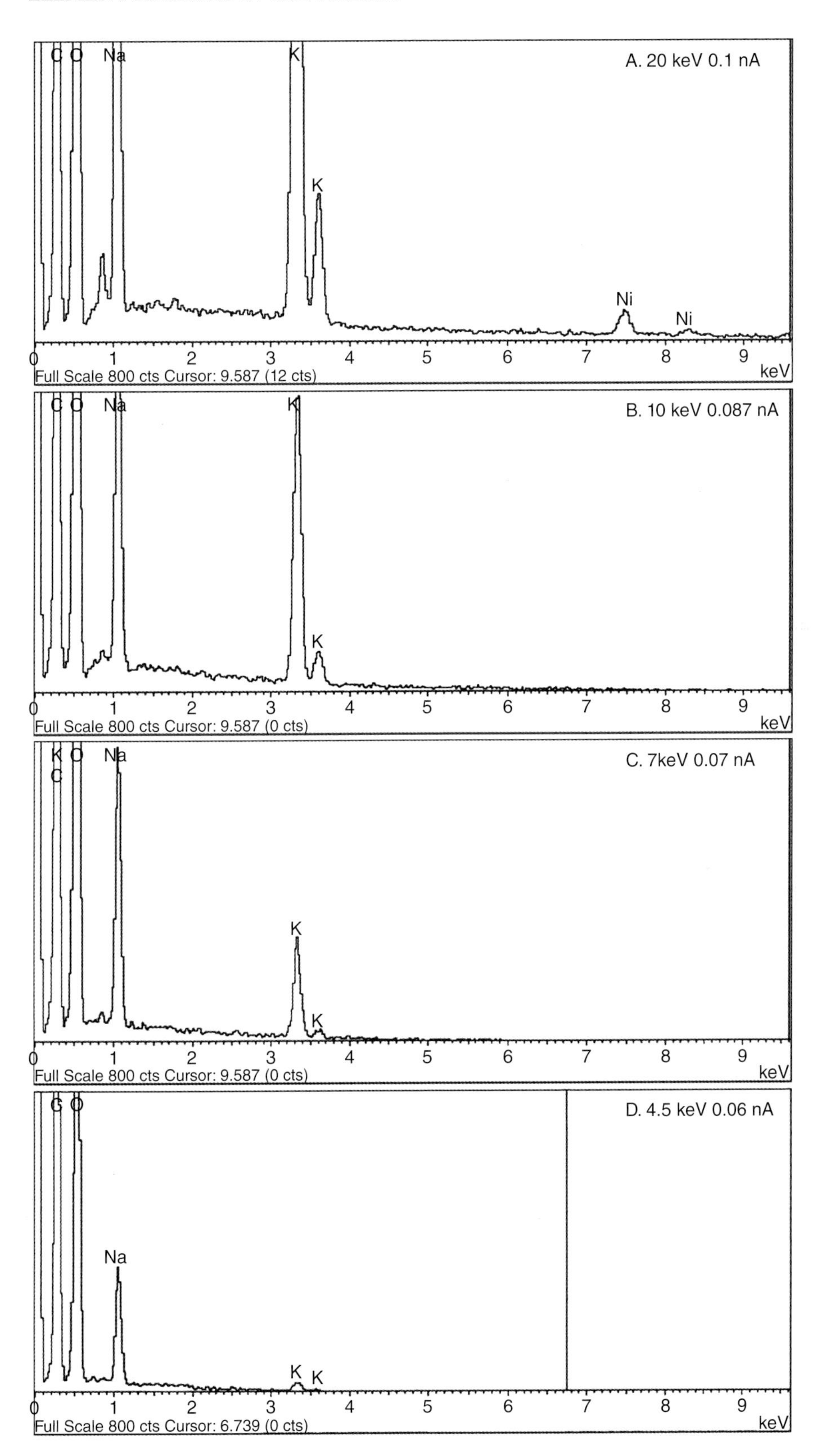
C O Na K K Ni Ni
A. 20 keV 0.1 nA
Full Scale 800 cts Cursor: 9.587 (12 cts)
keV
C O Na K K
B. 10 keV 0.087 nA
Full Scale 800 cts Cursor: 9.587 (0 cts)
keV
K C O Na K K
C. 7keV 0.07 nA
Full Scale 800 cts Cursor: 9.587 (0 cts)
keV
C O Na K K
D. 4.5 keV 0.06 nA
Full Scale 800 cts Cursor: 6.739 (0 cts)
keV

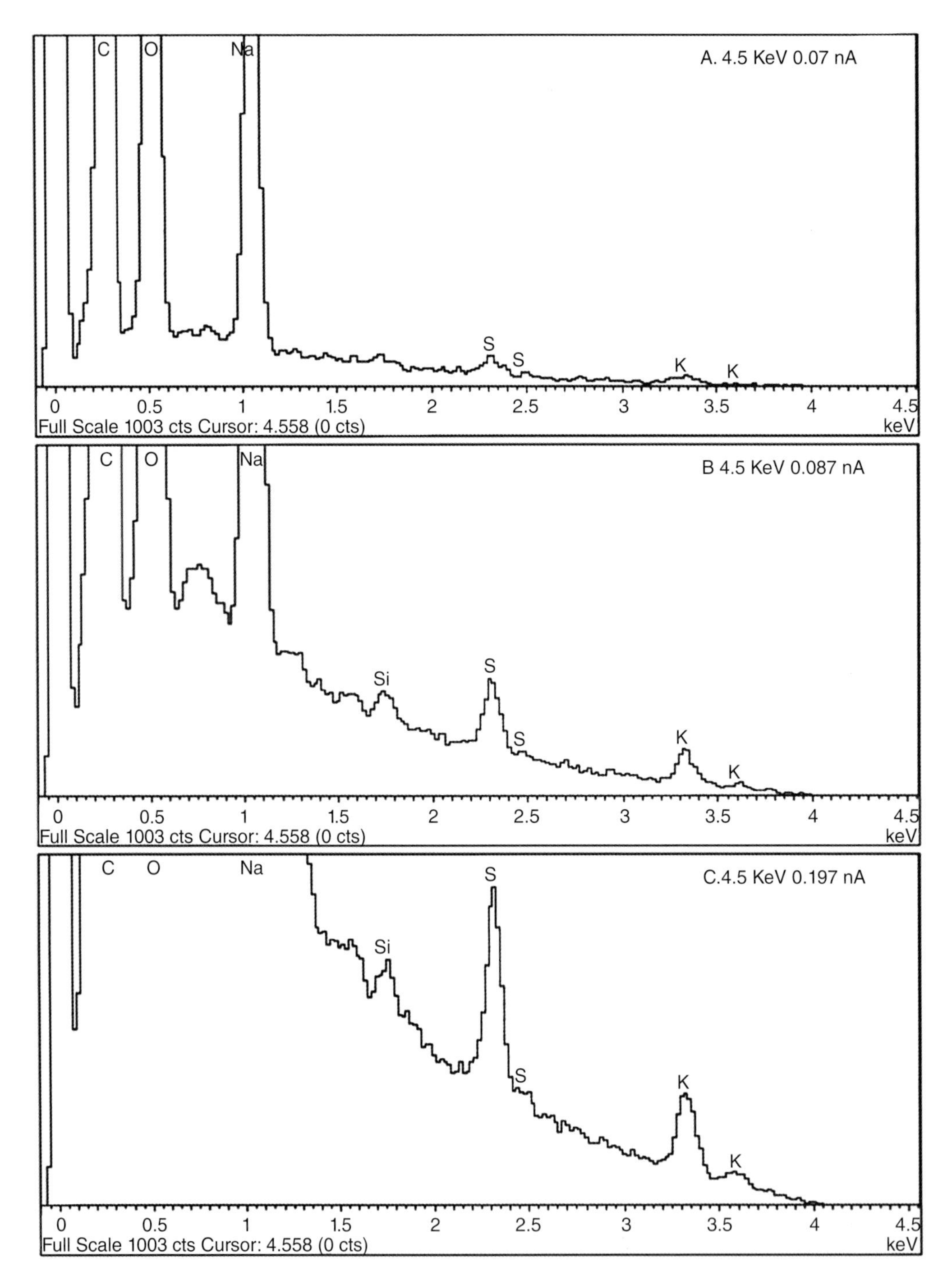

Figure 28.5 Spectra produced by analysis of sodium potassium tartrate crystals at the same voltage but with increasing beam current. (A) At 4 keV and 0.07 nA current production of K K_α X-rays is minimal. (B and C) As the beam current is increased there is an increase in the production of X-rays for all elements present in the spectrum so that even for K K_α a discernible peak is present even though the accelerating voltage is close to the critical excitation potential for this line.

the field emission source is the higher beam currents available at smaller spot sizes, which together with the more stable voltage (Section 28.5.2) makes these instruments ideal for low voltage operation. The user needs, however, to be aware that beam damage to the specimen increases with beam current, even in a conventional SEM operated at 5 keV structural damage has been reported to occur up to 800 nm below the surface of resin-embedded material (Scott 2011).

28.5.4 Surface Coating

At low voltages surface films can strongly affect analysis of materials science specimens (Statham 2006). However, for biological specimens, the depth of the beam specimen interaction is usually much greater than the surface coating layer so that the major effects of surface coating materials are the contribution of unwanted peaks to the spectrum and absorption of low energy X-rays emitted from the specimen (Marshall 1987).

28.6 PEAK IDENTIFICATION

The majority of microanalytical applications in biology have the purpose of identification of extraneous materials (see Chapter 14 by Gunning and Hauröder in this volume) and may have medical or legal implications (Ingram *et al.* 1999) so emitted X-ray peaks need to be identified correctly. Automatic peak identification routines supplied with commercial software should only be regarded as a guide to what a peak may possibly be and should not be regarded as being foolproof. Misidentification of peaks has been identified as a concern for materials scientists (Newbury 2005) and also occurs for spectra from biological specimens in which it is not unusual for automated identification routines to mislabel the Na K peak as Zn L, the $K_{K\beta}$ peak as $Ca_{K\alpha}$ or some of the Os M family as P K. The analyst needs to be sceptical about the results produced by their software and take the responsibility for peak assignment themselves. To do this there is a need to be familiar with spectrum artefacts such as incomplete charge collection, peak overlap that can alter peak shape and sum peaks and escape peaks that introduce peaks into the spectrum (Figure 16.6), although the latter are only likely to be encountered at high count rates. Full details of spectral artefacts can be found in microanalysis textbooks (Goldstein *et al.* 2003; Warley 1997). The steps in acquiring robust data begin even before the analysis of the specimen. The calibration of a detector system must be checked since, whichever peak deconvolution routine is used, any program will have difficulty in identifying peaks in a miscalibrated spectrum. Once this has been done, a spectrum acquired and an automatic peak assignment has taken place, the analyst should use their knowledge (and common sense) to check the results. Table 28.2 gives a series of relevant questions to help in checking peak assignment and Figure 28.7 shows a spectrum firstly mislabelled and then correctly labelled with the reasons for the peak assignments given.

28.7 QUANTIFICATION

In a spectrum the number of counts in an elemental peak depends primarily on the concentration of that element in the irradiated area. However, for a given concentration, the

Table 28.2 Check list to aid in peak identification

Is detector calibrated?
- If not perform detector calibration according to manufacturer's instructions.

Collect spectrum, perform automatic peak deconvolution.
- Look at peak assignment critically – remember this is only a guide.
- Only peaks that are statistically significant should be identified.
- If the spectrum was collected at high count rates with only few peaks present examine the spectrum for the presence of sum peaks and escape peaks.
- Identify peaks from extraneous sources, e.g. coating material.

For peaks assigned as K peaks:
- For peaks energy >3.0 keV are K_α and K_β peaks present?
- Are they in the correct ratio K_α:K_β 10:1 or greater?
- Are L or M lines present at lower energies?
- Are the peaks symmetric? A shoulder on a peak suggests peak overlap.

For peaks assigned as L peaks:
- If overvoltage is sufficient are corresponding K lines present at higher energy?
- Are M lines present at lower energy?
- Are members of family present in the correct proportions?

For peaks assigned as M peaks:
- If overvoltage is sufficient are corresponding L lines present?
- Are members of family present in correct proportions?

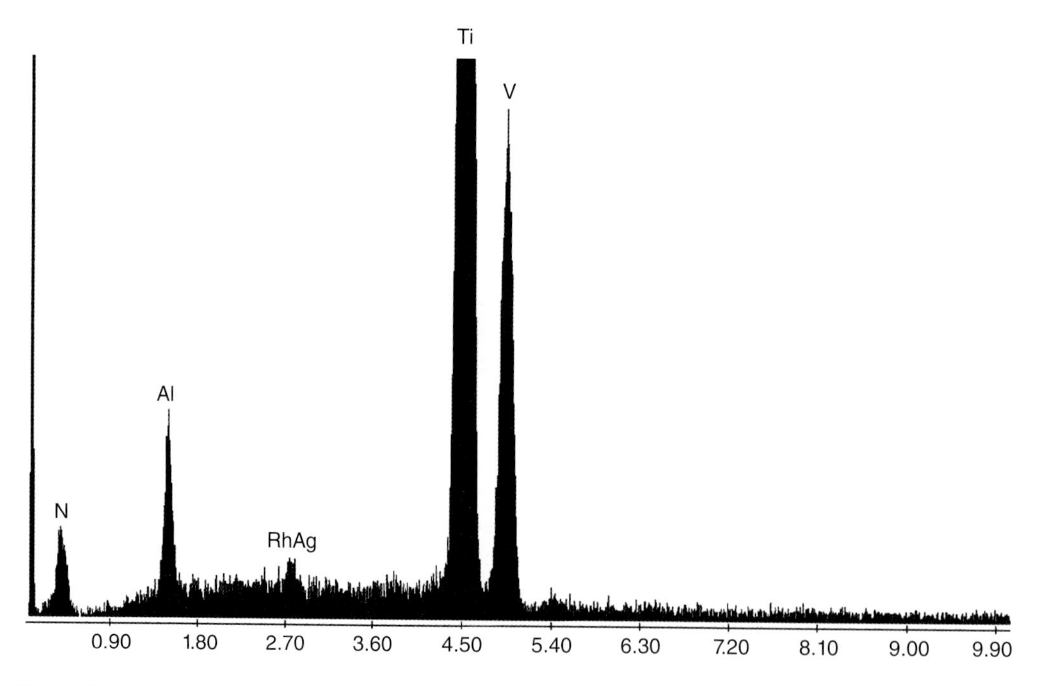

Figure 28.6 A spectrum produced by analysis of a dental implant screw that contains Al, Ti and V. Note the very high peak for Ti. Also the V K_α peak (4.95 keV) is beneath the Ti K_β peak (4.93 keV) so that identification of this element relies on the presence of a substantial V K_β peak. The peak at 2.8 keV labelled Rh is a Ti escape peak formed when incoming X-rays ionise Si atoms of the detector crystal and the Si X-ray produced escapes from the front surface of the detector. Escape peaks occur at high count rates and appear at 1.74 keV below the major element peak responsible for the original emission.

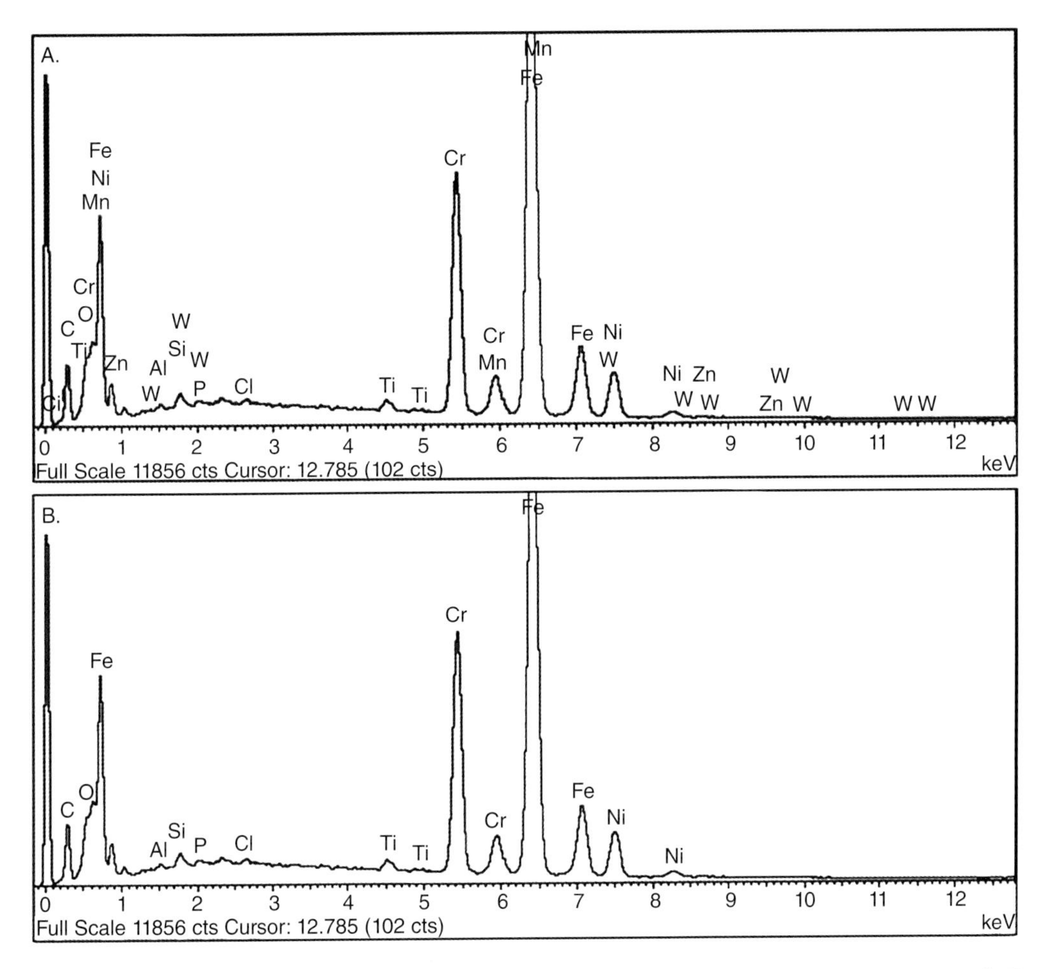

Figure 28.7 Spectra produced by the analysis of a dental implant screw containing Fe Cr Ti and Ni. (A) With peak labels after an automatic peak assignment a large number of labels are present since the software labels peaks that are likely to be present at a given energy position in the spectrum. (B) With peak labels correctly assigned. W is unlikely to be present since neither the M nor the L families of lines are consistently present in the expected proportions. Similarly, the peak assigned as Zn L is not accompanied by high peaks for Zn K_α or Zn K_β. The peak labelled as Mn is the Cr K_β peak, which is in the correct proportion to the accompanying Cr K_α.

number of X-rays generated differs for different elements (Section 28.3.1). In a bulk specimen the number of X-rays actually collected additionally depends on loss, or gain, due to interactions in the specimen as well as the efficiency of the detector itself. Quantitative routines are required to take these different factors into account. Quantification is available in most analytical software as a 'push button' option. There is a need to understand that the procedures embedded in software has been designed for materials science applications and may have underlying assumptions that are not valid for biological materials. If quantification is to be undertaken there is a need to be clear about the purpose of the investigation and to be sure that quantitative procedures will enhance the results obtained. The main methods for quantification, their advantages and disadvantages are described below.

28.7.1 ZAF

ZAF is program that has usually been supplied for SEM applications and is designed primarily for materials science specimens. In this procedure net peak intensities are corrected for loss within the specimen due to atomic number (Z), absorption (A) and X-ray fluorescence (F) on X-ray yield. One criticism of this routine for biological specimens, that the matrix composition cannot be measured (Boekestein *et al.* 1983), dates from the days of Be window detectors and is not true for thin-window detectors. However, with ZAF there is an assumption that the specimen has a smooth, polished surface, is examined normal to the electron beam and is of homogeneous composition. These requirements are not met with the majority of biological specimens, which present a rough surface that affects the acquired spectrum through absorption effects (Newbury and Ritchie 2015). A suitably smooth surface can be obtained by polishing a block of resin-embedded material but the presence of resin may cause ambiguity in the results obtained (Warley 2015). ZAF also does not perform well for low atomic number elements and should be avoided unless the specimen has a high mineral content.

28.7.2 XPP

The XPP (extended Pichou and Pouchoir) alogarithm (Pouchou and Pichoir 1991) was introduced to provide better performance than ZAF for both light and heavy elements. XPP also allows the analysis of tilted specimens. XPP has been extensively tested with materials science specimens (Newbury and Ritchie 2014; Pouchou 2002) and the program can be used with spectra exported from any commercial system in MSA format; it is available free from: http://www.cstl.nist.gov/div837.02/epq/dsta2/index.html.

28.7.3 Phi-Rho-Zed

Phi-Rho-Zed uses ionisation function curves to obtain a combined atomic number and absorption correction factor, which is then used to correct net peak intensities. This method has been exploited by Marshall and Xu (1998), Marshall *et al.* (2012), and Marshall (1987) for the study of frozen-hydrated specimens (Section 28.2.3). Comparison with standards is used to obtain results as mmol/kg wet weight, which can be converted to the physiologically relevant mmol/L after measurement of local water concentration achieved from measurement of O concentration before and after freeze-drying (Marshall and Xu 1998; Marshall *et al.* 2012; Marshall 1987). Quantification depends on the measurement of net peak intensities so a stable beam current is required.

28.7.4 Continuum Normalisation

The continuum-normalisation method of (Hall and Gupta 1982) is based on using the ratio of net peak counts to continuum counts obtained from a peak-free region of the spectrum as a measurement of mass of the element of interest per unit mass of specimen. Quantification is achieved by comparison with previously prepared standards (Warley 1990). The advantage is that the method is independent of fluctuations in beam current since these affect peak and continuum to the same extent. Strictly, this method is not applicable to bulk specimens, since it was derived for use with thin sections with the assumption that neither absorption nor fluorescence of X-rays occurs, but has been used for bulk biological specimens by some workers (Zs-Nagy *et al.* 1977). Continuum normalisation is the method of choice for analyses using a FEGSEM operated in STEM mode where thin section criteria hold.

28.7.5 Peak-to-Local Background

The method most frequently used for quantification of biological bulk specimens is the peak-to-local background (background under the peak, P/B) method, which was introduced to cope with specimens with uneven surfaces or particulate material (Statham 1979; Boekestein *et al.* 1984). The method is self-correcting for surface roughness, since both peak and background originate from the same energy region of the spectrum and should suffer absorption to the same degree. However, this assumption has been questioned (Gauvin and Lifshin 2000) and, in addition, it has been pointed out that the procedure fails if either electrons or X-rays escape from the lateral surface of the specimen (Tylko 2013). Despite this, P/B has proved a robust method for quantification, the advantage being that, in common with continuum normalisation, P/B is independent of variations in beam current and analytical time.

28.8 SUMMARY

The major limitation for EDS in biology using either SEM or FEGSEM is the specimen itself. There is a requirement for specimen preparation without disturbing the localisation of the elements of interest. Exact procedures vary from specimen to specimen and need to be determined experimentally. Previously the technique was also limited by difficulty in detecting elements present at low concentrations. The situation has, however, changed with the introduction of SDD detectors and elements present in trace concentrations have been detected in materials science specimens (Newbury and Ritchie 2015). It remains to be seen whether this can be achieved with biological materials.

For bulk biological specimens consisting largely of low atomic number material, the FEGSEM is unlikely to offer much advantage over conventional SEM when operated at higher voltages (15–30 keV), due to the large interaction volumes generated. There are two areas, however, in which FEGSEM will have increasing application. One is low voltage investigations. The stable voltage small spot size and higher beam current capabilities of the FEGSEM, coupled with the increased sensitivity of SDD detectors, offer distinct advantages for the analysis of low atomic number specimens. It may be that lack of application in this area to date has been due to limitations in previous instrumentation. The second area is operation of FEGSEMs in the STEM mode, giving a viable alternative to TEM for delivering high resolution EDS (Hondow *et al.* 2011). In some ways the wheel has turned full circle since this is the configuration (without the field emission source) adopted by Hall and Gupta (1982) in their pioneering work developing quantification for biological EDS.

ACKNOWLEDGEMENTS

The authors would like to thank Dr Clair Collins of Oxford Instruments Microanalysis Group for helpful comments on the manuscript and Dr Dale Newbury of NIST, Gaithersburg, for supplying reference 40 prior to publication. AW would like to thank Professor Antonio Campos, Faculty of Medicine, University of Granada, Spain, where much of this chapter was written.

REFERENCES

Alaminos, M., Sanchez-Quevedo, M.C., Muñoz-Avila, J.J., Garcia, J.M., Crespo, P.V., Gonzalez-Andrades, M. and Campos, A. (2007) Evaluation of the viability of cultured corneal endothelial cells by quantitative electron-probe X-ray microanalysis. *Journal of Cellular Physiology*, 211, 692–698.

Aronova, M.A. and Leapman, R.D. (2012) Development of electron energy-loss spectroscopy in the biological sciences. *MRS Bulletin*, 37, 53–62.

Boekestein, A., Stadhouders, A.M., Stols A.L.H. and Roomans, G.M. (1983) Quantitative biological X-ray microanalysis of bulk specimens: An analysis of inaccuracies involved in ZAF-correction. *Scanning Electron Microscopy* 1983, II, 725–736.

Boekestein, A., Thiel, F., Stols, A.L.H., Bouw, E. and Stadhouders, A.M. (1984) Surface roughness and the use of a peak to background ratio in the X-ray microanalysis of bio-organic bulk specimens. *Journal of Microscopy*, 134, 327–333.

Boothroyd, B. (1964) The problem of demineralisation in thin sections of fully calcified bone. *Journal of Cell Biology*, 20,165–173.

Brown, A. and Hondow, N. (2013) Electron microscopy of nanoparticles in cells, in *Frontiers of Nanoscience* (ed. Huw Summers), vol, 5, Elsevier, pp. 95–120.

Burgess, S., Li, X. and Holland, J. (2013) High spatial resolution energy dispersive X-ray spectrometry in the SEM and detection of light elements including lithium. *Microscopy and Analysis Compositional Analysis Supplement*, 27 (4), S8–S13.

Camus, P. (2011) Low beam-energy energy-dispersive X-ray spectroscopy for nanotechnology. *Microscopy and Analysis*, 25, S19–21.

Canny, M.J. and Huang, C.X. (1993) What is in the intercellular spaces of roots? Evidence from the cryo-analytical scanning electron microscope. *Physiologia Plantarium*, 87, 561–568.

Clode, P.L., Stern, R.A. and Marshall, A.T. (2007) Subcellular imaging of isotopically labelled carbon compounds in a biological sample by ion microprobe (nanoSIMS). *Microscopy.Research and Technique*, 70, 220–229.

Dvorackova, J., Bielnikova, H., Kukuttschova, J., Peikertova, P., Filip, P., Zelenik, K., Kominelli, P., Uvivova, M., Pradna, J., Cemakna, Z. and Dvovacek, I. (2015) Detection of nano- and micro-sized particles in routine biopsy material pilot study. *Biomed. Pap. Med. Fac. Univ. Palacky Olomouc Czech Repub.*, 159 (1), 87–92.

Echlin, P. (2001) Biological X-ray microanalysis: The past, present practices and future prospects. *Microscopy and Microanalysis*, 7, 211–219.

Echlin, P. and Taylor, S.E. (1986) The preparation and X-ray microanalysis of bulk frozen hydrated vacuolate plant tissue. *Journal of Microscopy*, 141, 329–348.

Fernandez-Segura, E., Cañizares, F.J., Cubero M.A., Campos, A. and Warley, A. (1999a) A procedure to prepare cultured cells in suspension for electron probe X-ray microanalysis: Application to scanning and transmission electron microscopy. *Journal of Microscopy*, 196, 19–25.

Fernandez-Segura, E., Cañizares, F.J., Cubero, M.A., Warley, A. and Campos A. (1999b) Changes in elemental content during apoptotic cell death studied by electron probe X-ray microanalysis. *Experimental Cell Research*, 253, 454–462.

Frey, B., Brunner, I., Walther, P., Scheidegger, C. and Zierold, K. (1997) Element localisation in ultrathin cryosections of high-pressure frozen ectomycorrhizal spruce roots. *Plant Cell and Environment*, 20, 929–937.

Gauvin, R. and Lifshin, E. (2000) Simulation of X-ray emission from rough surfaces. *Mikrochimica Acta*, 132, 201–204.

Goldstein, J.I., Newbury, D.E., Joy, D.C., Lyman, C.E., Echlin, P., Lifshin, E., Sawyer, L. and Michael, J.R. (2003) *Scanning Electron Microscopy and X-ray Microanalysis*, 3rd edn, Springer, New York.

Hall, T.A. and Gupta, B.L. (1982) Quantification for the X-ray microanalysis of cryosections. *Journal of Microscopy*, 126, 333–345.

Hardt, M. and Plattner, H. (1999) Quantitative energy dispersive X-ray microanalysis of calcium dynamics in cell suspensions during stimulation on a subsecond timescale: preparative and analytical aspects as exemplified with *Paramecium* cells. *Journal of Structural Biology* 128, 187-199

Hondow, N., Harrington, J., Brydson, R., Doak, S.H., Singh, N., Manshian, B. and Brown, A. (2011) STEM mode in the SEM: A practical tool for nanotoxicology. *Nanotoxicology*, 5, 215–227.

Huang, C.X., Canny, M.J., Oates, K. and McCully, M.E. (1994) Planing frozen hydrated plant specimens for SEM observation and EDX microanalysis. *Microscopy Research and Technique*, 28, 67–74.

Ingram, P., Shelburne, J., Roggli, V. and LeFurgey, A. (1999) *Biomedical Applications of Microprobe Analysis*, Academic Press, San Diego and London.

Joy, D.C. (1998) The efficiency of X-ray production at low energies. *Journal of Microscopy*, 191, 74–82.

LeFurgey, A., Davilla, S.D., Kopf, D.A., Sommer, J.R. and Ingram, P. (1992) Real-time quantitative elemental analysis and mapping: Microchemical imaging in cell physiology. *Journal of Microscopy*, 165, 191–223.

Marshall, A.T. (1987) Scanning electron microscopy and X-ray microanalysis of frozen-hydrated bulk samples, in *Cryotechniques in Biological Electron Microscopy* (eds R.A. Steinbrecht and K. Zierold), Springer-Verlag, Berlin, Heidelberg, pp. 240–257.

Marshall, A.T. and Wright, O.P. (1991) Freeze-substitution of scleractinian coral for confocal scanning laser microscopy and X-ray microanalysis. *Journal of Microscopy*, 162, 341–354.

Marshall, A.T. and Xu, W. (1998) Quantitative elemental X-ray imaging of frozen-hydrated biological samples. *Journal of Microscopy*, 190, 305–316.

Marshall, A.T., Clode, P.L., Russell, R., Prince, K. and Stern, R. (2007) Electron and ion microprobe analysis of calcium distribution and transport in coral tissues. *Journal of Experimental Biology*, 210, 2453–2463.

Marshall, A.T., Goodyear, M.J. and Crewther S.G. (2012) Sequential quantitative X-ray elemental imaging of frozen-hydrated and freeze-dried biological bulk samples in the SEM. *Journal of Microscopy*, 245, 17–25.

Mauritz, J.M.A., Seear, R., Esposito, A., Kaminski, C.F., Skepper, J.N.,Warley, A., Lew, V.L. and Tiffert, T. (2011) X-ray microanalysis investigation of the changes in Na, K and hemoglobin concentration in *Plasmodium falciparum*-infected red blood cells. *Biophysical Journal*, 100, 1–8.

McCully, M.E., Shane, M.W., Baker, A.N., Huang, C.X., Ling, L.E.C. and Canny, M.J. (2000) The reliability of cryoSEM for the observation and quantification of xylem embolisms and quantitative analysis of xylem sap *in situ*. *Journal of Microscopy*, 198, 24–33.

Newbury, D.E. (2005) Misidentification of major constituents by automatic qualitative energy dispersive X-ray microanalysis, a problem that threatens the credibility of the analytical community. *Microscopy and Microanalysis*, 11, 545–561.

Newbury, D.E. (2006) Electron-excited energy dispersive X-ray spectrometry at high speed and at high resolution: Silicon drift detectors and microcalorimeters. *Microscopy and Microanalysis*, 12, 527–537.

Newbury, D.E. and Ritchie, N.W.M. (2014) Quantitative X-ray microanalysis of low atomic number elements by SEM/SDD-EDS with NIST DTSA II: Carbides and nitrides and oxide, *Oh My! Microscopy and Microanalysis*, 20 (Suppl. 3), 702–703.

Newbury, D.E. and Ritchie, N.W.M. (2015) Performing elemental microanalysis with high accuracy and high precision by scanning electron microscopy/silicon drift detector energy dispersive X-ray spectroscopy (SEM/SDD-EDS). *Journal of Materials Science*, 50 (2), 493–518.

Ortega, R., Deves, G. and Carmona, A. (2009) Bio-metals imaging and speciation in cells using proton and synchrotron radiation X-ray microspectroscopy. *Journal of the Royal Society Interface*, 6, S649–S658.

Porter, A.E., Gass, M., Muller, K., Skepper, J.N., Midgley, P.A. and Welland, M. (2007) Direct imaging of single-walled carbon nanotubes in cells. *Nature Nanotechnology*, 2, 713–717.

Pouchou, J.L. (2002) X-ray microanalysis of thin surface films and coatings. *Mikrochimica Acta*, 138, 133–152.

Pouchou, J.-L. and Pichoir, F. (1991) Quantitative analysis of homogeneous or stratified microvolumes: Applying the model 'PAP', in *Electron Probe Quantitation* (eds K.F.J. Heinrich and D.E. Newbury), Plenum Press, pp. 31–75.

Rasch, R., Cribb, B.W., Barry, J. and Palmer, C.M. (2003) Application of quantitative analytical electron microscopy to the mineral content of insect cuticle. *Microscopy and Microanalysis*, 9, 152–154.

Reynolds, J.L., Joannides, A.J., Skepper, J.N., McNain, R. Schurgers, L.J., Proudfoot, D., Jahnen-Dechent, W., Weisberg, P. and Shanahan, C.M. (2004) Human vascular smooth muscle cells undergo vesicle-mediated calcification in response to changes in extracellular calcium and phosphate concentrations: A potential mechanism for accelerated vascular calcification in ESRD. *Journal of the American Society of Nephrology*, 15, 2857–2867.

Roomans, G.M. (1999) X-ray microanalytical studies of epithelial cells with reference to cystic fibrosis, in *Biological Applications of Microprobe Analysis* (eds P. Ingram *et al.*), Academic Press, San Diego and London, pp. 315–337.

Scott, K. (2011) 3D elemental and structural analysis of biological specimens using electrons and ions. *Journal of Microscopy*, 242, 86–93.

Scott, K. and Ritchie, N.W.M. (2009) Analysis of 3D elemental mapping artefacts in biological specimens using Monte Carlo simulation. *Journal of Microscopy*, 233, 331–339.

Skepper, J.N. (2000) Immunocytochemical strategies for electron microscopy: Choice or compromise. Invited review. *Journal of Microscopy*, 199, 1–36.

Skepper, J.N., Karydis, I., Garnett, M.R., Hegyi, L., Hardwick, S.J.,Warley, A., Mitchinson, M.J. and Cary, N.R.B. (1999) Changes in elemental concentrations are associated with early stages of apoptosis in human monocyte-macrophages exposed to oxidised low density lipoprotein: An X-ray microanalytical study. *Journal of Pathology*, 188, 100–106.

Statham, P.J. (1979) Measurement and use of peak-to-background ratios in X-ray analysis. *Mikrochimica Acta*, Suppl. 8, 229–242.

Statham, P.J. (2006) Practical issues for quantitative X-ray microanalysis in SEM at low kV. *Microscopy Today*, 14 (1), 30–32.

Thirion, S., Troadec, J.D., Pagnotta, S., Andrews, S.B., Leapman, R.D. and Nicaise, G. (1997) Calcium in secretory vesicles of neurohypophysial nerve endings: Quantitative comparison by X-ray microanalysis of cryosectioned and freeze-substituted specimens. *Journal of Microscopy*, 186, 28–34.

Thompson, K. (2013) Is energy resolution still an important specification in EDS? *Microscopy Today*, 21 (4), 30–34.

Tylko, G. (2013) Analysis of biologically-derived small particles – searching for geometry correction factors using Monte-Carlo simulation. *Microscopy and Microanalysis*, 19, 56–65.

Tylko, G., Banach, Z., Borowska, J., Niklinska, M. and Pyza E. (2005) Elemental changes in the brain, muscle and gut cells of the housefly *Musca domestica* exposed to heavy metals. *Microscopy Research and Technique*, 66, 239–247.

Warley, A. (1990) Standards for the application of X-ray microanalysis to biological specimens. Invited review. *Journal of Microscopy*, 157, 135–147.

Warley, A. (1997) *X-ray Microanalysis for Biologists*, vol. 16, Practical Methods in Electron Microscopy (ed. A Glauert), Portland Press, London.

Warley, A. (2015) Development and comparison of the methods for quantitative electron probe X-ray microanalysis analysis of thin specimens and their application to biological material. *Journal of Microscopy*, 50 (2), 493–518.

Warley, A. and Skepper, J.N. (2000) Long freeze-drying times are not necessary during the preparation of thin sections for X-ray microanalysis. *Journal of Microscopy*, 198, 116–123.

Zapotoczny, S., Jurkiewicz, A., Tylko, G., Anielska, T. and Turnau, K. (2007) Accumulation of copper by *Acremonium pinkertoniae*, a fungus isolated from industrial wastes. *Microbiological Research*, 162, 219–228.

Zierold, K., Tobler, M. and Muller, M. (1991) X-ray microanalysis of high-pressure and impact frozen erythrocytes. *Journal of Microscopy*, 161, RP1–RP2.

Zs-Nagy, I, Pieri, C., Guili, C., Bertoni-Freddari, C. and Zs-Nagy, V. (1977) Energy dispersive X-ray microanalysis of the electrolytes in biological bulk specimen. 1. Specimen preparation, beam penetration and quantitative analysis. *Journal of Ultrastructural Research*, 58, 22–33.

29

Image and Resource Management in Microscopy in the Digital Age

Patrick Schwarb[1], Anwen Bullen[2], Dean Flanders[3], Maria Marosvölgyi[4], Martyn Winn[5], Urs Gomez[1] and Roland A. Fleck[6]

[1]*Imagic Bildverarbeitung AG, Glattbrugg, Switzerland*
[2]*UCL Ear Institute, University College London, UK*
[3]*Friedrich Miescher Institute for Biomedical Research, Basel, Switzerland*
[4]*arivis AG, Business Unit arivis Vision, Rostock, Germany*
[5]*STFC Daresbury Laboratory, Daresbury, Warrington, UK*
[6]*Centre for Ultrastructural Imaging, King's College London, London, UK*

29.1 INTRODUCTION

For scientific imaging modalities the transition from film-based media to digital acquisition has provoked a paradigm shift regarding the quantity and complexity of acquired data. In the pre-digital era, many research groups used to acquire images on film slides, to be directly used for presentations or scanned for publications. As visual media, the original slides could be easily and quickly reviewed with a light box. In a few specialist cases, notably the macromolecular imaging (structural) community, digitisation of film data was routine and essential for the post-processing of data, leading to the generation of 3D structures.

'We are still in transition'. From today's perspective, hardcopies of images have many disadvantages; we are now happy to have overcome challenges of sharing results, storing prints, assigning metadata, etc. However, many sites struggle to fully update their labs to readily support digital standards. The challenge is mainly due to high costs for specialized equipment. Though costs are not the only reason for the slow transition to the fully digital lab, rather it is missing know-how for comprehensive IT solutions, incorporating all aspects of handling the vast amount of digital data from acquisition to storage. IT solutions also

Biological Field Emission Scanning Electron Microscopy, First Edition.
Edited by Roland A. Fleck and Bruno M. Humbel.

require a culture shift, with all members of a lab committed to inputting data and metadata in a timely manner and following standard procedures.

29.2 RESOURCE MANAGEMENT

With the cost and complexity of the equipment used in nearly all aspects of research comes the need for effective resource management. In order to have effective resource management one must be aware of some key concepts: resource discovery, usage optimization, monitoring of utilization, data management, and project management. In addition, the resource management system must be easy to use, and researchers as well as core facilities must derive direct benefit from using it.

In order to implement these concepts into practice a consortium called 'Open IRIS' (integrated resource and information sharing) was established to develop a solution to meet these needs. The aim of the consortium is to establish a tool that can be used freely to facilitate resource sharing and help optimize resource usage, in this way enabling researchers to discover and share resources in order to optimize their research. In making the tool free it is hoped that it will then be widely adopted and further facilitate resource usage. The consortium also helps to embed best practices directly into the tool that other labs and facilities can benefit from (billing policies, usage restrictions, statistics reporting, system design, etc.).

29.3 RESOURCE DISCOVERABILITY

Recognizing that multiple specialist tools are commonly needed in a researcher's daily work, it is essential to have a mechanism that can be used to discover any resources within a group, within an organization, or at neighbouring organizations. In addition, the spectrum of resources can range from a microscope to a 3D printer or a clean room. This is greatly facilitated by using a cloud hosted solution for resource management, which then gives the following benefits:

1. No need to have multiple local versions of resource lists to maintain and support.
2. Facilitates consistency of data by using consistent identifiers for resources, projects, researchers and organizations across labs and facilities that may span multiple organizations.
3. Facilitate collaborations and consortiums by enabling sharing across organizational boundaries.
4. Simplifies integration of institutional logins, as well as the ability to support social logins (e.g. Google, GitHub, LinkedIn).

However, to facilitate resource discovery one must be able to control visibility and access to resource providers and resources. This must include support for fine grained visibility and access settings that support users, organizations, groups, departments, projects and communities, as well as the optional ability to make information on resources or resource providers publically available.

In addition, one must make the system appealing, useful and simple for normal labs, as well as sophisticated enough for core facilities, so you have the broadest set of resources for discovery. Examples of computational clouds that provide user friendly services for biology include the Cloud Infrastructure for Microbial Genomics (www.climb.ac.uk) and the Cyber-infrastructure for Data Management and Analysis (www.cyverse.org). At the European level, Instruct (www.structuralbiology.eu) provides a mechanism for transnational access to high-end facilities in structural biology. However, resources must remain mindful of evolving regulations around data sharing (e.g. EU General Data Protection Regulation).

29.4 USAGE OPTIMIZATION

There are essentially two ways to optimize resource usage, which are by billing for resource usage or having a way to enforce booking restrictions of resources.

1. Billing. Billing for resource usage has a two-fold purpose: on the one hand it is to recover the costs of the usage of the equipment and, on the other hand, to ensure people only use the amount of resources they need. Billing has the major disadvantage of being very resource intensive in terms of effort on all sides by taking the time of researchers, facility administrators and finance groups. In addition, as academic research is still heavily reliant on one-time large purchases for equipment and generally groups have less funding for running costs, this can also be troublesome in terms of finding the funding. In addition, billing can potentially impede creative research. In order to effectively bill, charge calculations must be able to leverage complex rules based on:
   a. People. Produce prices based on rates and discounts associated with person, group or organization in order to achieve charges based on the financial relationship of a person to a facility.
   b. Time. Price calculations should support time of day and day (weekends and holidays). This can be used to encourage people to use resources during non-peak hours in order that such times are more heavily utilized.
   c. Resources. Ability to assign scale and a level of detail to prices at a per resource level, often referred to as data granularity.
2. Usage restrictions. Typically users are expected to behave fairly when using resources, but this is very difficult in reality when doing planning. The concept of 'fair share scheduling' has existed for many users with cluster computing, where policies are enforced via the job scheduler, but has not had much penetration into areas of instrument usage. However, this technique can be used in resource booking by implementing the concept of 'usage restrictions' where the booking system can be used to enforce certain rules that will ensure fair usage and optimize its usage. Some example rules are:
   a. Max booking duration. This is a rule in a booking system that defines the maximum length of a booking of a resource.
   b. Booking into the future. How far bookings can be done into the future, for example two weeks into the future to encourage people not to book a resource further in advance than they can effectively plan.
   c. Booking deletion policy. Not allowing the last minute cancellation of a booking, which then does not allow others to plan for available free slots coming available at the last minute.
   d. Booking limits. This can be defined in terms of hours or bookings, and can be enforced on a forward rolling or calendar basis (e.g. daily, weekly, monthly).

In order to further optimize resources, these rules should allow selective enforcement based on:

(a) Day. For example, weekends and holidays may have no 'max booking duration' enforced in order to encourage users with very long sessions to use the resources during periods of very low demand. Longer or extended "booking duration" may also be made available in the evening (out with normal working hours) and overnight to encourage users with long sessions to use resources during these periods of lower demand, freeing resources up for light users during the day.
(b) Time. You may decide to allow a 'max booking duration' of more than 2 hours in the evening to encourage people to work in the evenings for long sessions.
(c) People. There may be groups or organizations that have paid for part of a resource, so you may not restrict their usage of resources in the same way as others. For example, you may limit groups or users to a certain number of hours on a resource per month, but you may not restrict a certain group that helped pay for the resource.

29.5 UTILIZATION MONITORING

Reporting of utilization is essential for resource optimization and determining where to invest. However, it can also have a side benefit of letting other users know which users of a particular resource have the most experience on it, so they may contact them for help with using the equipment. The types of reporting that can be useful are:

1. Simple resource usage on a single resource. This can be done in a manner that is easily accessible from the resource itself so other users can see who is also familiar with a resource. It can also be leveraged for email distribution lists for a resource.
2. Misuse reporting. This can be, for example, monitored on an instrument computer to report when users do not show up for their bookings, which then blocks the usage of the resource for others. This information can be displayed publically, sent in email to the user as well as their manager, and also used in normal reporting of a facility.
3. Usage distribution. More complex reports on usage by groups, users and organizations on one or more resources.
4. Utilization. This type of reporting can be used to calculate the percent utilization of a resource over a time period. This can also be displayed in terms of off-hours usage (evening, weekends and holidays) and day time usage. In addition, this can be calculated over time to determine the effectiveness of an investment. This type of reporting will also capture downtime due to instrument failure and give a measure of the reliability of the resource.
5. Heatmap. This can help expose in a fast visual way what resources are most heavily used in a facility or by a user, group or organization.
6. Active booking. Report the number of booking sessions in a given period of time to get an impression of the activities in a facility.
7. Mean usage. You may want to determine the mean usage of a resource to determine the overall load on a resource.

These are very important reports that can be generated, but just as important is the ability to filter for or exclude certain types of resource usage based on people (user, group or organization), time (office hours, weekends or holidays) and type of bookings (operator assisted, training, out of service, maintenance and regular usage).

29.6 DATA MANAGEMENT

Data management is not often linked to resource management, but by requiring users to enter certain information before and after using a resource you can produce a digital trail that can be used to enforce data annotation and produce data management plans and reports in an automated manner: breadcrumb to link data, resource identifiers, personal identifiers (e.g. ORCID) and organizational identifiers (e.g. OpenAIRE). This linkage is made in the use of large central facilities such as synchrotrons. The SynchWeb/ISPyB Laboratory Information Management System is closely coupled to specific visits to a beamline, and covers pre-visit sample data, in-visit monitoring of data collection and post-visit data analysis.

29.7 PROJECT MANAGEMENT

The resource management system must also be able to support project management concepts for large complex projects. It should include the ability to track time of individuals, milestones and Gantt charts to chart a project schedule. In addition, as research is increasingly cross-disciplinary and with the arrival of techniques such as correlative microscopy it is important that the resource management system can be used for a wide variety of technologies (microscopy, genomics, proteomics, etc.). This then allows for more effective project management, with the ability to share and link information across technology platforms or labs.

Another important aspect of project management is the ability to assign and monitor resources used for a project, which then can also be enforced on the logins of the resources with a local agent. This can be done via requests, which permit access to specific resources for a defined period of time to a specific person. These requests can then be linked back to a particular project, which can span several labs or core facilities. From the point of view of a user, tracking the resources used for a project helps with writing acknowledgements in publications and with future resource requests.

Software for particular science domains often comes with its own project management tools. For example, CCP4 (http://www.ccp4.ac.uk) for macromolecular crystallography has provided tools for organising project data and reviewing computational jobs since the introduction of its first GUI around 2000. The CCP-EM project (http://www.ccpem.ac.uk) for cryoEM in structural biology is developing an analogous system. The OMERO (https://www.openmicroscopy.org/site) project for light microscopy also provides many data management tools. Image data can be imported into such systems, but it is preferable if third party software can gain direct access to a Centre's image store.

29.8 IMAGE ACQUISITION

Image acquisition devices and processes found in light and electron microscopy are widely spread in scientific labs. Although standard imaging equipment is usually located near to the experiments in the lab, more advanced techniques such as advanced light microscopy or electron microscopy are mostly located in specialized imaging centres, as expertise in advanced preparation is easily available. During acquisition a wide range of equipment-related metadata is stored, indispensable for further processing, analysis and presentation. These associated metadata are typically hidden in proprietary image file formats and can only be accessed by their assigned applications, or are even lost when converting supplier-specific formats to standard formats such as JPG or TIFF (e.g. where additions to the MRC format by FEI and Gatan were non-standard and thus lost during subsequent processing; see Cheng *et al.*, 2015).

Specifically in core facilities there is an increasing demand for linking additional project-, experiment- and user-related information to the acquired data, such as owner/start/date, fixation/staining and producer/acquisition date. As these image and associated data generated by various lab equipments from different suppliers are typically not compatible and the collection processes are not synchronized, device- and process-independent coordination and standardization of data collection is crucial for efficient post-processing and handling.

During the last decade acquisition equipment has improved dramatically. This leads to an imbalance between acquisition and processing time. Today's systems are capable of acquiring terabytes of data in a very short time. Just viewing these datasets absorbs hours and days of user time and computer power. For this reason, image acquisition needs fast and reliable viewing tools, allowing users to check its quality for additional acquisitions.

Having image and metadata linked and centrally available, time-critical image acquisition can be realized in the lab environment, whereas visualization and documentation can be done offline on mobile devices or desktop computers at any time and location.

29.9 MULTIMEDIA DATA IN SCIENCE

Generation of digital image data today ranges from simple digital photography to highly sophisticated recording of multidimensional light and electron microscopy data. Many of the digital acquisition devices in high-end microscopy are still using non-standardized file formats, requiring supplier-specific software applications for viewing and post-processing. This may be acceptable for certain communities, but the vast majority of users want to share and retrieve images with standard software tools and are hampered by the fact that even with a high level of commitment from the community no common scientific data exchange format, as we know it from medical (DICOM) or other environments, has been realized to this day.

29.10 BENEFITS OF DIGITAL MANAGEMENT SYSTEMS

As a consequence, complex and large 3D/4D datasets are created in proprietary formats and measured in gigabytes, terabytes and even petabytes. Managing these data effectively is extremely challenging. The challenge is due each image set being associated with the experiment's metadata, the operating parameters applied to the device used to acquire the images, the acquired images themselves and the image processing workflow. In many cases, the experimental process is separated from the image. Essential image-related information is scattered, typically in different media such as handwritten notes, Word® and Excel® documents as well as propriety image files, and thus is easily lost. Reviewing data, usually stored in multiple propriety file formats with associated supplier-specific software suites, also presents a challenge, as the file must be opened and viewed using the appropriate hardware and software components. These challenges are even increased by IT efforts to centralize specialized equipment in core facilities (Figure 29.1). For this reason tools for the simplification of these processes are of significant value in research imaging environments. These tools must also ensure data security and provide auditing capabilities. To address

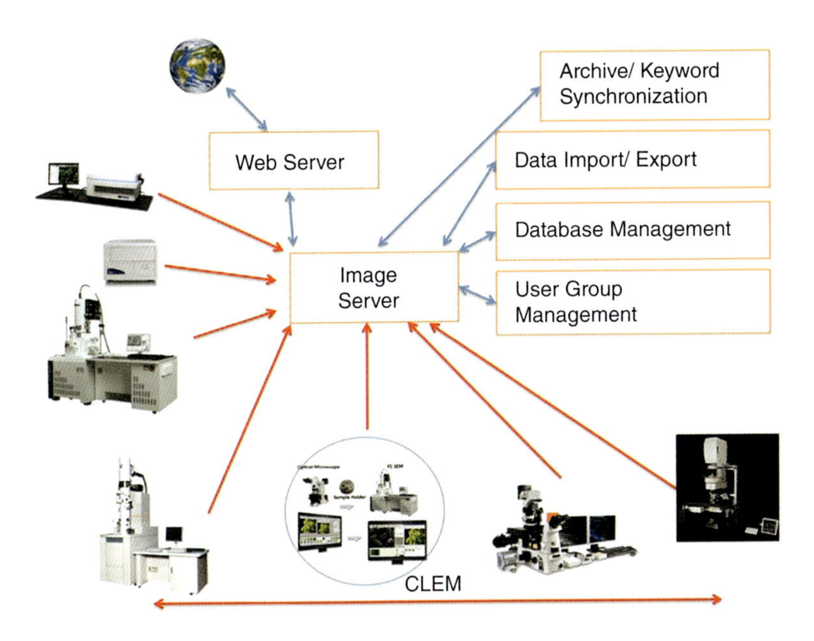

Figure 29.1 An example of a present-day research imaging environment with multiple specialized technologies feeding data to a central digital data management resource. The project pools core imaging technologies into a single data management system. The goal is to create the incentives and the unified structure required to provide a common database, sufficiently flexible and scalable to manage all imaging research workflows of the lab. This strategy also efficiently supports the overall IT concept.

this challenge, we propose the development of a Life-Science-focused Imaging Core Facility Information Management System, specifically designed for linking images to experimental metadata. Typically the five 'w' have been recognized to be important to an image dataset to make it unique (What, When, Where, Who, Why).

29.11 IMAGE DATABASE – A SYSTEM-RELEVANT COMPONENT IN SCIENCE

Although at first glance a sound server technology as well as a fast network seems to be sufficient for handling most problems in digital imaging, a closer look shows that it is not. Even if data storage and backup might ensure data security, a system with distributed data in multiple, supplier-specific files is hard to be searched for specific information.

Just imagine, how would you gather this information just from file names and then search the image server for these files? How would you retrieve the required information from these files without loading them individually into the proprietary applications they have been created with? Who else but the creator would then know where additional, experiment-specific information has been saved? Due to image manipulation, tracking all changes in an image processing workflow has already been identified as important in specific data areas, for example pharmaceutical and medical arenas. The traceability of data and how it has been processed is likely to become increasingly important to mitigate against the risk of inappropriate data manipulation in scientific research.

In the latest incarnation of an Image Database, raw as well as processed image data linked to associated metadata becomes mandatory for the efficient use of acquired data.

29.12 IMAGE DATABASE – AN INTEGRATED TOOL IN TODAY'S LABORATORY WORK

Research funding organizations ubiquitously recognize the importance of data being freely and readily available and that research organizations should provide mechanisms to facilitate and encourage data sharing across the bioscience community. Furthermore, good scientific practice requires/entails/prompts that data should be retained for ten years after the completion of a research project.

Despite these clear requirements, implementation of such systems into a working laboratory is often problematic (Figure 29.2). Biological scientists are not typically trained

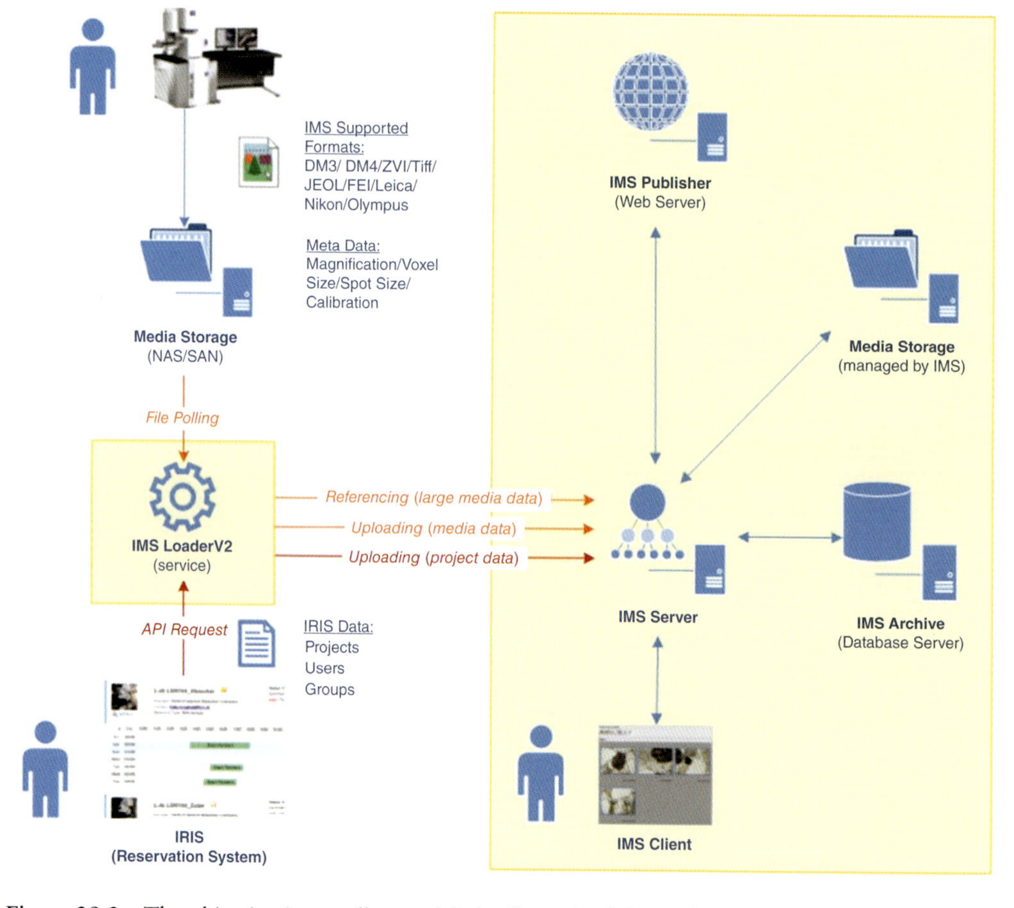

Figure 29.2 The objective is to collect and link all required data relating to an image and to manage user access and data exchange. However, this may be a challenging task. An image is the result of a number of complex acquisition steps (specimen preparation) applied to a sample (cell, tissue, organism), acquisition conditions (instrument-dependent) and post-processing (filtering/alignment). Meeting all these requirements, yet keeping it easy for the user to handle, must be the goal of this process.

in IT and may not be comfortable with complex tools, or may not see understanding computational systems as their top priority. Coordination at a national level can ease the task by ensuring that computational problems are solved once, rather than by each lab individually. The Collaborative Computational Projects, such as CCP-EM for cryoEM in structural biology (Wood *et al.*, 2015), exist to provide computational support at a national level for specific scientific domains. The CCPs have a strong emphasis on training, which can include courses on data management and available tools.

29.13 FROM TECHNICAL LIMITATIONS TO SOLUTIONS

For decades IT has tried to catch up with ever new demands for larger datasets and faster access by improving technology. Server power as well as storage capacity have increased dramatically over the last couple of years. Network speed is constantly rising as well, but network load, data acquisition speed and the acquired data volume are rising as well.

As a consequence, technology alone cannot cope with today's increase in speed and data volume – other solutions are sought after. This is the driving force for the current Big Data agenda, where the issues are often described in terms of the four Vs. In the current context, these could be listed as Volume (size of datasets), Velocity (speed of data acquisition), Variety (data from different instruments) and Veracity (uncertainty caused by loss of crucial metadata). We will look at some of them in more detail. There are three approaches to keep the data volume under control.

29.14 DATA VOLUME ON ACQUISITION

It is likely that most of the data we produce in a lab will not be of long-term value and could be deleted immediately. Saying this, sometimes it is still very difficult to quickly assess the value of the dataset right after acquisition. Analogous to the proposal of Library of Congress (USA) (Scola, 2015), who announced in 2010 that it intended to preserve digital heritage by acquiring Twitter's entire archive of tweets and make it available to researchers, the project now grapples with how to manage an archive that amounts to something like half a trillion tweets. In scientific imaging the near-exponential growth in the rate that digital data is generated, a succession of recent examples of digital fraud in the scientific literature and no obvious mechanism for discriminating between valuable and worthless data is a major consideration, compounded by the ease at which digital data can be deleted and digital formats and software become redundant. Digital data management is a critical question for science.

An intermediate solution is thus required that allows for relatively cheap storage of large volumes of data and a resource that accommodates the read/write speeds necessary for processing of data. One solution is a scratch area on a data server. A typical hierarchical data storage would consist of a local disk or solid state for fast i/o, mounted hard disks for normal work and tape for long-term storage. This allows the scientist to work on large data volumes for a limited period of time. After that period data would be deleted or moved to a slower backup medium. However, even if storage space is getting cheaper all the time, IT needs to back up heaps of data frequently, which in the end may be a bottleneck.

The disk requirements depend on the bit-depth of the data and the ability to losslessly compress image stacks (McLeod *et al.*, 2018). The bit-depth should be appropriate to the dynamic range of the data, and whether the microscope is running in counting or integrating mode. Compression schemes, which may be integrated into container formats such as TIFF and HDF5, vary in the space saving achieved and the CPU required to compress or uncompress images. They are generally more efficient for integer data, such as obtained from detectors in counting mode where many of the pixels will have identical values, in particular zero. Some image processing, such as gain correction or sub-pixel alignments in motion correction, produce floating point data which can compromise compression efficiency. As multidimensional microscopy supplier-specific file formats are still common, the decision of keeping raw data or converting them to generic formats is an issue. The Open Source Community has been pioneering in the approach to come up with a generic solution. However, even if a couple of suppliers are supporting this initiative and some of them even allow exporting datasets to such generic formats, most of them still continue to promote their proprietary formats.

Unfortunately for most lab environments, this means redundant storage of raw data from all involved equipment, as proprietary data can only be viewed and post-processed using proprietary applications and viewers.

As long as the community is not successfully pushing a common format this situation will not improve and each individual site needs to find its own solution according to its resources. One of the roles of CCP-EM is to provide a unified voice for the community, to help establish standards and to press for their adoption by companies and service providers.

29.15 REVIEW OF DATA

Reviewing large datasets can be quite a challenging issue, as in most cases the whole data volume has to be uploaded and viewed by a local proprietary application. This can even be compounded by changes to proprietary software as a programme develops through a series of version changes. A powerful image database, combining experiment metadata with acquisition process information, is helpful to quickly assess the datasets regarding their value for storage and further processing. For this purpose, fast previews of large datasets can also be of further help on slow networks and standard client hardware.

Transfer of large datasets across the network is a major issue. A number of parallel data transfer protocols exist, for example Globus GridFTP, which can be used for transferring data collected at the Diamond synchrotron, Aspera Connect used by the EBI Electron Microscopy Pilot Image Archive (EMPIAR) or iRODS used by CyVerse. In extreme cases, physical posting of hard drives may in fact be the most efficient and cost-effective solution.

29.16 STORE ONCE, VIEW MULTIPLE TIMES

One of today's big challenges will be the single data storage on a central location. Sending complete large datasets multiple times through a network is no option, as even in fast network environments expected live-time display speed typically cannot be provided. Using streaming technologies, however, allows us to store gigabyte-large datasets just once on a server and view them even on small devices such as tablets, smartphones or any standard desktop computer (where an application can run on the server, converting datasets to a smaller 2D picture, which can be down-sampled, compressed, etc., and then streamed. This concept relies on today's well-established technologies used by the big players providing data cloud solutions and streaming services. In addition, other solutions like web-based client/server technologies providing image data stored on a server over common web browsers are thinkable. A commercial solution for this is currently being developed by arivis. This system is even able to render large 3D datasets for interactive viewing in a standard web browser.

The ability to transport data effectively is vital where large datasets require significant post-acquisition processing, for example in the case of 3D datasets or large panoramic views. Where such post-processing is required it is important to define protocols for the post-acquisition handling of data and to understand the steps from image acquisition to 3D volume reconstruction.

29.17 3D IMAGE PROCESSING

FEGSEMs are increasingly used for 3D imaging applications, using specialized methods. The most common of these, Array Tomography (including ATUM), FIBSEM and Serial Block Face SEM, have been previously discussed in other chapters (see Chapter 21 by Templier and Hahnloser, Chapter 22 by Bullen, Chapter 23 by Genoud, Chapter 24 by Giannuzzi and Chapter 26 by Kizilyaprak *et al.* in this volume). The data collected by these methods is often significant in size and must usually be aligned and processed to produce a complete 3D volume. These techniques are capable of producing data in the terabyte range and the amount of data is only likely to increase over time. A recent paper reconstructing a 400 μm × 600 μm × 280 μm region of the mouse visual thalamus acquired 100TB of data, which took 9 months to align (Morgan *et al.*, 2016). Although this is currently an extreme example of volume imaging it is a good example of what is possible and the level of computing that may be required to effectively handle such datasets. To handle such large amounts of data specialized computing facilities are often required, but smaller datasets may be handled by existing software packages on single machines. For processing such datasets two basic steps are required, image alignment and 3D reconstruction.

The majority of 3D data generated by FEGSEM methods will require some image alignment. Although some of these methods are designed to minimize the alignment required, the effects of sample charging, thermal changes and other changes inherent in the imaging process can cause shifts in position that must be corrected before 3D reconstruction can begin. This may be achieved through several software solutions. Often there are software tools included in the acquisition and management software; for example, Gatan Microscopy Suite (Gatan, USA) and Atlas 5 (Zeiss, Germany) both contain tools for automatic alignment and drift correction. Other software packages, both commercial and free, may also be used. Amira (FEI, USA), arivis, Fiji (Schindelin *et al.*, 2012), and Imaris (Schneider, Rasband and Eliceiri, 2012) (Bitplane, Switzerland) image packages contain tools for image registration and alignment, and it may also be carried out using ImageJ (Abramoff, Magalhaes and Ram, 2004), which has a variety of plugins for image alignment, or the eTomo programme in the IMOD package developed for electron tomography by The Boulder Laboratory for 3-D Electron Microscopy of Cells at the University of Boulder, Colorado (Kremer, Mastronarde and McIntosh, 1996). Good image alignment is vital for accurate 3D reconstruction and therefore it is important to choose an alignment solution that works well for the data collected. It is often advantageous to use several tools in order to allow fine tuning of alignment, particularly where images impede some automatic alignment algorithms by their resolution, quality or subject.

Once alignment is completed, 3D models of the sample may be generated in a variety of ways. For fast visualization of the sample, where detailed analysis is not required, an intensity projection of the image stack can be a useful tool. This method will produce an overview of the complete stack and is particularly useful when tissue architecture encompassing very large areas must be examined. It is also useful for assessing the quality of a sample and alignment before proceeding to more time-intensive reconstruction methods. Intensity projections may be generated using a number of image processing software suites including Imaris, Amira, arivis and ImageJ. However, such intensity projections are less informative when tissue is particularly dense or when high-resolution information is required.

29.18 3D VOLUME PROCESSING

3D reconstructions of EM data have traditionally been carried out by segmentation of the image stack, defining objects either manually or automatically by defining contours on images, or 'slices' of the 3D image stack. These contours are then used to generate 3D meshes representing the objects in question. Quantitative information on shape, size, orientation and relationships between structures can then be drawn from these meshes. The process of producing such models can be highly time consuming and care must be taken to ensure that orientation, such as the 'handedness' of helical structures, are not affected by image processing or segmentation. In recent years efforts have been directed towards increasing the automation of image segmentation, in both light and electron microscopy. These efforts have now produced several tools in standard use, but EM poses particular problems for the automatic segmentation of volumes. EM is predominantly a bright-field imaging technique that produces complex images with large numbers of grey values, which can hamper many of the automatic segmentation processes used for other imaging modalities such as fluorescence microscopy. Despite this, several tools have been developed and incorporated into image analysis software packages. Much segmentation is still carried out in a manual or semi-automated manner, either by human oversight of semi-automated segmentation or by the use of tools to speed up manual segmentation, interpolate contours or define contours from existing meshes. With the use of these techniques segmentation, particularly of regularly shaped objects, may be achieved relatively quickly.

Segmentation and reconstruction is often the best method to understand complex structures in 3D datasets and to appreciate how such structures are related. However, the process is time consuming and labour intensive, and therefore for many laboratories it is difficult to produce large quantitative datasets over multiple samples. The technique is also not well suited for rapid assessment of potentially interesting structures. Alternative techniques may therefore be used for fast quantitation of 3D datasets. Stereology techniques are often a useful adjunct to reconstruction and can be used to assess a feature found by reconstruction (for example distribution of a structure) across multiple cells. Such analyses will lose resolution from being performed on point sampled datasets rather than complete reconstructions, but where used carefully and with a well-formed question can produce useful data from many cells without requiring full reconstruction. Stereology and reconstruction tools are also often available in the same imaging packages, so it is also possible to overlay stereology grids on to reconstructed objects or treat a stereology gridset as a 3D object that may be sliced across multiple axes. This can be particularly useful for assessing organelle distribution and we have used point counting stereology for this purpose (Bullen *et al.*, 2015). A stereology grid still requires a human expert to classify the points, but data can be generated relatively quickly compared to reconstruction, and therefore it is possible to classify a relatively large number of cells with a much lower time and labour cost.

The requirements for image management, from data acquisition to processing and 3D reconstruction, are becoming increasing vital as imaging volumes increase in size

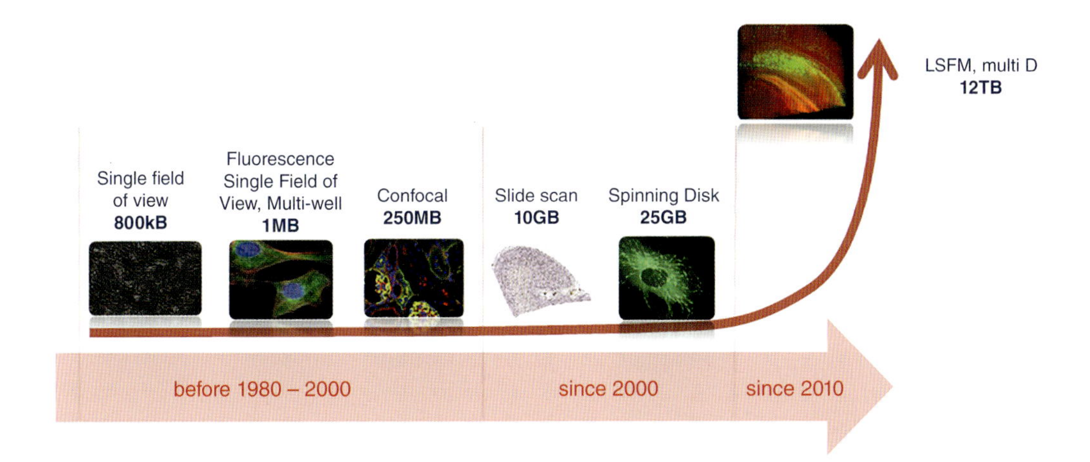

Figure 29.3 The requirements for image management, from data acquisition to processing and 3D reconstruction, are becoming increasing vital as imaging volumes increase in size and complexity.

and complexity (Figure 29.3). Therefore computing infrastructure must be able to grow and adapt to the changing needs of users, microscopy staff and regulatory bodies. The implementation of defined workflows and standard protocols for the storage, transfer and processing of data will greatly improve the consistency of data handling and data security, ease of acquisition and processing, and the quality of final processed data.

REFERENCES

Abramoff, M.D., Magalhaes, P.J. and Ram, S.J. (2004) Image processing with ImageJ. *Biophotonics International*, 11 (7), 36–42.

Bullen, A., West, T., Moores, C., *et al.* (2015) Association of intracellular and synaptic organization in cochlear inner hair cells revealed by 3D electron microscopy. *Journal of Cell Science*, 128 (14), 2529–2540. DOI: 10.1242/jcs.170761.

Cheng, A., Henderson, R., Mastronarde, D., Ludtked, S.J., Schoenmakerse, R.H.M., Shortb, J., Marabinif, R., Dallakyana, S., Agardg, D. and Winn, M. (2015) MRC2914: Extensions to the MRC format header for electron cryo-microscopy and tomography. *Journal of Structural Biology*, 192 (2), 146–150.

Kremer, J.R., Mastronarde, D.N. and McIntosh, J.R. (1996) Computer visualization of three-dimensional image data using IMOD. *J. Struct. Biol.*, 116, 71–76.

McLeod, R.A., Righetto R.D., Stewart, A., Stahlberg, H. (2018) MRCZ — A file format for cryo-TEM data with fast compression. *J. Struct. Biol.*, 201, 252–257.

Morgan, J.L., Berger, D.R., Wetzel, A.W. and Lichtman, J.W. (2016) The fuzzy logic of network connectivity in mouse visual thalamus. *Cell*, 165, 192–206.

Schindelin, J., Arganda-Carreras, I., Frise, E. *et al.* (2012) Fiji: an open-source platform for biological-image analysis. *Nature Methods*, 9 (7), 676–682.

Schneider, C.A., Rasband, W.S. and Eliceiri, K.W. (2012) NIH Image to ImageJ: 25 years of image analysis. *Nature Methods*, 9 (7), 671–675.

Scola, N. (2015) http://www.politico.com/story/2015/07/library-of-congress-twitter-archive-119698.html#ixzz3hvwXEAFm.

Wood, C., Burnley, T., Patwardhan, A., Scheres, S., Topf, M., Rosemane, A. and Winn, M. (2015) Collaborative Computational Project for Electron Cryo-Microscope, *Acta Crystallographica*, D71, 123–126.

30

Part 1: Optimizing the Image Output: Tuning the SEM Parameters for the Best Photographic Results

Oliver Meckes and Nicole Ottawa
Eye of Science, Reutlingen, Germany

30.1 IMAGE ADJUSTMENTS

30.1.1 Sharpness/Astigmatism

The first step for good SEM images is a clearly adjusted beam. Lens alignment (rotation centre) and astigmatism have to be set in an optimal position. The astigmatism has to be checked before each shot, since sample charging can cause distortion of the beam. This is often visible as astigmatism. The image should be adjusted and focused at 2 to 4 times higher magnification than the photoscan because the recording resolution is often far beyond the monitor display resolution (see Figures 30.1 to 30.3).

30.1.2 Brightness/Contrast

Many SEMs have “Auto Contrast Brightness”. This function cuts the image pixel intensities white (lights) and black (shade). Therefore the number of individual pixels at the maximum brightness (white) and maximum shade (black) fall below a defined percentage of the total range of pixel intensities (white, range of grey scales, black). This means that if an image contains a few bright areas these are set to white while a few dark ones are defined as absolute black. As a result, there is a loss of information in the “darks” and “lights” in the image. Images with a low signal (small spot, low keV or similar) cannot be adjusted at all, because the necessary information is masked by the “noise”. The comparable problem of loss of information and/or lack of adjustment occurs with the function “Auto Focus”.

Biological Field Emission Scanning Electron Microscopy, First Edition.
Edited by Roland A. Fleck and Bruno M. Humbel.

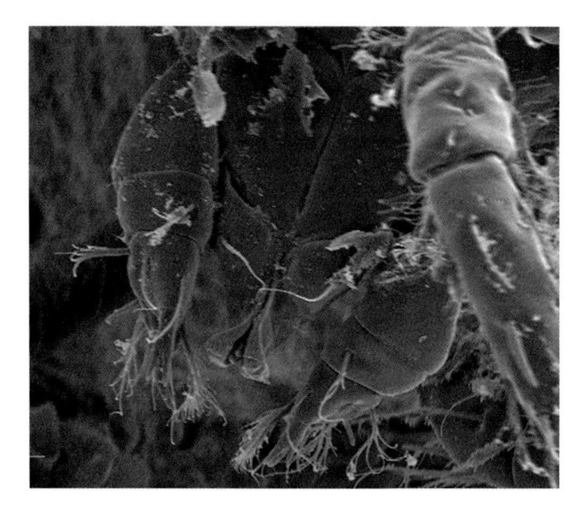

Figure 30.1 Focused on the monitor, 521 pixels.

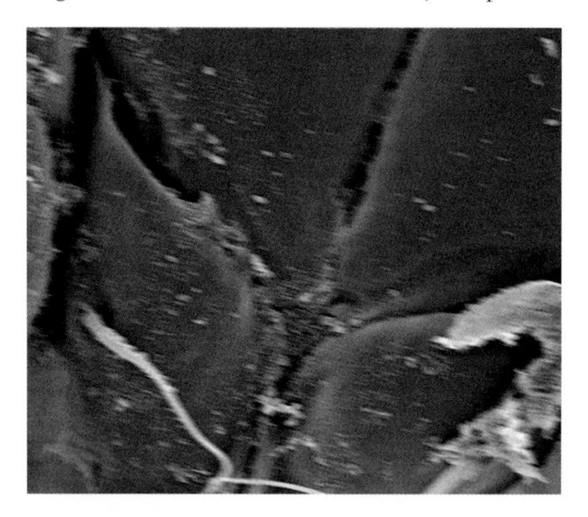

Figure 30.2 Excerpt from a 2000 pixel scan; the astigmatism is clearly visible.

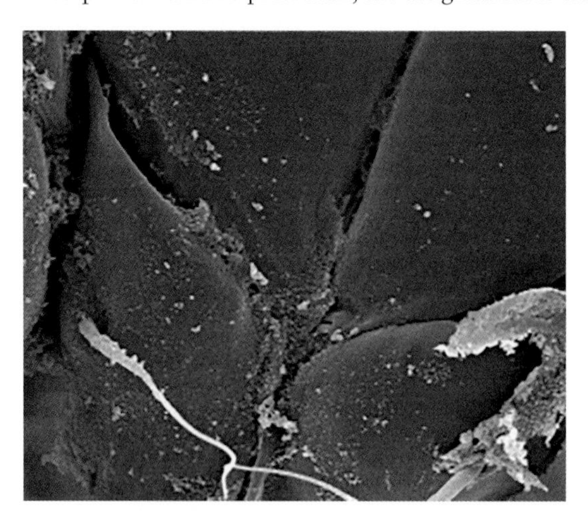

Figure 30.3 After "zooming-in" and focusing, the photoscan is sharp on the pixel level.

Figure 30.4 Result of "Auto Contrast Brightness" is that the contrast is adjusted too high: the histogram shows peaks to the upper and lower limit of the grey scale of the image. High peak: white, peak on basement: black. There are areas without information, e.g., at the top right, the leg of the mite.

A better way to address the accurate reproduction of the subject as an image with minimal loss of information is by using the histogram function. In this way an image can be exactly adjusted in brightness and contrast. Thus, ensuring an accurate reproduction of all relevant parts of the image can be before the photo scan is applied (Figure 30.4). When working with a sensitive specimen (i.e., cryo SEM or at high magnification with an increased risk of sample contamination; see Chapter 12 by Tacke *et al.* in Volume I and Chapter 18 by Walther in this volume), the pre-photo adjustments should not be directly performed on the scene of the the photo itself. Instead, pre-photo adjustments should be performed on a sacrificial area close to the photo area to prevent damage to the photographic field during whilst performing the adjustments. Once all adjustments are complete the operator may switch back to the field of interest to complete the photo scan. A final check of sharpness might be required at this stage.

As a "rule of thumb" it is better to take a picture with too low (for 16-bit) rather than with too high a contrast. A low contrast image can be compensated for in post-processing. When a SEM photo is recorded with too high a contrast, the missing information/structure is lost and no post-processing can retrieve it (Figures 30.5 to 30.7).

Figure 30.5 Correct contrast.

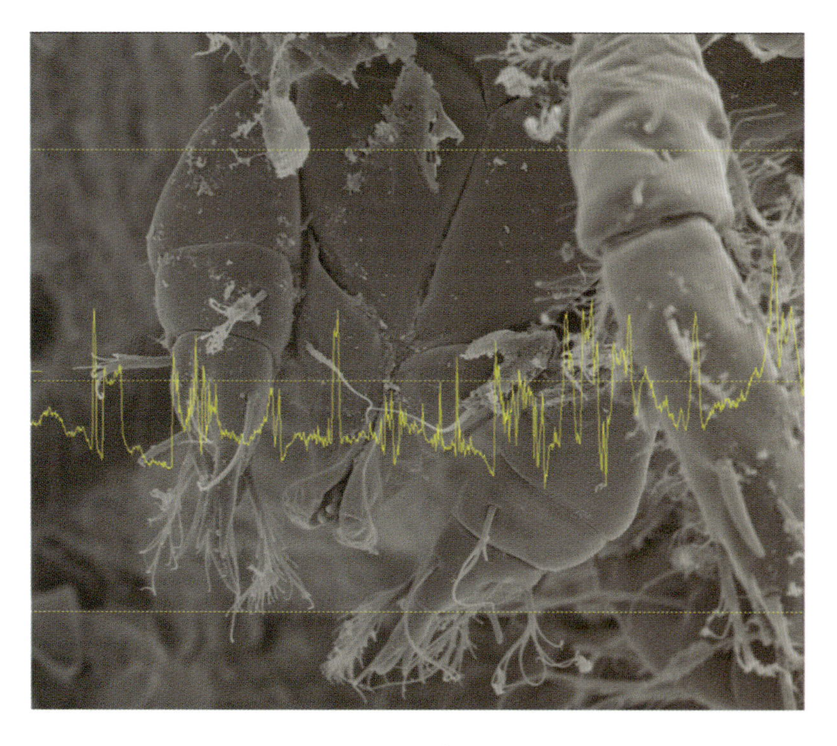

Figure 30.6 Too low contrast.

Figure 30.7 Too high contrast.

30.1.3 Integration Time/Noise Reduction

The last step to be optimized is to ensure that the scanning speed of the photo scan guarantees a noise-free picture. The integration time of the photo scan needs to be set on "fast scan mode". The image should appear clear, the edges sharp and surfaces smooth. If this is not the case, one has to select a higher (longer) integration time. By default, we use 3 to 30 μs/pixel, which results in a 3 to 6 minute scan at 4000 pixel-wide (UltraHD) pictures (according to approximately 1 min at 2000 pixels and 30 seconds at 1000 pixels width).

Frame- or line-averaging can also be used to reduce noise. However, these methods introduce a time-gap between the finally averaged measurements. Image shift or specimen shrinkage may occur and reduce the quality or sharpness of the final image. Averaging can be effectively used if the specimen is stable as it reduces the risk of charging. Other ways to get a noise-free picture could be to use larger spot sizes or wider apertures or higher voltages. The result in all possibilities will be the same: a stronger signal, which allows the photographer to decrease the contrast and therefore get less noise (see Figures 30.8 and 30.9).

However, these conditions are only true for low magnifications. While working with high resolution, small spot sizes and apertures are necessary; otherwise the above-mentioned imaging conditions could lead to a reduction in the resolution.

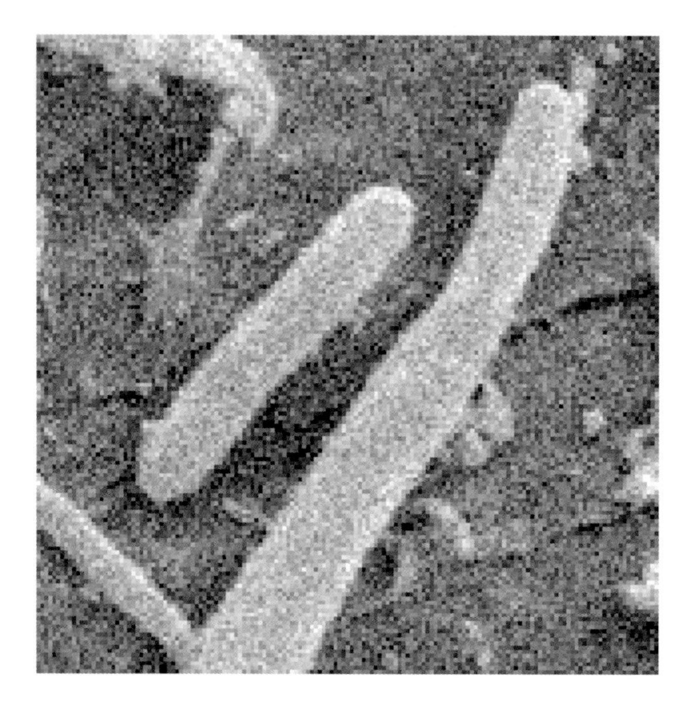

Figure 30.8 Short integration time, noise.

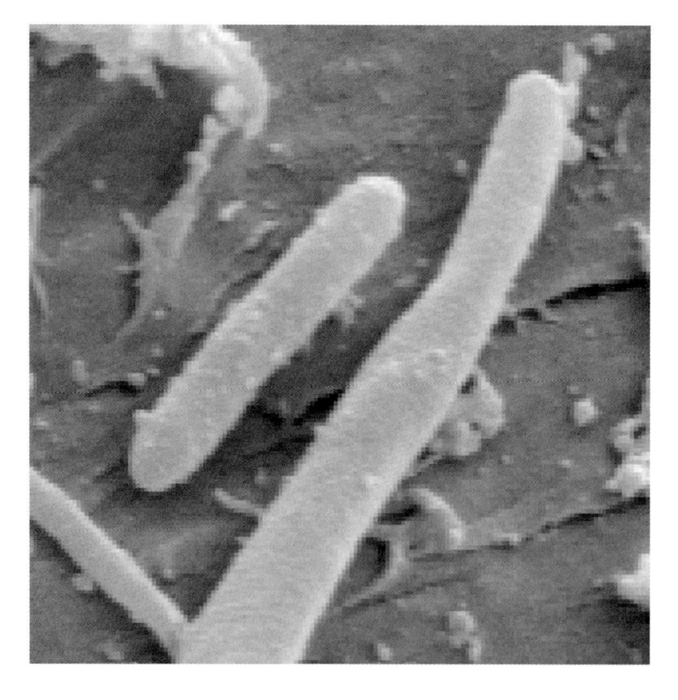

Figure 30.9 Longer integration, noise-free image.

30.2 EMPTY MAGNIFICATION AND USEFUL SCAN SIZE

When working near the resolution limit of the microscope, with over 100 000× magnification, a photo scan size of 500 or 1000 pixels width will be sufficient for a photo scan. The clipping at pixel level (Figures 30.11 and 30.12) shows that the sharpness of the selected scan size can be beyond the limit of the microscope resolution. The sharpness is not at "pixel-level", but in the range of more than 4 pixels; therefore these conditions only generate more pixels and higher damage of the specimen. We define this as "empty magnification", a magnification that shows no new details compared to a picture with lower resolution (see the markers in Figures 30.11 and 30.12).

An example calculation for a SEM with 1 nm resolution, remembering that 0.1 mm is the limit of resolution of the human eye:

$$1 \text{ nm} \times 100\,000 \text{ magnification} = 0.1 \text{ mm on the screen or paper}$$

$$0.1 \text{ mm} \times 1000 \text{ pixels} = 10 \text{ cm image width}$$

To get a clear view on the computer screen or paper, about 3 pixel points are needed to recognize and clearly define the resolution, e.g., 1 nm needs ~0.3 nm per pixel (see Figures 30.10 to 30.12).

A persistent beam may cause contamination or damage to the specimen (see Chapter 12 by Tacke *et al.* in Volume I). In addition, the sample can start drifting (moving under the beam) and adulterate the image (e.g., reduce image sharpness). It is, therefore, preferable to choose small scan sizes in combination with a higher integration time when working at high resolution. For example, a scan with 500 pixels in width pollutes the specimen to the same extent as a scan with 2000 pixels in width with a 4 times faster integration time, but results in a substantially reduced noise in the final picture.

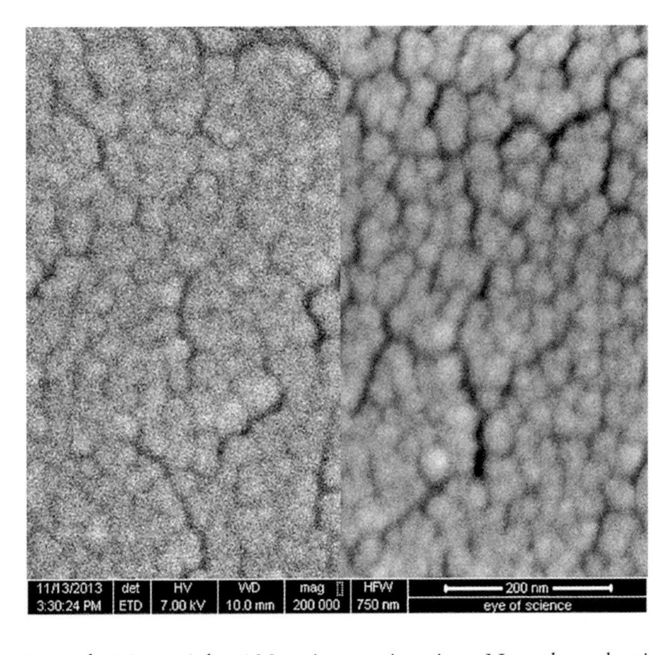

Figure 30.10 Left: 10 μs, right: 120 μs integration time. Note the reduction of noise.

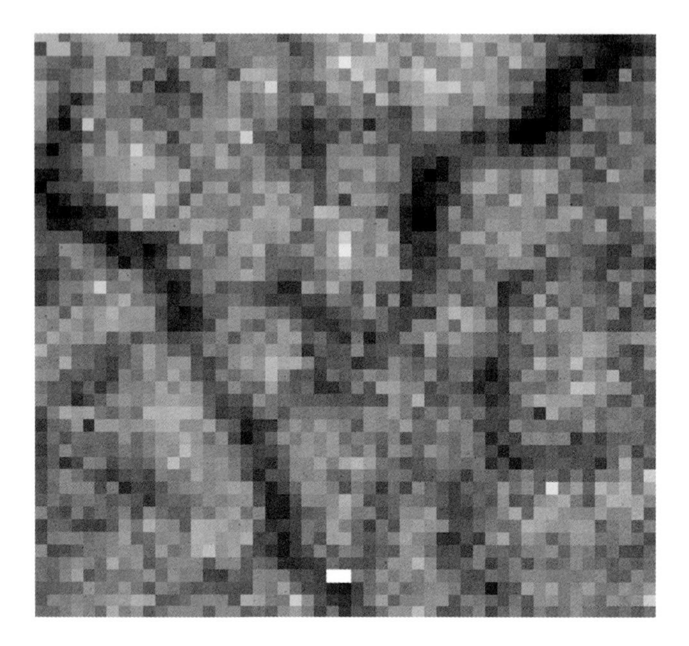

Figure 30.11 Cut-out of a 1000 pixel scan: two pixels define the object's edges.

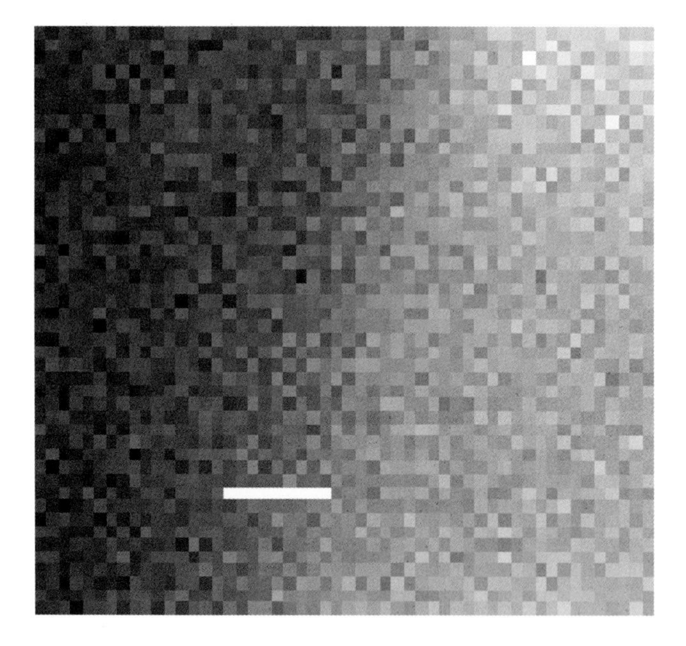

Figure 30.12 Cut-out of a 4000 pixel scan at the same magnification: 8 pixels define the object's edges. There is no new information.

30.3 SAVE: 8 BIT, 16 BIT, AND THE WHOLE IMAGE FORMATS ...

A modern, digital SEM scans all images in very fine grey scale (many thousand to several millions). When saving the image, this usually is reduced to 8 bits (256 grey levels). Some image formats also allow saving the very fine gradations (e.g., 12, 16, or 32 bit according to 4096, 65 536, and approximately 2 million grey levels respectively). The more grey levels, the larger is the size of the digital image.

30.3.1 The Various Image Formats

Excerpts from Figure 30.13 in different compression rates and formats illustrate the resulting artefacts and are shown in Figures 30.14 to 30.17.

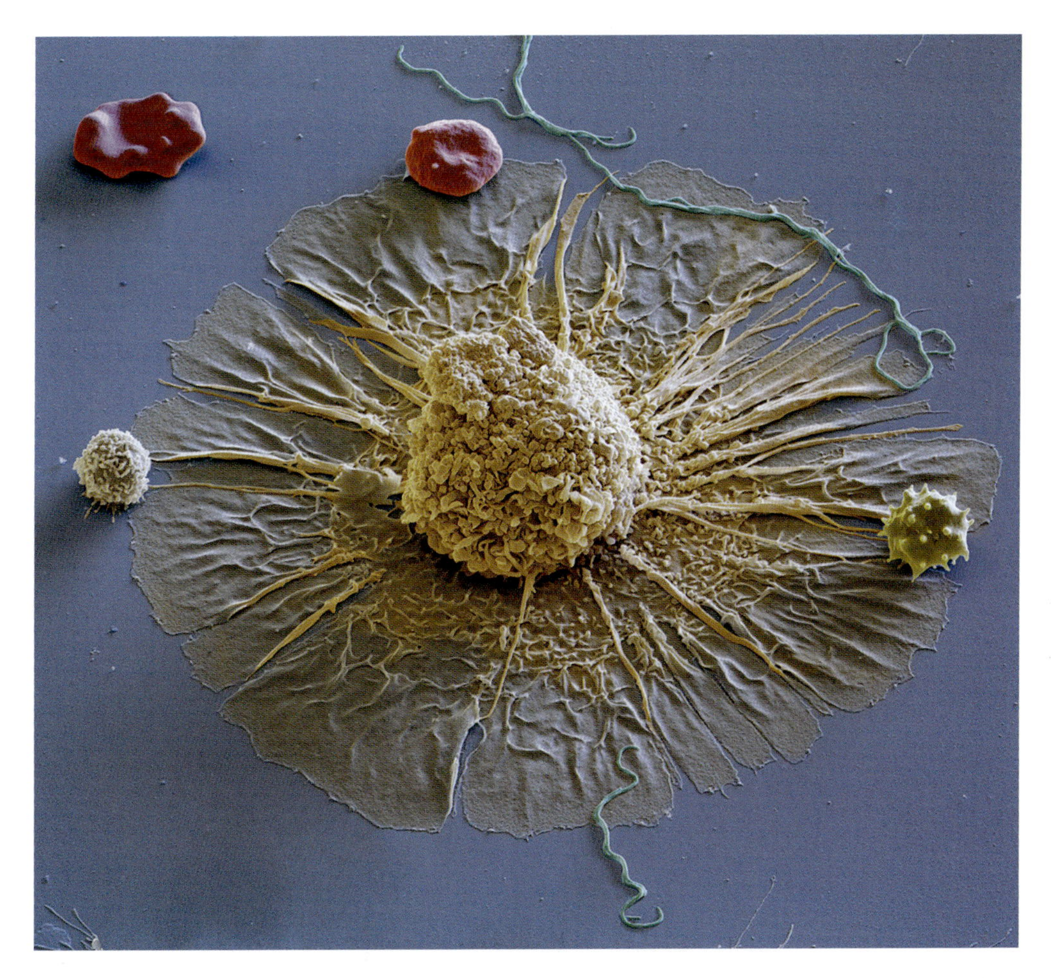

Figure 30.13 Coloured SEM image of red blood cells (red), macrophage (beige), granulocyte (white), and a deformed red blood cell (echinocyte, yellow); the long curved objects are *Borrelia* bacteria. Scaled to 1300 × 1200 pixel image size, magnification 2812:1.

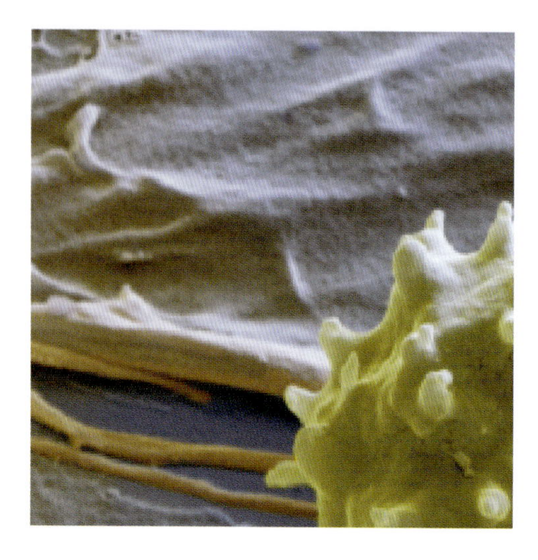

Figure 30.14 JPG level 12TIFF/PNG storage: no visible artefacts.

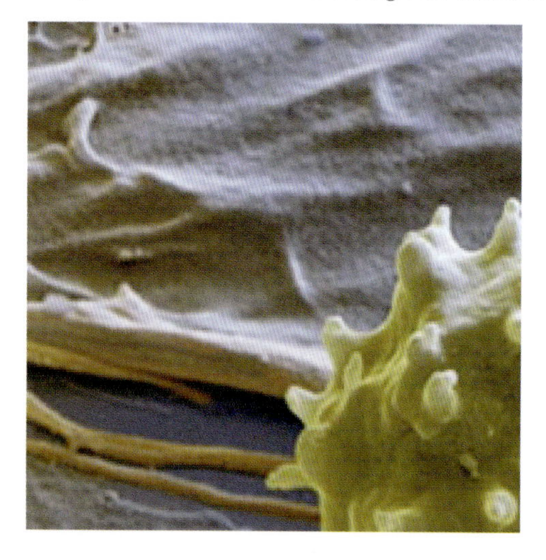

Figure 30.15 Stored JPEG level 6 "medium": colour and line artefacts become visible.

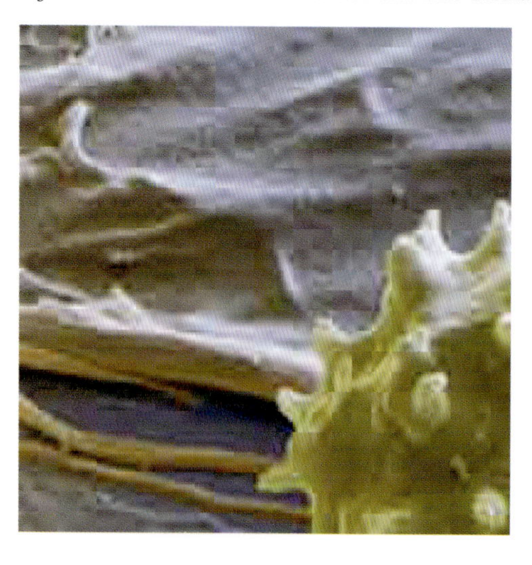

Figure 30.16 JPG level 1 "minimal" 8×8 pixel squares spoil any details.

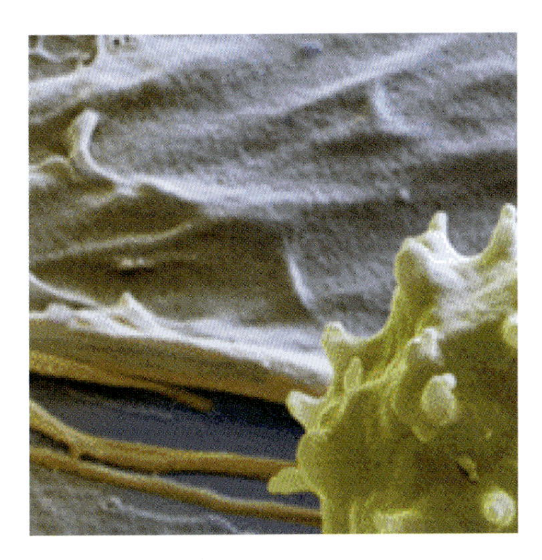

Figure 30.17 BMP, GIF, 256 colours: steps from one to another colour are clearly visible.

30.3.1.1 TIFF (Tagged Image File Format) (Figure *30.14*)

This format is very diverse. Some programme-specific, complex additional information can be saved (layers, fonts, alpha channels, 16 + 32 bit images, etc.). Files are saved 1:1, image size = file size. Using LZW, a lossless compression is possible, which reduces the file size to 50–80% of the original size – if the image is noise-free (LZW is a lossless compression of digital images with an algorithm developed by Lempel, Ziv, and Welch). Due to the low compression capability, it is not Internet suitable and is not supported there. Metadata can be embedded (information about photo data, image caption, copyright, etc.). It is the most common format for professional applications (photographers, pre-press).

30.3.1.2 JPEG (Joint Photographic Expert Group) (Figures *30.15* and *30.16*)

JPEG is a compression format. It can only be used with images in 8 bit (256 grey levels). The compression level can be selected and the resulting file size is, depending on the compression rate and image content, at 2% to 30% of the uncompressed file. In the highest quality level, the artefacts are practically not visible (Levels 10–12). At stronger compression rates distracting lines or squares (8×8 pixel) can appear in the image. JPEG is the standard format for data transfer and is used almost exclusively on the Internet. JPEG is not suitable when the images are processed further because at every change and storage, new artefacts may occur. In this case, the “original” should be obtained as a TIFF.

30.3.1.3 BMP (Bitmap) (Figure *30.17*)

This is a very simple graphic format that allows low compression. It is rarely used and is not suitable for scientific documentation because metadata cannot be added.

30.3.1.4 GIF (Graphics Interchange Format) (Figure *30.15*)

This format supports only 256 colours. After analysing, the colour scale is reduced to a "bitmap" of 256 colour. This is no RGB picture. This format is useless for colour pictures and colour reproduction. It can be used for greyscale pictures as it supports 256 levels of grey. Occasionally it is used in the Internet, especially because it allows animations to play. It is useless for scientific documentation, because metadata cannot be attached.

30.3.1.5 PNG (Portable Network Graphics)

"Successor to GIF" has suitable advanced possibilities for scientific documentation and also metadata can be embedded. It does not support animations. According to our tests the compression offers the same compression rate such as the TIFF LZW compression (depending on the motif 100% to 30% of the original size). It is less suitable than JPEG for exchanging data.

Part 2: Post-Processing of the Photomicrograph

Representative images highlight the variety of biological subjects available for colour rendering. The following is a small selection for inspiration with details of the conditions used to collect the raw images for subsequent colouring.

Lung capillaries are a network of capillaries in an alveolus (air sac) of the lung. The red blood cells are clearly visible through the cell walls of the blood vessels. Pneumocytes are coloured in yellow. Combination of SE and BSE signals, coloured, magnification 1620:1.

Section through epithelial cells of the bronchus. The cells are covered with cilia. Rhythmic movements of the cilia serve to move mucus and trapped particles away from the gas-exchanging parts of the lung. Combination of SE and BSE signals, coloured, magnification 10800:1.

Surface of a rice leaf covered with protective wax. Combination of SE and BSE signals, coloured, magnification 2250:1.

Freeze fracture through the leaf of *Atropa belladonna*, the deadly nightshade. The fracture clearly shows the cells containing chloroplasts (darker green), cores (red) and vesicles containing tannine, polyphenol, alkaloids, etc. (orange). Combination of SE and BSE signals, coloured, magnification 1530:1.

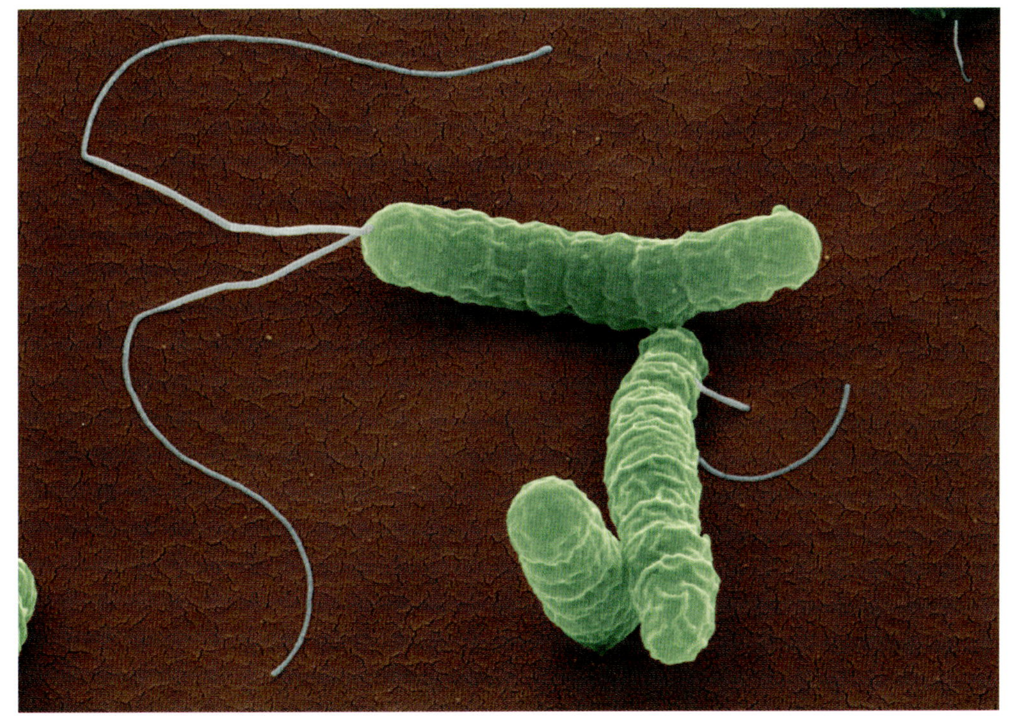

Helicobacter pylori is a bacteria that causes gastritis and is also the most common cause of stomach ulcers. Secondary electron signal, coloured, magnification 27000:1.

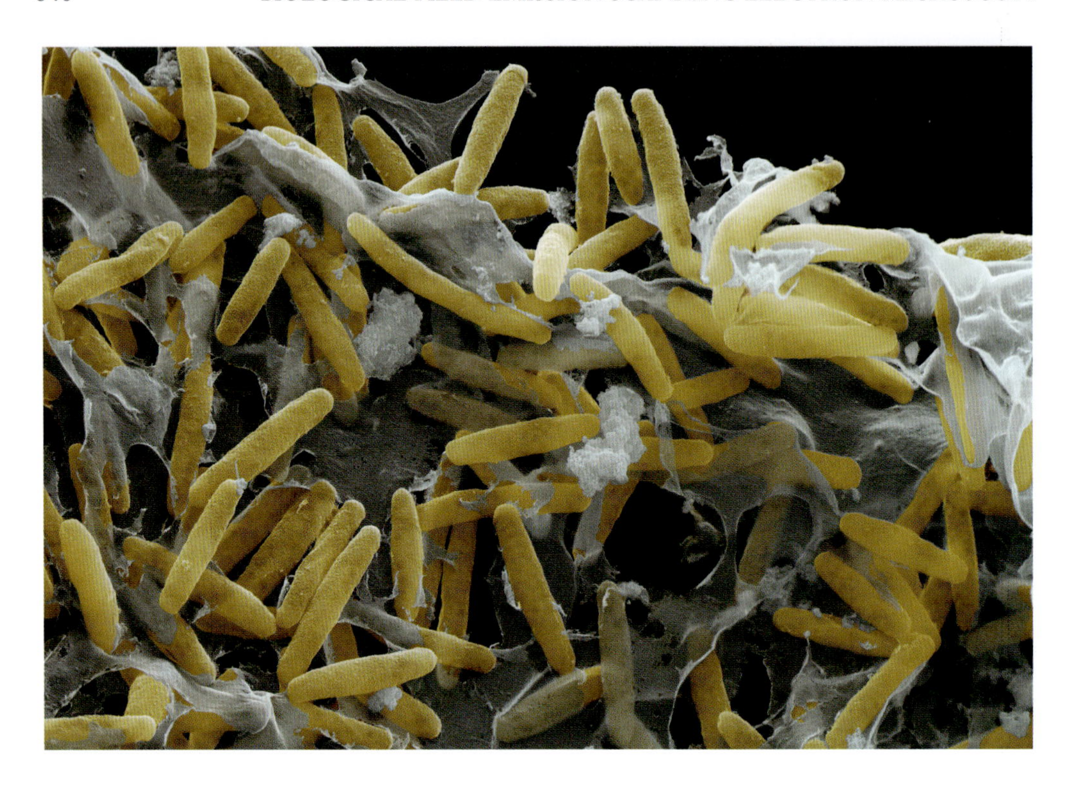

Myxococcus xanthus are gram-negative, rod-shaped bacteria that exhibit various forms of self-organizing behaviour as a response to environmental cues. Myxobacteria have the largest known prokaryotic genome. Combination of SE and BSE signals, coloured, magnification 9000:1.

30.4 OPTIMIZATION AND COLOURIZATION OF SEM IMAGES

In this multimedia world a conventional black and white scanning electron microscopy (SEM) representation (photomicrograph) of a cell or tissue can often be overlooked by the casual observer. With digital imaging techniques and colouring it is possible to breathe life into these digital images, which are then better able to engage adults and children alike and to educate non-specialists about the complexity and beauty of the living world.

To generate these coloured images is, however, not trivial and specialist knowledge of image processing is required to successfully perform the colouration. A professional colouring strategy is therefore required for SEM micrographs. From a sharp well-composed black and white SEM image (see Part 1 of this chapter), the following steps need to be followed to build the final coloured image.

From your FEG SEM image metadata, information relating to pixel resolution can be found. This baseline knowledge of the image is essential if post-image processing is to be performed effectively and appropriately. It is important to consider the file format used to record and store digital images. Not all file formats retain 100% of the original image. Instead, to reduce digital storage demands and simplify transfer of data files, formats can compress original data (see Part 1, Figures 30.13 to 30.17).

30.4.1 Common Digital Formats

TIFF (Tagged Image File Format) is a popular format for high colour-depth images, along with JPEG and PNG. TIFF is a flexible, adaptable file format for handling images and data within a single file, by including the header tags (size, definition, image-data arrangement, applied image compression) defining the image's geometry. A TIFF file, for example, can be a container holding JPEG (lossy) and PackBits (lossless) compressed images. A TIFF file also can include a vector-based clipping path (outlines, croppings, image frames). The ability to store image data in a lossless format makes a TIFF file a useful image archive, because, unlike standard JPEG files, a TIFF file using lossless compression (or none) may be edited and re-saved without losing image quality (if you have the enhanced scientific metadata you might need the extended Photoshop version to avoid loss of critical metadata). This is not the case when using the TIFF as a container-holding compressed JPEG. Other TIFF options are layers and pages.

JPEG (Joint Photographic Experts Group) is a commonly used method of compression for digital images, particularly for those images produced by digital photography. The degree of compression can be adjusted, allowing a selectable trade-off between storage size and image quality. JPEG in higher quality settings typically achieves 10:1 compression with little perceptible loss in image quality.

PNG (Portable Networks Graphics) is a raster graphics file format that supports lossless data compression. PNG was created as an improved, non-patented replacement for Graphics Interchange Format (GIF) and is the most used lossless image compression format on the Internet.

There are also scientific formats (e.g., DM3, DM4 (Gatan, Pleasanton, USA), and MRC), which are voxel formats. In most cases these can be readily converted to a .tiff or other common digital format using freely available software (e.g., ImageJ [1], Fiji [2], IMOD [3]).

30.4.2 Filters

Noise (find the tools in the menu "Filter" of Photoshop). Despite optimal signal strength and integration time used to acquire a SEM dataset often a percentage of noise remains visible in the image (Figure 30.18). This can be largely removed in post-production. Imaging programmes (like Photoshop) offer a wide variety of filters that need to be examined closely in their effect. To evaluate the effect of the filter it should be applied at 200% size image (1 image pixel = 4 screen pixels). This is achieved by selection of the appropriate zoom level for the displayed image.

Remove noise. Enhances clearly defined edges and blurs everything having low contrast. This leads to the loss of fine details and often distorts the image content (Figure 30.19).

Reduce noise. This filter works like "Remove noise" but allows for fine-tuning of the filter strength. Disadvantages are that if this filter is applied too strongly, it can lead to graphical distortions, object structures at pixel level get lost, and in the case where several adjoining dark or bright pixels are detected as lines they can be outlined incorrectly (Figures 30.20 and 30.21). Note that JPEG file format artefacts are intensified by this filter.

Gaussian blur. A slight noise will be removed effectively by filtering with a 0.5 pixel radius. This results in very smooth surfaces. At very sharp areas of the image it leads to a blurring of fine structures and a little loss of brilliance may occur (Figure 30.22). Severe noise has to be filtered with more than 1 pixel diameter but will result in a strong blur (Figure 30.23).

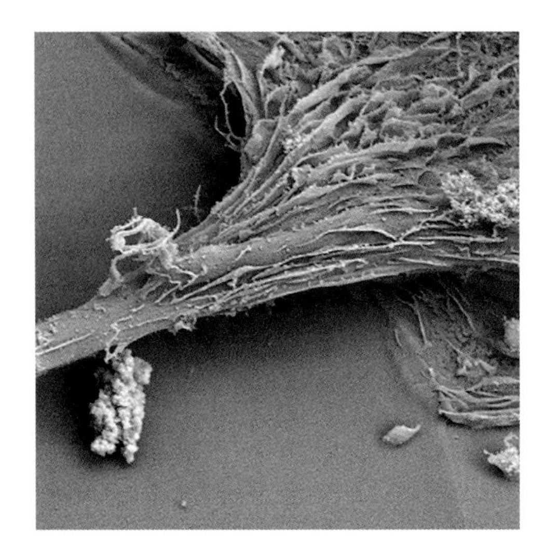

Figure 30.18 Original image, noise clearly visible in background.

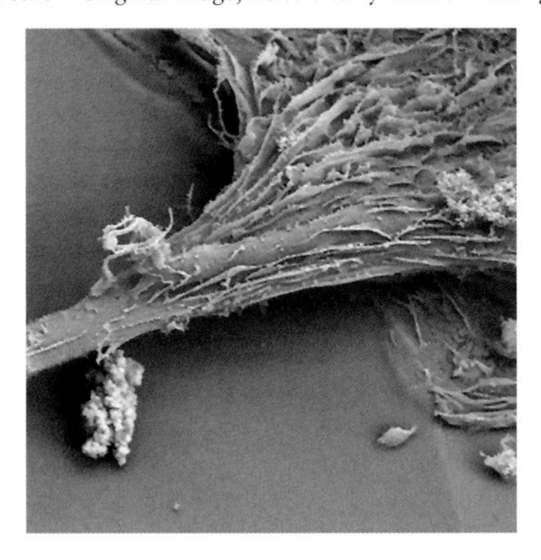

Figure 30.19 Remove noise.

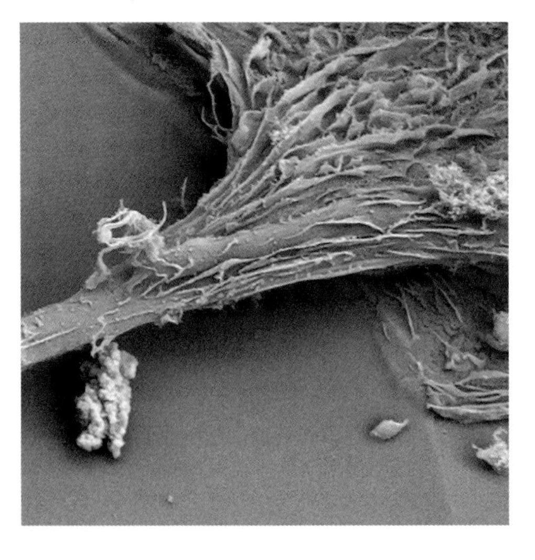

Figure 30.20 Reduce noise, level 2.

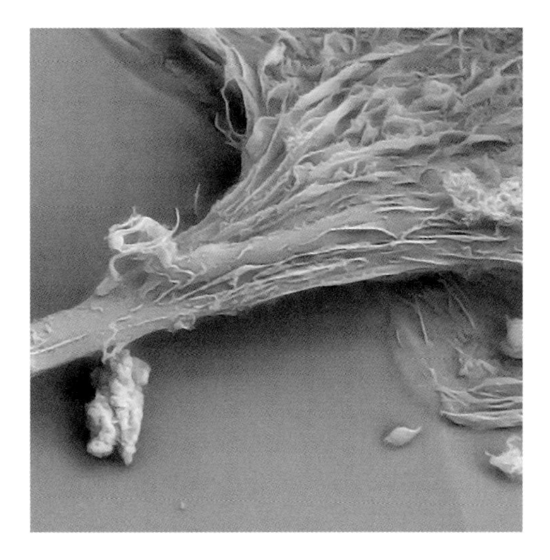

Figure 30.21 Reduce noise, level 10.

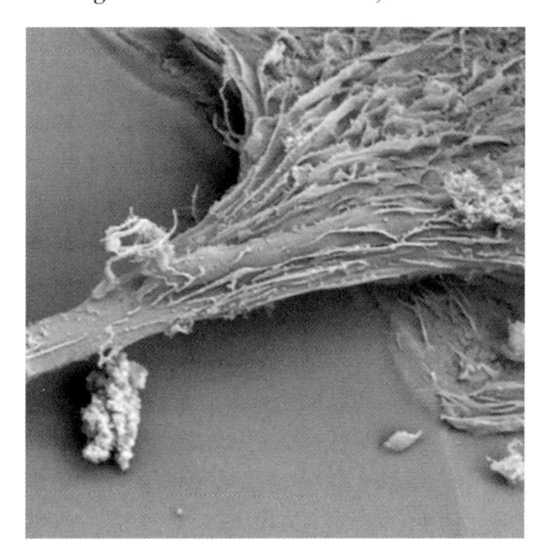

Figure 30.22 Gaussian blur, 0.5 pixel.

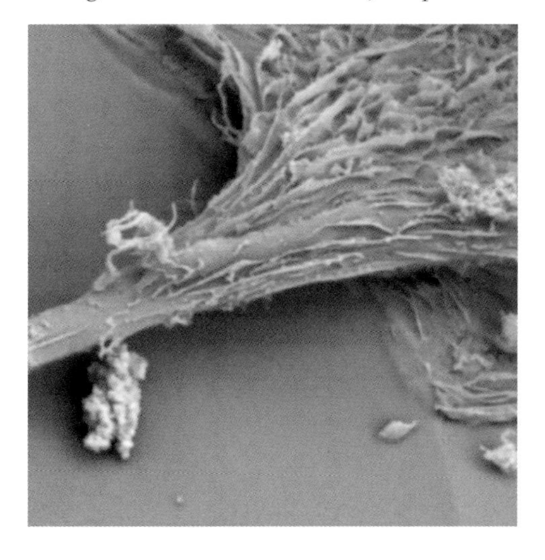

Figure 30.23 Gaussian blur, 1 pixel.

Keep in mind the fact that you need to avoid noise by scanning a SEM image so you don't need to filter in post-production and you will loose nothing (see Part 1 of this chapter, "Optimizing the Image Output").

Sharpen. An image that was already optimally focused in the SEM requires no sharpening in general. The filter "Sharpen" usually principally reinforces noise.

Unsharp mask. In low-contrast objects this filter can bring a significant improvement. The filter examines differences in brightness in the image and reinforces it. The effect can be controlled in intensity and radius, where 1 pixel in diameter and 100% intensity yields the same result as the "Sharpen" filter. A 1–5 pixel radius and 100% or even more produce unsightly dark and light edges to object contours. For the use at SEM images it will be interesting to set a 6–12 pixels radius and to regulate the strength only to 20–40%. The images will then appear clearer, sharper, and more defined (Figures 30.24 to 30.26).

Contrast Enhancement (Menu: Image > Adjustments > Shadows/Highlights). Images that have very large dark or almost white areas can be improved by the function Shadow/Highlight. This feature increases the contrast in dark or bright areas and makes it easier to see the details. This tool allows very fine tuning. It should be used with the "show more options" switched on. The default parameters are useless. An "Amount" of a few percent and only 2–5 pixels "Radius" give the best results. With "Tone" you may control the scope of the affected brightness levels. If you only want to regulate, for example, dark parts, switch the "Amount" of "Highlights" to zero. Also remember the "Black Clip" and "White Clip" which should also be on zero.

Figures 30.27 and 30.28 show the contrast enhancement using the shadow/highlight function. Improved image detail is shown following filtering with a 2–5 pixel radius, highlights set to zero, and black clip and white clip set to zero.

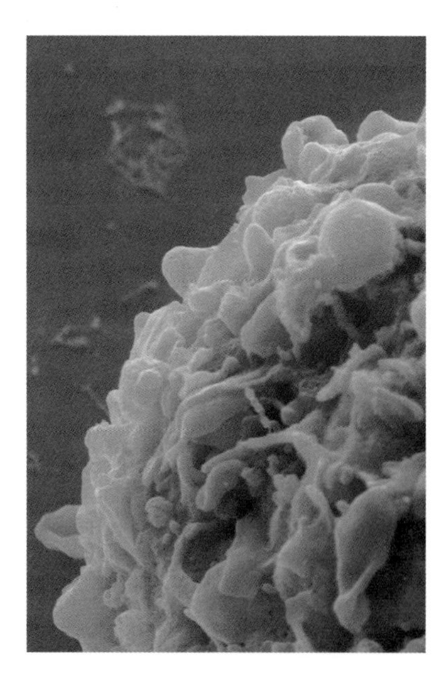

Figure 30.24 Original image.

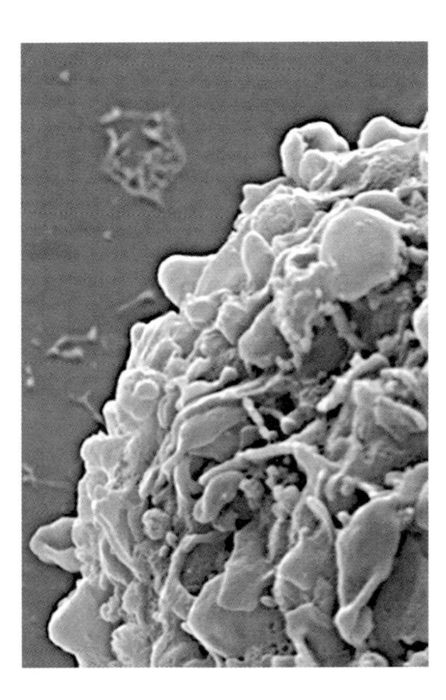

Figure 30.25 Unsharp masking at 4 pixels, 200%. A dark contour around the cell is visible; information in the brightest areas get lost.

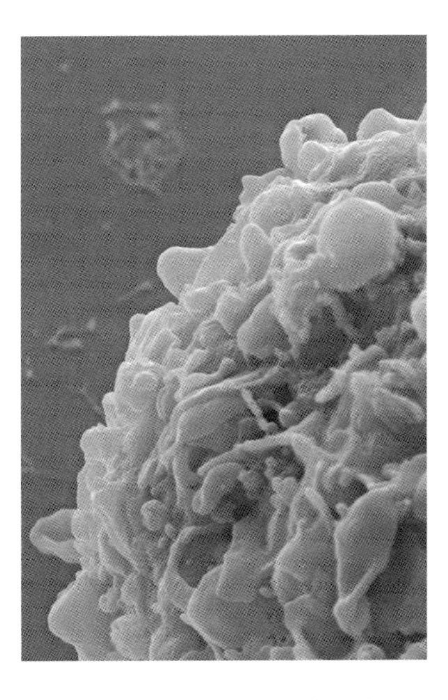

Figure 30.26 Unsharp masking at 8 pixels, 40% clearer than the original image, no visible artefacts.

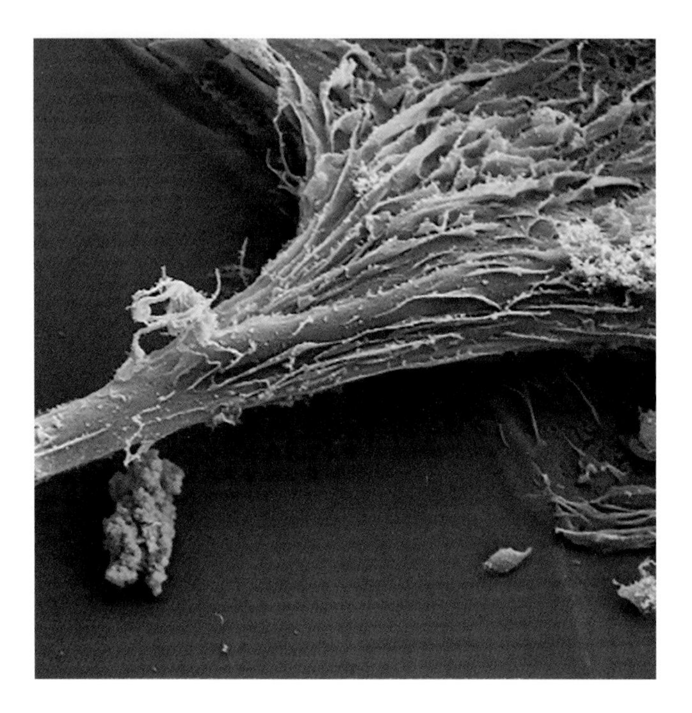

Figure 30.27 Original Image.

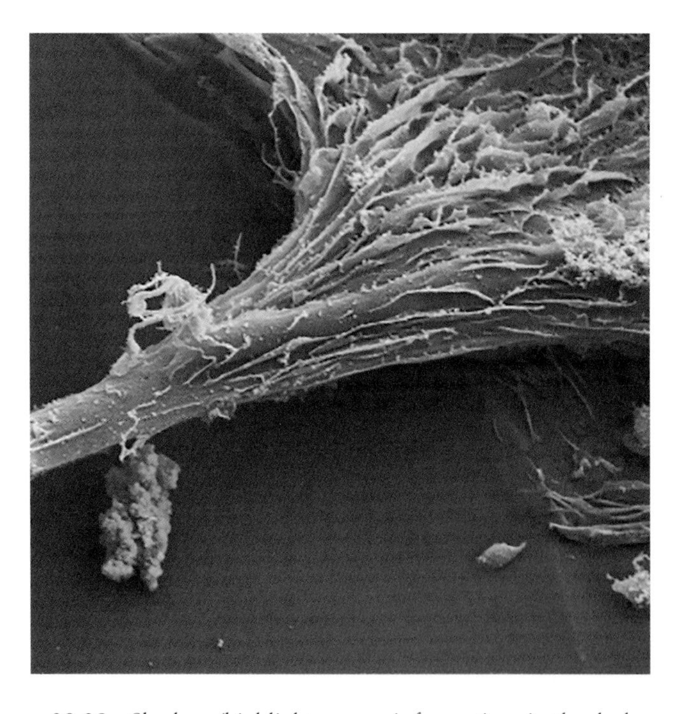

Figure 30.28 Shadows/highlights: more information. in the dark areas.

30.5 COLOURING OF SEM IMAGES

Colour photographs have a higher attentivity than black and white images. Also different types of content can be described better and easier using colours. An exact reworking of black-and-white to a colour image can only be achieved with a lot of manual work and patience. Although image editing programmes have "automatic" exemption tools, these mostly only work well with colour images. When working on an image consisting only of light–dark contours, these programmes quickly lose the boundaries between the object and background.

In order to create accurate masks, an "Artpad" (i.e. Wacom) is essential. Instead of the mouse a pen is used on its surface. The image editing programme converts the "drawing" with the pen directly to pixel graphics. These pens have even a tilt and pressure sensitivity, which can be programmed to affect size, opacity, or form and therefore allow very precise work.

30.5.1 Creating Masks

Photoshop or similar programmes offer tools that can be matched specifically to the needs of the postcolouration of black-and-white photographs. There are many ways to do this, with each also capable of being combined. Some of the more common of these are explained here (see Figures 30.29 to 30.36).

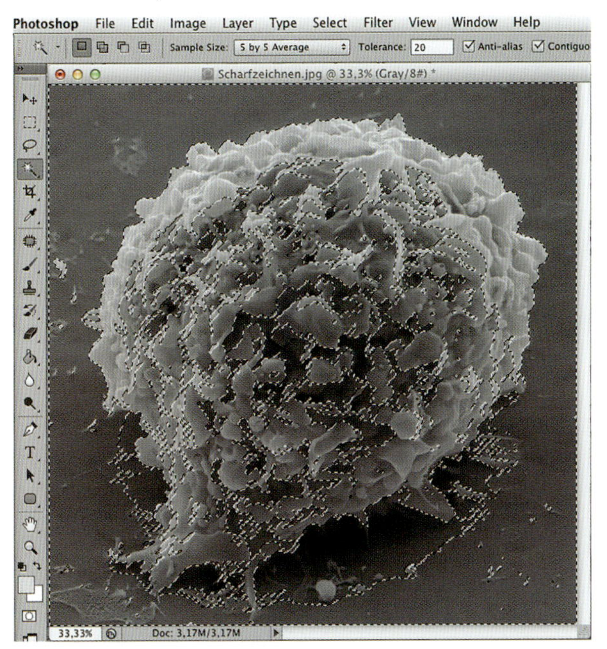

Figure 30.29 **Magic Wand (W).** This tool selects areas of equal brightness. The tolerance (how many different grey scales are addressed) can be redefined for each click. The tool can be exploited if the object has an even background and the object we want to highlight is significantly lighter or darker.

In areas of shade, where the background and object brightness switch, the wand fails as a selection tool.

Here the upper right corner (grey background) was selected with the Magic Wand tool. By holding "shift key" different selections are summarized.

In the lower part of the cell the grey values partly have the same values as the background; the selection extends to these areas.

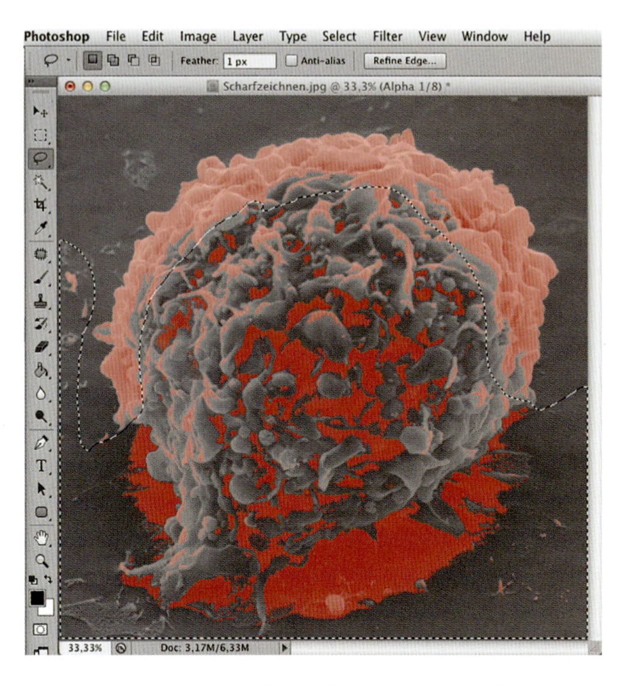

Figure 30.30 **Converting a selection to an alpha channel.** In the "Channels" window, the selection can be saved: "save selection as alpha channel". This alpha channel is a greyscale image (but is shown in red overlying) that also can be converted back to a selection again "load channel as selection". See screenshot below.

Lasso Tool (L). Now, too much selected areas will be deleted. In the "Channels" window the "Alpha 1" is activated. Using the lasso tool encircle the area that is faulty and clear it. (Menu + ←, the background colour must be set to white.). Rework is done best with the brush tool.

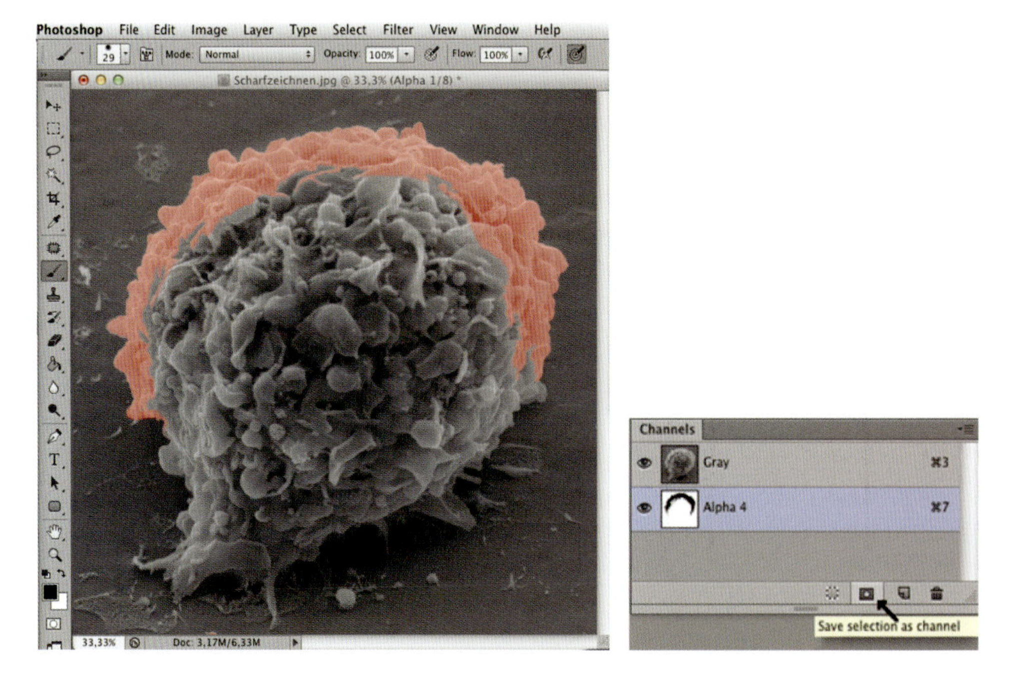

Figure 30.31 After deleting our image looks like this. Only the good outline of the alpha layer is left.

Figure 30.32 The detailed view shows that our mask is different from the picture. The mask is pixel-sharp; the underlying back and white image has smoother transitions.

Figure 30.33 **Gaussian Blur.** With a Gaussian Blur the softness of the alpha channel is adjusted to the contour of the original. The filter can be applied here to the whole alpha channel.

Caution in images with different focus. Here you have to start blurring the sharpest areas of the image. For further blur use the Blur-tool for reworking by hand (drop icon from the tool palette). In this case, set at 30%.

Figure 30.34 **Brush-Tool (B).** The lower area of this cell will be edited by hand on. The outline of the cell is traced thoroughly and completed with brushes matching in both size and softness.

Figure 30.35 **Paint Bucket Tool (K).** The complete outline of our object is now masked. Using the paint bucket tool the whole cell is backfilled (the bucket tool in the toolbox, G). The tolerance must be set to 245–250 here for a complete fill.

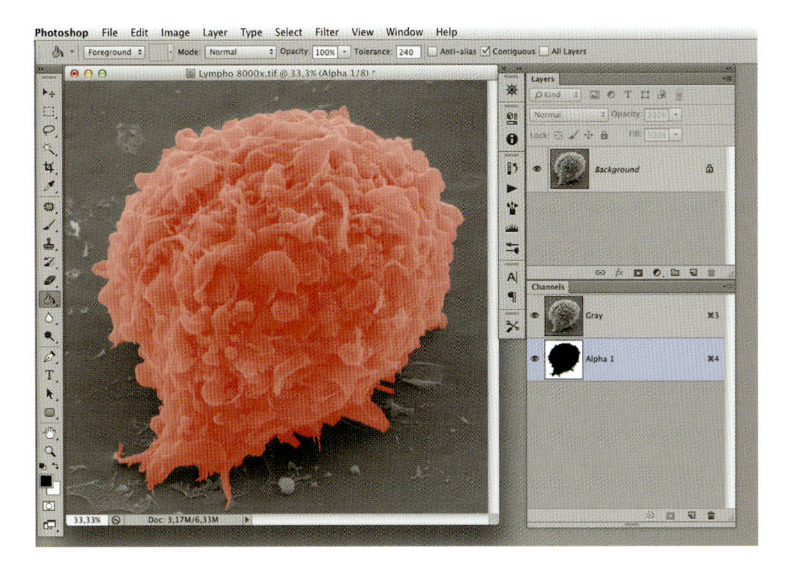

Figure 30.36 Final inspection at 200% and the first big step is done. The file should now be saved as a TIFF or Photoshop file.

30.5.2 From the Black-and-White Image with Alpha Channel to Colour Image

Figure 30.37 Now the file is converted into a colour image: Image > Mode > RGB. The image is visually black and white, but is now made up of three colour channels that still have all the same values.

Load Selection: drag the thumbnail of the alpha 1 channel in the Channels palette down to the dotted rectangle, or on the menu: Select > Load Selection > Alpha1.
The "ants" begin to run around our cell.

We want to give the white blood cell a beige tint and a blue to the background for better contrast.

Figure 30.38 Click on “RGB”–Channel to activate the image. Continue with: Menu Image > Adjustments > Colour Balance. A window with colour sliders opens. Here the colour matching of our cell is finely adjusted in the shadows, midtones and highlights; confirm with OK.

Also, the background should get a colour. This is the reverse of our selection: Menu: Select > Invert. The background can be coloured in the same manner.

Our alpha channel is no longer required and is dragged into the trash in its palette.

The file is saved under a new name; the black-and-white version is maintained for the making of another colour variation.

This was an example of a very simple masking. For complex images different masks can accumulate in numbers up to a dozen, if, for example, different bacterial species are to be distinguished or an insect should be displayed in all different shades of nature. An example is shown in Figure 30.39.

Figure 30.39 This image showing a jumping plant louse (*Psyllidae*) was coloured using nine Alpha layers for the different body parts as: wings, antenna, eyes, legs, patterns of the back, and so on. Finally, it was composed with a back-scattered electron scan to give an orange illumination from top right. This technique will be described in the next section. A full view image is shown in Figure 30.40.

Figure 30.40 Psyllids, also known as jumping plant lice, are sap-sucking insects that are a major agricultural and garden pest. This is a top view on head and thorax of *Cacopsylla picta* showing the compound eye, ocelli, antenna and the typical colours of a young imago. Magnification: 108:1.

30.5.3 Blending Various Detector Signals

As seen in the picture above, the psyllid is illuminated by an orange light from the right. This lighting enhances the spatial effect of the image. The backscattered electron signals of the SEM can be used for this lighting effect.

The image was generated by the three detector images, shown in Figures 30.41 to 30.43.

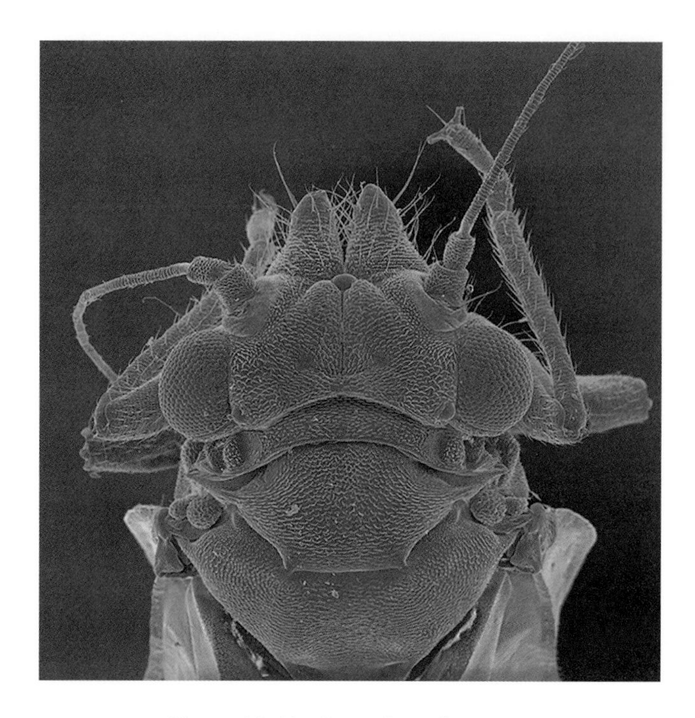

Figure 30.41 Secondary electrons.

Figure 30.42 Backscattered electrons 1.

Figure 30.43 Backscattered electrons 2.

Open the black-and-white files in Photoshop and then insert the backscattered electron images using "copy" and "paste" into the secondary electron file. Photoshop automatically creates new levels and places them "over each other" (see Figure 30.44). (Note that the pictures have to be set on "RGB" or "Greyscale". Layers are not supported in "bitmap" or "indexed colour" mode.)

Using the pop-up menu in the layers palette, you can select with the blending mode in which way the images are to be superimposed. The default blending mode is "Normal 100%". In the normal mode, the image is opaque to the underlying layer. We select "Layer 1" and set the pop-up menu of this layer to "Lighten". Now that the brightness of the selected image is added to the brightness of the underlying, both are fully visible. The same is set at the middle level. Usually the picture now appears to be too bright. Layers 1 and 2 mostly have to be adjusted to be much darker.

For colouring the lights, the change of "image/mode" to "RGB" is also required here. Then both BSE layers can be coloured. The quickest is with: Image > Adjustments > Hue/Saturation (shortcuts are shown in the menu; they differ in programme-version and the operating system) (see Figure 30.45). In this controller window we click on "colourize" (bottom right) and can then use the "Hue" and "Saturation" to select the desired colour. At the secondary electrons image (background layer) you can then do the differentiated colouring as described above.

Figure 30.44 Adding layers to the SEM image.

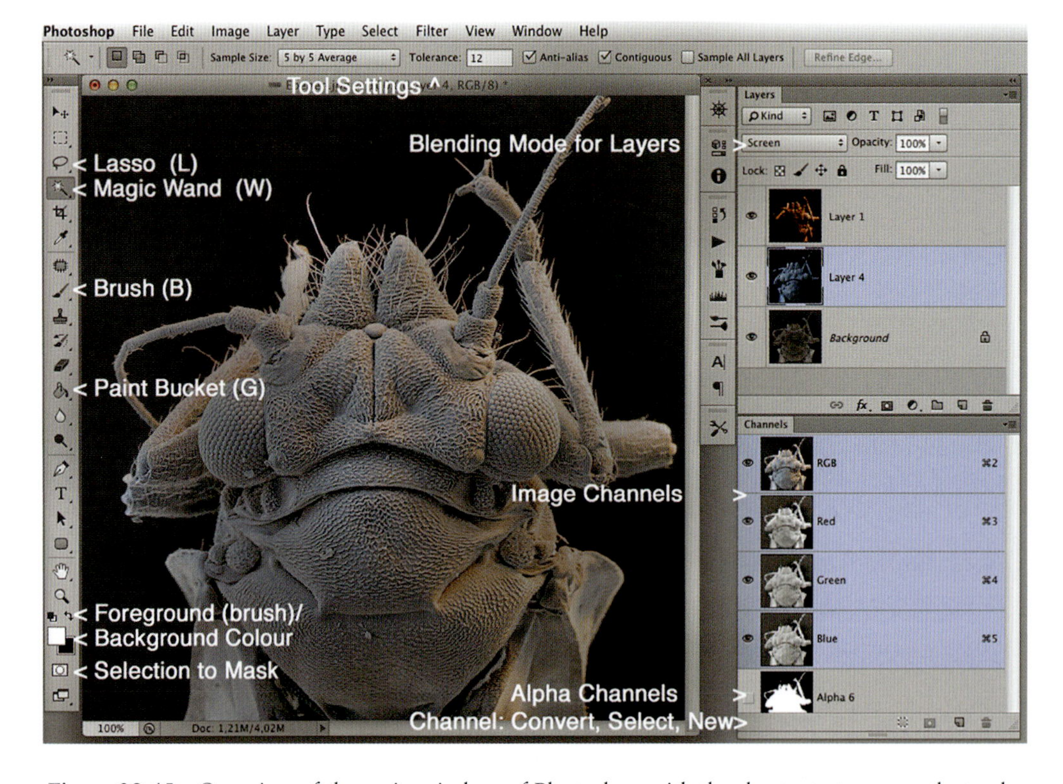

Figure 30.45 Overview of the main window of Photoshop with the shortcuts to access the tools.

30.6 CONCLUSION

As you can see, it is no longer a miracle to get a good black and white SEM image (see Part 1, Figures 30.1 to 30.9). However, it is also no longer a mystery as to how to achieve a good coloured SEM image, although it depends on the amount of time and patience you want to put into a single picture. Less noise together with a good contrast brings out a good black and white image and makes post-processing easier. The decision of how many masks you do and how fine the degree of colouring you want with your image soon adds up and increases the amount of patience you need for this enterprise. However, if the masks are precise and the work is done, the result is worth it even if the time spent in digital imaging doubles the time applied to sample preparation and its imaging at the microscope.

REFERENCES

Schindelin, J., Rueden, C.T., Hiner, M.C., *et al.* (2015) The ImageJ ecosystem: An open platform for biomedical image analysis. *Molecular Reproduction and Development*, PMID 26153368.

Schindelin, J., Arganda-Carreras, I., Frise, E., *et al.* (2012) Fiji: An open-source platform for biological-image analysis. *Nature Methods*, 9 (7), 676–682, PMID 22743772.

Kremer, J.R., Mastronarde, D.N., and McIntosh, J.R. (1996) Computer visualization of three-dimensional image data using IMOD. *J. Struct. Biol.*, 116, 71–76.

31

A Synoptic View on Microstructure: Multi-Detector Colour Imaging, nanoflight®

Stefan Diller
Scientific Photography, Wuerzburg, Germany

31.1 INTRODUCTION

Images have directly influenced some of sciences most powerful statements, not as words, rather as a universally understood language expressing the creative genius of Leonardo DaVinci to Rosalind Franklin's X-ray diffraction images of DNA (i.e. the X-ray diffraction image of DNA 'Photograph 51'), which in themselves led to the discovery of the double helix structure of DNA by Watson and Crick (Watson *et al.* 1953).

Physically correct imaging in electron microscopy means imaging in grey values or, more precisely: luminosity, which is a direct function of the electron signal coming from the specimen as a whole in transmission electron microscopes (TEM) or from a point on a surface in scanning electron microscopes (SEM). The microworld is not colourless – it becomes increasingly less colour saturated as we progress in magnification from the macro to the micro and on to the submicron scale. However, approaches to add colour to an SEM image are necessary as there is no way to directly record natural colours of the specimen in the electron microscope. This is because the de Broglie wavelength of the accelerated electron is more than a thousand times smaller than visible light. This small wavelength is, however, critical to defining the diffraction limit and hence magnifying power of an electron microscope when compared to microscopes using visible, ultraviolet or infrared light. As such, all colouring as applied to an SEM image is in fact an aesthetic interpretation by the image editor (see Chapter 30 by Meckes and Ottawa in this volume).

Biological Field Emission Scanning Electron Microscopy, First Edition.
Edited by Roland A. Fleck and Bruno M. Humbel.

Presently, many of the images used in print or broadcast media are in colour and are either generated by the simple translation of individual grey values to different colours and/or by an artist or scientist performing considerable photoediting; masking and colourization of the original image with programs like Photoshop, ImageJ and Gimp (see Watson *et al.* (1953), Diller(b), Diller(a) (2013) and Chapter 30) (Figures 31.1 and 31.2) or using a multi-signal detector approach like Scharf succeeded in adopting as a technique in 1991 (Scharf(a)) (Figure 31.3). In this case, each detector collects a signal and 'sees' a different solid angle of the specimen, which are then employed to contribute different colours to the whole image. The need for colour is most often attributed to making a scientific observation accessible to a non-scientific audience.

Figure 31.1 Large white cabbage eggs. Courtesy of Martin Oeggerli using three principal colours, generated using false colour rendering with Photoshop.

Figure 31.2 Reflective material utilizing a strongly contrasting red:green colour palette. Image courtesy of Steve Gschmeissner, generated using false colour rendering with Photoshop.

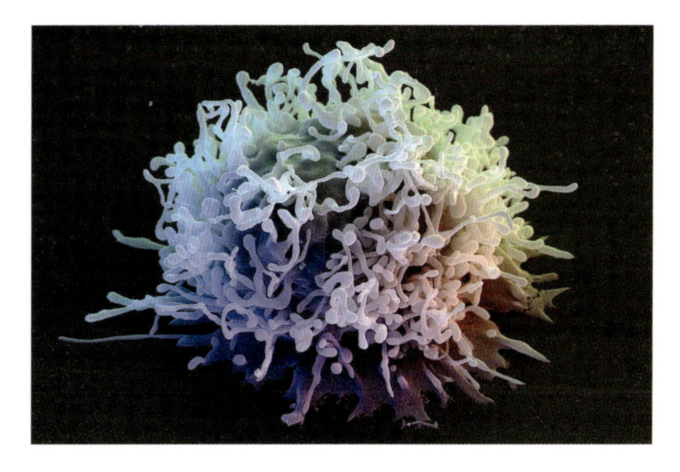

Figure 31.3 Human lymphocyte cell. Courtesy of David Scharf, 2013, generated using a multisignal detector approach.

31.2 BLACK AND WHITE VERSUS COLOUR IMAGING

Most of the scanning electron microscopes (SEMs) of today have more than one electron signal detector, for example secondary (SE) and backscattered detectors (BSE) located in lower and in-lens collection geometries (detectors are reviewed extensively in Chapter 30). These detectors can acquire electron signals either in parallel or sequentially. This leads to the possibility to easily colourize and visually enhance images (Figures 31.4 and 31.5) (Diller(a)).

Technically this colourization depends on the detectors seeing the specimen from different solid angles and/or using different signals like secondary or backscattered electrons or cathodoluminescence with each assigned a different colour. Colours mix because the solid angles of the detectors have some degree of overlap. Using secondary electrons to easily colourize small volumes or edges and backscattered electrons for topography or Z-contrast provides the possibility of enhancing greyscale ranges and hence colour transformation of the imaging signals.

31.2.1 Electron Microscopists and Coloured Imaging

Using colours to enhance the perception of the specimen and structures on the specimen is a powerful tool when employed to communicate imaging from a scientific source to the general public. A professional scientist or photographer must first search and acquire scientifically correct but also aesthetically appealing images. These two rules aid in making an image accessible to a non-expert viewer extending the impact of scientific imaging and aesthetics of the microworld to the wider public (Diller(a)).

However, these images when acquired from sources like the SEM are still images. Recently there has been an increasing interest in 3D imaging in electron microscopy (Fleck 2015) and commercial media outlets have begun to use short digitally synthesized cartoon films to represent biological processes. These are not generated directly from an imaged sample; rather they are computer-generated images (CGI). Alternatively, if one were instead to create a video showing movement directly from an otherwise fixed immobile object the resulting

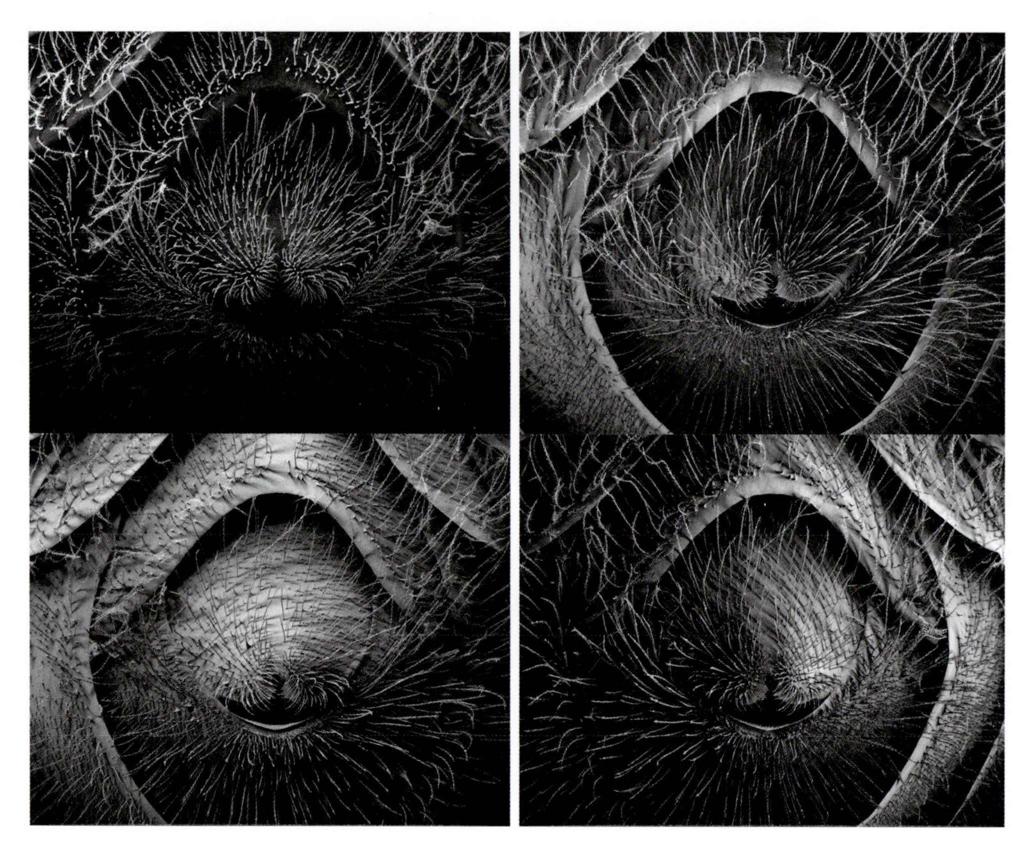

Figure 31.4 Honeybee anus: individual detector signals (images in the upper row made with secondary, the other images with backscattered spot detectors) which are first arbitrary coloured and then combined to generate a single colour contrasted image (see Figure 31.5).

Figure 31.5 Honeybee anus: multiple detector image of mixed colour signal from four separate detectors.

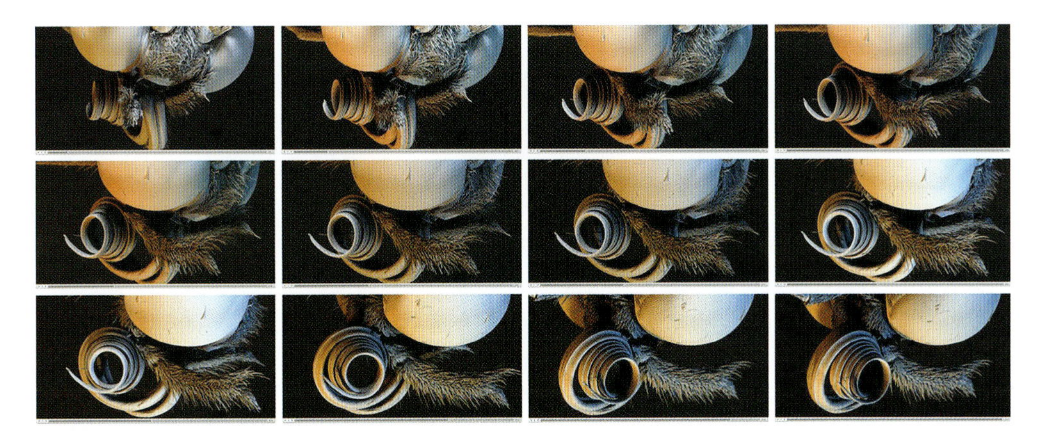

Figure 31.6 Example of an SEM movie as a series of still screenshots showing the progressive single camera angle rotation of a butterfly head.

film would be attractive and potentially bring the viewer closer to the 'living' specimen. This is the purpose of the nanoflight® technique (Figure 31.6).

In the SEM there is no easy way to acquire images with substantial detailed and consistent coloration whilst simultaneously moving a specimen under the electron beam and retaining different electron intensities necessary to provide surface detail. This has led to 'abstract' colouring and lighting to be adopted as a colouring solution. Starting from a greyscale base allows a pallet of colours to be applied to an image. When combined with movement, similar strategies to those adopted by cinematography directors can then be employed to enhance the viewer's perception of an image.

31.2.2 Colour

Painters, photographers and cinematographers exploit colour and the psychological and physical effect different colours and colour pallets can impart on us. These effects are often subtle and influence the viewer without them being consciously aware. In film colours are often employed as a device within a story, to create a sense of harmony or tension within a scene or to bring attention to a key point in a frame.

To help defining combinations of colours that are considered pleasing a colour wheel can be employed (Figure 31.7) (Itten 1961, 1970). A simple colour wheel comprises 12 colours based on a subtractive red-yellow-blue RYB colour model. Many more complex examples can be readily found with the aid of an internet browser. Primary colours are red, yellow and blue, secondary colours green, orange and purple (each are a combination of two primary colours) and tertiary colours are a mix of primary and secondary colours. On the colour wheel, two colours located on opposite sides are described as a complementary pair. Pairing complementary colours is the most common way to use two colours together, for example orange and blue (or the blue/green tertiary colour called teal). By paring a warm colour (orange) with a cool colour (blue) a high contrast and vibrant result is produced. These colour temperatures, red, yellow, orange and brown (warm colours), blue (cool) and green and purple (in-between), when combined have a subtle impact on an image and are able to highlight features within an image. If you consider a photograph, warm colours appear to come forward and cool colours recede, meaning that if a warm coloured subject is located

Figure 31.7 An example of a basic colour wheel to aid the selection and matching of complementary and split-complementary colour schemes.

within a cooler coloured background, the subject will stands out. This effect is often seen in photography, for example, where a strongly coloured flower (red rose) is set against the cooler background (green leaves), an effect that may also be effectively applied digitally to SEM images (Figure 31.2). To highlight the application of complementarity colours in film, examples are discussed with reference to commercially available films from which screen shots can be readily found using an internet search engine. In film the approach of employing strongly contrasted complementary colours has been wonderfully employed by Jean-Pierre Jeunet in the film *Amelie*. Here he pairs red and green with the main character Amelie, often represented in red against green, providing both high colour contrast but focusing the viewer's attention on Amelie herself. This is an adaptation of the photographer's strategy of creating a photograph with a high colour contrast. For example, by framing colours that are on opposite sides of the wheel, red flowers in a green field (e.g., poppies in a green field), creates this strong colour contrast. In films generated from the SEM a similar colour palette can be employed with remarkable effect (Figure 31.8).

A more subtle use of contrast between two colours is often seen in film where orange and teal are employed as a complementary pair. A ready example of this is in David Fincher's film *Fight Club*, where teal is used to push characters into the shadows and orange to highlight structures to increasing the perception of depth and draw the viewer's eye to different segments of the image. The same strategy is seen in television crime dramas and if the reader specifically looks for these cinematography 'tricks' they can be readily found. The effect has widely been used in the SEM movies (Figure 31.9). A different approach to the use of colours is a split-complimentary colour scheme when two colours next to the opposite are selected (three colours). This maintains the high contrast of a complementary pair but in a softer, less tense way (Figure 31.1). This effect may be seen in cinema in the Coen Brother's film *Burn After Reading*, which uses a pallet of red, green and teal. An alternative, three colour combination is when three colours evenly spaced around the wheel are selected. With this combination one colour will be dominant with the others accenting the dominant colour to provide a vibrant feel. The effect can be striking and can be seen in Jean-Luc Goddard's

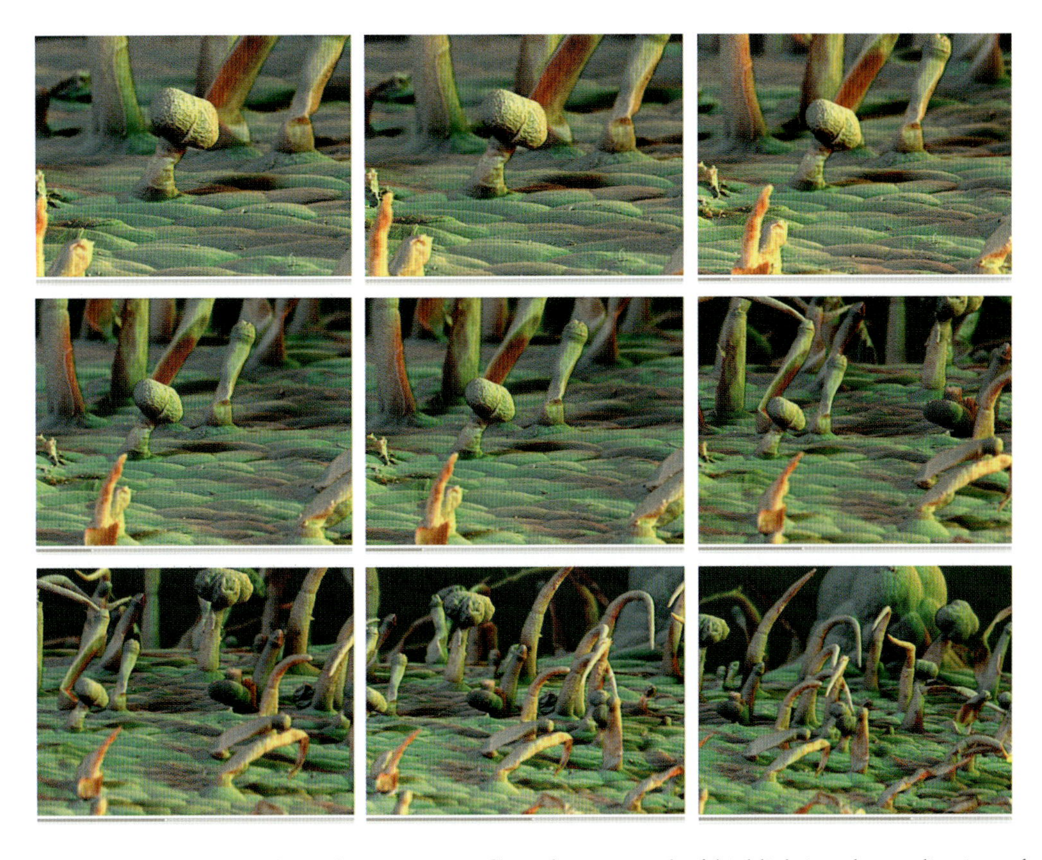

Figure 31.8 Still screenshots from an SEM film of a tomato leaf highlighting the application of strongly contrasting complementary colours (red/green).

Figure 31.9 Still screenshots from an SEM movie highlighting the use of colour (orange against a teal background) to emphasize the structure of interest. The lower two plates use a strong contrast between the warm orange and cooler teal to emphasize the foot process and bring it forward in the image. This is different from the upper two images with a flatter appearance. In the movie, the effect is to progressively bring the object of interest to the attention of the viewer.

1964 film *Pierrot Le Fou*, which makes use of a colour scheme of red, blue and green. Powerful colour strategies can be found in Sam Mendes' *Skyfall*, Alejandro González's *Biutiful*, Ridley Scott's *Blade Runner* and Éric Lartigau's *L'Homme qui Voulait Vivre sa Vie*; in each case the use of colour is employed to enhance the impact of scenes on the viewer. For a wide range of examples of colour themes employed in film refer to the Movies in Colour archive created by the graphic designer Roxy Radulescu (Radulescu).

Returning to the colour wheel, colours sitting next to each other on the colour wheel match well and can create an overall harmony in a colour palette. As they are close to one another they don't have the contrast and tension of a complementary colour palette. In photography, these colour combinations are common in landscapes as they are often found in nature. A pleasing impact can be achieved if one colour is permitted to dominate, a second to support and a third (with blacks, whites and grey tones) to accent. Consider the often seen, yet visually arresting photographs of a floor of bluebells in lush spring woods, which convey a tranquil mood reinforced by the dominance of blue and green colours.

31.2.3 Cinematographers Camera Tricks

Detailed models have been employed in film for years and by combining these models with illusions or 'tricks' the cinematography director can convince an audience that a small model is in reality a massive structure. Consider the opening scene to *Star Wars* where George Lucas creates the impression, using motion control, that a massive space ship is 'flying over' the audience. Combined with sound, this view of a small model creates the impression of scale and oppressive power. Many of these tools can be applied to films generated in the SEM. Motion control enables precise control of, and if necessary repetition of, camera movements, facilitating special effects photography. The process can involve filming several elements using the same camera motion and then compositing the elements into a single image. Common applications of this process include shooting with miniatures, either to combine several miniatures or to combine miniatures with full-scale elements, approaches that can be readily translated to the microfilm SEM world.

Lighting can add dimension and depth to an image. In the case of an SEM, movie lighting is introduced through the luminosity. Soft, even lighting which gives a gradual change in contrast will tend to flatten images, while lighting with harsher shadows with large abrupt changes in contrast a "fast falloff" gives the illusion of depth.

Parallax creates depth and makes a moving image more dynamic than a still image. It does this by imparting an impression of motion to the shot by moving the camera relative to the subject. This makes the subject's position/direction appear to change dependent on the viewing angle (if you cover one eye and focus on an object, then move to cover the other eye the object will appear to move because each eye provides a different viewing angle). In film, moving the camera and moving subjects creates the impression of activity (e.g. motion).

Things that are closer to our eyes obscure objects that are further away and in film occlusion of one object by another can be used to create depth. Tilting the camera up or down is a simple camera technique. Each can be mimicked in the SEM with a motorized stage and carefully mounted sample. Well-executed tilting, combined with action and coordination between the camera movement and the action is a simple way to engage the viewer. Stephen Spielberg uses the technique extensively in the black and white film *Schindler's List* which is filmed in black and white, so drawing extensively on the use of lighting to influence mood. Most often the sequence is acquired with a straight up or down tilt. Panning is the horizontal equivalent of tilt and in its most elegant form is a horizontal left or right sequence.

Again, this is a technique often seen in films by Stephen Spielberg to draw the viewer into the story. To combine tilting and panning a diagonal movement is the best.

Zoom shots are a powerful tool and highly accessible in an SEM. In cinema the director Ridley Scott uses the technique to make the sequence effective and 'macabre' by recording absolutely smooth and slow smooth zooms. In cinema tracking shots are complicated. The camera must be mounted on a dolly and moved along tracks. In the SEM the eight-axis stage and fixed detector position can be readily used to create a similar effect, with the sample tracking with respect to a fixed detector position. For a dynamic tracking shot foreground objects must be located between the camera and the main subject. These enhance parallax, with objects closer to the frame appearing to move faster than those more distant to the viewer, with occlusion creating a sense of depth.

Due to parallax, anything behind the subject (i.e., appearing to be in the distance) will be moving across the frame more slowly and therefore contribute less to the feeling of motion. However, if a fast sideways tracking shot is used an effect is achieved whereby the various planes in the background move at different velocities due to their varying distance from the camera (https://vimeo.com/144734321 (from second 45 on)). This effect can be seen on a very near 'fly-by' of some T-cells, with a focus-shift finally to the foreground. Since the specimen had been moved mostly radially, the further away that part is from the movement centre the more radial velocity that gets than the nearer one.

In the film *Gladiator* Ridley Scott has the main character gallop across a barren countryside, with no foreground objects in this apparently simple shot; the feeling of motion and scale is enhanced by the background. Focal length is important in tracking shots and in the SEM can be varied through detector selection and working distance. Stanley Kubrick would commonly film from a single point perspective with a symmetrical field and a cantered subject of interest and is seen to great effect in the film 2001: *A Space Odyssey*. A sample in an SEM may be treated in much the same way as a film director treats a model and using detectors in the SEM in much the same way as a director uses the camera it is possible to create a visually arresting moving image of the sample able to defy the physical limitations of the samples actual size and fixed, immovable state (Figure 31.6).

For readers interested in employing cinematography techniques a useful guide to composing scenes, describing various shots employed in film is available from Heiderich. In addition Bordwell provides a historical archive of examples of various cinematography techniques as they have been employed from 1918 to the present day (Bordwell 1997). Perhaps inspiration may be readily drawn from older, silent black and white movies, which are perhaps closer to the images generated by the SEM.

31.2.4 Motion Picture and the SEM

David Scharf is a scientist, a photographer and an artist, specializing in scanning electron microscope (SEM) imagery. For over 40 years he has been an innovator in the science, technology, methodology and photography of this uniquely fascinating form of imagery. He invented the Multi-Detector Colour Synthesizer in 1990, patented it in 1991 (Scharf (b), Scharf 1993). and was the first to bring controlled colouring to topographic and compositional information in SEM imaging. A Multi-Detector Colour Synthesizer uses multiple secondary electron detectors, which share the electron emission from a sample to collect different contrast/luminosity data dependent upon their orientation to the sample. Each signal from the detector is given a different colour. Colour selection is arbitrary and preferably selected based upon complementary colours, as described in the colour wheel (Figure 31.7).

The signals have the R-G-B format and are then converted from an analogue to a digital signal and combined. The colour complexity and shading enhance perception of depth and interpretation of the subject's topography.

Three secondary detectors simultaneously collecting signals from a single sample to provide multiple 'lighting' views of the object (Figure 31.10). When colour encoded and combined a highly aesthetically pleasing image is generated, with the colour contrast enhancing the visual interpretation and topography of the SEM subject. The perception of depth is often enhanced and can provide a dramatic 3D effect (Figure 31.3).

In 1980 Scharf recorded the first insects alive and moving in the SEM using real-time video recorded to S-VHS. In 1991 Scharf started using his colour system in recording real-time video, which he later upgraded to digital video. David Scharf created the first SEM movie sequences in 1996 leading to the production of the movie *THE HIDDEN DIMENSION*, in IMAX 3D and also the first HDTV sequences. In 2001, Scharf received an EMMY for his work on the National Geographic Television Documentary, *THE BODY SNATCHERS* (National Geographic Documentation). Data for these movies was acquired using a Digiscan digital SEM scan controller with the capacity to support multiple simultaneous electron signal detectors with the microscope and stage controlled by Digital Micrograph (Gatan, Pleasanton, USA).

Being a professional photographer working mostly in the history of arts, Diller started experimenting with electron microscopes in 1984, trying to adapt his macroscopic view to the unseen spaces beyond the resolution of the human eye. This has developed to the production of high definition videos in the SEM, inspired by the idea of flying over a landscape that is beyond the limits of the human eye in much the same way as in movies filmed from a helicopter, visually impressive landscapes are used to attract the viewer's interest.

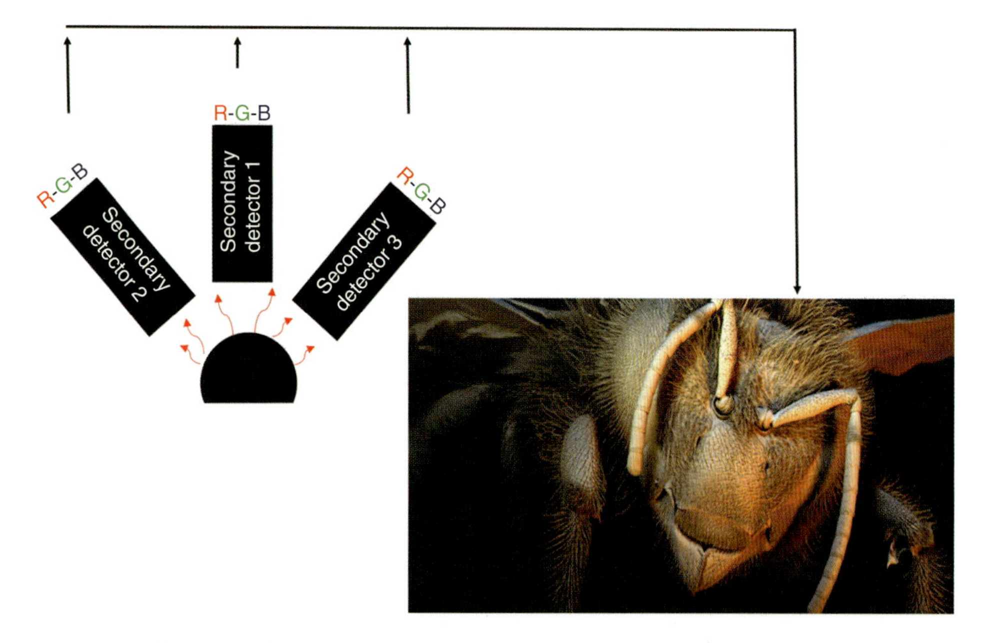

Figure 31.10 Schematic of a three secondary electron detector multidetector colour synthesizer, where the secondary electron signals scattered from an object are recorded at three different collection angles, converted to three different arbitrary colours and combined to generate a coloured image with complex contrast, an impression of 3D that is enhanced by the application of colour.

Hardware to allow ‘fly through’ movies of the nanoworld have been developed on a LaB_6 SEM (525 SEM, FEI, USA). This LaB_6 SEM has proven to be a powerful tool source, but lacks the resolution and depth that a FEGSEM would bring to the image. It was replaced in 2017 with a TESCAN MIRA3 FEGSEM (Tescan, Česká Republika), finally allowing 3D nanoflights to be recorded without the need to tilt at each frame position for left/right image pairs, since the 3D beam setup of the MIRA3 will do this by moving the electron beam within the microscope column. On the FEI 525 SEM additional hardware had been developed to make this old SEM remote controllable, the most important ones being focus and zoom magnification, followed by detector gain, black level and colours from each detector to be mixed into one final image. On the MIRA3 these remote capabilities are supplied by integrating commands used within the SHARK Remote API. To do this, some new extensions will need to be written within the nanoflight.creator software.

To the existing specimen movement in the SEM in X-Y-R dimensions five new axes had been added with a Kleindiek E5AT eucentric piezo sub-stage (Figure 31.11) (Kleindiek, Reutlingen, Germany). In 2015 these two stages were superseded by a newly designed fully encodered eight-axis piezo stage (SmarAct GmbH, Oldenburg, Germany) (Figures 31.12 and 31.13). This stage overcomes the limitations of earlier stages; stage 1 had three encodered axes (x, y, rotation), substage 2 five axes (Gonio, x, y, z, rotation 2) whereas only the two rotational axes had been encodered, the other linear axes being unencodered piezo stepper axes with which we were not able to completely linearize movement, resulting in the stage departing from the programmed path of movement.

This hardware, comprising different parts from different manufacturers, is made to work together through the development of digital control software, which allows precise ‘nanoflights’ to be taken around a specimen located in an SEM. Nanoflight® mostly employs solid angle detector mixing, often with different SE and BSE detectors. Presently, the highest specification instrument employs eight selectable detectors. Each is able to collect a signal that can be directly converted to a colour, which is then employed to

Figure 31.11 Kleindiek E5AT Piezo Substage with two encodered movements out of five (Kleindiek, Reutlingen, Germany).

Figure 31.12 Eight-axis closed-loop encodered piezo stage made by SmarAct GmbH, Oldenburg, Germany.

colour small volumes (edges are populated with data from secondary electron signals and topography/flat areas with backscattered signals) (Figures 31.4 and 31.5) (see Chapter 30).

In conclusion, these adaptations allow the operator to 'fly around' microstructures and create inspiring movies of this fascinating nanoworld. Some of the resulting nanoflight SEM movies can be seen on http://www.nanoflight.info or on the Vimeo nanoflight channel https://vimeo.com/channels/nanoflight.

31.3 MATERIALS AND METHODS

31.3.1 Nanoflight Hardware System Setup

- Philips/FEI 525 large chamber SEM with two SE detectors (180° angle difference) or current specification MIRA3 with eight-axis stage (Figure 31.13).
- SmarAct eight-axis fully encodered piezo stage divided into a 3D and a 5D substage (Figures 31.12 and 31.13).
- DISS 5 scanning and image acquisition system.
- Three backscattered diode detectors with 120° angle difference directly mounted below the final lens.
- Three backscattered diode detectors with 120° angle difference mounted 'off-axis' on wires below the final lens ca. two inches away from the beam centre to get structure-enhancing shadows in the imaging.
- Pointelectronic Videoprocessor to select and control four out of eight detectors for image acquisition.

Figure 31.13 Current generation nanoflight hardware, a MIRA3 FEGSEM with eight-axis SmarAct sample stage.

- Heilandelectronic boards mounted within original Philips/FEI electronics to remote focus, zoom magnification and beamshift.
- Control computer (running on Windows) to control all the connected SEM and additional hardware functions with the software package 'nanoflight.creator'.

To achieve an aesthetically pleasing movement of a specimen it is essential that the SEM is equipped with a very accurate (μ resolution) specimen stage movement with controllable very small steps (individual movements) of the specimen stage. The accuracy needed is directly dependent on the field-of-view (FOV) to be imaged. Most motorized stages in current SEMs might not be able to deliver these specifications when operating below a FOV of some hundred micrometres wide.

The 3D X-Y-R substage is based on customized crossed roller bearings with integrated optical encoders. The closed loop resolution is about 1 nm, which leads to a repeatability of about ± 180 nm for the complete travel of ± 75 mm. On top of the substage a compact 5D sample stage can be mounted. The first rotation is realized by an SR-4513 element with a closed-loop resolution of 15 μ°. The following linear positioners are based on SmarAct's crossed roller bearings SLC series, providing a repeatability of ± 30 nm and a closed-loop resolution of about 1 nm.

The smallest closed-loop rotary positioner SR-2013 is equipped with a stub-holder for positioning the sample within the SEM beam focus and has a closed-loop resolution of about 25 μ°. The available movements on the 5D substage are 360° for the first rotation, ±8 mm for the y axis, ± 10 mm for the x axis and ± 10 mm for the z axis and unlimited travel for the final rotation. All axes can be controlled using the standard SmarAct MCS controller (SmarAct GmbH, Oldenburg, Germany).

The exceptionally good repeatability in all stage coordinates will allow compucentrical movements, various translations of 3D paths for distinctive stereoscopic imaging and finally photogrammetric modelling of microstructures.

31.3.2 Nanoflight Software Setup

The 'nanoflight.creator' software is Windows-based and mostly written in C^{++} (National Geographic Documentation). First it had been intended as a proprietary software to work exclusively with the existing hardware at Diller's lab but the software architecture also allows it to be used with other remotely controllable instruments (for each new hardware an extension within nanoflight.creator needs to be written by the user) to take control of key parameters in the microscope (Figure 31.14). Since it is still 'work-in-progress' and not

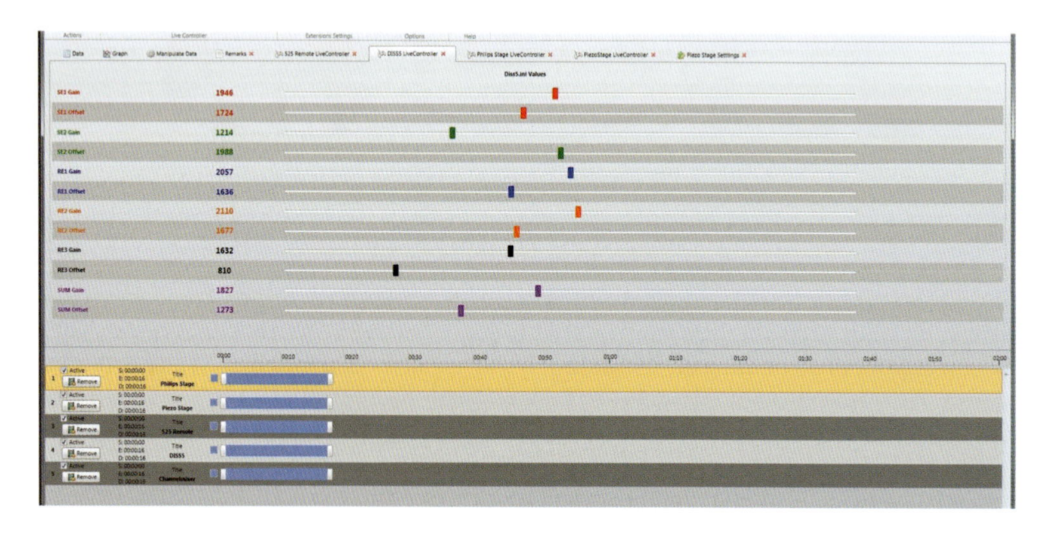

Figure 31.14 The nanoflight.creator extensions and value table for detector gain and offset.

without errors, it is not presently commercially available but we are striving to support SEMs using the SHARK Remote API from 2017 onwards.

The most demanding challenge for the development of the control software had been to make it modular, easily extendable for other remotely controllable SEMs or even for stereoscopic light microscopy in natural colours. The main program handles only data and user interface functions (GUI) and supplies a framework for all plug-ins used.

There are different levels of software modules: extensions for proprietary hardware like specimen stages, remote controlled functions on the SEM and so on; live controllers to control this hardware; add-ins like timelines to control the execution of programmed sequences within the overall time schedule; and pre-effect plug-ins like Autofocus and after-effect plug-ins like colour-correction or sharpness.

The 'nanoflight.creator' takes control of many parameters to be read from the attached hardware and for them to be sent back to this hardware during execution of the slow-scan frame sequence:

- Coordinates of the specimen position in eight axes (five linear, three rotators)
- Focus
- Beamshift for 3D imaging
- Detector gain and black level of up to eight detectors
- Detector colour and luminosity of up to eight detectors
- Number of frames between each acquired waypoint in the 'flightpath'
- Slow-scan start and adjustable delay depending on scan parameters (integration time, resolution, etc.)
- Value interpolation between the waypoints for the number of frames can be either various types of Catmull-Rom or linear
- Resulting splines of read-in values can be corrected/smoothed by dragging the spline curves at a point in time (Figure 31.15)

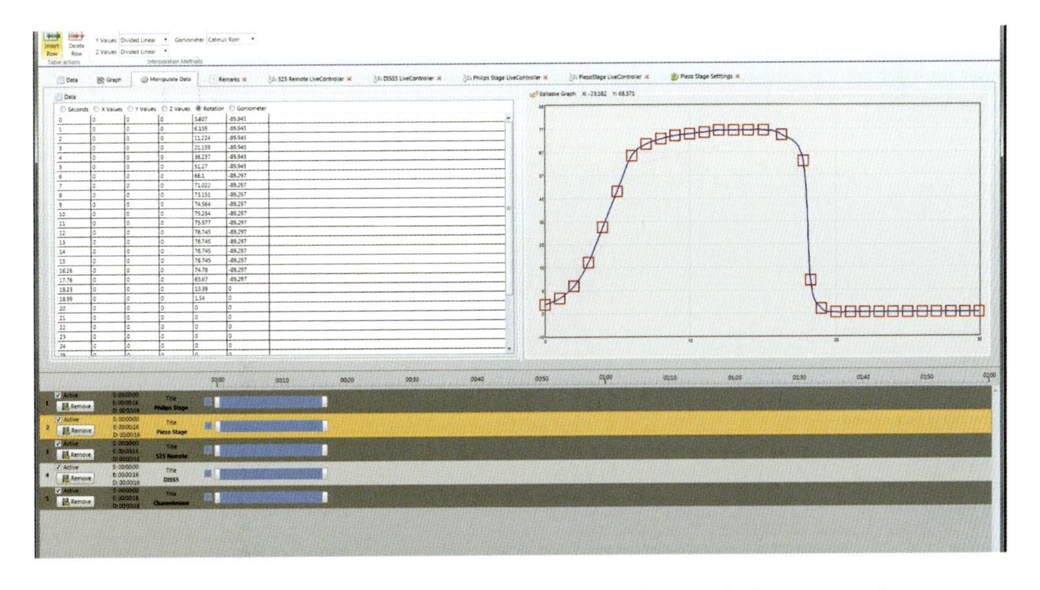

Figure 31.15 Menu for manipulating parameters in the nanoflight.creator software.

- Image data can be saved as a multichannel mixed colour image or as single grey scale images
- Preview function of the programmed sequence
- Stop/start at a new user defined frame-number

All these target-point function values are represented by vectors in a multidimensional space that stands for the settings of all axes used. Presently, the software uses eight axes with the new piezo-controlled encodered stage and therefore an eight-dimensional spline interpolation routine is needed. Within a normal image sequence more than forty hardware-specific parameters per image are thus sent to the system.

When preparing the movie, different target points or function values are selected with the live controllers in the software graphic user interface (GUI) (Figure 31.16) or the 3Dconnexion (3Dconnexion, Waltham, MA, USA) six axes controller, which can control the substage movements and all the remote values of the microscope and the scanning system (Figure 31.17) or within the specially adapted scanning system DISS5 GUI (Figure 31.18), a hardware/software solution to bring older SEMs up to date to active scanning and image acquisition of up to four detector channels. The DISS5 brings with it a script language we used to integrate its functions within the 'nanoflight.creator' software (Point Electronic GmbH, Halle, Germany).

To get the impression of smooth translations in the final movie, the transitions between the target points have to be interpolated by using splines. For this purpose, a special cubic Hermite spline is used, the so-called Catmull-Rom approach. The basic principle of creating a smooth curve through all data points (normally in two dimensions or three dimensions) had been generalized to arbitrary dimensions. Tangents at all points are calculated with respect to adjacent points to obtain a continuous path. Finally, 'nanoflight.creator' has a very extensive framework to facilitate programming of extensions for specific hardware

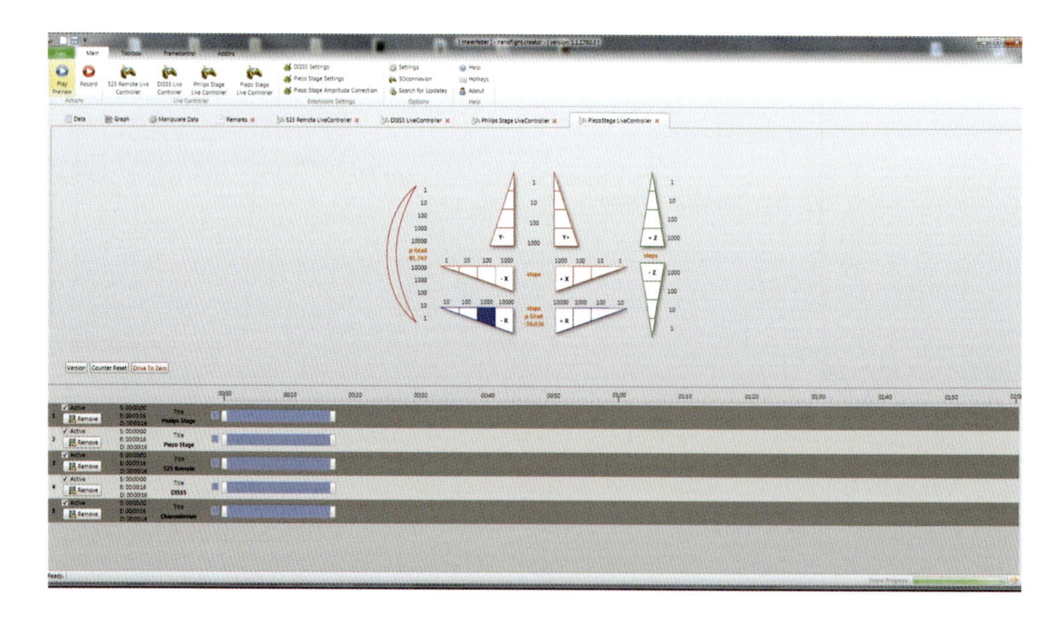

Figure 31.16 Kleindiek five-axis E5AT Substage Livecontroller in the nanoflight.creator software.

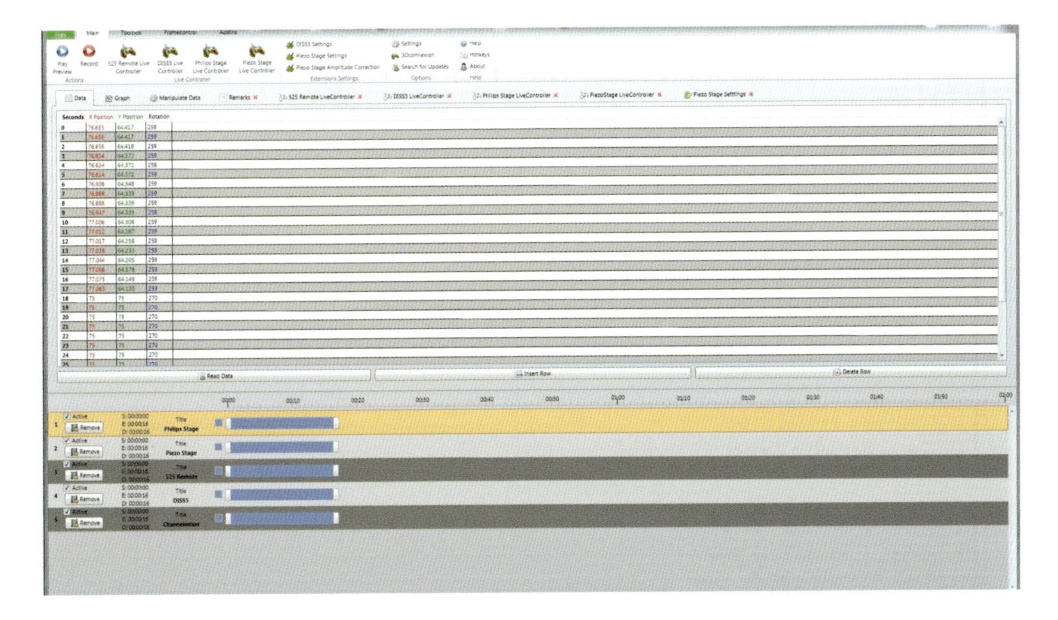

Figure 31.17 The extensions can be controlled by the 3D connexion Space Pilot.

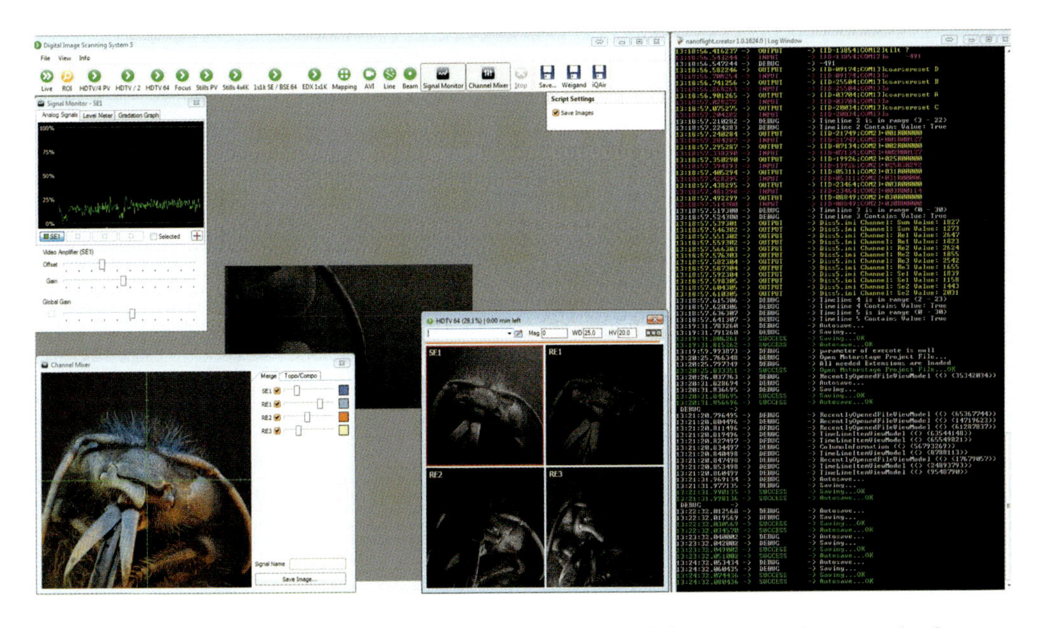

Figure 31.18 Scriptable DISS5 software with the possibility to mix colours on-the-fly.

and software. There is a comprehensive in-program help (Figure 31.19) for writing own extensions to fit your specific instrumentation.

The software 'nanoflight.creator' had been developed and compiled using Microsoft Visual Studio 2010 Service Pack 1 (Microsoft, CA, USA). To execute it needs the NET Framework 4. It is easily feasible to integrate a new hardware in 'nanoflight.creator'.

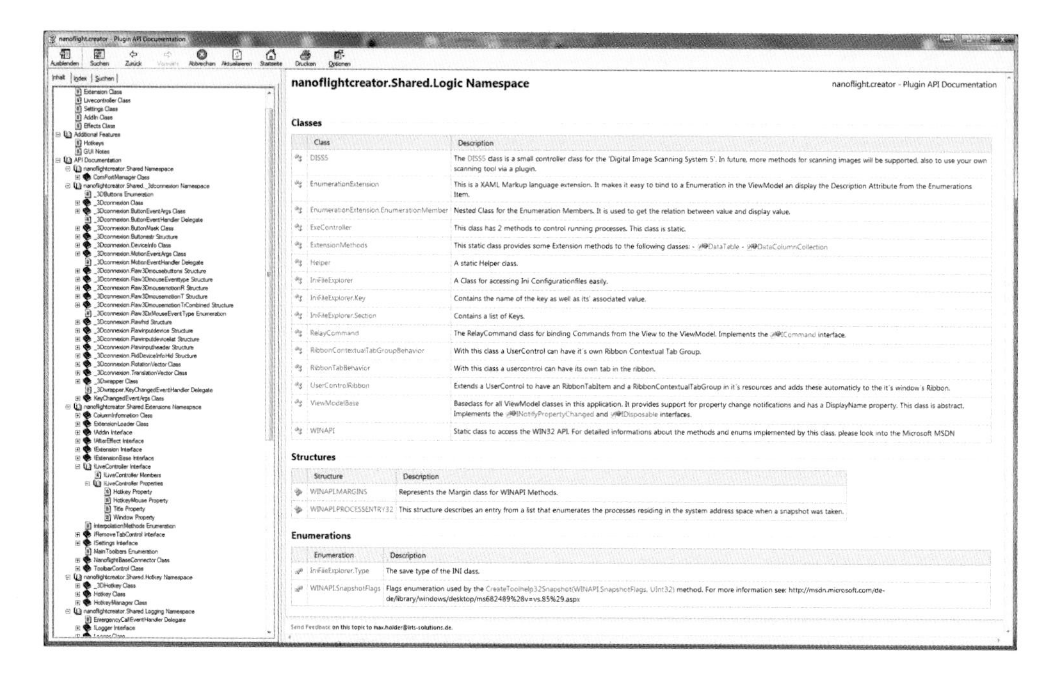

Figure 31.19 API documentation in nanoflight.creator.

To demonstrate the ability of nanoflight.creator' to integrate new hardware, this is an example extension for a specimen stage using the SHARK Remote API:

```
using nanoflightcreator.Shared.Logging;
namespace Extension.TESCAM_Example
{
public static class TESCAN
{
public static void StgCalibrate()
{
Logger.Instance.Log(LogType.Information, true, „TESCAN: StgCalibrate()");
}
public static void StgStop()
{
Logger.Instance.Log(LogType.Information, true, „TESCAN: StgStop()");
}
public static void StgMove(float vx, float vy, float vz, float vrota-
tion, float vtilt)
{
Logger.Instance.Log(LogType.Information, true,
string.Format(„TESCAN: StgMove({0}, {1}, {2}, {3}, {4})",
vx, vy, vz, vrotation, vtilt)); }
public static void StgMoveTo(float x, float y, float z, float rota-
tion, float tilt)
{
Logger.Instance.Log(LogType.Information, true,
string.Format(„TESCAN: StgMoveTo({0}, {1}, {2}, {3}, {4})",
x, y, z, rotation, tilt)); }
public static int StgIsBusy()
{
```

```
Logger.Instance.Log(LogType.Information, true, „TESCAN: StgIsBusy()");
return 0; }
public static int StgIsCalibrated()
{
Logger.Instance.Log(LogType.Information, true, „TESCAN: StgIsCalibrated()");
return 1; }
public static void StgGetPosition(out float x, out float y,
out float z, out float rotati- on, out float tilt)
{
float myX = 10f; float myY = 11f; float myZ = 12f; float myR = 120f;
float myT = 45f;
x = myX;
y = myY;
z = myZ; rotation = myR; tilt = myT;
Logger.Instance.Log(LogType.Information, true, string.Format(„TESCAN:
StgMoveTo({0}, {1}, {2}, {3}, {4})", x, y, z, rotation, tilt));
} }
}
```

The 'nanoflights' are still a 'project under construction' depending on the availability of modern hardware for nearly analogue viewing of the SEM sequences (e.g. high luminosity field emission microscopes with ultrafast scanning systems (DISS6 scanning system (Point Electronic GmbH, Halle, Germany)/detectors (ultrafast backscatter detector (PNDetector, Munich, Germany)) and beam blanking during stage movements).

31.4 CONCLUSION

Film and movement are powerful tools to attract the attention of the general public, children and others to the importance of science at the nanoscale. However, the tools developed to create the movies can also support wider remote access to instruments.

The nanoflights software is not commercially available, but can be accessed via the author. Future important developments include: developing an inverse kinematics calculation routine for the existing eight-axis piezo stage, developing an on-the-fly routine to grab and smooth the 'coordinate cloud' of the six-axis 3D connection space mouse, nanoflight.creator GUI remodelling, adding 2,5D modelling from 4 quadrant detectors and adding light imaging and Z-stacked movements with video or still-shot digital cameras. However, as a software suite it has considerable potential in creating visually arresting movies of samples mounted in the SEM. In addition, its power as an SEM remote control software suite proffers nanoflights forming a foundation for an open public microscopy resource.

ACKNOWLEDGEMENTS

The author is indebted for helping with this routine to Prof. Dr Hubert Mantz, University of Applied Sciences, Ulm. Grateful acknowledgments are also due to SmarAct GmbH, Oldenburg for sponsoring the new stage and Kleindiek Nanotechnik GmbH, Reutlingen, Germany for making the nanoflights possible at the very start. The 'nanoflight.creator' software was developed by Max Holder, IRIS Solutions GmbH, Wuerzburg, Germany and remains in development. It is presently only compiled to support the modified Philips/FEI 525 SEM at Diller's Laboratory and TESCAN SEMs using SHARK Remote.

REFERENCES

Bordwell, D. (1997) *On the History of Film. Harvard University Press.*

Diller(a) (2013), http://www.electronmicroscopy.info/media/nanoflight-poster_mc2013.pdf (accessed 25 September 2018).

Diller(a), www.electronmicroscopy.info (accessed 25 September 2018).

Diller(b), www.nanoflight.info (accessed 25 September 2018).

Fleck, R.A. (2015) *Electron microscopy in the 21st century.* Gray's Anatomy.

Heiderich, T., https://www.oma.on.ca/en/contestpages/resources/free-report-cinematography.pdf (accessed 25 September 2018).

Itten, J. (1961) *Kunst der Farbe*, Otto Maier Verlang, Ravensburg, BRD.

Itten, J. (1970) *The Elements of Color. A Treatise on the Color System of Johannes Itten based on his book The Art of the Color*, Van Nostrand Reinhold Company, New York, Cincinnati, Toronto, London, Melbourne.

National Geographic Documentation, http://www.films.com/id/17099 (accessed 25 September 2018).

Radulescu, http://moviesincolor.com/about (accessed 25 September 2018).

Scharf (1993). Multi-Detector Colour Synthesizer. US patent 5, 212, 383.

Scharf(a), http://www.scharfphoto.com/about/ (accessed 25 September 2018).

Scharf(b), http://www.electronmicro.com (accessed 25 September 2018).

Watson, J.D. and Crick, F. Francis (1953) A structure for deoxyribose nucleic acid. *Nature*, 171 (4356), 737–738.

Index

atomic number (Z), absorption (A) and X-ray fluorescence (F) on X-ray yield, 604
actin filament, 21, 39, 40, 202, 306, 307, 354, 579
Alexa Fluor 488, 470
Alpha bench-top series, 9
aluminium, 593
ambient pressure freezing methods, 168–169
Amira®, 507
antigenicity, 469
APEX, 504
Arivis®, 507
array tomography (AT), 130, 135
Aspera Connect, 620
Atlas 5 Array Tomography software, 490
ATUM. *see* automatic tape-collecting ultramicrotome (ATUM)
Auger electron spectroscopy (AES), 329
automatic tape-collecting ultramicrotome (ATUM), 468, 546
  device, 486
  mammalian cochlea section, 490–491
  sample preparation
    alignment and segmentation, 490–493
    imaging, 489–490
    mounting, 488–489
    sectioning, 487–488
    staining, 487
acquisition speed, 113
avidin–biotin complexes (ABCs), 470

backscattered electron (BSE), 2, 26, 87, 106, 109, 121, 183, 457, 499
  non-conductive biological samples, imaging of, 443
backscattered electron detectors (BSED), 500
BDM. *see* beam deceleration mode (BDM)
beam current, 4–5
beam deceleration, 42
  benefits for SEM imaging, 108
  development, 66–70
  effect, 67
  electron beam energy, 103
  principle, 85
  reduce beam energy, 107
beam deceleration mode (BDM), 85–87
beam penetration, 106–108
beam sensitivity, 324, 400–405
beam transparent, 277–278
biofilms, 448–449
biohazards material, 336–340
biological tissue specimen preparation
  dehydration considerations, 314–315
  fixation considerations, 312–314
  preparation methods, 315–317
biomaterials
  beam-sensitive, 324
  flow chart, 322
  sample preparation
    hard tissues and hard, 319–321
    process cycle, 320
    soft tissues and soft/semi-soft, 317–319
  water, 445–448
biomedical products, quality control, 336–340
biotinylated dextran amine (BDA), 469, 470, 475
block-face imaging, 130–132
broad ion beams (BIB), 182
BSE. *see* backscattered electron (BSE)
BSED. *see* backscattered electron detectors (BSED)

CAT. *see* correlative array tomography (CAT)
cathode lens, 85
cathodoluminescence (CL), 98
cathodoluminescent probes, 472
CCP4, 615
CCP-EM project, 615

Biological Field Emission Scanning Electron Microscopy, First Edition.
Edited by Roland A. Fleck and Bruno M. Humbel.

cellulose biosynthesis, 349
charge-coupled device (CCD), 496
charge free anti-contamination system (CFAS), 17
charge neutralization, 123
chemical fixation
  acrolein, 195
  action of fixatives, 203
  aldehydes
    formaldehyde (FA), 192–194
    glutaraldehyde (GA), 194–195
  amino acid of lysine, 193
  buffers, 203–204
  FIB-SEM approaches, 549–551
  malachite green, 198–199
  mixtures of fixtures
    buffered formaldehyde and picric acid, 201
    FA and GA, 200–201
    glutaraldehyde and osmium tetroxide, 201–202
    osmium tetroxide, 195–198, 202–203
    osmium tetroxide–potassium ferrocyanide staining, 202
    osmium tetroxide–tiocarbohydrazide–osmium tetroxide (OTO), 202
    ruthenium red, 199–200
    and tannic acid, 202–203
  SEM preparation
    critical-point drying (CPD), 205–206
    ionic liquids, 206–207
    resin, 208–209
    and room temperature preparation, 205–206
  specimen preparation, 590–591
  uranyl acetate, 198
  volume microscopy, 209–211
  water source, 204–205
chromatic aberration, 28, 42, 545
chromium, 593
circuit tracing, CAT, 463–465
Cloud Infrastructure for Microbial Genomics, 613
^{13}C-nuclear magnetic resonance (NMR) spectroscopy, 378
coating, HRSEM
  elevation angle, 286–288
  film thickness, 284–286
  signal enhancement, 280–282
  techniques, 282–284
  tungsten planar magnetron sputtering *vs.* electron beam evaporation (EBE), 284
cold-field emission (Cold-FE), 14, 44, 59, 108
condenser lens (CL), 57
confocal Raman microscope (CRM), 95–96, 98
conical tilt tomography, 576
conjugate array tomography, 466
continuum normalisation method, 604–605
contrast-enhanced DAPI channel, 473
convergent angle, 27–28
correlative array tomography (CAT), 135
  application, 476–478
  correlative light and electron microscopy, 462–463
  data acquisition
    EM, 474
    integrated LM/EM, 475–476
    LM, 472–473
    registration of LM and EM imagery, 474–475
  postembedding on-section immunohistochemistry
    antigenicity and/or fluorescence preservation of neuroanatomical tracers, 465, 469–470
    markers for postembedding on-section immunohistochemistry, 471–472
  sample preparation
    for circuit tracing, 463–465
    flat conductive substrate, 466–468
    flexible tape, 468
    for proteometric analysis, 465–466
  volumetric electron microscopic imaging, 462
  workflow, 463, 464
correlative light and electron microscopy (CLEM), 453, 462–463
correlative microscopy, 94–95
correlative volume imaging, 135–137
critical dimension SEM (CDSEM), 36
critical-point drying (CPD), 205–206
cryobiology
  electron microscopy observation, 234–227
  freeze drying, 232–234
  low temperature damage and injury, mechanisms of, 234–237
  temperature and condensed phases of water
    heterogeneous ice nucleation, 228–229
    homogeneous ice nucleation, 228
    irruptive recrystallisation, 231
    migratory recrystallisation, 231
    post-nucleation, 229–230
    spontaneous recrystallisation, 232
    supercooling, 227–228
    thawing, 231
    vitrification, 230
cryo-EDS, 165
cryo-electron microscopy (EM), 233, 355
cryo-FEGSEM
  beam sensitivity and coating, 400–405
  freezing, 399–400
  frozen-hydrated state, 592
  ice crystal formation, 399
  materials and methods, 409–411
  partial freeze drying, 408
  vitrification, 244–253
CryoFIB Lift-Out, 161–162

cryo-focused ion beam (FIB) milling, 137, 421–422
cryo-fracturing, 351, 400
cryo-immobilisation, 591–592
cryo-planing, 181–183, 405–408
cryo-plunging, 399
cryo-preparation
  beam sensitivity and coating, 400–405
  freezing, 399–400
  partial freeze drying, 408
cryopreservation
  biological systems, 239–244
  low temperature biology, 223–234
cryoprotectants, 237–239, 399
cryo rotate stages, 163
cryo-scanning electron microscopy
  ambient pressure freezing methods, 168–169
  coating, 183–184
    carbon coating, 160–161
    metal sputtering, 160
  column preparation chamber, 147–149
  cooling dewar, 149–150
  cryo-planing, 181–183
  fracturing, 157
  and freeze etching, 176–181
  freeze fracture (FF) technique, 176–181
  freezing methods
    ethane plunging, 152
    high-pressure freezing, 152
    propane jet freezing, 151
    slam freezing, 152
    slushed nitrogen freezing, 151
  high pressure freezing (HPF), 169–172
  imaging protocol for, 167
  JEOL FESEM, 55–56
  Leica EM VCT500 cryo-SEM, 185–186
  mounting methods
    edge, 154
    filter, 154
    hole, 155
    liquid film, 155–157
    rivet, 157
    surface, 154
  sample preparation, 150
  sublimation, 158–159
  techniques and equipment
    cryo-EDS, 165
    CryoFIB Lift-Out, 161–162
    cryo rotate stages, 163
    cryo-STEM, 164
    on-grid thinning, 161–162
    stage bias, 163–164
  types of cooling
    braid, 145
    gas, 145–147
  vacuum coaters *vs.* freeze fracture devices, 178–179
  vacuum cryo-transfer shuttle, 172–176
cryo-scanning transmission electron microscopy, 164
cryo-stage, 400
cryo-transfer, 400
custom antibodies, 470
Cyberinfrastructure for Data Management and Analysis, 613

data acquisition, 619–620
data management, 615
data storage, 620–621
desmotubule, 350
detection system, FESEM, 87–89
dextran, 469
3,3′-diaminobenzidine (DAB), 452, 470
DIGISCAN, 500
digital data management systems, 616–617
Digital Micrograph (DM), 500
diode sputtering (DS), 283
direct detection device (DDD), 496
direct imaging, 273
direct immunohistochemistry, 471
double-axis rotary shadowing (DARS), 250, 290, 353
dual-axis tilt tomography, 574, 576
dual silver enhanced colloidal gold-Alexa 488, 475
Durcupan, 555

elastic scattering, 1
electron backscattered diffraction (EBSD), 98
electron beam evaporation (EBE), 283, 284, 353
electron channeling patterns (ECPs), 11
electron cryo-tomography, 430
electron irradiation, 39, 160, 554
electron lens
  prealigned three-stage, 9
  principle, 27
electron optical system, 81
electron probe micro analyser (EPMA), 53
electrostatic voltages, 73
emission current density, 4
en bloc staining, 210, 487, 502, 504, 549–551
endoplasmic reticulum (ER), 350
energy dispersive analytics, 457
energy-dispersive X-ray (EDX)
  field emission gun scanning electron microscope (FEGSEM)
    low beam current, 598–601
    low excitation of X-rays, 598
    surface coating, 601
    voltage instability, 598
  microanalysis
    change control, 327–329
    contaminant identification, 329
    detection and analysis of foreign bodies and inclusions in tissue, 333–336

energy-dispersive X-ray (EDX) (*contd.*)
detection of asbestos and mineral fibres in lung tissue, 331–333
organic materials, 329–331
spectroscopy, 2, 251
systems, 10, 13
vs. wavelength dispersive spectroscopy, 594
energy filter, 66–70
energy-selective backscatter (EsB) detector, 120
environmental control module (ECM), 17
environmental scanning electron microscopy (ESEM), 17
bacterial biofilm on human vocal fold, 449
energy dispersive analytics, 457
gaseous secondary electron detector, 441–442
human skin in native-like state, 447–448
for hydrated biomedical samples investigation
biofilms, 448–449
correlative light and electron microscopy, 453
lipids, 449–450
living specimens, 454
specific staining, 452–453
tissue microanatomy, 451
tissue surfaces, 450
water in biomedical material, 445–448
hydrated samples, imaging of, 443–444
non-conductive samples, imaging of, 442–443
preparation steps
hydrated tissues, investigation of, 455
living specimens, investigation of, 454–455
Pseudomonas aeruginosa biofilm, 449
radiation damage, 444–445
secondary electron, 441–442
tissue cross-sections, 451
epon, 466, 555
ESEM. *see* environmental scanning electron microscopy (ESEM)
even scanning probe microscope (SPM), 98
Everhart–Thornley (ET)
SE detector, 37, 61, 62
type, 87, 92
extended Pichou and Pouchoir (XPP) alogarithm, 604
extraction voltage, 30
extractor, 417–418

F-actin atomic model, 582
F-actin filaments, 266, 269, 277, 280–282
FEGSEM. *see* field emission gun scanning electron microscope (FEGSEM)
FERA plasma, 80, 97
fibrous gauze wound dressing, 323–324
FIB/SEM EDX method
Everhart–Thornely detector, 536
for investigating interactions between biological samples and nanoparticles, 533–542
mouse urothelium
endosomes filled with $CoFe_2O_4$ NPs, 537–540
with FIB milling operation, 534–538
with MB49 cancer cells, 537
FIB-SEM Helios NanoLab 650, 558
field emission electron, 57
field emission gun (FEG), 5, 14, 103
field emission gun scanning electron microscope (FEGSEM), 233, 621
advantages, 357
automated acquisition, reconstruction, and analysis, 114
beam penetration, 106–108
biomedical products quality control, 336–340
cell biology, 347
contrast and signal-to-noise ratio, 108–109
energy dispersive X-ray microanalysis
change control, 327–329
contaminant identification, 329
detection and analysis of foreign bodies and inclusions in tissue, 333–336
detection of asbestos and mineral fibres in lung tissue, 331–333
organic materials, 329–331
energy dispersive X-ray spectroscopy
low beam current, 598–601
low excitation of X-rays, 598
surface coating, 601
voltage instability, 598
freeze-fracturing technique, 347
high resolution SEM imaging at low accelerating voltage, 104
imaging of biological material, 354
interaction volume in biological specimens, 595–597
multienergy deconvolution SEM (MED-SEM), 111
nuclear envelope, 352–358
nuclear pore complex, 345, 352, 355
optimal instrumental approach
fibrous gauze wound dressing, 323–324
gel networks, 325–326
nanocrystalline silver coated wound dressing, 323
resorbable polymer microspheres, 324–325
plant cell wall, 346–350
plasmodesmata, 350–351
quantification
continuum normalisation, 604–605
peak-to-local background method, 605
Phi-Rho-Zed, 604
XPP, *see* extended Pichou and Pouchoir 604

ZAF, *see* atomic number (Z), absorption (A) and X-ray fluorescence (F) on X-ray yield 604
sample preparation
dehydration considerations, 314–315
fixation considerations, 312–314
flow chart, 322
hard tissues and hard, 319–321
preparation methods, 315–317
process cycle, 320
soft tissues and soft/semi-soft, 317–319
serial block-face imaging (SBFI), 109–110
signal-to-noise ratio (SNR), 112
specimen preparation
chemical fixation, 590–591
coating material, 593
critical point drying, 590–591
cryo-immobilisation, 591–592
freeze-drying, 591–592
frozen-hydrated state, 592
resin-embedding, 590–591
scan speed, 113–114
spot size, 104–106
stability, 111–113
three-dimensional (3D) structures, 103
X-rays, 593–595
field emission (FE) noise, 44
field emission scanning electron microscopy (FE-SEM), 266
advantages, 81
assessment of cell and tissue architecture, 299
backscattered electron (BSE), 26
beam deceleration mode (BDM), 85–87
biological samples, imaging of, 89–90
commercialization of, 30–31
detection system, 87–89
development of in-lens, 32–34
essential for high-resolution, 160
examination of mammalian cells and tissues, 299–308
in food research
applications of food microscopy, 389–392
chemical fixation and dehydration, 392
cryo-preparation and observation, 386–389
facts and artefacts, 392–395
food microstructure, 385–386
freeze-fracturing and cryo-planning, 387
monoglyceride networks, 390–392
recrystallization temperature, 389
high sensitivity backscatter electron detector, 34
In-Flight Beam Tracing™, 79, 87
low accelerating voltage application, 34–36
low voltage electron beam, 300
optics and displaying modes, 81–84
resolution improvement effect, 42–43
retarding method and boosting method, 42
Schottky emission (SE) electron source, 43–44
sensitive samples, imaging of
at low energies, 90–91
low vacuum operation, 91–92
observation of biological samples, 92–94
techniques
correlative microscopy, 94–95
combined confocal Raman (CRM) integration, 95–96
FIB/SEM instrumentation, 96–99
ultra-high resolution microscopy, 84–85
ZEISS GEMINII® technology, 117
field emission SEMS, 57–59
field ion microscope, 5
FIJI, 507
flat conductive substrate, for correlative microscopy, 466–468
fluorescence light microscopy (FLM), 133
fluorescent markers, 471
focused ion beam (FIB), 94, 182
applications, 420–421
conventional milling geometry, 520
cryo-FIB milling, 421–422
for cryo-TEM lamella preparation, 423–426
Ga^+ ions, 519
imaging geometry
on FIB/SEM, 520
on single beam FIB, 520
ion beam
damage and artefacts, 419–420
generation and shaping, 417–419
and sample interactions, 417
ion-induced SE signal, 519
ion–solid interactions, 518
milled windows
cryo-microscopy and tomography, 426–431
cytoplasm, 426–428, 431
in situ mapping of macromolecular complexes, 429–431
primary neuronal cells, 428–430
milling, 417–418
particle–beam-induced deposition, 518, 519
principles, 417–420
with scanning electron microscope, 420–421
sputtering, 417
vacuum chamber, 517–518
versatility of lamella milling approach, 433
ZEISS GEMINI® technology, 119
focused ion beam-scanning electron microscopy (FIB-SEM)
chemically fixed mouse liver samples, 550–552
cross-sections and 3D tomography in, 99
FIB slice, 524–525, 527
geometry
of instrument, 557–558

focused ion beam-scanning electron microscopy (FIB-SEM) (*contd.*)
of resin block, 558–560
instrumentation, 96–99, 521–522
milled section of bone on dental implant, 523–524
mitochondrial network in mouse liver, 561
resin block, milling of, 556–557
resin formulations, 555
sample preparation
chemical fixation, 549–551
freeze-substitution, 551–554
high-pressure freezing, 551–554
resin embedding, 554–557
specimen geometries, 520–521
TEM specimen preparation with, 528–530
3D FIB-SEM tomography
with multi-signal SEM acquisition, 526–528
sectioning and imaging, 522–523
tomography, 132–133
for 2D sectioning and imaging, 520–524
yeast cells, 524
formaldehyde (FA), 192–194, 201, 463, 465
Fourier shell correlation (FSC), 00070:431
Fourier transform infra-red (FT-IR), 329, 331
freeze-drying
cryobiology, 232–234
cryo-immobilisation, 591–592
partial, 269–271
freeze etching, 176–181
freeze fracture (FF), 176–181, 252
freeze substitution (FS), 225, 551–554, 560
freezing methods, SEM
ethane plunging, 152
high-pressure freezing (HPF), 152
propane jet, 151
slam, 152
slushed nitrogen, 151

GAIA, 80, 96
gallium (Ga^+) ions, 419, 519
gaseous secondary electron (GSE) detector, 441–442
gel networks, 325–326
glutaraldehyde (GA), 194–195, 463, 465, 549, 551
gold particles, 41, 304, 305, 348, 353, 365, 370, 470–472, 477, 529

HAADF-STEM tomography, 546
Hard-Plus resin-812, 554
helium ion microscopy, 127
Heterocapsa circularisquama RNA virus, 579
heterogeneous ice nucleation, 228–229
high pressure freezing (HPF), 169–172, 226, 248–249, 551–554, 560
high-resolution scanning electron microscopy (HRSEM)
blurring, noise and artefacts
beam transparent, 277–278
contamination during imaging, 274–277
low dose imaging *vs.* SNR, 278–279
mass loss during imaging, 274
coating
elevation angle, 286–288
film thickness, 284–286
signal enhancement, 280–282
techniques, 282–284
tungsten planar magnetron sputtering *vs.* electron beam evaporation (EBE), 284
contrasting techniques for, 280–282
critical point drying (CPD) and freeze drying (FD), 267
high-vacuum cryo-transfer system, 271–273
macromolecular structure preservation, 266–269
molecular imaging, 292–294
partial freeze-drying, 269–271
work from metal coating, 290–292
high-vacuum cryo-transfer systems, 271–273
HM20, 555
homogeneous ice nucleation, 228
horseradish peroxidase (HRP), 504
HPF. *see* high pressure freezing (HPF)
human hair, in ESEM, 445
hydrated biomedical samples investigation, in ESEM
biofilms, 448–449
correlative light and electron microscopy (CLEM), 453
lipids, 449–450
living specimens, 454
specific staining, 452–453
tissue microanatomy, 451
tissue surfaces, 450
water in biomedical material, 445–448
hypoglossal (motor) nucleus, 478

image acquisition, 615–616
image database, 617–619
image output
adjustments
brightness/contrast, 625–629
integration time/noise reduction, 629–630
sharpness/astigmatism, 625
coloring
black-and-white image, 651–653
creating masks, 647–651
detector signals, 653–656
empty magnification, 631–632
fast scan mode, 629
formats
bitmap (BMP), 635

graphics interchange format (GIF), 636, 641
joint photographic expert group (JPEG), 635, 641
portable network graphics (PNG), 636, 641
tagged image file format (TIFF), 635, 641
optimization and colorization
digital formats, 641
filters, 641–646
photomicrograph, 637–640
post-processing, 627
and scan size, 631–632
image snapper, 83
immunoelectron microscopy (IEM)
cell wall components and syntase, 376–379
conventional electron microscopy (EM), 375
cytokinesis and septum formation, 378
freeze-fracture replica labeling method, 375
immunoglobulin G (IgG), 471
immunogold labelling, 304–307, 348
immunolabelling, 504
indium tin oxide coated (ITO) glass, 466–467
In-Flight Beam Tracing^{TM}, 79, 87
in-lens field emission scanning electron microscopy
development of, 32–34
high sensitivity backscatter electron detector, 34
JEOL FESEM, 59–61
low accelerating voltage application, 34–36
in situ cryo-FIB-lamella preparation technique, 423–426
integrated circuit (IC)
EBT-100, 19
ISI-ABT-Topcon, 19–20
MEA-3000, 18–19
orientated SEM, 18
SR-50, 19
intermediate lens (IML), 81, 83, 99
International Scientific Instruments (ISI) series, 10–11
inverse Radon transform, 570
ion beam imaging, 6
ion–solid interactions, 518
iRODS, 620
irruptive recrystallisation, 231

JEOL field emission scanning electron microscopy
beam system, 74–76
cryo-SEM, 55–56
development of beam deceleration, 66–70
energy filter, 66–70
field emission electron sources, 57
field emission SEM's, 57–59
in-lens field emission, 59–61
objective lens (OL), 61
semi in-lens, 61–64
development, 71–72
evolution, 64–66
super hybrid lens, 72–74
thermionic electron sources, 57
unique aberration, 70

Kapton tape, 487–488
Knossos, 507

lanthanum hexaboride (LaB_6) source, 4
Leica EM ICE high pressure freezer, 170
Leica EM VCT500 cryo-SEM set, 185–186
Leica VCT system, 422
light/electron microscope (LEM), 8
light microscopy (LM), 94–95
cellular architecture by, 103
cryo-correlative, 424
imaging, 98
polarized, 349
with SEM, 354
lipids, 449–450
liquid–metal ion source (LMIS), 417
living specimens, ESEM, 454–455
load-lock system, 98, 422
low vacuum secondary electron detector (LVSTD), 92
low-voltage scanning electron microscopy (LVSEM), 278, 363
Lucilia sericata, 454
LYRA, 80, 96

melt–refreeze, 229
metal coating, 183, 290–292
microcomputed tomography (microCT) methods, 138
migratory recrystallisation, 229, 231
milling
cryo-FIB, 182
and sputtering, 417
mini scanning microscope (MSM)
Alpha bench-top series, 9
DS130 (W) Tungsten
DS-130C LaB_6, 13–14
DS-130F field emission gun (FEG), 14
DS-130S, 14
ISI series, 10–11
SMS Bench-mounted series, 9–10
mini singlet oxygen generator (MiniSOG), 504
mitochondrial network, in mouse liver, 561
monoclonal antibody, 470
Monte Carlo (MC) simulations, 419–420
mounting methods, SEM
edge, 154
filter, 154
hole, 155
liquid film, 155–157
rivet, 157

mounting methods, SEM (*contd.*)
sample, 504–505
surface, 154
multiangle rotational shadowing (MARS), 250
multibeam SEM, 128–129
multicolor array tomography, 478
multi-detector colour imaging
black and white *vs.* colour imaging
cinematographers camera tricks, 666–667
electron microscopists and coloured imaging, 661–663
motion picture and SEM, 667–670
computer-generated images (CGI), 661
materials and methods
nanoflight hardware system setup, 670–672
nanoflight software setup, 672–677
multienergy deconvolution SEM (MED-SEM) techniques, 110–111
multimedia data, in science, 616

nanocrystalline silver coated wound dressing, 323
nanoflight®
hardware system setup, 670–672
software setup, 672–677
nanomedicine
FIB/SEM EDX method, 534–537, 540–541
materials and methods, 534–537
results, 537–540
neuroanatomical tracers, 465, 469–470
noise reduction, 279
non-conductive biological samples
backscattered electrons imaging of, 443
gaseous secondary electron imaging of, 442–443

objective lens (OL)
characteristics of, 14
depth mode, 81
and detector geometry, 29
in-lens type, 60
JEOL FESEM, 57, 61
of semi in-lens system, 36–37, 62
specifications of, 364
systems, 29
OMERO, 615
Open IRIS, 612
optimal instrumental approach, FEGSEM
fibrous gauze wound dressing, 323–324
gel networks, 325–326
nanocrystalline silver coated wound dressing, 323
resorbable polymer microspheres, 324–325
osmium tetroxide, 195–198, 549–551

partial freeze-drying, 269–271
particle–beam-induced deposition, 518, 519
peak-to-local background (P/B) method, 605
Phi-Rho-Zed, 604
photo-activated localization microscopy (PALM), 133
photomicrograph, 637–640
physical vapour deposition (PVD) technologies, 283
planarmagnetron sputtering (PMS), 283
plunge freezing, 423
point-spread functions, 106
postembedding multicolor imaging, 470–471
post-nucleation, 229–230
primary electron beam, 26
project management, 615
proteasome 20S core particle, 579
proteometric analysis, AT for, 465–466
Py-Knossos, 507

quality control, biomedical products, 336–340
Quanta 3D FIB-SEM, 523
Quantomix WetSEM capsule, 440
quantum dots, 471–472
quantum tunneling, 5
Quorum Polarprep system, 422

radial-gap, 85
Raman spectroscopy, 95, 329, 331
recrystalisation temperature, 233
reduced osmium–thiocarbohydrazide–osmium (ROTO) protocol, 487
regions of interest (ROIs), 473, 490
resin blocks, 502, 558–559
resin-embedding, 590–591
resolution gap, 567–568
resorbable polymer microspheres, 324–325
resource management, 612
discoverability, 612–613
optimization, 613–614
billing, 613
usage restrictions, 613
RhoANA, 492
root-mean-square (RMS) error, 475

SBEM. *see* serial block-face scanning electron microscopy (SBEM)
scanning electron microscope (SEM)
backprojection, 571–573
backscattered electron (BSE), 2
chemical fixation preparation
critical-point drying, 205–206
ionic liquids, 206–207
resin, 208–209
and room temperature preparation, 205–206
development with TEM, 8
electron source (cathode) for, 30
enhanced higher brightness sources, 4–5
incident beam electrons, 1–2

for ion beam imaging, 6
lanthanum hexaboride (LaB_6) source, 4
nanomanipulator with piezoelectric actuators, 456
objective lens systems and detector configuration, 29
principle, 1
resolution characteristic, 29–30
resolution mechanism of, 27–30
resolution *vs.* convergent angle, 27–28
scan raster, 2
signal-to-noise ratio, 3
structure and principle of, 26–27
thermionic emitter, 3–4
variable pressure (VP-SEM), 91
scanning probe microscopy (SPM), 80
scanning transmission electron microscopy (STEM), 13, 109, 121–122, 318
Schottky emission (SE)
cold field emission (CFE) electron gun, 44
electron source, 43–44
popularization of, 43–44
SEM, 81
Schottky type thermal field emission gun, 64
secondary electron (SE), 2, 38
coating, 183
coefficient, 571
detection modes, 499
enhancer, 68
signal generation, in ESEM, 441–442
secondary ions (SI), 98
semi in-lens FE-SEM
applications, 39–41
JEOL FESEM, 61–64
development, 71–72
evolution, 64–66
objective lens of, 36–37
signal detection system, 37–39
ultralow accelerating voltage, 43
semi in-lens type objective lens, 62
septation initiation network, 379
serial block-face imaging (SBFI), 109–110
serial block-face scanning electron microscopy (SBEM), 564–547
applications
cell biology, 508
material science, 509
neuroscience, 507–508
organism, 508
organs and tissues, 508
scanning electron microscopes
detection modes, 499–500
image acquisition, 505–506
microtome insertion, 500–502
post-processing and analysis of data, 507
pre-embedding immunolabelling, 504
sample mounting, 504–505
sample preparation, 502–504
variations, 502–503
serial sections with TEM, 496–498
serial section scanning electron microscopy (S3EM), 304
SIGMA (SS) series
environmental SEM, 17
integrated circuit
EBT-100, 19
ISI-ABT-Topcon, 19–20
MEA-3000, 18–19
orientated SEM, 18
SR-50, 19
measurement notation and bar, 20–22
vari-zone lens, 15
Alpha series, 17
conventional pin-hole design, 15–16
in-lens design, 17
through-the-lens design, 16–17
signal detection system
evolution of signal discrimination, 38–39
sensitivity of, 37–39
signal-to-noise ratio (SNR), 576, 578
FEG-SEM, 108–109
vs. low dose imaging, 278–279
scanning electron microscope (SEM), 3
silicon drift detectors (SDD), 594
silicon wafers, 36, 68, 130, 278, 304, 464, 466–468, 486, 489, 546
single-axis tilt tomography, 574, 576
single-particle reconstruction, 578–579, 581
singlet oxygen method, 472
SMS Bench-mounted series, 9–10
snorkel lens, 29, 85
sodium potassium tartrate crystals, 598–600
spherical aberration, 27
spontaneous recrystallisation, 232
sputtering, 419
and milling, 417
simultaneous iterative reconstruction technique (SIRT), 577
stochastic optical reconstruction microscopy (STORM), 133
supercooling, 227–228
super-resolution LM, 133
surface water, in ESEM, 443–444

TEM. *see* transmission electron microscopy (TEM)
TENEO VS (Volumescope), 501
TESCAN, 79–100
Texas Red, 469
thawing, 231, 235, 240–242, 398
thermionic electron, 4, 25, 30, 57
thermionic emitter, 3–4
thermoemission LaB_6, 81
3D electron microscopy (3DEM) techniques, 138
3D image processing, 621–622

three-dimensional field-emission scanning electron microscopy
- actin filament, 579
- three-dimensional ant reconstruction, 577
- applications
  - handedness, 581–582
  - resolution gap, bridging, 582–584
- backprojection, 570–571
- bacteriophage T4 tail, 579
- decision diagram, 569
- electron damage, 576, 578
- *Heterocapsa circularisquama* RNA virus, 579
- high electron-dose conical tilt tomography dataset, of metal-coated sample, 576, 578
- image formation in SEM, 571
- inverse Radon transform in two-dimensions, 570
- proteasome 20S core particle, 579
- protein monomers, 582–584
- sample preparation, 580
- secondary-electron escape probability distribution, 571–573
- SEM backprojection, 571–573
- single-particle reconstruction, 578–579, 581
- tomographic 3DSEM
  - resolution, 576–577
  - signal-to-noise ratio, 576
  - tilt strategies, 574–576
- tomographic reconstruction, 580
- voxel model, 571

3D volume processing, 622–623
3View®, 500–502, 507, 508
time-of-flight secondary ion mass spectrometry (TOF-SIMS)
- chemical mapping technique, 329
- FIB-SEM, 98
- setup and SPM, 80

TrakEM2 SIFT alignment algorithm, 473, 507
transmission electron microscopy (TEM)
- bone/coating interface, 528–529
- for chemically fixed and embedded yeast cells, 528–530
- coating/Ti substrate interface, 528–529
- collagen fibrils formed in bone, 529
- dental implants, 528–530
- development, 7–8
- electron beam, 496, 497
- human skin in native-like state, 447–448
- in-lens OL systems, 57
- limits, 497
- serial sections, 496–498
- tomography, 496–497
- two-dimensional projections, 496–497
- two-phase coating, 529

transmitted electrons (TEs), 87
Tyroglyphus casei, 454

ultra-fast scanning system, 83
ultrahigh resolution low-voltage scanning electron microscopy (UHR LVSEM)
- cell wall components, 367–371
- cell wall formation, 365–367
- resolution, 364–365

ultra-high resolution microscopy, 84–85
ultralow accelerating voltage, 43
ultralow-temperature low-voltage scanning electron microscopy (ULT LVSEM)
- freeze-fracture method of, 371–372
- HPF method, 371
- resolution of, 372
- septum formation, 372–375
- *in situ* observation of high-pressure frozen, 372

uranyl acetate staining, 487, 549

vacuum cryo-manipulation, 176
vacuum cryo-transfer, 173
variable pressure (VP-SEM), 91
vari-zone lens, 15
VAST, a program for computer-assisted manual space-filling segmentation and annotation 492
versatile cryo-transfer, 173
vitrification
- cryo-FEGSEM, 244–253
- temperature and condensed phases of water, 230

vitrified eukaryotic cells, 423
volume microscopy, 209–211
volumescope, 495
volumetric electron microscopic imaging, 462

WaferMapper, 489
wavelength dispersive spectroscopy (WDS), 55, 594–595
wavelength dispersive X-ray microanalysis (WDX), 327
Wehnelt cylinder, 3
Wuerchwitz mite cheese, 454

XEIA, 80, 97
X-ray microscopy (XRM), 138–139
X-ray photoelectron spectroscopy (XPS), 329
X-rays, 593

yeast cells, LVSEM
- high-pressure freezing technique, 363
- immunoelectron microscopy, 375
  - cell wall components and syntase, 376–379
  - cytokinesis and septum formation, 378
  - freeze-fracture replica labeling method, 375
- septum formation, 363

UHR LVSEM
cell wall components, 367–371
cell wall formation, 365–367
resolution of, 364–365
ULT LVSEM
freeze-fracture method of, 371–372
HPF method, 371
resolution of, 372
septum formation, 372–375
in situ observation of high-pressure frozen, 372
yttrium aluminum garnet (YAG) detector, 34

ZAF, *see* atomic number (Z), absorption (A) and X-ray fluorescence (F) on X-ray yield 604
ZEISS Efficient Navigation (ZEN), 134
ZEISS GEMINII® technology
biological imaging
advanced STEM imaging, 121–122
high-throughput and ease-of-use imaging, 123–126
imaging non-conductive, 122–123
optimum resolution, contrast, and surface sensitivity, 119–121
correlative microscopy
correlative array tomography (CAT), 135
correlative volume imaging, 135–137
cryo-FIB, 137
X-ray microscopy (XRM), 138–139
field emission scanning electron microscopes (FESEMs), 117
focused ion beam (FIB), 119
helium ion microscopy, 127
interface for easy correlation, 134–135
multibeam SEM, 128–129
SE and BSE, 122
3D imaging
array tomography, 130
block-face imaging, 130–132
FIB-SEM tomography, 132–133